Modern Statistics for Engineering and Quality Improvement

John Lawson
Brigham Young University

John Erjavec
University of North Dakota

DUXBURY

THOMSON LEARNING

Australia · Canada · Mexico · Singapore · Spain · United Kingdom · United States

DUXBURY

THOMSON LEARNING ™

Sponsoring Editor: *Carolyn Crockett*
Marketing Team: *Chris Kelly/Ericka Thompson*
Editorial Assistant: *Ann Day*
Production Editor: *Tessa Avila*

Permissions Editor: *Sue Ewing*
Manuscript Editor: *Cathy Baehler*
Print Buyer: *Nancy Panziera*
Printing and Binding: *Von Hoffman Graphics, Inc.*

For more information about this or any other Duxbury product, contact:
DUXBURY
511 Forest Lodge Road
Pacific Grove, CA 93950 USA
www.duxbury.com
1-800-423-0563 Thomson Learning Academic Resource Center

Printed in the United States of America

10   9   8   7   6   5   4   3   2   1

Library of Congress Cataloging-in-Publication Data

Lawson, John.
    Modern statistics for engineering and quality improvement / John Lawson, John Erjavec.
        p.   cm.
    ISBN 0 534-19050-2
    1. Engineering—Statistical methods. 2. Quality control—Statistical methods. I. Erjavec, John.
II. Title.
TA340 .L38 2000
620'.007'27—dc   00-046774

PREFACE

Intended Audience

This book is intended as a first course in statistics for undergraduate engineering majors (chemical, mechanical, electrical and manufacturing) or undergraduate statistics majors who intend to work in industry in areas of physical science and engineering. The only prerequisite for the course is college algebra, although calculus is required for some topics.

Objectives

Our goal in writing this book is to present a straightforward introduction to the statistical tools that we have found to be most useful to engineers and quality improvement specialists. We intend the book to be used for a one semester or two semester sequence and to serve as reference book for engineers on the job. We have each spent over 15 years working as statistical consultants in industry, and another 10 years in academia continuing that practice. We have spent countless hours consulting with engineers and chemists in R&D and manufacturing. We have a good appreciation for what statistical tools are most useful on the job, and we have attempted to limit this book to those topics. The topics we concentrate on are methods for collecting, analyzing and interpreting data in industrial studies. Tools such as experimental designs, control charts, and variance analysis are critical. We motivate the students to learn these techniques by showing their applications in industry.

Readers familiar with traditional textbooks in engineering statistics will find that we skip many of the topics that are found in those books, such as calculations of probability based on permutations and combinations and moment generating functions. We eliminate such background material, because we believe it is unnecessary for the practical application of statistics in an industrial environment, and for the continuity of this book. Leaving out marginal topics makes room for a more in depth coverage of the topics we feel are truly useful.

Brief Outline

The first two chapters of the book serve as an introduction and will stimulate interest by showing the need for statistical methods in industry.

Chapters 3–6 contain background material in probability and statistical theory. We have found through our experience that extensive details here are not required for learning the statistical techniques useful in industry. Therefore, the introductory material in Chapters 3–6 is brief. The topics in these foundation chapters are limited to the necessary prerequisites for the practical applications in the remainder of the book.

We present experimental designs as a key statistical tool in Chapters 7–15. These chapters play a central role in the book, because the importance of statistical methods in the planning and collection of data in industrial studies is crucial. Without the proper data, effective decision making is impossible. Experimental designs are the most powerful way of collecting the proper data in industrial studies.

The final chapters of the book discuss methods for studying and reducing variability. Variability reduction is the basis for quality improvement in industry. Chapters 16–18 present quality improvement tools such as variance characterization, control charts, and robust design studies.

Sample Course Outlines

With 18 chapters, there are more topics than can be covered in a one-semester class, but the instructor has flexibility in choosing practical topics for the last third of the class. For example, most instructors would spend 2/3 semester covering Chapters 1–10. In a class for chemical engineers, the last third of the course could cover Chapter 12 on screening designs and Chapters 13–15 on response surface methods, while a course for manufacturing engineers may instead cover Chapter 16 on measurement errors and Chapter 17 on quality control. Finally, a class for mechanical engineers may cover Chapter 11 on multiple level factorials, Chapter 12 on screening designs and Chapter 18 on robust engineering design. A two-semester sequence could cover the entire book at a more leisurely pace.

Computer Use and Problem Solving

The statistical computations that are discussed in the book are presented in worksheet formats that can be easily adapted to a modern spreadsheet program such as EXCEL®, LOTUS 123®, or Quatro Pro®. Spreadsheet programs are common tools available to engineers in the workplace, and therefore we emphasize the worksheet approach to computations. Of course, modern statistical software reduces the effort needed to carry out some statistical computations and graphics. We illustrate the use of a statistical package for some methods using MINITAB®.

Unique Aspects

The book is comprehensive enough to serve as a reference for working engineers. Some of the more practical aspects of the book that set it apart from others are described here. While covering some of the same basic experimental designs—such as 2^k, screening experiments and response surface experiments, discussed in other books—this book provides a framework or strategy that shows when each type of design should be used and how they can be used in a sequence to build engineering knowledge. While this book talks about the need for randomization in experimentation, as do most other books, it also shows, in a practical way, what to do when it is difficult or impossible to randomize. Because of difficulties in randomizing, approximately 25% of all industrial experiments turn out to be split plot experiments. These experiments are frequently mishandled and misinterpreted in industry. This book provides a simple way of setting up and analyzing split plot experiments, building on the framework for 2^k designs. This book doesn't just talk about power of tests in a simple case like the one sample test on the mean, but provides Wheeler's simple formula that can be used to determine how many replicates are needed in factorial experiments or the size of a fraction in screening experiments. Finally while other books present only hand calculation formulas or illustrate the use of special purpose statistical software to do calculations, this book makes it clear how all statistical calculations can be made on general spreadsheet programs that are usually available to students and practicing engineers.

This book reduces the percentage of pages discussing traditional probability and inference and devotes more pages to topics particularly pertinent in an industrial setting that can be truly considered engineering statistics. Many real examples from industry are presented in the book and emphasis is placed on interpretation of results, not just on calculations.

John Lawson
John Erjavec

CONTENTS

Part I
Introduction

Before we can begin a study of statistics, we need to give engineers and scientists incentive to study statistics and provide or review the basic quantitative tools that are used throughout the rest of the book. Chapters 1 and 2 provide an overview of where statistics fit into the scientific method, and how they are used in quality improvement efforts that are vital in industry today.

CHAPTER 1
The Scientific Method and Statistics

1.1 Introduction

Statistical methods are a collection of mathematical tools that have been proven to be indispensable to the modern engineer. In recent times, application of statistical methods by engineers in industry have brought about breakthrough improvements in quality, reliability, and cost of manufactured products. As a result of, many U.S. corporations have invested heavily in on-the-job training programs in statistical methods for their engineers, and have sought to influence academic institutions to integrate more practical training in statistics into the engineering curriculum[1].

What are the job functions of engineers, and in what ways are statistical methods valuable to them on the job? Engineers do a variety of things. They do so many things, in fact, they had to be divided into four major classifications: Chemical, Civil, Electrical, and Mechanical. But, broadly speaking we can say that all engineers design and develop new products, structures, or systems. They improve or modify past designs, and they build and test prototypes, tools, and manufacturing processes. They test, troubleshoot, improve, and maintain control over products and manufacturing processes, and they devise appropriate maintenance and service schedules. In all these functions, engineers must collect and analyze data[2], for good decisions can rarely (if ever) be made on the basis of opinion or unfounded theory.

The process of collecting and analyzing data is what statistics is all about. Therefore whether engineers have taken any formal courses in statistical methods, they will use statistics. There are good and bad strategies for both collecting and analyzing data. Good strategies are necessary for successful work, and have been developed by scientists over the past century. These good strategies employ statistical methods and they are used by effective scientists and engineers. In fact, these methods are an inseparable part of the scientific method.

This book teaches practical statistical methods for collecting and analyzing data that will be useful to engineers in industrial applications. In this chapter we will introduce the subject by discussing the role that statistics plays in the scientific method, and how the scientific method is used in engineering work, in an industrial setting. Next, we will describe some simple statistical methods that can be used in the hypothesis (theory) formulation stage of the scientific method. Finally, we will illustrate their application with some actual case studies from industry.

1.2 Statistics and the Scientific Method

Statistics has been called the science of science, or the science of learning from data. To better understand what that means, let's consider the scientific method in general as George Box has described it in his article "Science and Statistics" [3]. We use the scientific method to learn about a system we care about. We start by proposing a hypothesis or theory to describe the way the system behaves (the real but unknown situation, or state of nature). Next, we deduce the consequences of our theory and compare this to the known or available facts. If the facts do not contradict our theory, we feel more confident in it. Otherwise, we have to go back and modify our theory or hypothesis. Once our theory conforms to the known facts, we need to test it further by collecting additional data. This we do by experimentation. When our theory or hypothesis is contradicted by experimental data,

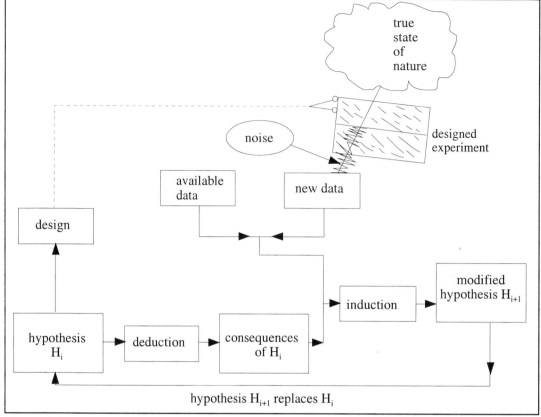

Figure 1.1   Data Generation and Data analysis.

we are again forced to go back and modify it. The whole idea is diagrammed in Figure 1.1 which is copied from Box[4]. In this picture, experimentation is represented by the window which allows us to peer into the true state of nature. The experimental data which we obtain is represented by the wiggly line coming through the window. The wiggle reflects the fact that experimental data cannot show precisely the true state of nature. Instead it is always clouded by experimental variability or noise.

The role of statistics, in the scientific method, comes in three places. First is at the initial stages where available facts are examined to postulate a reasonable theory. Here powerful descriptive statistical methods can summarize vast quantities of data, reveal the prominent features, and aid the researcher in hypothesizing the underlying mechanism or determining whether a proposed mechanism agrees with available facts. The second place where statistics is used in the scientific method is in the experimental stage. What experiments need to be conducted? What combinations of controllable variables would be the best to try? Statistical experimental designs, the subject of Chapters 7 to 15, offer the most efficient plans for conducting experiments. The third place where statistics are used in the scientific method is where new data or data obtained from experiments is compared to the consequence of proposed hypotheses. Here objective methods of statistical analysis reduce errors in conclusions and prevent human biases and optimism from clouding data interpretation.

What is the relevance to engineers? Whenever there is insufficient theory to predict or provide knowledge for the solution to a problem, trail and error or experimentation must be used to reach a workable solution. In industrial problems such as manufacturing troubleshooting or product design, empirical experimentation is often the only available approach for engineers to solve problems. Which experiments to perform (what data to collect) and how to interpret that data, are some of the major questions faced by any R&D or process-improvement study team. Statistical experimental designs help us answer the first question. They are extremely efficient and provide quantum gains over the alternative approach of studying one variable at a time. Statistical experimental designs are not expensive, at least not when compared to any other experimentation strategy aimed at giving the same information. It is exactly the application of statistical experimental designs that has been so successful in Japan.[5] It is not sparkling new manufacturing facilities or automation that has been the main contribution to the Japanese increase in quality and productivity. Rather, it is due to the many small gains made on a regular basis through the application of statistical experimental design techniques that have accumulated over the years. It has been reported[6] that at a Japanese company visited by a group of American researchers, 6000 experiments had been performed in one year using plans similar to those discussed in Chapters 7–15 and 18. In another company 48 multifactor experiments were conducted in the design and development of one product. The chief engineer at that company said that the use of planned experiments was the most important factor in developing a high quality product[7].

Data collected from statistically designed experiments lend themselves directly to statistical analysis of data, which is the third place where statistics is involved in the scientific method. Once a set of experimental data has been collected, how does one determine whether it confirms or contradicts his hypothesis or theory for explaining the situation? If ten scientists or engineers are given the same set of data, they may all come to different conclusions based upon their previous background and personal biases. However, if the data are subjected to statistical analysis, no subjectivity is involved. Statistical methods are founded in probability theory and are designed to be objective. Using them minimizes the possibility of bad conclusions caused by experimental noise in data. Statistical analysis methods that are appropriate for data collected through the experimental plans presented in this book are described and illustrated as needed.

In the remainder of this chapter we will describe three simple tools that are useful to engineers in formulating a scientific hypothesis in an industrial setting. Actual industrial case studies illustrate their use.

1.3 Tools Helpful in Formulating Hypothesis

1.3.1 Pareto Diagrams

When troubleshooting problem manufacturing processes, or redesigning a product or system, there are usually lots of available data or facts that need to be examined before embarking upon any new approach or solution. Proper organization of these available facts can greatly facilitate the process of speculating what changes or solutions may be fruitful. One simple tool used for this organization is the *Pareto Diagram*.

Pareto Diagrams were first introduced by J. M. Juran, a pioneer in the field of quality control. The purpose of this diagram is to determine the few vitally important aspects of a body of information, so that attention is properly focused on them rather than on the many trivial aspects. The diagram is constructed by taking a body of information or facts, dividing it into several classifications and then constructing a bar chart which clearly displays the category that is most abundant or important. For example, Table 1.1 below shows data on defective circuit boards classified by defect type. Figure 1.2 shows a Pareto Diagram of the same information. The Pareto Diagram is constructed by making a bar chart with one bar for each category. The heights of the bars can represent the frequency of each category, as they do in Figure 1.2, or the dollar cost, the time spent, or any other meaningful numerical value associated with each category. A line graph is often added to the diagram to show the cumulative total for each classification. The calculations are shown in Table 1.2 and the line is added in Figure 1.3.

Table 1.1 Defects in Circuit Boards

Type Defect	Count
Bad Part	1
Missing Part	3
Wrong Orientation	3
Improper Mount	27
Wrong Component	4
Damaged on Insertion	51
Lead Unconnected	12
Wrong Point of Insertion	8

The construction of a cumulative line is illustrated in Table 1.2 and Figure 1.3.

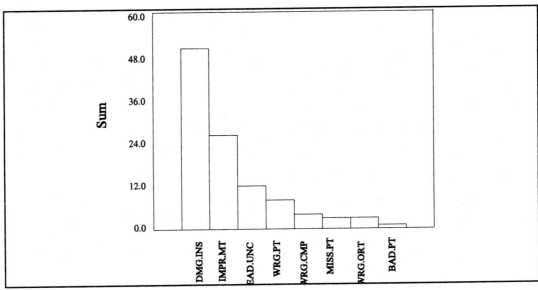

Figure 1.2 Defective Circuit Boards

Table 1.2 Cumulative Defect Count

Type Defect	Cumulative Count	%
Damaged on Insertion	51	47
Improper Mount	78	72
Lead Unconnected	90	83
Wrong Point of Insertion	98	90
Wrong Component	102	94
Missing Part	105	96
Wrong Orientation	108	99
Bad Part	109	100

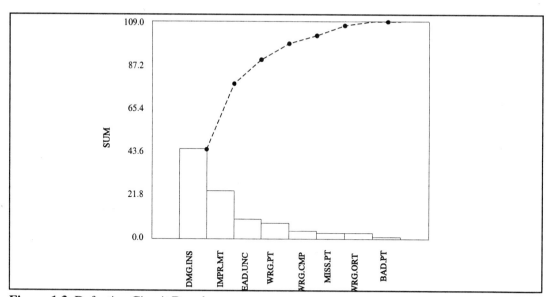

Figure 1.3 Defective Circuit Boards

The figure allows us to concentrate on the most common, most expensive, or most time consuming categories. For example, in Table 1.2 we can see that Damaged on Insertion, Improper Mount, and Lead Unconnected, account for 83% of all defective circuit boards. By focusing on the important aspects of a problem we are better able to hypothesize the causes and therefore potential cures for the problem. It would be a waste of time (at least at this juncture) to try to improve component quality, while trying to improve insertion and mounting appears critical.

If the circuit board assembly is a labor-intensive process, manufacturing engineers might want to know if operator skill must be increased or the task complexity decreased. Are certain mistakes more prevalent? Are these more prevalent mistakes made more often by less experienced operators, or could the circuit pack design be changed to make assembly easier and less error prone? Additional Pareto Diagrams could help to continue to focus attention in the appropriate direction. Defects could be classified by defect type, and later by operator experience level and successive diagrams could be constructed to lead the way to a proper hypothesis. For example, if only inexperienced operators are causing the more common defects, it would be hypothesized that a better training program could reduce errors. On the other hand, if experienced operators are making the common errors, it might be hypothesized that a better design should be devised.

Pareto Diagrams are also useful for presenting the results of experimental studies. For example, if a study were conducted to determine if additional training of operators might reduce the number of components damaged on insertion and the number of improper mounts, a before and after Pareto Diagram might be constructed.

1.3.2 Process Flow Diagrams

The available facts about how a process is actually being run or how a product is actually being used must be considered carefully before any changes (for improvement) are suggested or hypothesized. A good tool for clearly representing how a process is currently being run is the *process flow diagram*.

Process flow diagrams symbolically show the steps in a process in the order they are performed. For manufacturing processes these diagrams may already exist; but for many tasks that are performed in a business, the steps may not be so obvious and may never have been diagramed. Process flow diagrams may also be used to illustrate the way a product is used. Process flow diagrams are best when they are constructed by a team of individuals. What one person forgets about the process another will remember, and pooling the knowledge of a group always gives the most complete information. Finally, a process flow diagram should never be considered complete until all the people who actually do the work agree that the diagram accurately reflects the way the work is currently done. The existing process flow diagram for many
manufacturing processes may not accurately reflect the way things are currently done.

There are no specific rules about how process flow diagrams should look, but the symbols shown in Figure 1.4 are fairly universal. To construct a flow diagram, the team should think through the process and identify the major process steps. These can be written on a flip chart, blackboard, or other convenient surface. Next, storage points and decision points should be identified. Decision points may add greatly to the complexity of a process but it is critical that each be identified in describing the process. Often a process can be simplified by eliminating

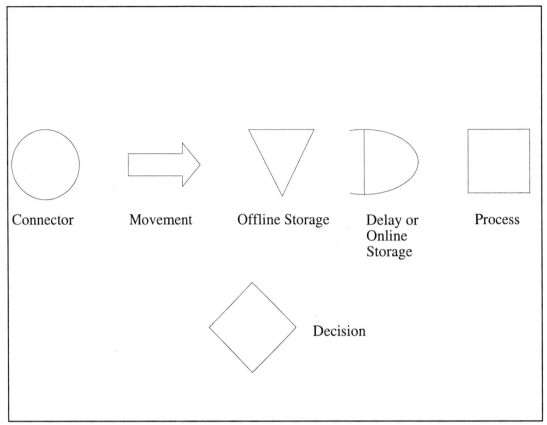

Figure 1.4 Process Flow Diagram Symbols

unnecessary decision points and storage points. The final step, as was already stated, is to check the flow diagram with all individuals actually carrying out the work see if it accurately reflects what is being done. If it does not, it should be modified to truly represent the current process.

Consider an example of the hand-assembly portion of a circuit board assembly plant. A process flow diagram of the production line may appear as Figure 1.5. If there are problems in the process, such as delays or an excessive number of defectives, the process flow diagram is a good first place to start looking for possible causes. The diagram, along with knowledge of existing problems, may stimulate thought and lead a team of individuals to guess or hypothesize possible causes and solutions to problems.

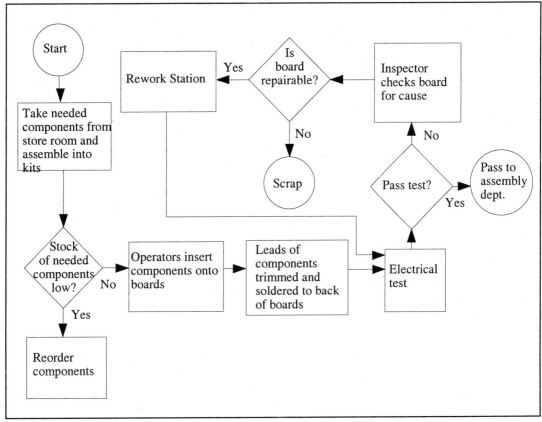

Figure 1.5 Process Flow Diagram of Manual Circuit Board Assembly

1.3.3 Cause and Effect Diagrams

The chances of hypothesizing, or proposing, an adequate solution to a problem are higher if brainstorming sessions are used where the opinions of many individuals who are involved, can be obtained. A good tool for summarizing brainstorming sessions on the cause of a problem, or proposed solutions to a problem is the *cause and effect diagram.*

The cause and effect diagram was originated by Kaoru Ishikawa, the famous Japanese quality control guru, while he was explaining to some engineers the myriad of factors that can affect the variability in a quality characteristic. Cause and effect diagrams are best constructed in team meetings, because the thoughts of one individual may stimulate ideas in others. The diagram tends to keep the discussion focused on enumerating causes of a problem, and the process of constructing the diagram usually proves to be very educational for everyone involved.

Cause and effect diagrams are constructed by writing the problem or concern in a box at the right. Next, a main arrow is drawn from the left pointing at the box, as shown in Figure 1.6(a). Then, stems representing the main classifications of causes are attached to the main arrow. Ishikawa has suggested organizing causes into four main classifications: Materials, Methods, Machines and Man,. Of course these classifications are not sacred and can be modified to be more appropriate for a particular problem when desired. However, the number of main branches should always be near four, never less than three or more than six, because that would represent poor organization. Adding Ishikawa's four classifications to the diagram results in Figure 1.6(b). Finally, specific things thought to be causes of the problem can be added as twigs to the main stems. For example, in a circuit board assembly plant, a cause and effect diagram addressing the problem of faulty circuit boards might appear as Figure 1.6(c).

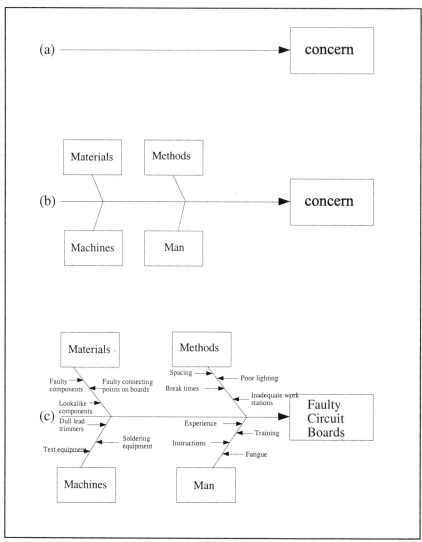

Figure 1.6 Cause and Effect Diagram Construction

Another way of organizing a cause and effect diagram is to use the pattern of a process flow diagram. In this version of the cause and effect diagram, the main arrow coming from the left shows the major process steps and decision points. The specific causes of the problem are then attached at the process step where they would occur. An example of a cause and effect diagram of this type for the hand-assembly portion of a circuit board assembly might appear as Figure 1.7.

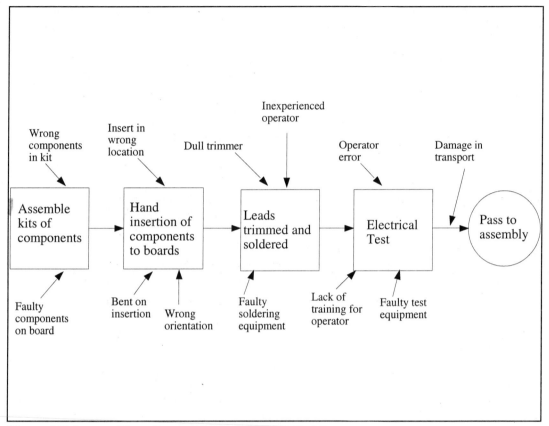

Figure 1.7 Process Flow Cause and Effect Diagram

Specific causes should be added to the diagram through group brainstorming. One convenient way this can be done is to record an idea from each person in the group until all ideas are exhausted. Using this technique, each person takes a turn and must either add one specific cause to the chart or pass.

Another way to stimulate ideas for specific causes, discussed by Ishikawa, is to ask "why?" five times. In other words, one could question the brainstorming group with: Why are circuit boards faulty? If one of the group's answers was because some components placed on the board do not function, the next question might be: Why don't some components function? If one of the answers to this question was because some components are not placed properly on the board, a logical next question might be: Why are some components not placed on the board properly? By asking why five times or until all ideas are exhausted many detailed causes can be discovered and added to the chart.

Cause and effect diagrams for the most part represent opinions or hypothesis. However, if certain causes are known and quantifiable, that information could be added to the chart with footnotes or other symbols. In this way, the cause and effect diagram can show the level of process knowledge.

After completion of the cause and effect diagram, solutions may be hypothesized for correcting the problem. As these hypothesized solutions are tested through experiments, the resulting information can be used to update the diagram. Causes found not to be important can be crossed out. Others that are demonstrated to be important can be highlighted or footnoted. Demonstrated solutions can be added to the appropriate causes of problems with footnotes, etc. In this way, the cause and effect diagram not only serves to collect and organize opinions about problems, but it can be used to guide and record the progress of research efforts aimed at solving the problems.

1.4 Industrial Examples

In this section we present two examples of the use of the simple tools shown in this chapter to help in proposing or hypothesizing the cause of real industrial problems. The first example discusses the improvement in a casting form design in the automotive industry, and the second illustrates the improvement in manufacturing consistency in the electronics industry.

1.4.1 Chrysler Corporation Kokomo Castings Plant[8]

New Process Gear (NPG), a division of Acustar that builds manual transmissions for Chrysler, discovered some of the most frequent problems they were experiencing were due to defective castings received from the Kokomo Plant. Process Quality Improvement (PQI) teams at the Kokomo plant were notified of the problems in regular weekly telephone conferences with their customer NPG.

To help hypothesize the cause of the defective castings, data on the defects supplied by NPG were summarized by Pareto Diagrams. First, the chart was made by grouping the defects into classes as shown in Figure 1.8. In this figure we see that 52.5% of the defects were due to housing porosity. Next, all defects due to housing porosity were classified as to the location of the defect, and another Pareto Diagram was constructed showing the percentage of the defects at each location. This is shown in Figure 1.9. After reviewing the Pareto Diagrams, and noticing that the majority of porosity defects were concentrated in one area of the casting, the PQI team hypothesized that the porosity defects could be reduced by making a design change to the casting forms. It was believed that the problem would be reduced by adding feeders to the casting form between the gate and the main area where the porosity defects were found.

The additional feeders were installed in the casting forms, and defect data continued to be collected and monitored by control charts (which are explained in Chapter 2). The overall defect rate decreased from 7.8% to 5.9% (a 24% decrease) and was sustained over the next month. The annual scrap savings projected from this small change to the casting forms was projected to be $40,000 annually. Data illustrating this improvement will be shown in Chapter 2.

The Pareto Diagram was extremely useful in summarizing a large body of facts or data on casting defects in a way that directed the PQI team's attention to one of the underlying causes.

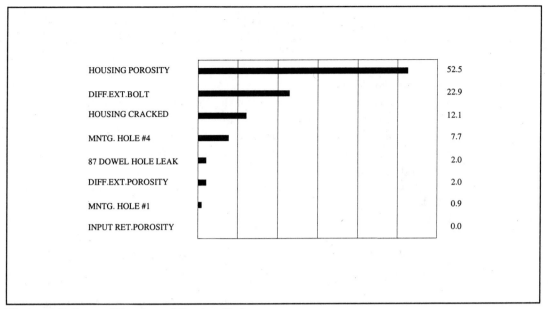

Figure 1.8 Pareto Diagram of Casting Defects

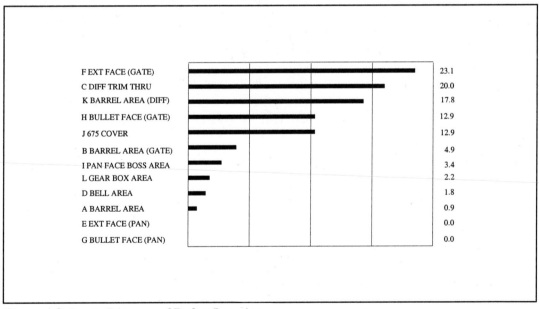

Figure 1.9 Pareto Diagram of Defect Location

Forming a scientific hypothesis about the cause of a process, or design problem, is always the first step in finding the solution. Facts, data, and simple diagrams such as the Pareto, are very helpful in formulating viable hypothesis.

1.4.2 Copperplating Ceramic Substrates

The Fixed Products Microcircuits Operation of the Communications Sector in Motorola Corporation copperplates hundreds of different hybrid circuit boards. As part of their 6σ program (a company-wide program aimed at reducing variability), they conducted a study to find ways to improve their manufacturing process by making the copperplating step more consistent and thereby decrease variability within different and similar ceramic substrate circuit boards[9]. By decreasing variability, the Microcircuits Operation could ensure that they could meet customer specifications with very high confidence.

One of the first steps taken in the study was to accurately determine what was being done in the process at present. Figure 1.10 is a simple flow diagram of the copperplating process. After reviewing the flow diagram, brainstorming sessions were held to hypothesize which process factors could affect plating thickness. The results of these brainstorming sessions are summarized in the cause and effect diagram in Figure 1.11.

Once the initial ideas were on paper, data from the process was examined in attempt to confirm the ideas hypothesized, or to come up with new ideas concerning the factors that affect plating thickness. After collecting data, performing experiments, and analyzing data to confirm or contradict the hypothesis listed in Figure 1.11, it was found that tighter control of process variables (such as temperature and surface tension) and relocation of the anodes in the plating cells reduced the range of variability copperplating by more than 66%. This allowed the company to meet their tight specifications. In addition, the more uniform copperplate layer allowed for a reduction in the average thickness and for substantial material cost savings. The actual data collected and the statistical summaries of the data that led to these improvements will be shown in Chapters 3 and 12, as the statistical methods that were used are described.

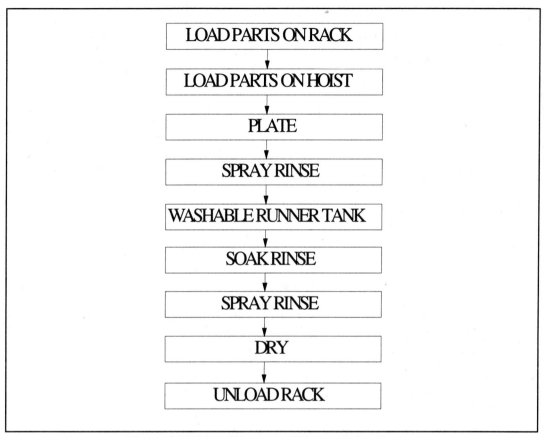

Figure 1.10 Flow Diagram of Copperplating Process

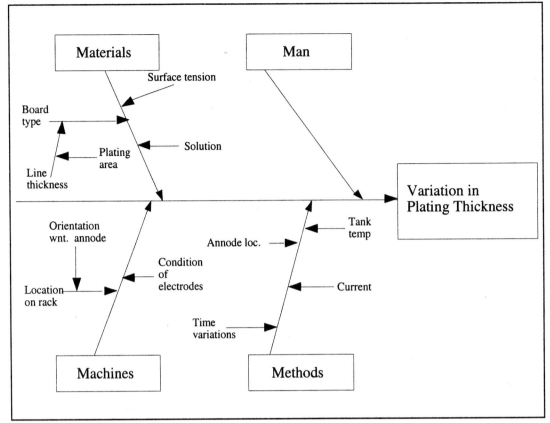

Figure 1.11 - Hypothesized Causes for Variation in Plating Thickness

1.5 Summary

The job of the engineer is to design and develop new products and processes, and modify and improve past designs. There are successful and unsuccessful strategies for doing this. The late Bill Hunter[10] proposed a four-step outline of a strategy for improving any design or process. His outline is as follows:

(1)   To improve anything there must be change. Without change we will have no better than the status quo.

(2)   To justify any change, we should have good data to serve as a rational basis for making this change. Without good data, decisions to change are often based on unfounded opinions and results are spotty at best.

(3)   Twin questions must then be addressed. What data should be obtained or collected, and how should that data be analyzed and interpreted to reach the most reliable decision about change?

(4)   Statistics is the discipline that addresses these twin questions.

Statistical methods are the embodiment of the scientific method, and are crucial to engineers as they design and develop new products and manufacturing processes, and build and test prototypes.

In this chapter we have described three simple tools that are useful for summarizing available facts and formulating scientific hypothesis. The remainder of the block will present methods for collecting and analyzing data and numerous practical examples in engineering work. Important concepts from this chapter are:

Job function of Engineers 3
The Scientific Method 3, 4
Role of Statistics in the Scientific Method 4, 5
Pareto Diagrams 5 to 9
Process Flow Diagrams 9 to 10
Cause and Effect Diagrams 11 to 14

1.6 Exercises and Review Questions for Chapter 1

1.1 Describe the scientific method in your own words and describe when it should be used in engineering.

1.2 Describe in your own words the areas where statistical methods are used in the scientific method.

1.3 Explain the difference between a hypothesis about how a system works and knowledge of how the system works.

1.4 In an injection molding process, a sample of plastic parts are sampled hourly. Defects are recorded. For the past month, the following numbers of defects have been recorded:

Defect Type	Number of Defects Found
Sinks	17
Scratches	30
Black Spots	100
Splay	26
Flow Lines	18

a) Present this data graphically in a way that would stimulate thinking about ways to reduce defects in the process.

b) Discuss the next step you would recommend to reduce defects.

1.5 Construct a flow diagram of the process you use in preparing for an exam in a quantitative subject such as statistics.

1.6 Construct a cause and effect diagram where you can list your hypothesis (or opinions) as to the reasons for differing grades earned by students in a statistics class. Use the main categories: Materials (i.e., text, etc.), Man (students), Methods (study habits etc), and Measurements.

1.7 The results of this exercise may help your to better focus you time and study habits.

    a) Classify or categorize the activities you normally do during your waking hours Monday through Friday (i.e., eating, exercising, attending this class, talking to friends, watching TV, etc.)

    b) Count or approximate the number of hours you spend on each activity during a typical week.

    c) Construct a Pareto Diagram of your results in b).

    d) Discuss the non-essential activities that you might be able to reduce. Which would benefit you most in terms of increased study time?

1.8 Collect data on the points you lost in a sport/game that you play/watch categorized by reason the points were lost. Construct a Pareto Diagram for the data, and draw conclusions about the major weakness(es) that are exhibited. For example, if you play ping-pong (or volleyball, tennis, etc.), keep track of 100 points that you lost by category (net ball that could not be returned, set up opponent for an easy "kill," over-hit a ball, missed an easy shot, etc.).

1.9 Construct a Pareto Diagram for a hobby that occasionally has poor quality, and draw conclusions. For example, if you frequently take photographs, review 100 poor pictures and categorize the problems (too dark/light, out of focus, part of the scene cut off, too wide/narrow view, poor composition, etc.).

1.10 Develop a process flow diagram for a manufacturing process you are familiar with. An actual commercial process (from experience through a co-op or summer job) would be ideal, but a home process such as cooking (e.g., baking chocolate chip cookies), washing (clothes, dishes), or maintenance is also acceptable.

1.11 Draw a cause-effect diagram for a quality problem that you have. Possibilities include low grades in a particular course, lack of quality in some hobby, etc.

References

1.   ABET Conference on Statistics in the Engineering Curriculum, Leesburg Va., May 1989.

2.   Bisgaard S.(1989) "Teaching Statistics to Engineers", *American Statistical Association 1989 Proceedings of the Section on Statistical Education*, Alexandria Va.

3.   Box G.E.P. (1976) "Science and Statistics" *Journal of the American Statistical Association*, Vol 79., p. 971.

4.   Ibid.

5.   Wu, Yuin, "The Unknown Key to Japanese Productivity", Reprint available from American Supplier Institute, Dearborn Michigan. Ibid

6.   Box, G.E.P., Kackar, R.N., Nair, V.N., Phadke, M., Shoemaker, A.C. and Wu, C.F.J., "Quality Practices in Japan", *Quality Progress*, March 1988, pp37–41.

7.   Ibid.

8.   "Kokomo Castings Plant," *Statistical Process Control*, Vol. 1, June 1988, published by Quality & Productivity Statistical Methods Office, Chrysler Motors.

9.   Delott, C. and Gupta, P. (1990) "Characterization of Copperplating Process for Ceramic Substrates", Quality Engineering, Vol. 2, No.3, pp. 269–284, Marcel Dekker, Inc.

10. Hunter, W.G., "Managing Our Way to Economic Success: Two Untapped Resources" Technical Report No. $, Center for Quality and Productivity Improvement, University of Wisconsin, Madison Wisconsin, 1986.

CHAPTER 2
Concepts of Quality Control

2.1 Introduction

In the last chapter we discussed the role of statistics in the scientific method and illustrated the ideas with two industry examples. In this chapter we discuss quality control, one of the main areas where the scientific method and statistics are used in engineering. Along with this discussion we will introduce two important statistical techniques that are used frequently in quality control and engineering, these are: quality control charts and experimental design plans.

2.2 Historical Perspectives of Quality Control and Engineering

Prior to the industrial revolution, the jobs of engineering, manufacturing, and quality control were all performed by one individual, the craftsman. He was solely responsible for the design and production of goods, and for satisfying his customer. After the industrial revolution, and with the division of labor, the engineering function was separated from manufacturing. Therefore, the only way the designer, or engineer, could ensure the quality (or functionality) of a final product that was made of interchangeable parts subject to manufacturing imperfections, was to place specification limits on the critical dimensions and characteristics of the components. Because of this, the industrial definition of quality has come to be associated with the satisfaction of engineering specifications.

With his 1911 treatise, *The Principles of Scientific Management*[1], Frederick Taylor introduced the ideas of work standards and wage incentives. These ideas were widely accepted in western management, and the emphasis of the manufacturing department quickly became production levels not quality. An outgrowth of Taylor's principles was the establishment of inspection or quality departments in manufacturing firms[2]. The inspection department became the policeman to ensure engineering specifications were met, and more often than not, adversarial relationships developed between the engineering department, the production-focused manufacturing department, and the policing inspection or quality department. Because the inspection department did little more than separate good (i.e., meeting the engineering specs) and bad parts, the idea emerged that you couldn't have both high quality and high productivity, and unfortunately this belief that better quality products cost more to produce is still quite prevalent today.

The benefit of producing high quality at a low cost was recognized early by some. This was never more apparent than when the Bell Telephone Company was in the process of trying to establish affordable nationwide telephone communications. High-quality components and switching systems were critical, yet high cost had to be avoided. Therefore, the telephone company became one of the early leaders in seeking improved methods for quality control in manufacturing. Walter Shewhart, a physicist at the Bell Telephone Laboratories, published his book, *Economic Control of Quality of Manufactured Product*[3] in 1931. In this book he presented the idea that through in-process inspection and charting of data, manufacturing personnel could easily recognize when simple process adjustments were necessary. These simple process adjustments would prevent poor quality parts from being produced, and would cost virtually nothing since they were normally being performed anyway, albeit at the wrong times. Shewhart showed a direct connection between the costs of producing high quality and the variability in process output. The variability in process output made inspection (and its associated costs) necessary. Variation in output is the cause of scrapped or reworked parts (internal failure costs) and customer dissatisfaction (external failure costs). Shewhart claimed these costs could all be reduced or avoided if process variability could be minimized. If that could be done inexpensively, high quality was possible at low cost.

2.3 Quality Control and Statistics

An essential component in Shewhart's ideas was a statistical tool called the *control chart*. This chart made it possible to identify when a production process was predictable and running in a state of statistical control, or when *assignable causes* were making the quality of the process output deteriorate. By running in a state of statistical control, Shewhart did not mean the process could be controlled to produce parts with an undeviating dimension, or to produce parts with some unvarying characteristic, for there will always be some variability in manufactured output. What Shewhart meant was that the variability in controlled processes would appear the same as variation in casino gambling games: unpredictable in the short run but following predictable long-run frequency patterns. Thus, Shewhart named the cause of variation in statistically controlled processes to be *chance causes*. The control chart was the tool that Shewhart introduced to help determine whether assignable causes, or merely chance causes were affecting the variation in manufactured output. When the variation in process output is due solely to chance causes, there is no need for operators to make adjustments, and any attempt to do so will only increase the variability of output and number of defective parts. When variation is due to assignable causes, the control chart will indicate that an adjustment is needed. By using the control chart to determine when adjustments should be made, the variability in process output will be reduced at virtually no cost.

Shewhart's ideas were so valuable, that when World War II broke out, the U.S. War Production Board sponsored thousands of eight-day seminars to teach defense contractors the use of control charts and other statistical tools for controlling quality[4]. A master teacher in many of these courses was W. Edwards Deming who had been tutored by Shewhart. Deming was an eloquent advocate of Shewhart's ideas and he did much to clarify and add to them. Deming called Shewhart's chance causes for variation *common causes* and he called assignable causes *special causes*. He emphasized that finding the appropriate cause of variation was the most essential step in reducing variation. Deming explained that special causes were things such as a worn tool in a metal removal process, or a machine out of adjustment. These were things that happened from time to time, but could be compensated for by local machine operators, if they had control charts to help them identify when action was necessary. Common causes, on the other hand, were things that affected all production output, such as poor lighting in the plant, variable sources of raw materials, and inadequate training or instruction for operators.

Deming expounded the idea that common causes were responsible for 85% of the variation and poor quality in industry, and that only by action of management could these causes be reduced. Local operators had no influence on the purchase of raw materials or plant layout design. But, he explained that as an outgrowth of Taylor's work standards much of western management had become comfortable in the role of judges. They sought to judge workers performance against some numerical standard, and blame them for all problems of low production rates and low quality. The common belief had become "if workers wouldn't make any mistakes, there wouldn't be any problems with quality." Management wasn't looking for ways to reduce common causes of variation, they blamed everything on special causes and worker errors. Since, as Deming indicated, 85% of all variation and quality problems are actually due to common causes, management's role as judges created impediments to communication between workers and management and robbed the workers of their chance for pride in their work. Who wants to come to work with variable raw materials, poor workstation design and inadequate training and then be blamed for all the problems? In a situation like this, workers become demoralized and barriers develop between workers and management which further reduce the quality of product and productivity of the firm. Deming explained that management was unaware that there were barriers, and that only through their action could the barriers be removed.

To help management remove barriers between itself and workers and promote cooperation in attacking variation (the real cause for poor quality and low productivity), Deming formulated his 14 points for management that are summarized in Table 2.1. These points focus management's attention on the chronic systems problems rather than errors by individuals, and on the use of teamwork, cooperation and scientific methods to detect and correct problems. In-depth explanations and examples of each of the points can be found in Deming's own book *Out of the Crisis*[5].

Table 2.1 Deming's 14 Points for Management

1. Create constancy of purpose for improvement of product and service

2. Adopt the new philosophy; we are in a new economic age created by Japan

3. Cease dependence on mass inspection to ensure quality

4. End the practice of awarding business on the basis of price tag alone

5. Improve constantly and forever the *system* of production and service

6. Institute training on the job

7. Adopt and institute leadership

8. Drive out fear

9. Break down barriers between staff areas

10. Eliminate slogans, exhortations, and targets for the workforce

11. Eliminate work standards and numerical quotas

12. Remove barriers that rob people of pride of workmanship

13. Encourage education and self-improvement for everyone

14. Take action to accomplish the transformation

After World War II, U.S. companies curtailed production of defense equipment and returned to producing consumer goods such as automobiles, appliances, and radios. With consumer savings and appetites for goods at high levels, many of the lessons learned during the war-production efforts were quickly forgotten. With widespread shortages of consumer goods, emphasis was on production, not quality. With most of Europe's and Asia's manufacturing facilities disabled from the war, U.S. manufacturers had a virtual monopoly in world markets, and their success further encouraged the Taylor style of management. In the U.S., Deming became a forgotten civilian hero of World War II.

Meanwhile, industry was devastated in Japan. Japan had no natural resources and many of its manufacturing facilities had been demolished in the war. Economists thought the West would have to feed the Japanese for decades to come. Under MacArthur's occupation forces, the Japanese even struggled to reestablish telephone communications. Poor-quality components and switching systems caused much difficulty. It was at this point that American experts such as Deming , Juran, and others were called to teach Japanese businesses the quality control techniques that were so successful during the war effort. Deming first visited Japan in 1947 to advise the Allied Forces in sampling techniques. He was invited back by the Japanese Union of Scientists and Engineers (JUSE) in 1950, 1951, 1952, 1955, 1960, and 1965 as a teacher and consultant to industry. His major message was not to engineers or production supervisors, but rather to the top management of companies. He taught foremost, the simple principle summarized in Figure 2.1[6].

| Improve Quality | -> | Less Rework and Waste | -> | Productivity Improves | -> | Capture Market with Lower Price and Better Quality | -> | Stay in Business | -> | Provide Jobs |

Figure 2.1 Simple principle taught by Deming

With nothing to lose, Japanese companies wholeheartedly adopted both Shewhart's view of process control, and Deming's admonition to eliminate barriers between management, workers, and the workers' right to pride of workmanship. The result was a resounding success. Japanese products have become world standards for quality. A prestigious, national quality award was established in Japan and named after Deming, who in 1960 was himself awarded the 2nd Order Medal of the Sacred Treasures, by the emperor. This is the highest honor Japan bestows on foreigners. However, in the U.S., Deming's message was still largely ignored until 1980 when NBC aired the white paper *If Japan Can, Why Can't We.* This documentary portrayed Deming and his message as the major catalyst for Japanese companies that had taken worldwide markets away from slumping U.S. counterparts in the auto and electronics industries.

Since 1980, companies from the U.S. and other countries have become aware of the need to improve quality of products and services in order to remain competitive, and a new definition of quality has evolved. In modern terms, quality no longer means conformance to standards or specifications, but rather it means pleasing the customer. Companies that have been successful in making a change to this new philosophy have developed an entirely new style of management that has been referred to in Japan as CWQC (Company Wide Quality Control) and in the U.S. as TQM Total Quality Management). Table 2.2 summarizes five assumptions or paradigms that guide TQM.

Table 2.2 Total Quality Management (TQM)

1. Pleasing customers (quality) is the number one priority and driver for all business activities. Continuous improvement in this area is necessary for prosperity of the firm.

2. Most (85% - 95%) of the problems in pleasing customers are systems problems and not the errors or faults of individuals.

3. Cooperation, teamwork, and synergy involving all individuals with some system knowledge is the best way to solve systems problems.

4. Scientific methods and statistical facts, rather than unfounded opinions, should be the basis for all decisions made by workers and leaders in order to improve work systems.

5. Because of points 2 and 3, TQM is necessarily a management system based on mutual trust, respect, and cooperation between workers and management.

Point 4 in Table 2.2 emphasizes the need for decisions based on statistical facts and scientific methods. When management and workers focus on processes rather than individuals, and communicate via the results of objective statistical analysis, rather than opinions, there is much less chance of misunderstanding and adversarial relations. Statistical methods such as control charts can in fact become communication tools to foster teamwork and cooperation on the solution to systems problems. In the past, management assumed all problems were special causes due to worker errors. That not only left the root causes for problems untouched, but also created barriers that prevented cooperation on company objectives.

The U.S. Department of Commerce, in 1988, established the Malcolm Baldrige National Quality Award (MBNQA) that is given annually to companies' in several categories to recognize them for outstanding quality management and achievement. Awards are given based on companies applications that describe their approach, deployment, and results in several areas that are called the Baldrige Award Criteria. The prime purpose of the award is to encourage competitiveness of U.S. firms through education. Winning companies are required to share information on their strategies in quality management, and many firms use the Baldrige Award Criteria as a checklist for their own self-assessments and improvement programs. Figure 2.2 illustrates the direction that firms move when they implement the Baldrige Award Criteria as their management philosophy. As companies increase their customer-driven focus they improve reliability, on-time delivery, and additional customer pleasing features. When companies improve their operational performance they improve in areas such as internal efficiencies, effective use of resources, and reduced costs. Improvement along both axes in Figure 2.2 naturally leads a company to become more competitive. TQM is being encouraged by automakers and the U.S. Department of Defense to be used by all their suppliers.

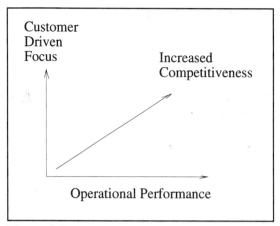

Figure 2.2: Results of implementing Baldrige Award Criteria

The need and benefits for companies adopting the TQM philosophy are threefold[7]. First, it is necessary for survival in today's competitive environment. A January 1992 special issue of *Business Week* reported that Japanese firms had captured 34% of the U.S. auto market. In 1979, Texas Instruments and Motorola were leading in worldwide sales of integrated circuits with a 57% market share. By 1990, they were down to less than 20% and Japanese suppliers NEC, Toshiba, and Hitachi were leading in sales worldwide. If American automakers, integrated circuit manufacturers, and others had not started to adopt a TQM management philosophy, similar to their Japanese competition, they may not have survived.

The second reason for adopting TQM is for profitability. A TQM management style leads to lower costs, higher customer acceptance, and increased profits and financial performance. The U. S. General Accounting Office issued a report in May 1991 titled "U.S. Companies Improve Performance through Quality Efforts ." This was a survey of the performance of companies that had applied for the Baldridge Award prior to 1991. The report showed that these companies reduced their cost of quality by an average 9% a year, and increased their market share 13.7% per year. Becker and Golomski[8] compared the revenues of the companies in the U.S. General Accounting Office survey, and others known to be using the TQM philosophy, to all U.S. manufacturing firms. They found that TQM firms increased revenue 8.3% per year on average compared to 4.2% on average for all manufacturing firms. In 1999, Hendricks and Singhal[9] compared winners of supplier and state quality awards based on Baldrige criteria to control companies from the same industries. They felt that award winners had demonstrated implementation of TQM, and found that the award winners substantially outperformed the controls in terms of operating income, sales, and asset growth over a 5-year period. With respect to shareholders, a 1995 news release from the United States National Institute of Standards and Technology (NIST) stated that quality management was a sound investment. NIST "invested" a hypothetical $1000 in each of the five publicly traded, whole company winners of the Malcolm Baldrige Award. The investment was tracked from the first business day in April of the year they won the award to Oct. 3, 1994. NIST found that investment in these companies outperformed the S&P 500 by 6.5 to 1, a 188 percent return on investment compared to 28 percent return for the S&P 500. Tai and Przasnyski[10] repeated this comparison in 1999, adjusting for risk and market movements, and found that, though not as spectacularly as had been shown in the NIST study, MBNQA winners did beat the S&P 500 index. They quoted Deming who said if customers are satisfied ultimately shareholders would be satisfied too.

The final reason for adopting a TQM philosophy is because it is the moral thing to do. Customers receive better product and service at a lower cost; companies stay in business and jobs are preserved; employees are treated with dignity and find more fulfillment in their work; shareholders receive greater return on their investment. Everyone wins and it's a better way of doing business. We would be remiss to do anything less.

Companies that adopt a TQM philosophy will necessarily use statistical methods. Table 2.3 lists the seven major categories or core values in the Baldrige Award Criteria. Category 2 is often referred to as the "brain center". To respond to this criteria companies must describe what data is collected in the company, how that data is analyzed, interpreted, shared, and used to make decisions regarding improvements in Customer Driven Focus and Operational Performance. This is practically the definition of what statistical methods are all about. To respond to Category 5 Management of Process Quality, companies must describe how they improve the design of their products, how they control and improve their manufacturing processes and support services, and how they manage and improve their supplier relations. In this book we will present the statistical methods of designed experiments and control charts. These are the most effective tools engineers can use to accomplish the core values required by Category 5. In the remainder of this chapter we will present three examples to illustrate the basic ideas of control charts and designed experiments, and show how they can be used in a TQM environment.

Table 2.3 Major Categories in Baldrige Award Criteria

1. Leadership

2. Information and Analysis

3. Strategic Quality Planning

4. Human Resource Development and Management

5. Management of Process Quality

6. Quality and Operational Results

7. Customer Focus and Satisfaction Results

2.4 Control Chart Examples

To better illustrate the concept of Shewhart control charts, we will show two industrial examples where these charts were used beneficially.

Example 1: The first example is taken from a company that manufactures high-speed wheel balancers that are used by auto tire dealers. Many of the parts for the wheel balancers are cylindrical in form and are produced on numerical controlled (NC) lathes. A particular part called the shank is illustrated in Figure 2.3, with one of its critical dimensions shown. Prior to using control charts to control this dimension, the NC lathe operators followed a typical procedure. They were to take one shank every 10 minutes and measure the critical dimension. If a measurement exceeded the upper specification limit, the operator adjusted the insert or replaced the worn tool and then made the adjustment. Since the major assignable cause for variation in this lathe operation was tool wear, a graph of measured dimensions during a typical manufacturing run looked like a classical tool wear curve, as shown in Figure 2.4.

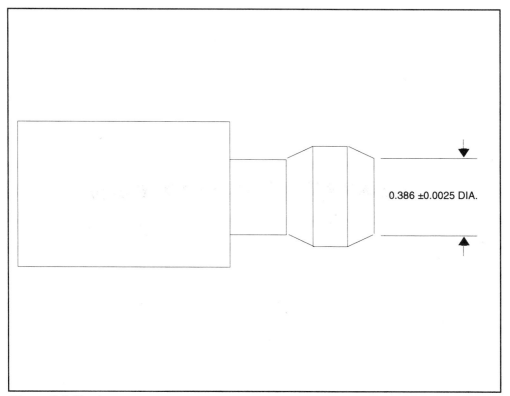

Figure 2.3 Shank

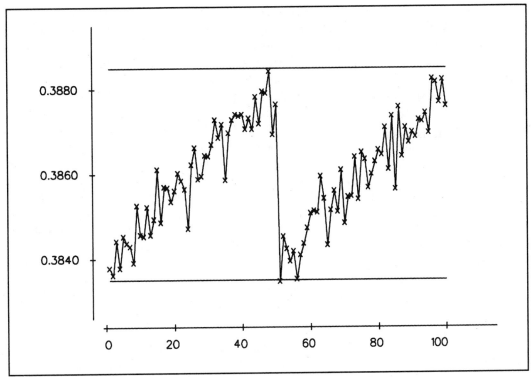

Figure 2.4 Tool Wear Curve in Measured Diameters

In Figure 2.4, which contains the upper and lower specification limits, it can be seen that the operator's procedure was producing parts with dimensions within specifications. Actually less than 1% were ever found to be out of spec. Initially, the manufacturing manager was resistant to changing the operator's procedure, and felt there was no need to use control charts or any other means to hold a tighter tolerance. However, there was waste associated with the current procedure.

End of the line sampling inspections were routinely conducted and indeed a small proportion (<1%) of parts were rejected. In addition, occasional problems occurred in assembly and testing of the final product. If parts didn't fit together well, because one was at the low end of its spec range and an adjacent part was at the high end of its spec range, the wheel balancer wouldn't function correctly during the final test after assembly. This necessitated a rework area for assembled products that didn't pass final inspection. Here they were partially disassembled, reassembled with different components by trial and error and then retested until they passed the final test. If all parts were manufactured closer to their nominal dimension, there would be no need for end of the line inspection or a rework area. There were direct and measurable labor costs for inspection and rework that could be avoided if a tighter tolerance could be held. There were other costs that were sometimes more difficult to quantify.

When the product was in the customers' use, wear would occur to internal rotating parts due to friction, and the balancer would then occasionally fail. To alleviate this problem the company maintained a nationwide fleet of repair vans that could be quickly sent to a customer's shop. If the balancer was under warranty, the company would cover the repair cost, otherwise the customer would. Wheel balancers with all components close to their nominal specifications would be more likely to remain in failure-free service longer than those balancers with some components near the specification limits at assembly. Field failures, whether under warranty or not, were costs the company had to absorb. Costs of maintaining the fleet of repair vans and costs of repairing units on warranty could be measured. However, the larger cost due to loss of repeat sales to unhappy customers may be difficult to quantify, yet real nonetheless.

Therefore, additional costs were being incurred because a tighter tolerance was not being held, but how could a tighter tolerance be maintained? No one is to blame for the problem except the management policy. Instead of a policy that called for operators to adjust or replace a tool when part dimensions exceed the tolerance limits, as shown in Figure 2.3, tighter limits could be defined. But, how tight could the limits be made before the natural variability due to common (or chance) causes would render them counterproductive? An appropriate control chart (a statistical tool) provides the answer to this question. The control limits on a control chart are based on the natural common cause variability. The limits are based on the normal distribution which is discussed in Chapter 3, and the details of calculating control limits are presented in Chapter 17. For now, just assume the limits have been established and are used to adjust the process. We will illustrate how control charts might be used if the operator's procedure was changed from measuring one part every 10 minutes to measuring five consecutively produced parts every 50 minutes.

Table 2.4 shows measurements of the dimension shown in Figure 2.3, from a hypothetical production run using the changed measurement procedure. Five consecutively produced parts were measured every 50 minutes, and the values along with ranges (maximum minus minimum) and averages are shown by time in the table. Since it is unlikely that tool wear, tool breakage, a slip in tool adjustment, or other assignable causes for variation would occur within a consecutive set of 5 parts, the range for each sample of 5 measurements is an indication of the variation caused by chance causes (or common causes). Figure 2.5, shows a Range control chart for the ranges in Table 2.4. The center line (0.001122) is the average range ($\bar{R}$), and the upper and lower control limits are $D_4\bar{R}$ and $D_3\bar{R}$ where the factors $D_4=2.114$ and $D_3=0.0$ are given in Appendix Table A.2 and will be explained in more detail in Chapter 17. In this figure, the random pattern of the points within the control limits indicates that there are no special or assignable causes of variability within the individual groups of measurements, and that the average range $\bar{R}=0.001122$ is a good measure of the common cause variability.

Table 2.4 Shank Dimensions

Time	Measurements					Range	Average
8:30	0.3839	0.3837	0.3844	0.3837	0.3843	0.0007	0.3840
9:20	0.3845	0.3843	0.3839	0.3851	0.3844	0.0013	0.3844
10:10	0.3846	0.3853	0.3845	0.3848	0.3859	0.0014	0.3850
11:00	0.3850	0.3857	0.3856	0.3852	0.3854	0.0008	0.3854
11:50	0.3861	0.3859	0.3856	0.3846	0.3860	0.0016	0.3856
12:40	0.3867	0.3859	0.3859	0.3863	0.3862	0.0008	0.3862
1:30	0.3868	0.3873	0.3868	0.3870	0.3856	0.0017	0.3867
2:20	0.3871	0.3873	0.3874	0.3872	0.3872	0.0003	0.3872
3:10	0.3872	0.3873	0.3870	0.3877	0.3870	0.0007	0.3872
4:00	0.3881	0.3879	0.3884	0.3868	0.3874	0.0016	0.3877
4:50	0.3836	0.3846	0.3842	0.3838	0.3840	0.0010	0.3840
5:40	0.3837	0.3841	0.3843	0.3846	0.3849	0.0012	0.3843
6:30	0.3853	0.3851	0.3859	0.3853	0.3841	0.0018	0.3851
7:20	0.3853	0.3856	0.3851	0.3860	0.3846	0.0013	0.3853
8:10	0.3856	0.3855	0.3864	0.3853	0.3863	0.0011	0.3858
9:00	0.3865	0.3857	0.3860	0.3862	0.3864	0.0008	0.3861
9:50	0.3866	0.3871	0.3861	0.3872	0.3854	0.0018	0.3865
10:40	0.3877	0.3865	0.3870	0.3866	0.3868	0.0012	0.3869
11:30	0.3870	0.3873	0.3872	0.3873	0.3868	0.0006	0.3871
12:20	0.3884	0.3882	0.3877	0.3881	0.3874	0.0010	0.3879

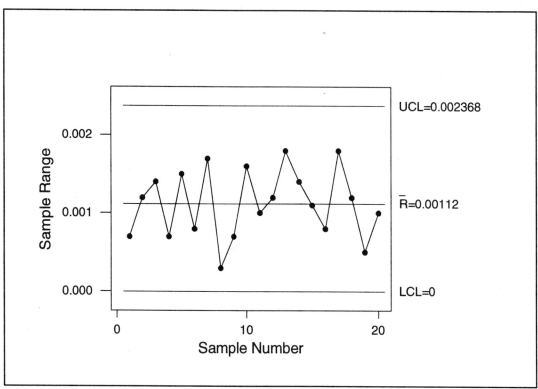

Figure 2.5 Range Control Chart for Table 2.4

Figure 2.6 is a control chart for averages. The points are the group averages shown in Table 2.3. The centerline is the overall average $\overline{\overline{X}}=0.38594$, and the upper and lower control limits are given by $\overline{\overline{X}} \pm A_2 \overline{R}$, where the factor $A_2=0.577$ is also given in Appendix Table A.2. The distance, $A_2 \overline{R}$, of the control limits from the overall average, is a multiple of $\overline{R}$ which represents the natural variability of the process due to common (or chance) causes. From this control chart, it can be seen that there are many points out of the limits indicating the special cause of tool wear. If the operators were to use the control chart effectively during production, they would make an adjustment at the first sign of a point going beyond the control limits. This would result in a chart similar to Figure 2.7.

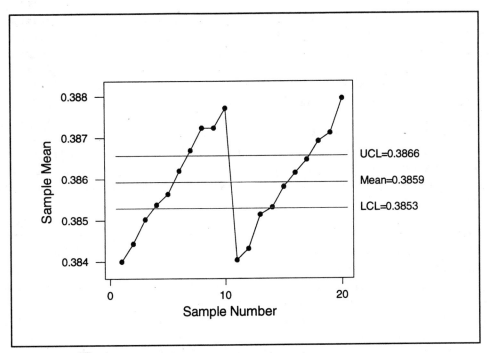

Figure 2.6 $\overline{X}$ Chart for Table 2.4

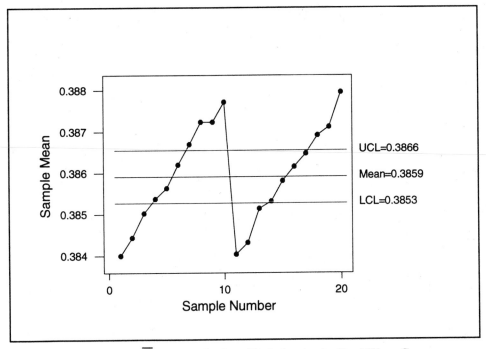

Figure 2.7 Using the $\overline{X}$ Chart to Reduce Variation in the Tool Wear Curve

The data used to construct Figure 2.7 would be similar to the data shown in Table 2.4. However, as the operator collected the data, he or she would plot the average for each set of consecutive measurements on the chart. Thus the chart would be developed as the data were collected. The control limits (based on previous data such as Table 2.4) would be placed on the chart prior to the operator collecting the current data. Each time a point exceeded the upper control limit, or any other criteria established by the production management, the operator would take action and adjust the tool setting. If the individual measurements from Figure 2.7 were graphed as in Figure 2.4, the result would look like Figure 2.8. There it can be seen that the individual measurements are now well within the specification limits. In fact, the overall variability has been reduced 50%, at virtually no cost. The operators make the same number of measurements and the same type of machine adjustments (although they do make them more frequently).

The control chart provides the guidance as to when the adjustments should be made. The control limits are based on the natural variability that would be expected due to piece-to-piece raw material differences and other common causes. If limits tighter than the control limits were used as a guide for making machine adjustments, the operators would be fooled by the common cause variation and actually adjust the machine too often. Adjustments made too often is what Deming has called *tampering*, and it can actually introduce more variation in the final dimensions in addition to causing extra costs due to machine down time. If wider limits than the control limits were used as a guide for making adjustments, wider variation as shown in Figure 2.3 would result. Thus the control limits provide the optimal guide, and failure to use them will cause any manufacturing process to run less efficiently than it otherwise could; just like driving a car that is out of tune. In harmony with the TQM philosophy, the use of control charts can help the company management to focus on the system or procedure that machine operators were following, rather than affix blame for problems on machine operators or assembly errors.

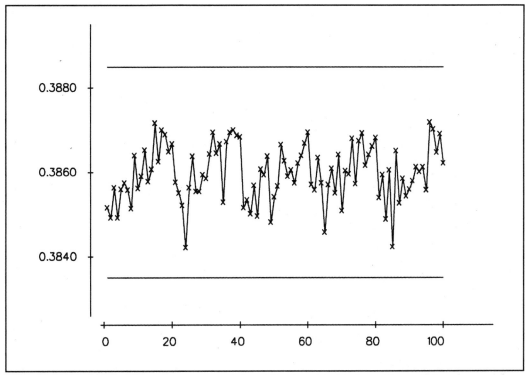

Figure 2.8 Individual Measurements and Specification Limits from Figure 2.7

Example 2: A second example of the use of control charts in industry is a continuation of the example in the Kokomo Casting Plant from Chapter 1. In Chapter 1, Pareto Diagrams were used to show that the majority of defects in the castings produced for NPG at the Kokomo plant were due to porosity. In addition to making Pareto Diagrams, PQI (Process Quality Improvement) teams at Kokomo made Shewhart control charts of counts of porosity defects. (note: the control charts presented here were simulated to represent the descriptions in reference 8 of Chapter 1. The actual data and charts were not provided in the article). Counts of porosity defects from 100 castings were plotted on a C-chart (where C denotes counts) in Figure 2.9. The center line on the chart is $\bar{c}$, the average defect count, and the upper and lower control limits are $\bar{c} \pm 3\sqrt{\bar{c}}$, which will again be explained in detail in Chapter 18.

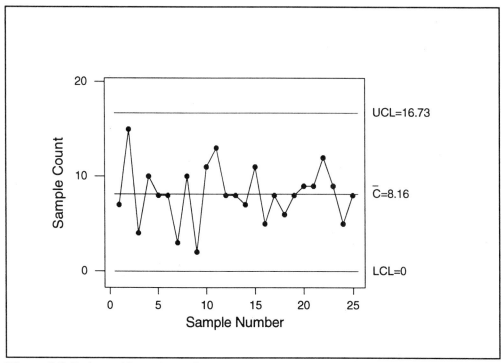

Figure 2.9 C-chart of Porosity Defects

The C-chart in Figure 2.9 shows a random pattern of points within the control limits. Since there are no points out of the limits nor are there any unusual trends, the variation in porosity defects is solely due to common causes (or, as Shewhart would call them, chance causes). This chart indicates that there are no assignable causes. This is different than the case in the previous example, where special causes such as a worn tool showed up as trends and points out of control on the control chart. Casting operators can make adjustments to the process (that might affect the porosity) like changing the pouring temperature, or varying their pouring technique. However, unlike the lathe operation in Example 1, it would be unproductive to ask operators to make an adjustment after a day with a higher than normal defect count (such as the second day on the chart). Since none of the points fall outside the control limits, there are no assignable causes. Making adjustments when there are no assignable causes would again be what Deming calls tampering, and may actually make the long run average defect count higher.

Since the chart indicates only common causes for porosity defects, the PQI teams at Kokomo looked for something that might affect porosity defects for all castings at all times, rather than trying to determine the specific cause for high defect days, or determine correcting adjustments. Thus, in examining the casting forms (used for all castings) and the location of defects as described in Chapter 1, they hypothesized that adding feeders to the casting form might decrease the number of porosity defects. But, hypothesizing a solution is just one step in the scientific process, the next step is to implement the change and collect more data.

The control chart is also an excellent tool for visualizing the effect of a process change, such as that undertaken at Kokomo. If data before and after the process change were plotted on a C-chart, with control limits determined from the before data, we would see a figure such as Figure 2.10. In this figure, we again see no points out of the limits. However, the string of consecutive points below the center line after incorporation of the process change is not random and represents an out of control signal. It gives strong evidence that the process change actually decreased the defect rate. If the points after the process change would continue to produce a random pattern within the limits, it would indicate that the process change had virtually no effect. We must rely on something like the chart to make a reliable conclusion. Simply comparing the average defect counts before and after a process change can be misleading without knowledge of the normal day-to-day variability that is displayed on the control chart.

The two examples above illustrate the basic idea of Shewhart's control charts, composed of a center line and upper and lower control limits. These charts are useful in separating the causes

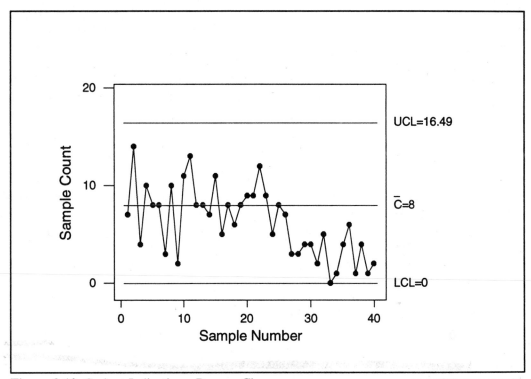

Figure 2.10 C-chart Indicating a Process Change

of variation between common causes that affect all production, and special causes that normally can be adjusted or corrected for by the local operator. This distinction helps to define the role and responsibilities of management and workers, and reduce tension between them in the workplace.

39

Only management can authorize changes that affect all production. For example, it is not the casting operator's job to redesign the casting form. Nor is it management's job to approve all tool adjustments on a lathe operation. Control charts can be used by operators to determine when to make an adjustment and when to leave the process alone. Control charts can help management to quantify what the process is currently producing, and to determine if process improvement efforts actually had a beneficial effect.

2.5 Quality Improvement in Engineering Design

Quality control is an activity usually associated with manufacturing. Quality, as Shewhart defined it, was inseparably connected to the reduction of variation, and most variation in part dimensions and function are assumed to be created during manufacturing. The reduction in special and common causes of variation are manufacturing related activities. However, shortly after World War II while working in the ECL (Electronics Communication Laboratories of Nippon Telephone and Telegraph Company, Japan's equivalent of the Bell Telephone Labs), Genichi Taguchi formulated a different viewpoint of quality control. Instead of reducing variation caused in manufacturing, Taguchi sought to counteract the influence of factors that introduce variation in the response by better engineering design.

Although the failure of many telephone switching systems could be traced to faulty or off-target components, in 1948 Japanese companies could not afford the process controls and instrumentation necessary to reduce common causes, hold tighter tolerances, and produce more uniform components. Taguchi was an engineer, working at the ECL, trying to develop reliable switching systems by trial and error using the poor quality components he had available. He learned from American advisors the statistical process control techniques of Shewhart, and the experimental design techniques that had been used extensively in United States and Great Britain for agricultural research and chemical process improvement and optimization. The experimental design techniques appeared more relevant to Taguchi's job than did control charts.

He realized that experimental design techniques had originated in agricultural experimentation. There they were often used to find hybrid varieties of crops that would be resistant to environmental factors such as drought or blight. Taguchi equated variation in poor quality switching components (over which he had no control in 1948) to the variation in environmental factors faced by the agricultural researcher. He used experimental design techniques to systematically examine different variations of circuit designs, in order to find one that would work consistently with the poor quality components he had available. He was successful in doing this and started what has now been called *off-line quality control*. Off-line refers to quality control activities conducted in the engineering design stage, rather than on the manufacturing line.

One of the early and most famous examples of off-line quality control was published in the book *Sangyu Furontea Mongatari* (*Frontier Stories in Industry*, published by Diamond Sha Publishing Company in Japan). This example took place in 1953, when the Ina Seitō Tile Company was having difficulty in controlling the size of their tiles coming out of the kiln. The cause of the size variation was traced to variation in the temperature gradient in the kiln (a common cause). Rather than trying to reduce the common cause of variation, which would have required expensive investment in instrumentation and controllers, the engineers tried to reduce the influence of the temperature gradient by experimenting with various tile clay formulations. They systematically varied different components in the clay formulation according to an experimental design plan. A portion of their results is shown in Table 2.5.

Table 2.5 Ina Seitō Tile Experiment Results

	Orthogonal Table L_8							Definition of Factors A, B and D			Number of Defectives in 100 Tiles
								Content of a certain lime(%)	Fineness	Kind of Agalmotolite	
Factor	A	B	C	D	E	F	G	1	2	4	
No. Column	1	2	3	4	5	6	7	1	2	4	
1	-	-	-	-	-	-	-	5	Coarser	Existing	16
2	-	-	-	+	+	+	+	5	Coarser	New	17
3	-	+	+	-	-	+	+	5	Finer	Existing	12
4	-	+	+	+	+	-	-	5	Finer	New	6
5	+	-	+	-	+	-	+	1	Coarser	Existing	6
6	+	-	+	+	-	+	-	1	Coarser	New	68
7	+	+	-	-	+	+	+	1	Finer	Existing	42
8	+	+	-	+	-	-	+	1	Finer	New	26

In this table, the first column (No.) represents an experimental tile formulation number. The second column (the Orthogonal table which is actually a group of 7 columns) represents an experimental plan in coded units. The third, fourth and fifth columns represent translations of the codes - and + from the first, second, and fourth columns of the Orthogonal array into actual levels of components in the tile formulations. In a sense, each row under columns 3, 4, and 5 represent the recipe for an experimental tile formulation. The last column represents the number of defective tiles (out of 100 measured) for each of the 8 experimental tile formulations tested. These results are represented graphically on a cube in Figure 2.11. In this figure it can be seen that when 5% of a certain lime was added to the formulation (back of the cube), the number of defective tiles was reduced to 12.5% (average of 16, 17, 12, and 6) from 35.5% (average of 6, 26, 42, and 68). This was accomplished without any investment in oven temperature controllers. Further confirmation experiments verified that this improvement could be sustained with this formulation change.

Judging that an increased percentage of the lime will in the future reduce tile size variation is called an *inference*. Normally an inference is drawn from experimental results based on a *decision rule* that is something like control limits in control charts. In this case, the decision rule would be any change in the average number of defective tiles (from one amount of a tile component to another) greater than 20.77% would be considered important. Fineness and type of Agalmotolite, the other two factors, would not be considered important since they only produced changes of 5.25% and 10.25% respectively. The inference that increasing the lime percent actually reduced tile variation was confirmed by continuing to run the process with the higher percentage of lime.

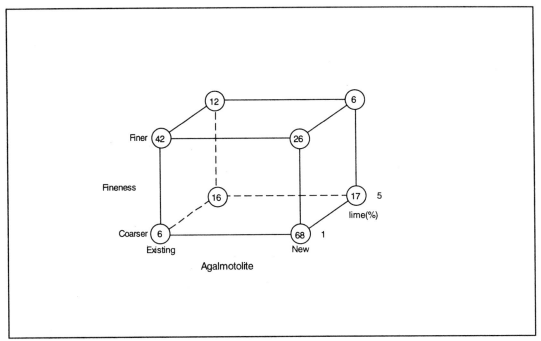

Figure 2.11 Tile Formulation Experiment

Many other examples of off-line quality control in Japan involved electric circuits. Although it was difficult to tightly control the nominal properties of electrical components such as resistors, capacitors, and transistors made in the early 50s, engineers successfully used designed experiment plans to find combinations of target (or nominal) values for each circuit component that would make the assembled circuits work, despite variations from the target in individual components. Taguchi called this off-line quality control activity *parameter design* since it involved designing the parameters of a circuit in terms of nominal values for each component. Japanese electronics companies used parameter design to modify existing circuit designs so that they would work reliably with very inexpensive and low quality components (ones which are highly variable from one to the next). They were so successful in producing reliable and inexpensive products, that by 1980 most U.S. companies were forced out of the consumer electronics market.

On a grant from a Japanese University, Genichi Taguchi visited the U.S. in 1980 to discuss his ideas about quality control in engineering. His stated purpose for his visit was to repay the U.S. for the assistance they had provided Japan in the area of quality control after the war. Among others, he visited the AT&T Quality Assurance Center in Homdale, N.J. There, with Taguchi's help, a case study was conducted in the design of parameters in an integrated circuit manufacturing process. This case study was very successful and resulted in a four-fold reduction in process variation, a threefold reduction in fatal defects, and a two-fold reduction in processing time after only a few weeks of experimentation.

American industry has recognized the value of using quality control techniques in engineering design. AT&T, The American Supplier Institute, and many others have extensive training courses in Taguchi Methods that are taken by literally thousands of working engineers in the U.S. each year. Many successful case studies are published in technical journals each year, and many more are kept secret for proprietary reasons. Taguchi Methods basically involve the use of experimental design techniques in the engineering stage in order to reduce the influence of uncontrollable variation upon product performance. Later in this book, we will present an in-depth review of experimental design concepts in Part III and IV (Chapters 7 - 15), and an in-depth review of quality control and Taguchi Methods in Part V (Chapters 16 - 18).

2.6 Summary

In this chapter we have described the interrelationships between engineering, quality control and statistical methods such as control charts and experimental design plans. In today's competitive marketplace, companies must produce high-quality products at reasonable costs if they are to remain in business. Engineers who design products and production lines to produce products need to know the established techniques for designing and producing quality products. These techniques include Total Quality Management (TQM) and statistical tools such as control charts and experimental design plans that are presented in this book. The following is a list of the key concepts and ideas presented in this chapter:

2.6 Exercises and Review Questions for Chapter 2

2.1 Briefly describe in your own words the contributions of the following individuals: Taylor, Shewhart, Deming, Taguchi, and the affects of their contribution on quality of manufactured products.

2.2 Briefly define or describe TQM, and the benefits of utilizing it.

2.3 a) Briefly describe your method of commuting from home to class.
   b) Describe some of the causes for delays in the time it takes you to make the trip.
   c) Classify each of the causes you list in b) as common causes or special causes and explain why.

2.4 Explain who should be responsible for eliminating special causes of variation in manufacturing processes, and who should be responsible for reducing or controlling common causes of variation.

2.5 Briefly describe why the following claim is false: "higher quality is always more costly"

2.6 Briefly explain the value of control charts. Use examples from the chapter if it helps you to explain.

2.7 What was the purpose for Deming's 14 points.

2.8 Briefly describe the Baldrige Award and give reasons why a company might apply for it or implement its criteria in business practices.

2.9 Find an article in a recent (last 12 months) "popular" magazine (e.g. Newsweek, Business Week, etc.) relating to quality control and/or quality management. Summarize the article in 100-150 words.

2.10 Look up Malcolm Baldrige, Genichi Taguchi, Kaoru Ishikawa, or W. Edwards Deming and write a 100-150 word summary of his accomplishments.

2.11 Find an article in a recent technical journal (e.g. Journal of Quality Technology, Quality Progress, etc.) on the use of control charts for on-line quality control or experimental designs used for off-line quality control. Summarize the article in 100-150 words.

References

1. Taylor, F. W., *Principles of Scientific Management*, (New York: Harper & Brothers,1911).

2.Juran, J .M., *Juran on Leadership for Quality: An Executive Handbook*, (New York: Free Press,1989), p. 4.

3. Shewhart, W. A., *Economic Control of Quality of Manufactured Product*, (New York: D. Van Nostrand, 1931). Reprinted by the American Society for Quality Control, Milwaukee, Wis.

4. "Quality Goes to War" *Quality Progress,* Vol. 24, No. 12 (1991)

5. Deming, W. E., *Out of the Crisis*, (Cambridge, MA: M.I.T. Press, 1989) .

6."Quality Goes to War," Ibid.

7. Coppola, A., "The Case for TQM, *RAC Journal,* Vol. 1, No 3, 4th Quarter 1993.

8.Becker, S. W., and Golmski, W. A., "TQM and the Organization of the Firm: Theoretical and Empirical Perspectives", *Quality Management Journal,* January, 1994 pp. 18–24.

9.Hendricks, K. B., and Sighal, V. R., "Don't Count TQM Out", *Quality Progress*, April 1999, pp. 35–42.

10.Tai, L. S., and Przanyski, Z. H., "Baldridge Award Winners Beat the S&P 500", *Quality Progress*, April 1999, pp. 45–51.

Part II
Basic Tools

In this section we present some of the basic descriptive and inferential statistical methods that will be used throughout the rest of the book. Chapter 3 is a brief overview of probability theory that is the basis for inferential statistics. Chapters 4 and 5 present descriptive and graphical tools for examining data. Chapter 6 is an overview of some basic tools of statistical inference.

CHAPTER 3
Theoretical Background: Probability and Statistics

3.1 Introduction

In this chapter we will lay the theoretical foundations for decision rules used in the scientific method. When data are collected to confirm or reject some preconceived hypothesis, a decision is reached by knowingly or unknowingly following some rule for comparing the data to the expected results. Decision rules of this type are commonly used by engineers for: 1) deciding when to leave a manufacturing process alone and when to make adjustments to correct an apparent problem and, 2) deciding when to believe certain product design changes or manufacturing process alterations resulted in improved performance based on limited test data. Formal decision rules used for these situations are control chart limits for monitoring process performance and confidence (significance) limits for judging test results.

One of the problems in establishing decision rules for dealing with manufacturing data and experimental test results is inherent variability in data. Tests conducted under the same conditions and manufacturing runs made at the same process settings, usually do not give identical results. The results differ due to chance causes from many unknown sources. One goal of experimentation, research, and quality control in manufacturing is to separate assignable causes for variation from chance or random causes. The engineer would like to know if the *seemingly* improved product is a result of his changes to the production process (or product design), or the sum of many chance causes that normally fluctuate in production and product use.

Probability theory has shown that chance variations follow definite patterns. Initially, mathematicians developed probability theory to help gamblers evaluate their betting strategies. Later researchers in the sciences have learned to use the laws of probability to separate the effects of chance influences and identify the effects of their manipulated variables. We at this point introduce the basic ideas of probability theory which will help us to establish limits for control charts and draw inferences from data resulting from experiments.

3.2 Probability and Random Variables

The throwing of dice or the tossing of a coin are examples of *random trials* in that they can be performed in exactly the same way at the same place and yet will yield unpredictably different results, solely due to what Shewhart would call chance causes. A random trial is one whose result is unpredictable in the short run, but forms a definite pattern in the long run. For example, a single toss of a coin is unpredictable, but after a large number of tosses approximately 50% will be heads. Consider the random trial of tossing five coins simultaneously: penny, nickel, dime, quarter, and half dollar. A numerically measured outcome of the random trial, such as the number of heads, is termed a *random variable*, i.e., one whose value is unpredictable. The *sample space* is defined to be the set of all possible outcomes of a random trial, often expressed in terms of values of the random variable. For example, in the random trial of tossing five coins simultaneously with the random variable being the number of heads, the sample space would be {0, 1, 2, 3, 4, 5} which represents all possible numbers of heads that could result. An *event* is any one (or combination) of outcomes of a random trial. Like the sample space, events are often expressed in terms of values of the random variable, for example the event of getting three or more heads in the toss of five coins is the set {3, 4, 5}.

Two random trials, or two random variables, are said to be *independent* if the outcome or value of the first in no way increases the predictability of the second. For example, knowing the value of the first of two thrown dice in no way helps to predict the second.

The *probability of an event* is a number between 0 and 1 which represents the certainty with which the event will occur (1 if completely certain, 0 if impossible). Suppose we have a coin, which is evenly balanced and which will not stand on edge. With such a coin, the chances of tossing a head or a tail are identical and one or the other will certainly turn up. The probability of getting either a head or a tail on a toss is 1, and there is a 50% chance of a head and a 50% chance of a tail. These facts can be expressed by:

$$P(H) + P(T) = 1$$
$$P(H) = P(T) = 0.5$$

which is read as:

"The probability of a head plus the probability of a tail is 1 (or certainty)" and "The probability of a head equals the probability of a tail which equals 0.5."

The first bar chart in Figure 3.1 symbolizes this situation. There are only two possible outcomes (or events), heads or tails, and the probability of each (the height of the bar) is the same, namely 0.5. We can consider the random variable to be the number of heads (0 or 1) and 0.5 is the probability that the random variable assumes either 0 or 1.

If we flip two coins there are four possible outcomes, two of which are equivalent in terms of the random variable (number of heads). See Table 3.1.

Table 3.1 Possible Outcomes for Two Coins

1st Coin	2nd Coin	Number of Heads
T	T	0
T	H	} 1
H	T	
H	H	2

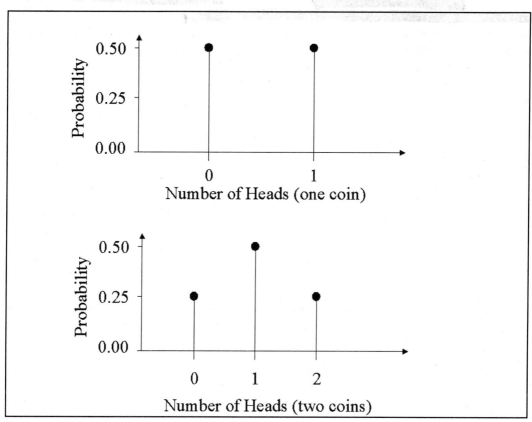

Figure 3.1 Bar Charts of Probability

A *joint event* or *intersection event* is defined to occur when two events both occur, or is defined to be the set of outcomes that is common to both events. For example if we call A the event of getting a head on the first coin (a penny) and B the event of getting a head on the second coin (a nickel), then AB or A∩B is defined to be the joint event of getting a head on both coins. This corresponds to the last outcome in Table 3.1. Each of the four outcomes in Table 3.1 can all be seen to be different joint events defined in terms of events for the 1st coin and events for the 2nd coin. If the individual events are independent, the probability of joint events can be calculated by multiplying the probabilities of the individual events. Mathematically we say that:

$$P(AB) = P(A \cap B) = P(A) \times P(B). \qquad (3.1)$$

Which in words says, "The probability of the joint event AB (which is the same thing as the intersection of the two events, A∩B) is equal to the product of the probability of event A and the probability of event B." Equation (1) is called the *multiplication rule* for independent events.

Since the outcome of one coin toss is independent from another, we can use the multiplication rule to find the probabilities of the four individual events shown in Table 3.2. Let us tabulate the probabilities of all of the possible outcomes:

Table 3.2 Probabilities of Joint Events

1st Coin		2nd Coin		Joint	
Event	Prob.	Event	Prob.	Event	Prob.
T	.5	T	.5	(T, T)	.5×.5 = .25
T	.5	H	.5	(T, H)	.5×.5 = .25
H	.5	T	.5	(H, T)	.5×.5 = .25
H	.5	H	.5	(H, H)	.5×.5 = .25

A more general multiplication rule for probabilities is:

$$P(AB) = P(A \cap B) = P(A) \times P(B|A)$$

which leads to the definition of the *conditional probability* of an event B occurring given that an event A has occurred. The conditional probability is denoted by $P(B|A)$ and is equal to $P(A \cap B)/P(A)$ by solving the equation above. This rule is useful for computing joint events in finite populations. For example, what is the probability of drawing two consecutive kings from a deck of 52 face cards? This can be written as $P(K_1 \cap K_2) = P(K_1) \times P(K_2|K_1) = (4/52)(3/51)$. Where 4/52 is the probability of drawing a king on the first card from a shuffled deck, and 3/51 is the probability of drawing a second king, given that one has already been removed from the deck. Conditional probabilities are used for evaluating sampling plans with small lots of items.

There is also an *addition rule* for probabilities. It can be used to find the probability of an event by summing the probabilities of the individual outcomes that constitute that event. For example the event of getting exactly one head from the toss of two coins is comprised of two individual outcomes in Table 3.2, namely (T,H) and (H,T). Therefore the probability of exactly one head is: .25 + .25 = .50. The addition rule can also be used to find the probabilities of events that are comprised of two *disjoint events*. Disjoint events are events that have no outcomes in common. For example, the event, A, of getting four or more heads in the toss of five coins is comprised of the outcomes (H,H,H,H,H), (T,H,H,H,H), (H,T,H,H,H), (H,H,T,H,H), (H,H,H,T,H) and (H,H,H,H,T). The event, B, of getting 1 or less heads is comprised of the outcomes (T,T,T,T,T), (H,T,T,T,T), (T,H,T,T,T), (T,T,H,T,T), (T,T,T,H,T) and (T,T,T,T,H). These two events are disjoint since they have no individual outcomes in common. The addition rule says that the probability of the combined event A+B (also known as the union of the two events, A∪B) is equal to the sum of the probabilities of the two disjoint events. Mathematically we write:

$$P(A+B) = P(A \cup B) = P(A) + P(B). \tag{3.2}$$

Equation 2 is the addition rule for disjoint events. In terms of the sets A and B defined above, this says that the probability of getting either 4 or more heads or 1 or less heads is equal to the probability of getting 4 or more heads plus the probability of getting 1 or less heads.

Using both the multiplication rule and the addition rule we can calculate the probabilities of the three events, defined in Table 3.3 in terms of the number of heads. These probabilities are shown in the second bar chart in Figure 3.1.

Table 3.3 Probabilities of Events

1st Coin		2nd Coin		Joint	
Event	Prob.	Event	Prob.	Event	Prob.
T	.5	T	.5	0 heads	.5×.5 = .25
T	.5	H	.5		
			}	1 head	.5×.5+.5×.5=.50
H	.5	T	.5		
H	.5	H	.5	2 heads	.5×.5 = .25

3.3 Discrete Random Variables and Distributions

3.3.1 Introduction

The number of heads in the toss of five coins is one example of a random variable. There are actually two types of random variables that will be useful to us. The first type, *discrete random*

53

variables, can only take on a finite, or countably infinite number of values. The second type of random variables are called *continuous random variables* and can take on any value in a segment (or the entirety) of the real number line. The number of heads is an example of a discrete random variable.

In industrial processes much of the data that can be collected can be thought of as observed values of random variables. For example, when an NC lathe process is in statistical control, the shank dimensions studied in Section 2.3 would be an example of a continuous random variable. When the casting process discussed in Section 1.4 is in statistical control, the number of porosity defects would be considered a discrete random variable. In this section we will discuss discrete random variables in detail.

Symbolically, we usually refer to a random variable with a capital letter such as Y or X. The probability that a discrete random variable takes on a particular value, y, can then be written as P(Y=y). Lower case letters are usually used to represent values that the random variable can take on. The function p(y) = P(Y=y) that assigns a probability to each value that the random variable can assume is called the *probability mass function* or *discrete probability density function*. The bar charts in Figure 3.1 are actually graphical representations of two specific probability mass functions. Values that the random variable can take on are represented on the horizontal axis of the bar chart. The probability that the random variable assumes a specific value is the height of the bar and is represented by the probability mass function, p(y). Probability mass functions have two properties. The first is $0 \leq p(y) \leq 1$, or probability values assigned to the possible values, y, that the random

variable takes on are always between zero and one. The second property is that $\sum_{y} p(y) = 1$, or

the sum of the probability values assigned by the probability mass function over all the values, y, that the random variable can take on is equal to 1.

The function that assigns probabilities to events of the type $\{Y \leq y\}$, i.e., $F(y) = P(Y \leq y)$, is called the *cumulative distribution function* or *CDF*. From the CDF, we can directly determine the probability that the random variable will take on a value less than or equal to any value y. The distribution function has the property that $0 \leq F(y) \leq 1$, for any value y, since F(y) represents a probability. For discrete random variables, the CDF is a step function as shown in Figure 3.2, and

can be written in terms of the probability mass function as $F(y) = \sum_{k \leq y} p(k)$. Figure 3.2 shows the

CDFs for the number of heads in the toss of one or two coins that correspond to the probability mass functions in Figure 3.1. In the first graph in Figure 3.2, we see that F(y) is flat for $0 \leq y < 1$ since the probability of getting zero or less heads is equal to the probability of getting 1/2 or less heads which is equal to the probability of getting 3/4 or less heads, etc. This is true, of course because we can only observe zero or 1 head. The CDF has vertical jumps or discontinuities at the points where the probability mass function p(y) takes on non-zero values. When describing the behavior of a discrete random variable, we may use either the probability mass function or the cumulative distribution function, depending on which is more convenient for the purpose. Probability mass-distribution functions and their associated cumulative distribution functions are grouped into families of

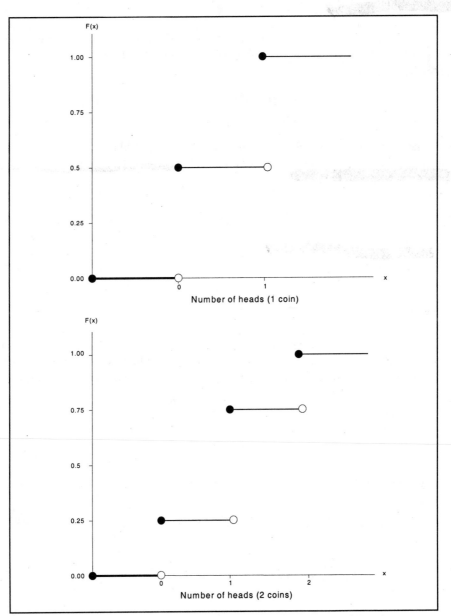

Figure 3.2-- Cumulative Distribution Functions for Number of Heads

functions called *distributions* for short. A distribution family consists of a single functional form for the mass function and cumulative distribution, but different values of constants called *parameters* for each member of the family. The shape of the mass function or cumulative function in a family of distributions will change as values of the parameters change.

3.3.2 Binomial Distribution

The probability mass function shown in Figure 3.1 and the cumulative distribution function shown in Figure 3.2, for the number of heads in the toss of two coins, can also be represented by the formula:

$$p(y) = \frac{2!}{y!(2-y)!}\left(\frac{1}{2}\right)^2 . \quad \text{for} \quad y = 0, 1, 2$$

where y is the observed number of heads. This formula is a special case of a family of probability mass functions called the *Binomial Distribution*. The general probability mass function for a Binomial Distribution is given by:

$$p(y) = \binom{n}{y} p^y(1-p)^{n-y}, \quad y = 0 \text{ to } n$$

(3.3)

and the cumulative distribution function is given by:

$$F(y) = \sum_{k=0}^{y} \binom{n}{k} p^k(1-p)^{n-k}, \quad y = 0 \text{ to } n \qquad (3.4)$$

where the random variable Y is a count of the number of times a failure occurs in n independent random trials, each of which results in either a success or failure; and p represents the constant probability of a failure in a single trial. The parameters of the Binomial Distribution are the constants n and p, which do have physical meaning. The symbol $\binom{n}{y} = \dfrac{n!}{y!(n-y)!}$ %Hp = COMB

IN PROB %

Example 1: Let us show that the binomial probability mass function obeys the

rule $\sum_{y} p(y) = 1.0$. First, the values y can take on are the integers from 0 to

n, therefore, for the binomial, $\sum_{y} p(y) = \sum_{y=0}^{n} \binom{n}{y} p^y(1-p)^{n-y}$, but

56

$$(a+b)^n = \sum_{k=0}^{n} \binom{n}{k} a^k b^{n-k} \quad \text{is the binomial expansion and is true for any}$$

constants a and b. Letting p=a, and (1-p)=b, we get

$$\sum_{y=0}^{n} \binom{n}{y} p^y(1-p)^{n-y} = (p+(1-p))^n = 1^n = 1.0 .$$

The Binomial Distribution is used to make predictions about the number of defective items that will be found in an inspected sample taken from a large lot of manufactured parts, and it is used to establish reasonable control limits for the number of defectives. The only things that have to be known is the parameter p, the long run proportion of defective items, and the parameter, n, the number of items inspected at a particular time.

Example 2: If samples of n=10 casted transmission housings were inspected during each hour's production and the long run proportion of defective housings (due to all causes) was p=6%, then we can use the binomial probability mass function to predict that the probability of zero defective housings on a particular hourly inspection would be:

$$\binom{10}{0} (.06)^0(.94)^{10} = 0.5386 .$$

A result that would be expected with regularity.

The Binomial Distribution is used in cases where the value the random variable can take on (which represents the number of defects), is restricted to integers between 0 and the number of items inspected. Figure 3.3 represents three different Binomial probability mass functions and their corresponding cumulative distribution functions defined by different values of the parameter p, with the parameter n fixed at 15. For each of these mass functions, the value the random variable can take on is restricted to be between n=0 and n=15. The shapes of the distributions can be seen to differ as the parameter p takes on different values. The top mass function, with p=.2, is said to be skewed to the right, and with this mass function it it is more probable to have small values arise. The second mass function, with p=.5, is symmetric. The third mass function, with p=.8, is said to be skewed to the left, and with this mass function it is more probable to have large values arise.

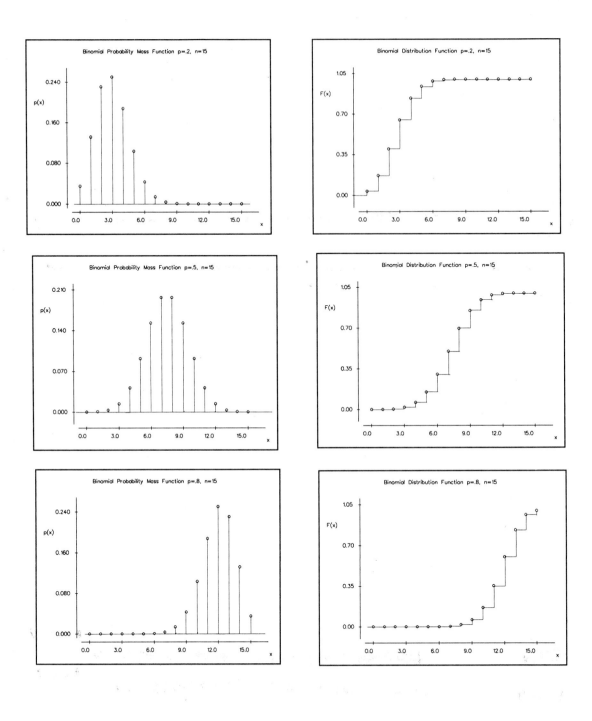

Figure 3.3 Binomial Distribution with n=15

58

Example 3: Using the Binomial probability mass function along with repeated use of the addition rule we can see that the probability of 4 or more defective housings in one hour's inspection would be a very rare event:

$$P(Y \geq 4) \quad = \quad \sum_{y=4}^{10} \binom{10}{y} (.06)^y (.94)^{10-y} \quad = \quad 0.002029 \ .$$

3.3.3 Poisson Distribution

In some quality control inspections, an item is not classified as either defective or not, but rather a count of defects is associated with each item. For example when an auto paint job is inspected, there could be scratches, runs, light application areas, and other classes of defects. Each paint job inspected may have 0, 1, or more defects. In this situation, there is really no upper limit, like n, to the number of defects that can be counted. A family of discrete probability mass functions-distribution functions that are used for making predictions and establishing limits in this case is the *Poisson Distribution*. The probability mass function for the Poisson Distribution is:

$$p(y) \ = \ \frac{e^{-\lambda} \lambda^y}{y!} \quad , \tag{3.5}$$

and the distribution function is:

$$F(y) \ = \ \sum_{k=0}^{y} \frac{e^{-\lambda} \lambda^k}{k!} \quad , \tag{3.6}$$

Where the only parameter λ represents the average number of defects per unit inspected. Again $0 \leq p(y) \leq 1$ for all y, and $\sum_{y=0}^{\infty} \frac{e^{-\lambda} \lambda^y}{y!} = 1.0$. The summation goes to infinity, since there is no upper limit to the values y can take on in this case.

Figure 3.4 shows three different Poisson Distribution probability mass functions indexed by their parameter λ. Theoretically the horizontal axis should extend to ∞, however the p(y) values are negligible for y>20. From the graphs, it can be seen that the Poisson probability mass functions are generally skewed to the right, but as λ increases they become more symmetrical.

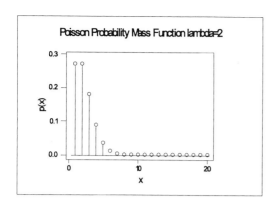

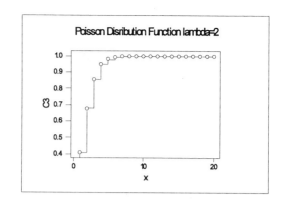

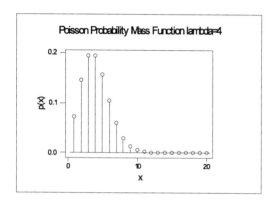

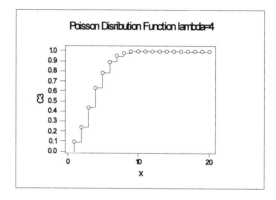

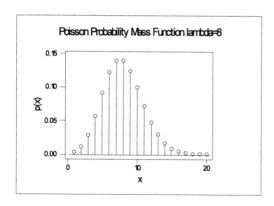

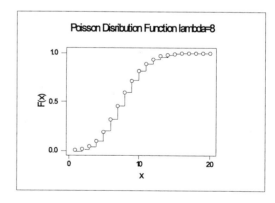

Figure 3.4 Poisson Distribution

Example 4: Insulated wire is being produced. A 100-foot section is inspected every 15 minutes for quality of the insulation layer. Historically the average number of defects per 100-foot section was 1.2. The probability of two defects in the next section inspected is $\dfrac{e^{-1.2}1.2^2}{2!} = 0.2169$, and the probability of

3 or more defects in the next section inspected is

$$P(Y \geq 3) = \sum_{y=3}^{\infty} \frac{e^{-1.2}1.2^y}{y!} = 0.1205$$

or $P(Y \geq 3) = 1 - P(y \leq 2) = 1 - \sum_{y=0}^{2} \frac{e^{-1.2}1.2^y}{y!}$

The Binomial and Poisson Distributions are just two of a large class of discrete probability mass functions that are useful in describing random variability in physical phenomenon. However, these two are useful for a wide range of industrial and engineering work.

3.4 Continuous Random Variables and Distributions

3.4.1 Introduction
Probability density functions represent probabilities for continuous random variables, which can take on any value in a continuum rather than a discrete set of values like the examples thus far shown. One example of a game of chance that produces a continuous random variable is the spinner example shown in Figure 3.5 below.

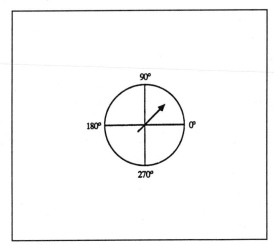

Figure 3.5-- The Spinner

When the spinner is spun, the arrow stops pointing anywhere from $0°$ to $360°$ and it can take on any value in the continuum. The resulting angle is the random variable, Y. Since Y can take on an infinite number of values in the range from $0°$ to $360°$, it is impossible to define a probability mass function, p(y). Instead, a *probability density function* is defined. The probability density function is the smooth curve, f(y), represented by the straight line in Figure 3.6. The function, f(y), does not represent the probability of specific values for y, but rather gives the relative probability of events occurring. The probability of a specific angle is defined to be zero, and the probability that the result falls in some interval of angles is calculated as the area under the curve. The distribution function, F(y) shown in the bottom of Figure 3.6, is defined as the cumulative area under the curve from the left up to the point y. In terms of the density function

$$F(y) = \int_{-\infty}^{y} f(y)dy \qquad (\text{ alternatively } f(y) = \frac{d}{dy}F(y) \text{) }, \text{ and it is a smooth, non-decreasing curve}$$

for a continuous random variable.

Example 5: For the Spinner the distribution function is:

$$F(y) = \int_{0}^{y} \frac{1}{360}dy = \frac{y}{360} \qquad \text{for } 0 \le y \le 360,$$

$$= 0.0 \qquad \text{for } y < 0,$$

$$= 1.0 \qquad \text{for } y > 360.0$$

Two properties of probability density curves are $f(y) \ge 0$ and $\int_{-\infty}^{\infty} f(y)dy = 1.0$, which is to say the density function is greater than or equal to zero for all values that the random variable can take on, and the area under the curve defined by the density function is 1. Probabilities for specific events are determined by taking the area under the density curve in the interval that represents the event, or by taking the difference of the distribution function at the two endpoints of the interval. Probabilities of events such as $\{Y \le 270°\}$ can be read directly from the graph of the distribution function.

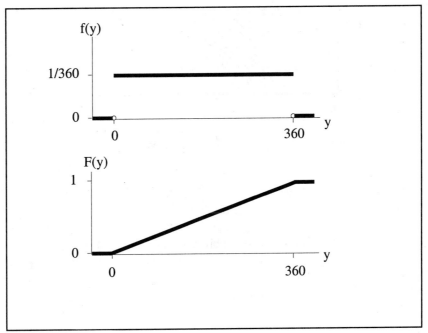

Figure 3.6 Density and Distribution Function for Spinner

Example 6: The probability that the spinner stops between 90° and 180° is 0.25, which is the area under the curve in this interval in Figure 3.7. Or it can be computed as the difference between
F(180°)-F(90°)=(180/360)-(90/360)=90/360=.25 .

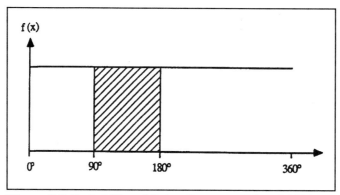

Figure 3.7 Probability of Being Between 90° and 180°

For Example 6, the density function is a straight line, since it is equally likely that the spinner may stop at any point on the circle. But in general, probability density functions can take on a variety of shapes, as dictated by the physical circumstances or by the game of chance which they represent. Some common families of probability density-distribution functions are the *Uniform Distribution*, the *Exponential Distribution*, the *Normal Distribution*, and the *Lognormal Distribution*. Each of these distributions is a whole family of curves indexed by parameters like the discrete distributions we discussed earlier. In this section, we will describe each of these families of density curves and some common situations where they would be used for predictions and inference.

3.4.2 Uniform Distribution

The random variable representing the result of the spinner is one example of a Uniform Distribution. The Uniform Distribution represents the situation where the random variable, Y, can take on any value in an interval completely at random. If the interval of values the random variable can take on is from a to b, then

$$f(y) = \begin{cases} 0 & \text{for } y<a, \\ \dfrac{1}{b-a} & \text{for } a\leq y\leq b, \\ 0 & \text{for } y>b. \end{cases}$$

(3.7)

and

$$F(y) = \begin{cases} 0 & \text{for } y<a, \\ \dfrac{y-a}{b-a} & \text{for } a\leq y\leq b \\ 1 & \text{for } y>b \end{cases}$$

(3.8)

64

The parameters of the Uniform Distribution are the endpoints a and b.

For the spinner example, a=0 and b=360. Computer and calculator random number generators, such as RAND, generate values of uniform random variables on the interval 0 to 1. Probabilities are easy to calculate for uniform random variables since they can be represented by the areas of rectangles.

3.4.3 Exponential Distribution

The exponential distribution is used to represent the interval in time or space between completely random events. Exponential random variables take on only positive values, since negative intervals of time, etc. are not defined. The density function for the exponential distribution is:

$$f(y) = \frac{1}{\theta}e^{-\frac{y}{\theta}} \quad \text{for} \quad y \geq 0 \tag{3.9}$$

and the distribution function is:

$$F(y) = 1 - e^{-\frac{y}{\theta}} \quad \text{for} \quad y \geq 0 \tag{3.10}$$

where the parameter, θ, represents the long run average value for Y. Figure 3.8 shows a graph of an exponential density and distribution function with $\theta=5$.

Example 7: If breakdowns in repairable production equipment occur completely at random, with an average interval between breakdowns of 10 months, then the exponential distribution can be used to predict the probability of the next breakdown. $\theta = 10$, and the probability that the next breakdown will occur in 6 months or less would be:

$$\int_0^6 0.1e^{-0.1x}dx = 1 - e^{-.1(6)} = 0.451$$

and the probability that the next breakdown will occur in less than one year is:

$$\int_0^{12} 0.1e^{-0.1x}dx = 1 - e^{-.1(12)} = 0.699$$

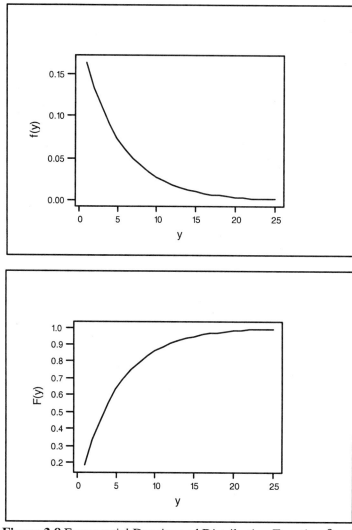

Figure 3.8 Exponential Density and Distribution Function $\theta = 5$

3.4.4 Normal Distribution

Certain probability densities have so much importance in statistics that areas under the curve have been tabulated for future reference. One such distribution is the Normal, or bell-shaped, Distribution. This distribution is useful for describing variability in industrial measurements such as lengths or weights. Natural variation in living organisms and their characteristics also tend to follow a Normal Distribution. For example, the bar charts of measurements in Figures 3.9 and 3.10

can be closely approximated by a normal distribution curve.

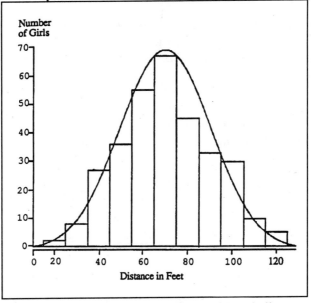

Figure 3.9 Normal Curve Fit to Data of Baseball
Throws for Distance by First-Year High School Girls

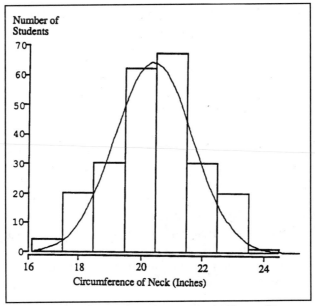

Figure 3.10 Normal Curve Fit to Neck Circumference
of 231 Male College Football Players

67

Each normal density curve is completely defined by two parameters, the mean (average) and the standard deviation. The mean of a normal density is usually represented by the greek letter μ, and the standard deviation by the greek letter σ. These parameters are illustrated in Figure 3.11, which is a graphical representation of the density curve. The mean is the value at which the density curve peaks and about which the distribution is symmetrical. It is thus a measure of central tendency (the middle of the range of y). The normal curve has points of inflection one standard deviation above and below the mean, and nearly all of the area under the curve (99.994%) is included between $\mu \pm 4\sigma$. Thus, the standard deviation of the density is a measure of its dispersion (the range of y). This figure shows the areas under a normal curve which correspond to various intervals on either side of the mean.

Mathematically, the normal density is given by the function:

$$f(y) = \frac{1}{\sqrt{2\pi}\sigma}\, e^{-\frac{1}{2\sigma^2}(y-\mu)^2}$$

(3.11)

where:

$f(y)$ = height of the curve
μ = mean
σ = standard deviation

The cumulative distribution function for the Normal Distribution has no closed form and is usually expressed as the probability that the normal random variable y is less than or equal to some value a, or:

$$\Phi(a) = P(y \le a) = F(a) = \int_{-\infty}^{a} \frac{1}{\sqrt{2\pi}\sigma} e^{-(y-\mu)^2/2\sigma^2} dy \qquad \textbf{(3.12)}$$

Areas under the normal curve can be found by numerical integration. Most statistical programs such as MINITAB, spreadsheets such as LOTUS 123 and even some handheld calculators have built-in routines to calculate areas under the normal curve for specified values of μ and σ.

The Normal Distribution is, of course, a family of density functions which differ by their centers (μ's) and their spreads (σ's). Figure 3.12 shows two normal densities with different means, and different standard deviations. The Normal Distribution with mean $\mu=0$ and standard deviation $\sigma=1$, is called the Standard Normal Distribution. The standard normal random variable is denoted by the symbol Z, and the distribution function for the Standard Normal is denoted by $\Phi(z)$. The distribution function for any Normal Distribution can be evaluated through the Standard Normal Distribution using the standardization formula $F(y) = \Phi(z)$, where $z = \frac{y-\mu}{\sigma}$ (i.e., subtract the mean and divide by the standard deviation). Therefore, one table of areas under the Standard Normal

Distribution can be used to evaluate areas under any Normal Distribution.

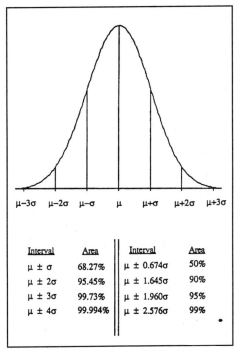

Interval	Area	Interval	Area
$\mu \pm \sigma$	68.27%	$\mu \pm 0.674\sigma$	50%
$\mu \pm 2\sigma$	95.45%	$\mu \pm 1.645\sigma$	90%
$\mu \pm 3\sigma$	99.73%	$\mu \pm 1.960\sigma$	95%
$\mu \pm 4\sigma$	99.994%	$\mu \pm 2.576\sigma$	99%

Figure 3.11 The Normal Distribution

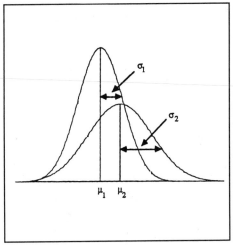

Figure 3.12 Two Different Normal Densities

Table A.3 in Appendix A gives the tail areas (i.e., $1-\Phi(z)$) of the Standard Normal Distribution. To use this table, find the value of z in the margins of the table and look up the tail area in the body of the table. For example, to find the area under the standard normal density to the right of 1.05, look along the row labled 1.0 and under the column labled x.x5 to read the value 0.1469. Since the normal curve is symmetric, the area to the left of -1.05 is the same 0.1469. Consider the following example which shows how to use Table A.3 to look up an area under a Normal Distribution without mean zero and standard deviation 1.

> **Example 8:** If the NC lathe process of Chapter 2 was in statistical control producing shanks with a mean dimension of $\mu=0.386$, and a standard deviation of $\sigma=0.0008944$, then the percentage of parts expected to fall within or outside the specification limits of 0.3835 to 0.3885, could be found by integrating Equation (11) numerically (no closed solution is possible), or by using Table A.3 in Appendix A.

Note: The appendix table shows the area under the *standard normal* density with mean zero and standard deviation 1. To use the table we must convert values to z-scores with the formula $z=(x-\mu)/\sigma$.

Converting the specification limits in Example 8 to z-scores we subtract the mean and divide by the standard deviation as shown below:

Lower Spec Limit: $\qquad z_{LSL} = \dfrac{0.3835 - 0.386}{0.0008944} = -2.795$

Upper Spec Limit: $\qquad z_{USL} = \dfrac{0.3885 - 0.386}{0.0008944} = 2.795.$

The table shows the tail area beyond a given z-score. To use the table we would locate the first two digits of our z-score 2.7 in the first column under z. Next we look to the right to the column x.x9 and read off the tail area 0.0026 from the body of the table. Actually, since our z-score is 2.795, we should interpolate between 2.79 and 2.80, but both these z-scores show the same tail area of 0.0026. As the figure at the top of the table indicates, the tail areas are the same for both positive and negative 2.79, since the curve is symmetric. Therefore the area outside the interval -2.79 to 2.79 is found by the addition rule $0.0026+0.0026 = 0.0052$ and represents the probability of a part being outside the engineering specifications. These calculations are shown graphically in Figure 3.13. The probability of falling within the engineering specs would be $1-0.0052 = 0.9948$, since the area under the entire curve must be one.

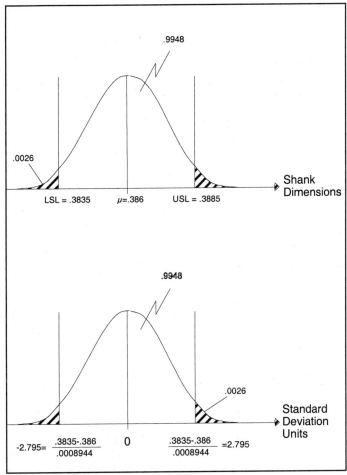

Figure 3.13 Probability of Being Within Engineering Specs

3.4.5 Lognormal Distribution

The Lognormal Distribution is a family of skewed density functions that are related to the Normal. It is often used to model industrial measurements that can span orders of magnitude (i.e., 1 to 10 to 100, etc.). An example of a Lognormal density function is shown in Figure 3.14.

71

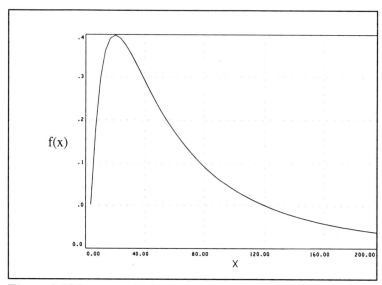

Figure 3.14 Lognormal Distribution with $\mu=3$ and $\sigma=1$

The Lognormal can be derived from the Normal Distribution, in the following way. If the random variable, Y, has a Normal Distribution density with mean, μ, and standard deviation, σ, then a random variable, $X=e^Y$, follows the Lognormal density with density function:

$$f(x) = \frac{1}{x\sigma\sqrt{2\pi}} \, e^{-(\ln(x)-\mu)^2/2\sigma^2} \tag{3.13}$$

The parameters of the Lognormal distribution are usually written as $\tilde{m}=e^\mu$, and σ. $\tilde{m}$ is the median or 50th percentile of the distribution. The Lognormal random variable is strictly positive, and its density is usually skewed to the right. If a random variable, X, follows a Lognormal Distribution, and a random variable, Y, is defined as $Y=\log_e(X)$, then the random variable Y will follow the Normal Distribution. This can be shown using the distribution functions of the random variables X and Y.

If the random variable X follows the Lognormal Distribution with parameters μ and σ, then is the density is

$$f_x(x) = \frac{1}{x\sigma\sqrt{2\pi}} e^{-(\ln(x)-\mu)^2/2\sigma^2},$$

and $F_x(x) = P(X\leq x) = \int_0^x f_x(x)dx$ is the distribution function of X. But, if $Y=\ln(X)$, then

$$F_y(y) = P(Y\leq y) = P(\ln(X)\leq y) = P(X\leq e^y) = F_x(e^y)$$

72

since $\log_e$ is a monotonic function. (Note: The subscripts x and y are used here to distinguish the density and distribution functions of X and Y.)

The density function of Y can be found by differentiating the distribution function to get:

$$f_y(y) = \frac{d}{dy}F_y(y) = \frac{d}{dy}F_x(e^y)$$

$$= f_x(e^y)e^y$$

$$= \frac{1}{e^y\sigma\sqrt{2\pi}}e^{-(\ln(e^y)-\mu)^2/2\sigma^2} \cdot e^y$$

$$= \frac{1}{\sqrt{2\pi}\sigma}e^{-(y-\mu)^2/2\sigma^2}$$

but this is the density of a Normal Distribution. The following example shows how the Lognormal Distribution can be used in practice.

Example 9: Time to failure of a critical component in a piece of manufacturing equipment is a random variable and follows a Lognormal Distribution with $e^\mu = \tilde{m} = 50$ hours and $\sigma = 1.5$. What is the probability of no failures in a two-week (i.e., 80-hour) period?

$$P\{NoFailure\} = P\{Y > 80\} = P\{lnY > 4.382\}$$

$$= P\left\{\frac{lnY-ln50}{1.5} > \frac{4.382-3.912}{1.5}\right\}$$

$$= P\{Z > .3133\} = .3771$$

If a factory contains five identical and independent pieces of manufacturing equipment, what is the probability that none will be idle during a two-week period due to failure of the critical component?

$$P\{None\ Fail\} = [P\{No\ Failures\ on\ one\}]^5$$

$$= .3771^5 = .007625$$

3.5 Expected Values and Moments

In a game of chance or a random trial, it is understood that the results of individual trials are unpredictable. The number of tails in five tosses of the coin or the position of the spinner after one spin cannot be predicted. However, by the very nature of the probability density and distribution functions we have defined, we can predict the long run average of repeated random experiments or the relative frequencies with which certain events occur.

The long run average of a random variable is called the *expected value* or *mean* of that random variable. This quantity is of importance to gamblers who are interested in whether or not they will win or lose in the long run when they play a certain strategy in a game of chance. The mean or expected value is also of interest to engineers who want to know the average dimension (or any other characteristic) of items made by a process that is operating in a state of statistical control. Mathematically we represent the expected value of a discrete random variable as the weighted sum $E(Y) = \sum_y y \cdot p(y)$, and for continuous random variables as an integral $E(Y) = \int_{-\infty}^{\infty} y f(y) dy$. Graphically we can think of this as the center of gravity of the one dimensional probability density function or probability mass function as shown in Figure 3.15.

Example 10: Consider the tossing of five coins. The expected value of the random variable Y (which represents the number of heads) is given by:

$$E(Y) = \sum_{y=0}^{5} y \binom{5}{y} \left(\frac{1}{2}\right)^y \left(\frac{1}{2}\right)^{5-y}$$

$$= \left(\frac{1}{2}\right)^5 \sum_{y=0}^{5} y \binom{5}{y}$$

$$= \left(\frac{1}{2}\right)^5 [\ 0 \cdot 1 + 1 \cdot 5 + 2 \cdot 10 + 3 \cdot 10 + 4 \cdot 5 + 5 \cdot 1\ ]$$

$$= \frac{80}{32} = 2.5$$

In other words, on the average when five coins are tossed 2.5 heads will result.

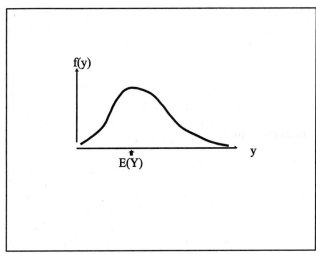

Figure 3.15 A Continuous Density Function

Example 11: Considering the continuous spinner, the expected value of the degrees, Y, will be:

$$E(Y) = \int_0^{360} y\left(\frac{1}{360}\right) dy = \left[\frac{1}{360}\left(\frac{y^2}{2}\right)\right]_0^{360}$$

$$= \frac{360}{2}$$

Therefore, the average of a large number of spins should be about $180°$.

      For most families of distributions, the expected value can be expressed as a function of the distribution parameters. For example, Table 3.4, shows the expected values of the distributions we have presented.

      The nth central moment of a density or mass function is defined to be the expected value of the nth power of the difference from the mean, i.e., $E(Y-E(Y))^n$. The second moment gives us information about the dispersion in a density or mass, as the expected value gives us information about the center. Higher moments give us other information about the shape of the distribution, such as skewness. In fact, knowledge of all the moments n=1, 2, ... would be equivalent to knowing the density or mass function equation. For our purposes the second moment is most interesting.

      The second central moment, $E(Y-E(Y))^2$, is called the variance of the distribution, denoted by Var(Y), and represents the average squared distance from the center (expected value or mean) of the density or mass function. It directly gives us information about the dispersion. The standard deviation is the square root of the variance. In a Normal Distribution, the standard deviation is the distance from the mean to the point of inflection, as shown for the two normal densities in Figure 3.12. Table 3.4, shows the relationship between variance, standard deviations, and distribution parameters for the specific distributions we have presented in this chapter.

Table 3.4 Expected Values, Variances and Standard Deviations for Common Distributions

Distribution	Parameters	Expected Value $E(Y)$	Variance $E(Y-E(Y))^2$	Standard Deviation
Binomial	n,p	np	np(1-p)	$\sqrt{np(1-p)}$
Poisson	λ	λ	λ	$\sqrt{\lambda}$
Uniform	a,b	$\dfrac{a+b}{2}$	$\dfrac{(b-a)^2}{12}$	$\sqrt{\dfrac{(b-a)^2}{12}}$
Exponential	θ	θ	θ^2	θ
Normal	μ, σ	μ	σ^2	σ
Lognormal	μ, σ	$e^{\mu+\frac{1}{2}\sigma^2}$	$e^{2\mu+\sigma^2}[e^{\sigma^2}-1]$	$\sqrt{e^{2\mu+\sigma^2}[e^{\sigma^2}-1]}$

3.6 Sampling Results

Unfortunately, in most real-life situations not associated with games of chance, we do not know the density function or characteristics of the population with which we work. The expected value and standard deviation are unknown. We must get along with a sample of data (observed values of a random variable). Our task is to estimate the mean or expected value and standard deviation of the unknown density from a relatively small amount of data. We don't want to be taking an infinite number of measurements or running an infinite number of random trials to get our answer. Here, probability theory provides us with a reasonable approach.

We define a random sample $Y_1, Y_2, ...Y_n$, from a distribution $f(y)$ as independent random variables each having the same distribution. In a practical sense, $Y_1, Y_2, ...Y_n$ are the results of n repeated trials of a game of chance. $y_1,...y_n$ represent the observed values of the random variables.

If the density function, $f(y)$ or probability mass function $p(y)$ is unknown, then the expected value or long run average result, $\mu = E(Y)$, is also unknown. But, if we have observed results we can use

the sample average $\bar{y} = \left.\sum_{i=1}^{n} y_i \middle/ n\right.$ as an estimate of μ. The sample average, $\bar{Y}$, is not a constant

like μ, but is a random variable that will vary from one random sample to another. Probability theory gives us insight to the behavior of sample means.

3.6.1 The Law of Large Numbers

The *Law of Large Numbers* is a theorem that shows what happens to the sample average as our sample size, n, becomes large. In a way, it corresponds to what most people intuitively sense as the law of averages. The theorem states that the sample average, $\bar{Y}$, of a random sample of n, can always be made as close to $\mu = E(Y)$ as desired by increasing n, the number of values in the sample. For example, with the toss of a fair coin, let the random variable equal 1 if a head is tossed, and 0 if a tail is tossed. Then, for obvious reasons, the expected value of this random variable is 1/2. As shown in Figure 3.16 if the average of several consecutive tosses (these are, of course, independent random variables) is plotted, the line will always converge toward 1/2 as the number of tosses becomes large.

Therefore, if we have one two-headed coin in a set of five or a loaded die, we can estimate the expected value of the random variable (even though we don't know $p(y)$) by repeatedly performing the random trial (coin toss or die throw) and averaging the results we get. As long as we have a large number of trials (100 or more) our sample average will be reasonably close to the true expected value. This theorem justifies our use of the sample average and standard deviation calculated in Chapter 2 for the Shank Dimension control chart.

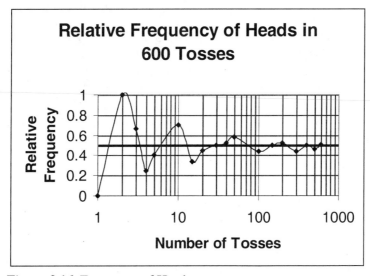

Figure 3.16 Frequency of Heads

3.6.2 The Central Limit Theorem

Another theorem in probability theory which has wide application in modern statistics is the Central Limit Theorem. The *Central Limit Theorem* states that no matter what the density or mass function of a set of independent random variables is, the distribution of their average will become closer and closer to the bell-shaped Normal Distribution as the number of random variables averaged increases. For example, in Figure 3.17 the probability mass function for the average of several tosses of a die can be seen to quickly approach a bell-shaped curve.

Not only do averages of independent random variables tend to follow a bell-shaped or Normal Distribution, but it can also be noticed from Figure 3.17 that averages are more closely clustered about their mean than individual values. This can be expressed mathematically by the fact that the distribution of sample means of all possible samples of size n from any population having mean, μ, and standard deviation, σ, will be approximately normal with mean, μ, and standard deviation $\dfrac{\sigma}{\sqrt{n}}$.

Example 12: In the probability mass function for the random variable that represents the number of dots on a die, shown at the top of Figure 3.17, the mean and standard deviation are $\mu=3.5$, and $\sigma=1.708$. But, by the Central Limit Theorem, the distribution of the average number of dots on four die is approximately normal with mean $\mu=3.5$ and $\sigma = 1.708/\sqrt{4} = 0.8539$. Therefore, if we toss four die repeatedly and call the average of the four dice our random variable, we would expect 3.5 as a long run average. Although we cannot predict the result of a particular throw of four dice, the Central Limit Theorem, and properties of the Normal Distribution, tell us that we should expect about 68% of the results to be within 3.5 ± 0.8539(one standard deviation from the mean) and 95% of the results to be $3.5 \pm 2(0.8539)$.

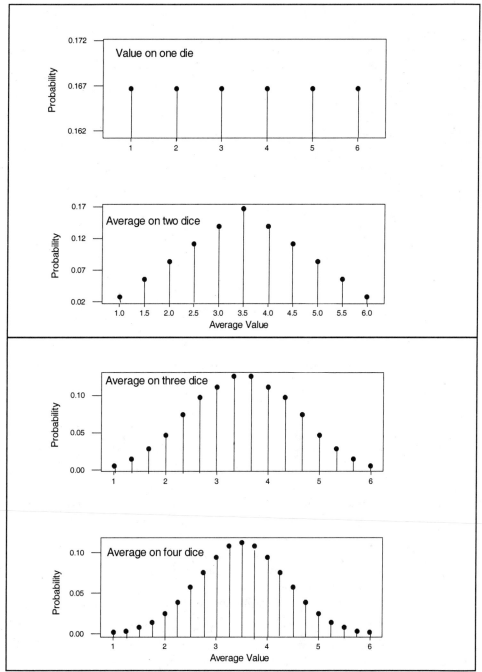

Figure 3.17 Distribution of the Average

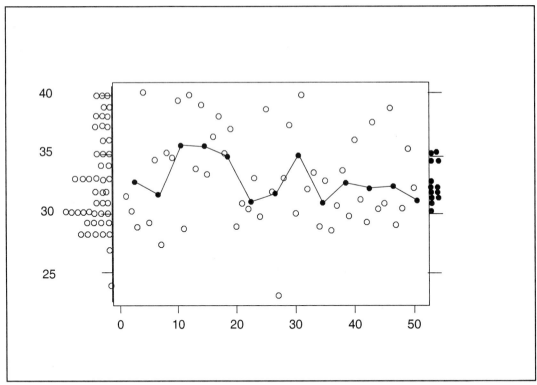

Figure 3.18 Consistency of Averages

For typical industrial data the situation is depicted in Figure 3.18. The open circles represent measurements made on the output of an industrial process. The solid circles represent averages of four consecutive points on the graph. At each end of the graph we see the distribution of individual measurements represented as a dot plot (on the left) and the distribution of averages of four (on the right). We can see that averages are more closely grouped around the process mean, which confirms our feeling that we should have more faith in an average than a single measurement.

There are special forms of the Central Limit Theorem that relate to the Binomial Distribution and Poisson Distribution. Briefly they say that whenever the parameters n and p of the Binomial Distribution are such that $n \geq 30$ and $np \geq 5$, or the Poisson Distribution parameter $\lambda > 5$, then the density function, $f(y)$, for the Normal Distribution, with $\mu = np$ and $\sigma = \sqrt{np(1-p)}$, or $\mu = \lambda$ and $\sigma = \sqrt{\lambda}$, can be used to approximate $p(y) = P(Y = y)$. Sums of binomial or Poisson probabilities of the form $\Sigma p(y)$ can be approximated by areas under the normal curve $f(y)$. This is illustrated graphically in Figure 3.19, which shows the relative frequency of results from 20 coin tosses. The bar heights represent Binomial probabilities, and they can be approximated remarkably well by the normal density curve.

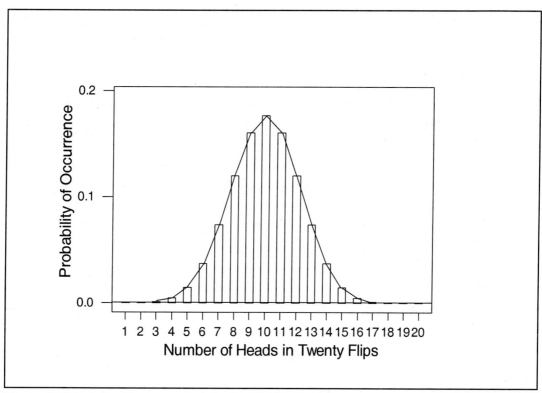

Figure 3.19 Relative Frequency Results for 20 Coin Tosses

Example 13: Suppose we want to find the probability that we would get more than 13 heads after tossing 20 coins. This can be calculated using the Binomial Distribution with n=20, and p=1/2 by summing the terms from 14 to 20

as: $\displaystyle\sum_{y=14}^{20} \binom{20}{y} \frac{1}{2}^{y}(1-\frac{1}{2})^{n-y} = 0.05766$

or approximated using the Normal Distribution with $\mu=20(.5)=10$, and

$\sigma = \sqrt{20(.5)(1-.5)} = 2.236$ by calculating the upper tail area beyond 14. To

do this we first determine the number of standard deviation units, 14, is from the mean i.e., ((14-10)/2.236)=1.78. Looking this value up in Appendix Table A.3 for the Normal Distribution, we read the upper tail probability of.0375.

The accuracy of the approximation can be increased by making a continuity correction, consisting of adding or subtracting 1/2 from the count before standardizing in order to make the area under the curve more closely approximate the area of the bars in Figure 3.9. For this example we would subtract 1/2 from 14 since the bar at 14 really extends from 13.5 to 14.5. Doing this our standardized value becomes ((13.5-10)/2.236)=1.56, and looking this value up in Table A.3 we read the tail areas 0.0594.

In practice, the Normal approximation to the Binomial is quite useful for developing control or decision limits for counts of defective items and similar industrial data.

3.7 Expected Values and Variances of Linear Combinations of Random Variables

If a random variable $Y=aY_1+bY_2$, where Y_1 and Y_2 are independent random variables, then the expected value of Y can be expressed as $E(Y)=aE(Y_1)+bE(Y_2)$ since:

$$E(Y) = E(aY_1+bY_2) = \int_{-\infty}^{\infty}\int_{-\infty}^{\infty}(ay_1+by_2)f_1(y_1)f_2(y_2)dy_1dy_2$$

$$= \int_{-\infty}^{\infty}ay_1f_1(y_1)\int_{-\infty}^{\infty}f_2(y_2)dy_2dy_1 + \int_{-\infty}^{\infty}by_2f_2(y_2)\int_{-\infty}^{\infty}f_1(y_1)dy_1dy_2$$

$$= a\int_{-\infty}^{\infty}y_1f_1(y_1)dy_1 + b\int_{-\infty}^{\infty}y_2f_2(y_2)dy_2$$

$$=aE(Y_1) + bE(Y_2)$$

Similarly it can be shown that the $Var(Y) = \sigma_Y^2 = a^2Var(Y_1)+b^2Var(Y_2) = a^2\sigma_{Y_1}^2+b^2\sigma_{Y_2}^2$.

In general if,

$$Y = \sum_{i=1}^{k} a_i Y_i \qquad\qquad (3.14)$$

where the $Y_i's$ are independent random variables, then ,

$$E(Y) = \sum_{i=1}^{k} a_i E(Y_i) \qquad\qquad (3.15)$$

and

$$Var(Y) = \sigma_Y^2 = \sum_{i=1}^{k} a_i^2 Var(Y_i) = \sum_{i=1}^{k} a_i^2 \sigma_{Y_i}^2 \qquad\qquad (3.16)$$

Example 14: This explains the consistency of means since $\bar{Y} = \sum_{i=1}^{n} \frac{1}{n} Y_i$ is a special case of the general formula $Y = \sum_{i=1}^{k} a_i Y_i$ with $a_i = \frac{1}{n}$ for i=1 to n. Therefore if all random variables Y_i have the same variance, σ^2, then $Var(\bar{Y}) = \sum_{i=1}^{n} \frac{\sigma^2}{n^2} = \frac{\sigma^2}{n}$, and the standard error of $\bar{Y}$ is $\sigma/\sqrt{n}$ as illustrated in Section 3.6.

Example 15: Using the same logic and deducing from the general formula, the expected value and variance of a difference of two means such as $\bar{Y}_1 - \bar{Y}_2$ can be shown to have expected value $E(Y_1) - E(Y_2)$ and variance $\sigma_1^2/n_1 + \sigma_2^2/n_2$. Thus it can be said that the mean of a difference is a difference of the means, but the variance of a difference is the sum of the variances.

3.8 Probability Based Decision Rules

When there is variability in data, all decision rules are subject to error. We can never expect to be right 100% of the time because random or chance caused by variability will occasionally fool us. However, if we base our decision rules on the understanding of probability theory, we can formulate objective rules that will minimize the chance for error, or at least let us calculate the chances of being wrong when we make judgments.

Probability-based decision rules are constructed in the following way. We hypothesize the situation to be the usual or expected. To reject the hypothesis in favor of an alternative, we must have evidence to the contrary. For example, in manufacturing we might hypothesize the process is running in a state of statistical control with variations due solely to chance or common causes. We would have to see unusual data like that shown in Figures 2.6 and 2.7, that were very improbable if the hypothesis was true, before we would reject this hypothesis. In research, we might hypothesize that the only difference in performance of prototype products based on differing design alternatives (or using different component materials), would be due to chance causes unrelated to the prototype differences. If data shows strong differences that are unlikely to be chance causes, we could have evidence to reject the hypothesis and assume the alternative (that the prototype differences caused the performance differences).

Consider the following example. If the NC lathe process for producing cylindrical parts, discussed in Chapter 2, was in a state of statistical control centered at the nominal dimension, then we would expect the average dimension of five measured parts to follow a

Normal Distribution with mean $\mu = 0.3860$ and standard deviation $\sigma_{\bar{x}} = 0.0002$. This is what

would be expected from probability theory. If the hypothesis were true, and the cutting tool was not worn or in need of adjustment, it would be very unlikely (probability<0.0027) for an average of five measurements to fall outside the interval 0.3854 to 0.3866. These limits are three standard deviations below and above the mean, respectively. If a result did fall outside these limits it would be much more likely that the cutting tool was worn or out of adjustment. Therefore these limits are established as control limits (or our decision rule) for deciding when to stop the process and make an adjustment, and when to let it run.

In the previous example, we outlined the rational for 3σ limits for control charts. In Chapter 6 similar quantitative decision rules will be developed for interpreting results of tests and research programs. Less structured graphical tools will also be developed in the next two chapters. These decision rules form a sound basis for interpreting data from production environments and industrial research programs.

3.9 Summary

In this chapter we have briefly outlined some important ideas in probability theory that will serve as the theoretical basis for many of the objective decision rules used for interpreting industrial and research data. The concepts of random trials, random variables, and probability distributions have been defined. Much of the variability witnessed in industrial data can be modeled with probability distributions that represent the behavior of games of chance like coin tossing and dice throwing. The following is a list of key concepts and terms used in this chapter:

3.10 Exercises for Chapter 3

3.1 Two fair dice are tossed. The dice are not connected and the result on one die in no way affects the result on the second.

   a) Use the multiplication rule to determine the probability of getting a sum of 2.

   b) Use the multiplication and addition rules to find the probability of getting a sum of 7.

3.2 A random variable, Y, represents the number of dots showing when one fair die is rolled.

   a) Write the probability mass function for Y.

   b) Write the cumulative distribution function for Y.

   c) Make a graph of both the probability mass function and the cumulative distribution function for Y.

3.3 Plastic parts are manufactured on an injection molding process. Some parts are classified as defective due to flashing, flow lines, incomplete fill, or other problems. The number of defective parts, Y, in a run of 1000 is a Binomial random variable with probability of defective, p=.005.

   a) Find the probability of no defective parts in a run of 1000.

   b) Find the probability of one or more defective parts in a run of 1000.

   c) Find the probability of two or more defective parts in a run of 1000.

3.4 Home tub and shower stalls are made of composite materials in a production process that involves laying up fibers, spraying resin, and molding. Defects such as micro cracks (in spider web patterns) often appear in the final products. Although these defects do not affect the performance of the product, they are unappealing to customers. The number of defects per unit, Y, is a random variable that follows the Poisson Distribution with $\lambda=1$.

   a) Find the probability that a tub and shower unit will have no defects.

   b) Find the probability that a tub and shower unit will have three or more defects.

   c) Find the probability that a production run of 100 units will have more than 40 defect-free units.

3.5 A simulation program for mating tolerances calls a subroutine that generates a random number that is equally likely to be any value between 2.49 and 2.51.

a) What is the probability that the subroutine generates a number between 2.492 and 2.498?

b) What is the probability that the subroutine generates a number greater than 2.504?

3.6 After studying control charts for a lathe process, engineers noticed that the time between adjustments to correct for tool wear followed approximately an exponential distribution with average time between adjustments $\theta = 3.94$ hours. If the lathe has just been adjusted, then:

a) What is the probability that another adjustment will be required in less than three hours?

b) What is the probability that it will be five or more hours until an adjustment is required?

3.7 Contact window forming is one of the more critical steps in fabricating state of the art CMOS integrated circuits. The purpose of the windows is to facilitate connections between the gate sources and drains. The diameter length of the window is a critical dimension which is produced by photo lithography and etching processes. The distribution of the diameters is Normal with mean $\mu = 3.5 \mu m$ and standard deviation $\sigma = .15 \mu m$.

a) Find the probability that a contact window diameter is between $3.3 \mu m$ and $3.6 \mu m$.

b) Find the probability that a contact window diameter is between $3.1 \mu m$ and $3.4 \mu m$.

c) Find the probability that a contact window diameter is greater than $3.6 \mu m$.

3.8 A continuous random variable Y with density function

$$f(y) = \begin{cases} .5y & \text{for } 0 \le y \le 2 \\ 0 & \text{otherwise} \end{cases}$$

a) Sketch the graph of the density function

b) Find $P\{Y \ge 1\}$

c) Find the cumulative distribution function F(y) and sketch the graph of this function

d) Find the mean or $\mu = E(Y)$ and indicate where it is on the graph you made in a)

3.9 On a graph, mark a horizontal scale from -10 to 10 and on this scale sketch the density function of the following distributions

    a) Normal with mean $\mu=0$, and standard deviation $\sigma=1$

    b) Normal with mean $\mu=0$, and standard deviation $\sigma=2$

    c) Normal with mean $\mu=2$, and standard deviation $\sigma=1$

What does this exercise tell you about the mean and standard deviation in relation to the shape of a Normal Distribution?

3.10 You are testing cheap ammunition which has 10% duds (ammunition which does not fire). How many pieces must you test fire (n), in order to be 95% sure of getting at least one failure (a dud)?

3.11 The number of burnt corn flakes in a 14 oz. box is well described by a Poisson Distribution with an average, $\lambda = 2.5$ per box.

    a) What is the chance of getting four or fewer burnt corn flakes in one box?

    b) What is the chance of getting five or more burnt corn flakes in one box?

3.12 In Wheel of Fortune, the segments of the wheel are 15 degrees in width. If two of the segments are Bankrupt, what is the probability of spinning a Bankrupt?

3.13 North Dakota has an average of two blizzards per year in the five month period of November through March.

    a) If the number of blizzards can be described by a Poisson distribution, what are the chances of having seven or more blizzards in the five month period?

    b) If the time between blizzards can be described by an exponential distribution, what are the chances of having at least four months without a blizzard?

3.14 Verify the mean and the standard deviation used in Example 12 for the tossing of a single die are: $\mu = 3.5$ and $\sigma = 1.708$.

3.15. The net weight of a box of corn flakes is well described by a Normal Distribution. The target mean is 14 oz. The machine that fills the boxes has an adjustable average fill weight, and a standard deviation of $\sigma = 1/4$ oz.

a) What should the 3σ limits be for the Shewhart control chart if you are plotting the weight of a single box?

b) What should the 3σ limits be for the Shewhart control chart if you are plotting the weight of an average of 12 boxes (one case)?

CHAPTER 4
Descriptive Tools

4.1 Introduction

When diagnosing the cause of production problems or seeking to draw conclusions from R&D experiments, the data are usually studied. Simple graphs of data and numerical summaries, such as averages and ranges, greatly help to simplify this task. Examining a mass of raw numbers, such as the shank measurements of Figure 2.3 in Chapter 2, would be a difficult task without a simple graph like a control chart to summarize the data and display the intrinsic characteristics. Graphs are also a great way of communicating results and recommendations to management, who must make expenditure decisions. A short explanation along with a graph usually makes the results of a study clear.

In this chapter we will describe some of the simple descriptive statistics and graphs we will use throughout the rest of the book for summarizing and displaying data. The rationale for these techniques is simpler than the basis for more formal techniques that use probability theory, and in some cases graphs and simple summaries are all that are needed to draw conclusions from data. In other cases where trends and differences are not so obvious, the formal statistical analysis explained in Chapter 6 is needed to draw conclusions.

To motivate the use of the descriptive techniques let us reconsider the two industrial examples presented in Chapter 1. In the first example quality improvement teams were attempting to reduce the number of defective cast transmission housings, and in the second example Motorola employees were attempting to reduce the variability in thickness of copper plate layers. If the manufacturing processes in these two examples were in a state of statistical control, the variables of interest (i.e., number of defective housings, and copper plate thickness) can be considered to be random variables. The number of defective housings would be a discrete random variable following the Binomial Distribution, and the copper plate thicknesses can be considered to be a continuous random variable following a Normal Distribution. We will refer to these examples as we illustrate the some of the descriptive statistics.

4.2 Data Collection and Descriptive Statistics

Data can always be assumed to represent numerical measurements or observations taken on a sample of items. The sample represents a subset of some larger population or universe of items for which we would like to draw inference about. There are different ways that sample data can be obtained in industrial and engineering studies. The two extremes might be called *observational studies* and *experimental studies*. In *observational studies* data is collected without manipulating or controlling the circumstances. For example, when control charts are being developed the first set of data is usually collected without making adjustments to the process. This is done so that the natural variability in the data can be observed. In *experimental studies*, on the other hand, every means of controlling the environment is employed that is known. For example, in the experiment conducted to reduce tile size variation, presented at the end of Chapter 2, every component of the clay was carefully metered and mixed, and controlled changes were made in the mix from one run to the next.

The purposes of observational and experimental studies are usually different. In observational studies, the intent is typically to describe some characteristic of a production process or product performance from a sample of data. For example, we might be interested in the reliability of a mechanical product, and therefore we might like to collect data to determine the mean time to failure, or the distribution of failure times in order to establish a reasonable warranty period. In experimental studies, on the other hand, the intent is usually to establish cause and effect relationships and find optimal conditions. In the tile size variation experiment, for example, the purpose was to see if varying clay components would *cause* a reduction in size variation.

Deming[1] further classifies observational studies into *enumerative studies* and *analytical studies*. In *enumerative studies,* the purpose is to describe a well-defined population of items. An example is when a receiving department samples a portion of raw materials from a shipment coming into a factory. The purpose is to make measurements on the sampled items to determine if the entire shipment will be acceptable for use. The purpose of *analytical studies,* on the other hand, is to describe an aspect of a population that is not well-defined in time or space. For example, when collecting data on the failure times of a mechanical part in order to establish a warranty period, the population of parts we would like to warranty have not yet been manufactured.

Although the purpose of most data collection is to draw inference about some larger set from a sample, descriptive statistical methods cannot directly help in this purpose. Descriptive statistical methods are simply used to describe the sample of data we have. In that respect the same descriptive methods may be used for data from experimental or observational studies, enumerative or analytical. Viewing descriptive statistics may give us some insight about the larger set or population from which our data was obtained, but to actually draw inference we need the probability-based tools to be described in Chapter 6.

4.3 Graphical Descriptive Statistics for Visualizing Data

Graphical methods such as line graphs, dotplots, histograms, stem leaf diagrams, scatter plots, and boxplots are useful for helping us to visualize the information or meaning in a set of data.

4.3.1 Line Graphs

Line graphs are a simple way of displaying data in relation to time or direction. Most simple spreadsheet programs have the ability to construct these graphs. They are constructed by marking a vertical axis for the data and a horizontal axis to represent time or space. The purpose of making these graphs is to examine the data for trends or patterns. Financial data, such as quarterly earnings or daily stock prices, are usually shown on this type of graph so that increasing or decreasing trends can be visualized.

Figure 4.1 is an example of a line graph of weight measurements of samples taken across the web of a non-woven fabric. The horizontal axis represents the direction across the web, as shown in Figure 4.2, rather than time. In the manufacturing process that produced the fabric, the wet fibers were sprayed on the web, and then dried and wound around a spool. There was concern because the strength and thickness of the final towels was inconsistent. Process changes were being contemplated, but the simple line graph of weights of samples taken across the width of the web was consistent when data were collected during different runs, and indicated that the nozzles that sprayed the fibers were not calibrated.

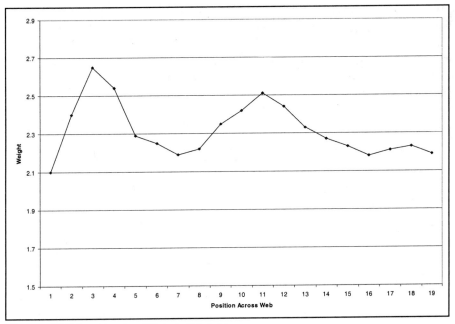

Figure 4.1 Line Graph of Fabric Weights

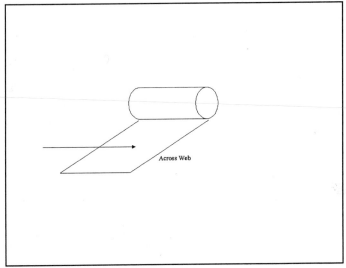

Figure 4.2 Nonwoven Fabric Web

4.3.2 Dotplots

When there is no trend or pattern in data taken over time or along a direction, line graphs are unnecessary. However, it is still useful to graphically summarize the variability in the data. A simple graph that is particularly suited to do that is the dotplot. A dotplot is essentially a line graph with the time dimension removed. The dotplot is constructed by making a horizontal scale that will include the smallest and largest values in the data set, and then simply placing a dot above the scale where each data point falls. If two data values are the same (or the same within the scale of your graph) put two dots, one on top of the other, on the same spot on your scale. A dotplot of the six values 106.968, 113.526, 96.027, 89.752, 105.961, 105.960 is shown below in Figure 4.3.

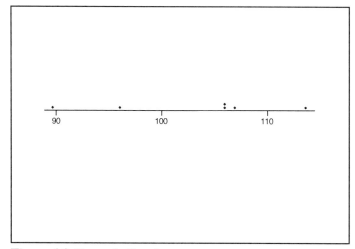

Figure 4.3 Example Dotplot

Although some statistical software like MINITAB have procedures to generate dotplots, they are easy to construct by hand with paper and pencil, and for that reason they are used quite often with small data sets. Dotplots are also quite compact. Several can be drawn above each other to compare several sets of data. It is easy to recognize outlier data points on dotplots and to compare the location of the center and the variability or spread between two or more data sets.

4.3.3 Histograms

When there is a larger amount of data (i.e., 50 values or more), dotplots will become very busy looking with many dots stacked on top of each other. A histogram is better way to represent the data in this case. A histogram is a bar chart that in appearance is similar to a probability mass function. The horizontal axis is broken up into intervals, and the heights of the bars above each interval represent the number of observed data values in each interval. If the data represent observations of a discrete random variable, such as the number of defective transmission housings, each interval would be defined to contain exactly one integer value. On the other hand when the observed values are of a continuous random variable, we must decide how to divide the range of the data into intervals. Example histograms are shown in Figure 4.4 and Figure 4.5 which summarize a sample of data from the production lines for transmission housings and copperplated ceramic substrates. Notice the histogram in Figure 4.4 has narrow bars indicating discrete values are being summarized, while Figure 4.5 has wide bars indicating the interval contains values of a continuous variable.

Normally histograms are constructed automatically by statistical computer programs such as MINITAB. Even graphic hand calculators such as the Texas Instruments TI85 or the Hewlett

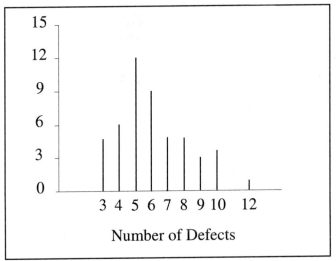

Figure 4.4 Transmission Housing Defects

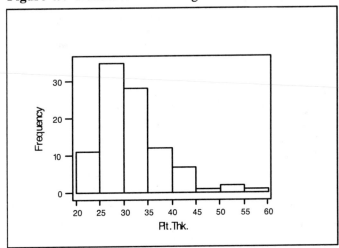

Figure 4.5 Plating Thicknesses

Packard HP48 series have built-in programs to construct histograms. If you construct a histogram of continuous data by hand, or use a spreadsheet such as LOTUS 123, you need to have a few guidelines for dividing the range of the data into intervals. The number of intervals that should be in a histogram of continuous data depends upon the number of observed data points you have. A reasonable number is given by rounding, to the nearest integer, the value obtained from the formula[2]

$$I = [1.0 + 3.3\log_{10}(N)] \tag{4.1}$$

where I is the number of intervals and N is the number of data points to be summarized. When constructing the intervals and class boundaries, the problem of having data values fall exactly on the boundaries can be avoided by defining all intervals to include their upper boundary, but not their lower boundary. Another way to avoid the problem is to make the range slightly wider than the range of the data, and carry one more decimal in the class boundaries than in the data points.

Table 4.1 Copper Plate Thicknesses

28.6	29.2	39.7	25.7	44.2	21.1	23.0
29.8	31.9	36.0	28.9	27.0	29.5	31.8
25.6	26.1	26.6	39.2	22.8	33.5	29.6
33.0	29.8	34.3	31.0	32.8	25.2	22.1
36.0	39.4	30.1	28.2	27.5	22.5	27.1
40.6	29.6	26.4	40.7	57.3	28.8	34.8
30.6	28.6	31.4	36.8	25.6	30.3	34.0
30.2	26.2	22.0	28.9	35.5	22.7	26.9
28.5	42.8	26.8	29.8	29.2	32.8	25.2
22.4	22.5	30.7	36.9	33.6	29.6	33.1
44.0	31.2	35.0	35.1	39.7	50.1	43.6
24.4	33.4	25.0	33.4	39.4	34.1	30.6
27.1	29.3	27.4	32.6	31.9	33.4	23.8
53.5	25.5	41.9	31.5	45.0	32.0	

Example 1: Consider the data in Table 4.1 that represent a sample of copper plate layers summarized in Figure 4.5. Since there are 97 data points, a reasonable number of intervals for a histogram would be $8 \approx [1.0 + 3.3\log_{10}(97)] = 7.56$. The maximum data point is 57.3 and the minimum is 21.1. The range 57.3-21.1=36.2 is divided by the number of intervals, 8, to get the interval length, 4.525. Normally this would be rounded up to the same number of significant digits as the original data or less, i.e., 4.6 or 5. By rounding the interval length up, the range covered by the histogram is increased, and therefore the first interval should start somewhat below the minimum value in the table (i.e., at 21, instead of 21.1 using an interval length of 4.6, or 20 rather than 21.1 using an interval width of 5.0).

If the first interval starts at 21.0 with an interval width of 4.6, some data values (such as the 25.6 in the first column and third row of Table 4.1) will fall exactly on the boundary between two intervals. To avoid this problem, the number of significant digits in the starting value can be increased to one more than is recorded in the data, i.e., make it 21.05 rather than 21.0, or count a value as belonging to an interval only if it is greater than or equal to the lower boundary and strictly less than the upper boundary.   Next we list the intervals and count the number of data values falling into each. Below are two alternative sets of intervals and counts for the data in Table 4.1. The first set uses a starting value of 20.0 and an interval of 5.0 and counts all data points falling on the boundary as belonging to the upper interval. The second set uses a starting value of 21.05 and a interval of 4.6 and avoids the problem of data values falling on the boundary. Figure 4.5 was produced using the intervals and counts from the first set.

	Set 1		Set 2	
Interval	**Count**		**Interval**	**Count**
20-25	12		21.05-25.65	17
25-30	34		25.65-30.25	31
30-35	28		30.25-34.85	26
35-40	12		34.85-39.45	10
40-45	7		39.45-44.05	8
45-50	1		44.05-48.65	2
50-55	2		48.65-53.25	1
55-60	1		53.25-57.85	2

Histograms of data can be used to get a rough idea what the density or probability mass function that generated the data looks like. They can tell us whether the distribution is symmetric or skewed, and roughly what the mean (or center of the data) is. In Figure 4.6 are some example histograms that illustrate symmetry, skewness, and bimodality. If there are odd values in a data set that appear like they do not belong, they may stick out like a sore thumb on a histogram. For example, the class at the far right in Figure 4.7 contains only one value. This should be a signal to recheck the data point. It could be a measurement or recording error.

Histograms are also useful for before and after comparisons of data. Figure 4.8 is a comparison of the plating thickness data before and after Motorola implemented tighter process controls. The figure clearly indicates the reduction in variability that resulted. Figure 4.9 illustrates what a before and after comparison of transmission housing defects might look like. Here it can be seen that the change to the casting form clearly reduced the average (center) number of defects.

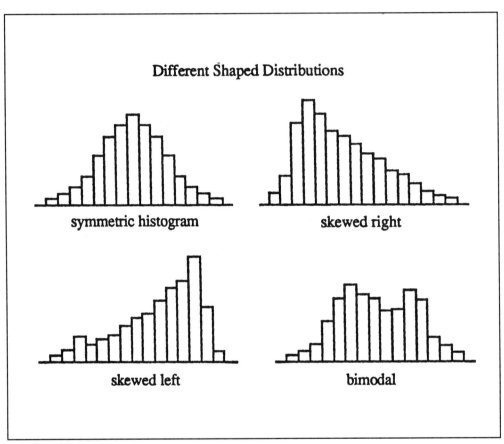

Figure 4.6 Histograms Display Different Shapes

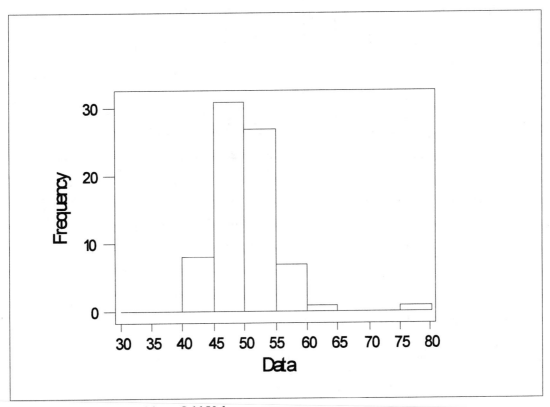

Figure 4.7 Histogram with an Odd Value

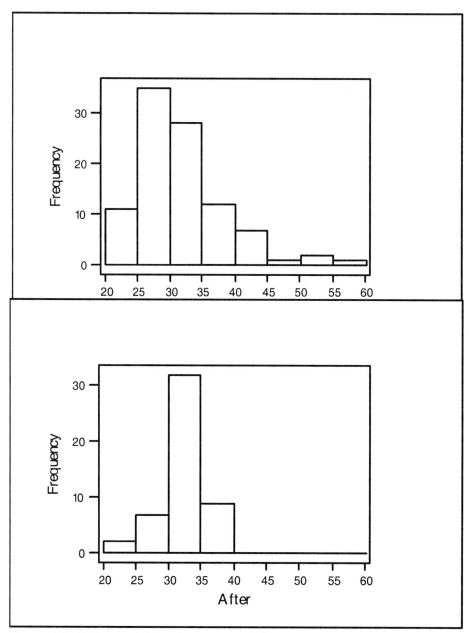

Figure 4.8 Before and After Comparison of Copper Plate Thickness

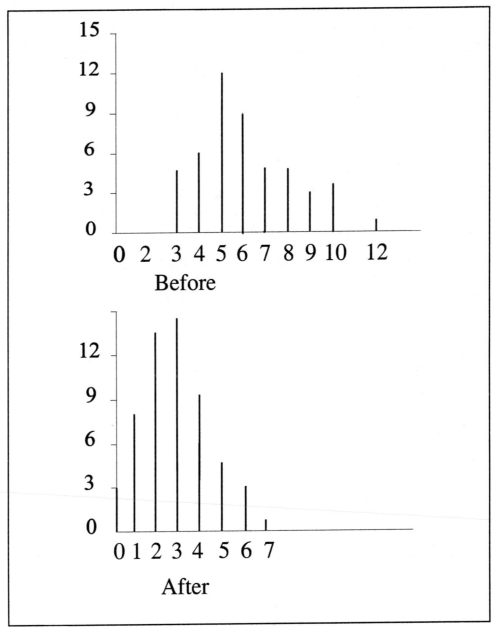

Figure 4.9 Before and After Comparison of Number of Defective Castings

4.3.4 Boxplots

A boxplot is similar to a histogram in that it shows the shape of the distribution of data. However, it is more compact and therefore better when you want to compare several groups of data. A boxplot is constructed by ordering the data from smallest to largest and then locating the smallest value or minimum, the largest value or maximum, the 25th percentile, the 50th percentile (or median), and the 75th percentile. The 25th percentile is a number such that 25 percent of the values in the data set are less than this value. Similarly the 50th and 75th percentiles are numbers that are greater than 50 percent and 75 percent of the values in the data set, respectively. The boxplot is drawn by making a box, the bottom of which starts at the 25th percentile, and the top of which is at the 75th percentile. A horizontal line cuts through the box at the 50th percentile. The width of the box is usually arbitrary, however if several data sets are being compared the width can be drawn proportional to the number of values in each data set.

From the top and bottom of the box, lines called *whiskers* are drawn. The whisker at the top of the box extends to the largest value in the data set that is within 1.5 times the interquartile range (the difference between the 75th percentile and the 25th percentile) from the 75th percentile. Likewise, the whisker at the bottom of the boxplot extends from the bottom of the box (i.e., the 25th percentile) to the smallest value within 1.5 times the interquartile range from the 25th percentile. Individual data points that are beyond the end of the whiskers are usually odd data points like the one indicated in Figure 4.6. These odd values, sometimes called *outliers*, are indicated on the boxplot with a single asterisk.

Table 4.2 Data for Example Boxplot

Data	Data Sorted	Rank
6.524	3.125	1
5.311	4.414	2
5.214	5.214	3
6.174	5.311	4
3.125	5.553	5
4.414	5.874	6
5.553	5.901	7
5.874	6.174	8
6.247	6.247	9
5.901	6.524	10

Example 2: For an example of constructing a boxplot, consider the data in Table 4.2. The first column in the table is the original data. The second column is the data sorted from smallest to largest and the third column is the rank order of each value. The 50th percentile or median is the middlemost number between if there are an odd number of values. If there are an even number of values as in Table 4.2, the 50th percentile is the average of the two middlemost numbers. For the data above, this would be the average of 5.553, the 5th largest value, and 5.874, the 6th largest. We

usually use the notation $y_{.50} = \dfrac{(5.553 + 5.875)}{2} = 5.714$. Likewise, the 25th

percentile would be a number that is 25% down from the top. Its rank is calculated by dividing the number of data values (plus 2) by 4 (i.e., (n+2)/4 where n= the number of data points). Since there are 10 values in Table 4.2 the rank of the 25th percentile would be (10+2)/4=3.0. The 25th percentile itself is, in this case, 5.214. In general, if the calculated rank is not an integer value the 25th percentile is found by interpolating between two values. The rank of the 75th percentile is calculated as (3n+2)/4. For the data in Table 4.2 this would be 8.0, and the 75th percentile is 6.174. The interquartile range is then the difference of 6.174-5.214=0.960. The whisker at the bottom of the box will extend from the 25th percentile (5.214) to the smallest number greater than 5.214-1.5(0.960)=3.774. This means the smallest value 3.125 will be represented as an outlier on the plot and the whisker will extend to the next to smallest value, 4.414. The whisker at the top of the box will extend from the 75th percentile, 6.174, to the largest value less than 6.174+1.5(0.960)=7.614. All values are less than this so the upper whisker will extent to the maximum value of 6.524. The example boxplot is shown in Figure 4.10.

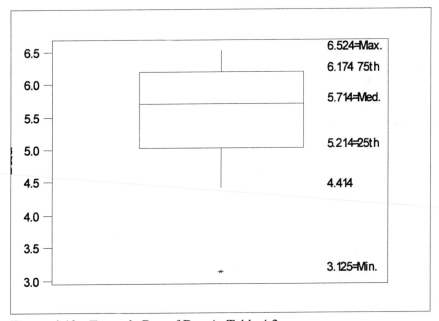

Figure 4.10-- Example Box of Data in Table 4.2

103

Symmetry or skewness of distributions can be seen as easily with a boxplot as it can with a histogram. Figure 4.11 below shows boxplots of left skewed, symmetric, and right skewed data. This should be compared to similar pictures shown for histograms in Figure 4.6.

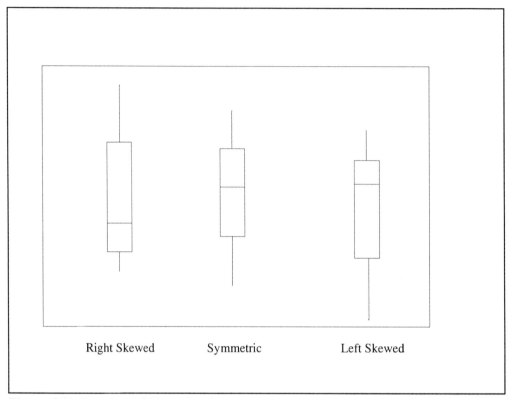

Figure 4.11 Boxplots of Skewed and Symmetric Data

Boxplots are most valuable for comparisons of different sets of data. It would be difficult to compare more than two or three histograms on the same scale, but 10 or even 20 boxplots can easily be squeezed on one graph for comparison. In the copperplating example from Chapter 1, Motorola employees were trying to reduce the variability in plating thickness. To do this they ran the plating cell at four different combinations of process variables they hypothesized influenced plating thickness. (These four combinations were actually chosen according to a fractional factorial experimental design to be described in Chapter 12.) Boxplots of the plating thicknesses in each of the four experimental runs are shown in Figure 4.12. The length of the boxplot from top to bottom and from the end of the top whisker to the bottom whisker are measures of variability. It can be seen from the figure that Combination 2 had the least variability in plating thickness.

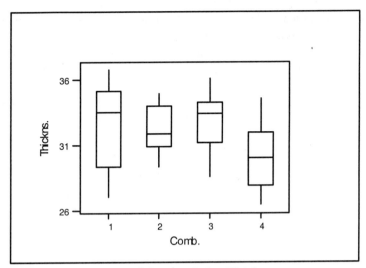

Figure 4.12 Boxplots of Copperplating Thickness

Most statistical software programs can create boxplots, however general spreadsheet programs like QUATTRO PRO® or Microsoft EXCEL® will not.

4.3.5 Stem Leaf Diagrams

Another graphic that is easy to construct by hand is the stem leaf diagram. Construction of the stem leaf diagram is similar to that of the dotplot, but the scale is more coarse and the actual data values rather than dots are placed on the diagram. The scale for the stem leaf diagram is made by breaking the numbers between two adjacent digits.

Example 3: Consider the data in the first column of Table 4.1. The values in this column range from 22.4 at the low to 53.5 at the high. Breaking the numbers between the tens place and the ones yields a scale comprised of 4 stems, namely 2, 3, 4 and 5. The stem leaf diagram is drawn by writing the stems in a vertical column with a vertical line drawn to the right. Next, the digits to the right of the break point are added to the stems as leaves. The example is shown below in Figure 4.13.

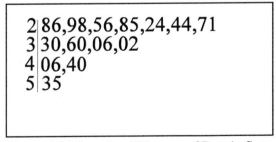

Figure 4.13 Stem Leaf Diagram of Data in first Column of Table 4.1

Sometimes the leaves are rounded off to one digit thereby eliminating the need for the commas between them. When the right number of stems are selected, the stem leaf diagram takes on the appearance of a histogram with the added advantage of having the actual data values displayed at the same time. In other words, from Figure 4.11, it is easy for us to copy down the original numbers 28.6, 29.8, 25.6, 28.5, 22.4, 24.4, 27.1, 33.0, 36.0, 30.6, 30.2, 40.6, 44.0, and 53.5.

If we were to create a stem leaf diagram of all the data in Table 4.1 using the four stems shown in Figure 4.11 it would not be very informative. With 97 data values classified into only four stems, the graph would be too short and wide to allow us to visualize anything. To give some flexibility in the number of stems, each digit to the left of the vertical line can again be broken into substems. For example, to create 8 rather than 4 stems we could use the eight lines:

$$
\begin{array}{r|}
2* \\
2\bullet \\
3* \\
3\bullet \\
4* \\
4\bullet \\
5* \\
5\bullet \\
\end{array}
$$

where leaves beginning with the digits 0-4 are placed on the line with the *, and leaves beginning with the digits 5-9 are placed on the line with the •. An example using this scale for the data in the first column of Table 4.1, again, is shown below in Figure 4.14.

$$
\begin{array}{r|l}
2* & 24,44 \\
2\bullet & 86,98,56,85,71 \\
3* & 30,06,02 \\
3\bullet & 60 \\
4* & 06,40 \\
4\bullet & \\
5* & 35 \\
5\bullet & \\
\end{array}
$$

Figure 4.14 Stem Leaf of Data
in 1st Column of Table 4.1
Using 2 Substems

To get even more stems or categories, the stem digits can be broken into five substems represented by the symbols:

$$
\begin{array}{c}
*| \\
t| \\
f| \\
s| \\
\bullet|
\end{array}
$$

where, * is for leaves beginning with the digits 0-1, t is for leaves beginning with the digits 2-3, f is for leaves beginning with the digits 4-5, s is for leaves beginning with the digits 6-7, and • is for leaves beginning with the digits 8-9. Figure 4.15 shows a stem leaf diagram of all of the data in Table 4.1, rounding the leaves off to a single digit. Compare this figure to the histogram shown in Figure 4.2. It gives essentially the same picture but is constructed by hand with a pencil and paper with about the same amount of effort as recopying the list of numbers.

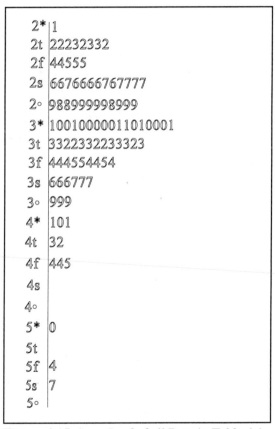

```
2* | 1
2t | 22232332
2f | 44555
2s | 6676666767777
2∘ | 988999998999
3* | 10010000011010001
3t | 3322332233323
3f | 444554454
3s | 666777
3∘ | 999
4* | 101
4t | 32
4f | 445
4s |
4∘ |
5* | 0
5t |
5f | 4
5s | 7
5∘ |
```

Figure 4.15 Stem Leaf of all Data in Table 4.1, Leaves Rounded to One Digit

Another valuable use for stem leaf diagrams is for sorting numbers from low to high. Recall this had to be done to get the ranks and find the percentiles of the numbers in Table 4.2 in order to construct a boxplot. We will again require sorting numbers to create probability plots that will be described in Chapter 5. Statistical software, general purpose spreadsheet programs, and some calculators such as the HP48 series have built-in sort algorithms. But if you are without one of these tools and have to sort a set of data by hand, it can be a tedious task. First you have to look through the list of numbers to find the minimum, copy it to the top of the new list and cross it out on the original list. Then repeat this process over and over until the data is sorted. All in all this requires you to go through the list of numbers many times. Constructing a stem leaf diagram requires that you go through your list of numbers only once, and when you are finished the numbers are "almost" sorted. For example in Figure 4.12, only the leaves on the second and third stems ended up out of order. Since the number of leaves on each stem is small, these can almost be arranged in your head to produce the sorted list 22.4, 24.4, 25.6, 27.1, 28.5, 28.6, 29.8, 30.2, 30.6, 33.0, 40.6, 44.0, 53.5. Whenever you have to sort 10 - 20 numbers by hand, constructing a stem leaf diagram first will usually reduce the time and frustration considerably.

4.3.6 Scatter Plots

The graphical representation of data we have presented thus far only describes one measured characteristic. If the purpose of collecting data is to examine correlation or cause and effect relationships between two measured characteristics, then a different type of graph is needed. When data are collected in pairs, the scatter plot is the graph we use to study relationships. Figure 4.16 is an example scatter plot of gas mileage versus weight on 1979 model cars (taken from the Splus data set autos-obtained from U.S. Government EPA Statistics and the April 1979 issue of *Consumer Reports*).

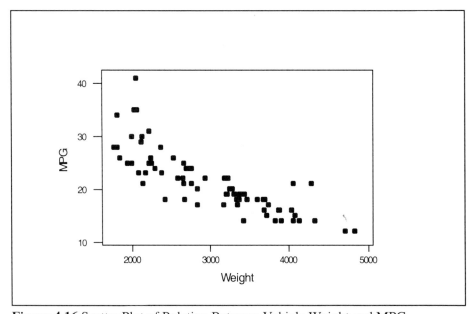

Figure 4.16 Scatter Plot of Relation Between Vehicle Weight and MPG

In this data set several variables were measured on 74 cars models. Each point on the graph represents the pair of variables (X=weight, Y=MPG) for a specific car model. From this graph we can see that gas mileage is better for lighter cars.

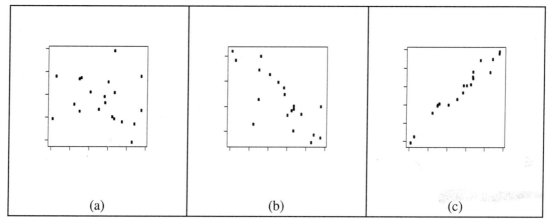

(a)   (b)   (c)

Figure 4.17 Scatter Plots Indicating Little, Moderate Negative, and Strong Positive Relations

Figure 4.17 (a) is an example of a scatter plot that shows little relation between the variable plotted on the X and Y axes, while (b) shows a moderate negative relationship and (c) shows a strong positive relationship. By positive relationship, we mean as X increases Y tends to increases, and vice versa for negative relationships.

4.4 Numerical Descriptive Statistics for Summarizing Data

Numerical descriptive statistics are often used when a set of data must be summarized in words for an oral or written report where graphics cannot be incorporated. We will discuss two main types of summary statistics. The first is a single number summary of the location or center of a set of data, and the second is a single number summary for the variability in a set of data. We have already used some of these statistics in Chapters 1 and 2, so the description here will be brief.

4.4.1 Location

The indicator of location (or center) of a data set that we will use most frequently is the arithmetic mean or average. It is a direct estimate of the expected value defined in Chapter 3. It is represented by placing a bar over the symbol used for the data, and is calculated by summing all the data, then dividing the sum by the number of values summed. If the data values are represented by y_i, for i=1 to n, where n is the number of data points, the usual notation to represent the mean would be $\bar{y} = \dfrac{\sum\limits_{i=1}^{n} y_i}{n}$. Sometimes the median or 50th percentile is also used as an indicator of the center

in a set of data. The median is especially useful when the data have a skewed distribution like Figure 4.5, or where there are outliers as shown in Figure 4.7 or 4.10. The median is not pulled away from the middle of the data by a few odd values in the extreme tails, and can be more representative of the center of the data than the mean. However the mean is used in the formal statistical decision rules based on probability theory, and therefore will be used frequently in the remainder of this book.

Since the intent of the quality improvement teams at the Kokomo Plant was to reduce the number of defective castings, as shown in Figure 4.9, comparison of the means of the before and after data on number of defects per week would be a useful and concise summary. The mean number of defects before the improvement effort was 6.14 compared to 2.89 after the mold redesign.

4.4.2 Variability

The purpose of the Motorola project, discussed in Chapter 1, was not to change the average or mean thickness of copper plate layers, but rather to reduce the variability in the thicknesses.

4.4.2.1 Range

One of the simplest measures of variability, that we have already presented, is the range. The range is simply the difference between the smallest and largest values in a set of numbers. The range is the most common measure of variability used when constructing control charts, like those shown in Chapter 2. One problem with using the range as a measure of variability is its sensitivity to odd values or outliers in the data. As the number of values in a data set increase, the likelihood of one odd value also increases and the range becomes an unreliable measure of variability. For example, the two data sets presented in the boxplots in Figure 4.14 are identical except for the largest value, yet the range of Data Set 1 is 33.36 while the range of Data Set 2 is 51.37. If only the range were presented, without the accompanying figure, it could be very misleading because the data sets are identical except for one value.

For small sample sizes, however, the range is an efficient measure of variation. For this reason, the range has been traditionally used for constructing control charts where subgroup sizes

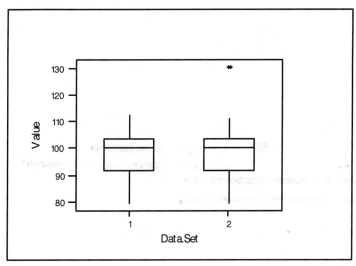

Figure 4.14-- Boxplots of Data Sets 1 and 2

are small (i.e., 3 to 7).The range can be converted into an estimate of the standard deviation (which is discussed next) using tabled factors d_2 (Table A.2) that are used in the construction of control charts. Examples of the use of the d_2 factors will be given in Chapters 16 and 17.

4.4.2.2 Variance Standard Deviation

The sample variance of a set of numbers is represented by the symbol, s^2, and is calculated by the following formula:

$$s^2 = \frac{\sum_{i=1}^{n}(y_i-\bar{y})^2}{(n-1)} = \frac{\sum_{i=1}^{n}y_i^2 - \frac{\left(\sum_{i=1}^{n}y_i\right)^2}{n}}{(n-1)} \tag{4.2}$$

It is a direct estimate of the population variance, σ^2, defined in Chapter 3, and can be thought of as the average squared distance from the arithmetic average in a set of numbers. Table 4.3 shows an example of equation (4.2) to compute s^2 using the data in the first column of Table 4.1. All statistical programs have the ability to calculate this quantity directly. Most scientific calculators and spreadsheet programs also have this equation built-in. However, some like the @VAR function in LOTUS 1-2-3 actually calculates $((n-1)/n)s^2$ which is the population variance for a finite population. If you are using a program or a calculator, you need to make sure you are getting the correct quantity. Check your program or calculator you are using with the data in Table 4.3. If you get the incorrect variance (i.e., the population variance of 67.365), you can correct it by multiplying by $n/(n-1)$.

When describing a set of numbers, calculating the variance, s^2, is usually just an intermediate step. The number that we actually use most often to describe the variability in a set of numbers is the sample standard deviation s. The sample standard deviation of a set of numbers is represented by the symbol, s, and is simply the square root of the sample variance, s^2. This, again is a direct estimate of the population standard deviation, σ, defined in Chapter 3, and can be thought of as the average distance from the arithmetic average in a set of numbers. We can visualize the standard deviation on the Normal Distribution (in Figures 3.11 and 3.12) as the distance from the mean to the point of inflection on the curve. The larger the standard deviation, the wider the Normal curve will be.

Table 4.3 Example Variance Calculation

Index of Values i	Data y_i	Squared Data y_i^2
1	28.6	817.96
2	29.8	888.04
3	25.6	655.36
4	33.0	1089.0
5	36.0	1296.0
6	40.6	1648.36
7	30.6	936.36
8	30.2	912.04
9	28.5	812.25
10	22.4	501.76
11	44.0	1936.0
12	24.4	595.36
13	27.1	734.41
14	53.5	2862.25
Sums:	454.3	15,685.15

$$s^2 = [15,685.15 - \frac{(454.3)^2}{14}]/(14-1) = 72.547 \quad : \text{Sample Variance}$$

$$((n-1)/n)s^2 = [15,685.15 - \frac{(454.3)^2}{14}]/14 = 67.365 \quad : \text{Population Variance}$$

Table 4.4 below shows a comparison of all the variability estimates we have discussed using the before and after copper plate thickness data that was shown graphically in Figure 4.8. The after measures of variability are smaller indicating a more consistent plating thickness, as could be seen in Figure 4.8.

Table 4.4 Comparison of Variability Measures for Copperplating Data

Measure of Variation	Measurements before Process Improvement	Measurements after Process Improvement
Standard Deviation(s)	6.872	3.493
Range (R)	36.20	15.000

4.5 Summary

In this chapter we have described some graphical and numerical descriptive statistics. These descriptive statistics will be used throughout the rest of the book to present and summarize data from observational and experimental studies. Some of the simple graphical tools presented in Chapters 1 - 4 have been so useful in quality improvement studies that they have been given a special name. They are called, in the quality improvement literature, the seven tools, which are:

(1) Flow diagrams
(2) Cause and effect diagrams
(3) Pareto diagrams
(4) Line graphs
(5) Scatter plots
(6) Histograms
(7) Control charts

The important concepts and descriptive statistics you should be familiar with are:

1.Deming, W. E.,*Quality Productivity and Competitive Position*, (Cambridge, MA: MIT Press, 1982).

2. Sturges, H. A., "The Choice of A Class Interval ," *Journal of the American Statistical Association*, Vol. 21, (1926), pp. 65–66.

4.6 Exercises for Chapter 4

4.1 Consider the following table of data which represent the drained weights of size 2½ cans of tomato in purèe (taken from E. L. Grant, *Statistical Quality Control*, McGraw Hill Book Company 1946).

```
22.0  22.5  22.5  24.0  23.5  20.5  22.5  22.5  23.0  21.5
20.0  20.5  23.0  22.0  21.5  21.0  22.0  22.0  23.0  22.0
22.5  19.5  22.5  22.0  21.0  23.0  23.5  21.0  22.0  20.0
19.0  20.0  22.0  20.5  22.5  21.5  20.5  19.0  19.5  19.5
21.0  22.5  20.0  22.0  22.0  21.5  23.0  22.0  23.0  18.5
20.0  19.5  21.0  20.0  20.5  19.0  21.0  21.0  21.0  20.5
19.5  20.5  21.0  20.5  21.0  20.0  21.5  24.0  23.0  20.0
22.5  19.5  21.0  21.5  21.0  21.5  20.5  22.0  21.5  23.5
19.0  21.5  23.0  21.0  23.5  21.0  20.5  19.5  22.0  21.0
20.0  23.5  24.0  20.5  21.5  22.0  20.5  21.0  22.5  20.0
```

Use your calculator, or computer to generate a histogram of the data.

4.2 The data below came from a voltage-endurance test of an insulating fluid, and represents time to breakdown at 35kV.

   16, 33, 41, 87, 93, 98, 134, 258

   a) Calculate the median, 25th percentile, and 75th percentile of the data.

   b) Construct a boxplot of the data.

   c) Calculate the sample mean, $\bar{y}$, of the data.

4.3 The data below represent conveyor speeds in cm/sec.

8.1, 7.7, 7.4, 5.8, 7.6, 6.8, 7.9, 6.3, 7.0, 8.0, 8.0, 8.0, 7.2, 6.0, 6.3, 6.7, 8.2, 8.1, 6.6, 6.5 8.5, 7.4, 7.2, 5.6, 6.3

Construct a stem leaf diagram of the data.

4.4 The data below are the test scores of 25 students on two consecutive tests in their Engineering Statistics course. The average grade on both exams was 80.

  a) Construct a stem leaf plot for Test 1 scores and a second plot for Test 2 scores.

  b) Construct a boxplot for Test 1 scores and a second box plot for Test 2 scores.

  c) Comment on whether or not you feel the students did equally well on both tests, and explain your conclusion.

Student	Test 1	Test 2
1	73	78
2	68	76
3	79	85
4	82	81
5	84	83
6	85	81
7	79	86
8	78	79
9	87	89
10	87	86
11	74	78
12	62	40
13	71	78
14	81	85
15	90	93
16	88	85
17	76	82
18	84	82
19	81	79
20	84	84
21	91	90
22	72	48
23	78	78
24	84	91
25	82	83

4.5 The data to the right represent the number of times a particular Web page on the Internet was accessed during a one-minute interval (between 2:00 and 3:00 p.m. EST).
  a) Draw a histogram for the data.
  b) What distribution best describes the data? Determine the parameters of that distribution from the data.
  c) Draw the probability mass function for the distribution from Part (b) on top of the histogram from Part (a).

3	3	3	1	4
1	3	6	5	1
7	5	5	1	6
1	0	4	2	2
5	0	0	3	3
2	2	1	2	3
3	0	3	2	3
4	3	7	0	4
3	4	4	4	3
4	8	4	2	1

4.6 The data to the right represent the length of time (in seconds) between accesses of a particular Web Page on the Internet (between 2:00 and 3:00 pm EST).
  a) Construct a stem leaf diagram of the 50 data points.
  b) Construct a boxplot of the 50 data points.
  c) Draw a histogram for the data.
  d) What distribution best describes the data? Determine the parameters of that distribution from the data.
  e) Draw the probability density function for the distribution from Part (d) on top of the histogram from Part (c).
Note: For this continuous distribution, the f(y) plot will require a separate Y-axis.

14	78	17	27	42
9	44	41	34	4
12	1	23	1	20
28	5	12	58	17
40	25	10	3	58
16	21	8	1	10
3	1	45	5	34
14	7	4	2	14
48	14	15	70	4
5	4	30	1	1

4.7 The data to the right represent the concentration of fifty batches of product (which has a nominal concentration of 50%).
  a) Construct a stem leaf diagram of the 50 data points.
  b) Construct a boxplot of the 50 data points.
  c) Draw a histogram for the data.
  d) What distribution best describes the data? Determine the parameters of that distribution from the data.
  e) Draw the probability density function for the distribution from Part (d) on top of the histogram from Part (c).
Note: For this continuous distribution, the f(y) plot will require a separate Y-axis.

50.7	50.7	50.3	51.5	51.0
51.3	51.0	51.6	50.4	51.2
51.8	51.2	50.6	50.7	51.7
50.2	52.0	51.3	50.8	50.8
51.6	50.9	51.4	50.6	50.0
51.1	50.5	51.2	51.1	51.0
50.6	50.8	50.4	51.6	51.2
51.2	50.2	50.7	51.2	50.1
50.5	51.4	51.0	51.4	52.1
51.7	51.6	50.3	51.5	50.3

CHAPTER 5
Probability Plots

5.1 Introduction

In the last chapter we discussed graphical and numerical descriptive statistics for displaying and summarizing data. Descriptive statistics can be used when the data come from observational studies or experimental studies. However, as we stated before, descriptive statistics can only summarize the data we have. To use data for constructing more formal decision rules, like those discussed in Chapters 1 - 3, we need a probability model or probability distribution like those discussed in Chapter 3.

How do we know what probability density function, f(y), or equivalently what cumulative distribution function, F(y), would best represent a set of data? This question can be answered using graphs of the data similar to the graphical descriptive statistics used in the last chapter. The histogram, boxplot, dotplot, and stem leaf diagram all give us a rough idea what the density function, f(y), is like. They can show us if it is symmetric like a Normal Distribution, or skewed like a Lognormal or Exponential Distribution. We can gain an even more precise picture by summarizing data to approximate the CDF curve, F(y), rather than the density function, f(y).

5.2 The Empirical CDF (ECDF)

Recall from Chapter 3 that the CDF, $F(y) = P\{Y \leq y\}$. If we have a data set $y_1, y_2, \ldots y_n$, we can approximate the $P\{Y \leq y\}$ by counting the proportion of values in our data set that are less than or equal to y. We can do this simultaneously for many values of y by constructing a graph.

Consider the data from the first column of Table 4.1 that was used previously to illustrate stem leaf diagrams and dotplots in Chapter 4. This data is reproduced in sorted order in the second column of Table 5.1. The first column of Table 5.1 are the data order numbers or ranks. The usual notation is as follows: if the unsorted data values are represented by the symbols y_i, for i=1 to n, then the sorted values, or order statistics, are represented by the symbols $y_{(i)}$, where i is the rank that ranges from 1 to n. The third column of the table are the proportion of data values less than or equal to $y_{(i)}$. It was calculated by dividing the rank numbers in the first column by the total number of values (i.e., n=14 in this table). These proportions, in the third column, are not the best estimates of the probabilities $P\{Y \leq y_{(i)}\}$ that we can find. For example, the proportion of values less than or equal to the largest value 53.5 is 1.0, but it is unrealistic to assume that the probability of producing a copperplating thickness greater than 53.5 is zero. In fact, with a larger sample of data (i.e., all of Table 4.1) we do observe values greater than 53.5. Also the proportion of values less than or equal to the 7th value 29.8 is .50, but we know the median would be the average of the 7th and 8th values (29.8 and 30.2).

A better estimate of the $P\{Y \leq y_{(i)}\}$ is found by subtracting ½ from the rank numbers (i-½) before dividing by the number of values (n). This calculation is shown in the 4th column of Table 5.1. Notice that in this column the estimate of $P\{Y \leq 53.5\}$ is not 1.0, and the number y such that $P\{Y \leq y\} = 0.50$ falls halfway between the 7th and 8th values as the median. The plot of $y_{(i)}$ versus $\hat{F}_i = (i-\frac{1}{2})/n$ (i.e., 2nd versus 4th columns in Table 5.1) is called the empirical cumulative distribution function (ECDF). Figure 5.1 shows a graph of these columns. The connecting lines are added to facilitate interpolating percentiles. For example as shown in the graph, the 25th percentile is found by drawing a horizontal line from .25 on the vertical scale to the connecting line. Then a vertical line is dropped from the connecting line to the horizontal scale where the 25th percentile is read.

Table 5.1 Data from Column 1 of Table 4.1

Rank i	Sorted Data $y_{(i)}$	Proportion $\leq y_{(i)}$	Estimate of $\hat{F}_i = P\{Y \leq y_{(i)}\} = (i - \frac{1}{2})/n$
1	22.4	1/14=.0714	.5/14=.0357
2	24.4	2/14=.1428	1.5/14=.1071
3	25.6	3/14=.2142	2.5/14=.1786
4	27.1	4/14=.2857	3.5/14=.2500
5	28.5	5/14=.3571	4.5/14=.3214
6	28.6	6/14=.4286	5.5/14=.3926
7	29.8	7/14=.5000	6.5/14=.4643
8	30.2	8/14=.5714	7.5/14=.5357
9	30.6	9/14=.6429	8.5/14=.6071
10	33.0	10/14=.7143	9.5/14=.6787
11	36.0	11/14=.7857	10.5/14=.7500
12	40.6	12/14=.8571	11.5/14=.8214
13	44.0	13/14=.9286	12.5/14=.8929
14	53.5	14/14=1.000	13.5/14=.9643

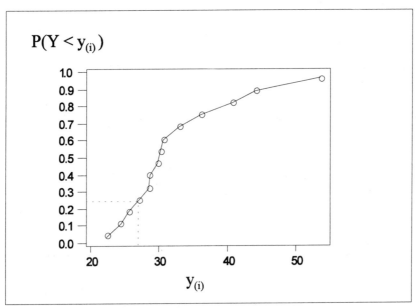

Figure 5.1 ECDF Plot of Data from Table 5.1

From a set of data, the ECDF plot gives us a graphical picture of what the cumulative distribution, F(y), looks like, just as a histogram gives us a picture of what the density function, f(y), looks like. The shape of the ECDF plot can help us to guess what the equational form of the cumulative distribution function, F(y), is. We can do this by remembering what the shapes of CDFs look like for various common distributions. For example, Figures 5.2 and 5.3 show the CDF plot for a Uniform Distribution and a Normal Distribution. The CDF plot for data from a Uniform Distribution, in Figure 5.2, is a straight line between the limits of extreme low and high values the data can take on. This fact is due to the form of the cumulative distribution function given in Equation 3.8 in Chapter 3. On the other hand, the equation for the cumulative distribution function for a Normal Distribution, is a more complicated integral equation given in Equation 3.12, and the shape of the CDF curve for a Normal Distribution is the symmetric sigmoid curve shown in Figure 5.3.

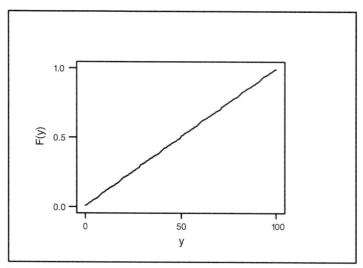

Figure 5.2 CDF Plot for Uniform Distribution

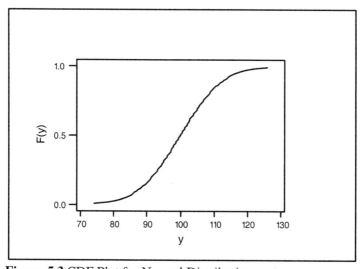

Figure 5.3 CDF Plot for Normal Distribution

122

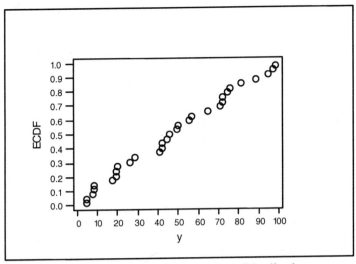

Figure 5.4 ECDF Plot Data from Uniform Distribution

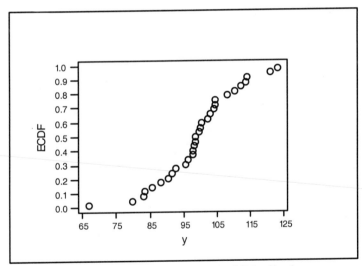

Figure 5.5 ECDF Plot Data From Normal Distribution

Figures 5.4 and 5.5 show ECDF plots of data from a Uniform Distribution and a Normal Distribution. From these ECDF plots it is easy to see which data set came from a Uniform Distribution and which one is from a Normal Distribution.

Look at Figure 5.1. What distribution does the data appear to be from? The data are obviously

123

not from a Uniform Distribution, because the points do not fall on a straight line. However, it is more difficult to judge by eye whether the data fall close to a symmetric sigmoid. This problem can be avoided by using a specially scaled graph paper called Normal Probability paper. It has nonlinear scaling at the top and the bottom of the vertical scale so that a symmetric sigmoid drawn on this graph paper appears as a straight line. For example, Figure 5.6 is the data from Table 5.1 replotted on Normal Probability paper. The nonlinear scale can be seen by noticing the distance between .99 and .999 at the top of the vertical scale is the same as the difference between .5 and .8, near the middle of the scale, and the difference between .001 and .01 at the bottom of the scale. The straight line is added to the points in the figure to help in visually judging how close the points come to a straight line. In this figure it can be seen that the data from Table 5.1 do not fall along a straight line but a curved or bent line on the Normal Probability paper. This indicates the data comes from a right skewed distribution as will be explained in Section 5.4.

Preprinted Normal Probability paper can be purchased at most stationery or college bookstores where engineering graph paper is sold. Some statistical programs like MINITAB have programmed procedures to produce plots on Normal Probability paper. However, as we will show in the next section, we can produce an equivalent Normal Probability plot on arithmetic graph paper, by transforming the data points to be plotted (rather than the vertical scale of the graph paper). The option to make Normal Probability plots on arithmetic graph paper allows us to make the same graphs using popular spreadsheet programs like LOTUS 1-2-3® or Microsoft EXCEL®.

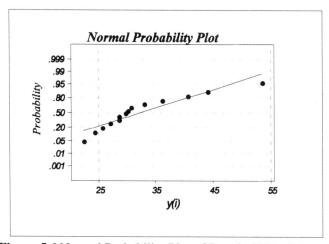

Figure 5.6 Normal Probability Plot of Data in Table 5.1

124

5.3 Making Probability Plots on Arithmetic Graph Paper

5.3.1 Probability Integral Transform

Consider the ECDF plot for data from a Uniform Distribution shown in Figure 5.4. This plot was made on arithmetically scaled graph paper. If the data is from a Uniform Distribution, the points in this plot tend to fall along a straight line. This is due to the special form (i.e., Equation 3.8) of the Uniform CDF. However, a linear set of points on arithmetic graph paper is what we would like to see for Normal Probability plots. There is a theorem that will allow us to accomplish this. The theorem says essentially this: If the cumulative distribution function, CDF, for a random variable, Y, is $F_Y(y)$, then there exists a transformation $u=h(y)$, such that the distribution function of U, $F_U(u)$, is uniform on 0 to 1. Furthermore, the transformation equation is $h(y)=F_Y(y)$.

This theorem follows from the fact that:

$$F_U(u) \;=\; P\{U \le u\} \;=\; P\{\,F_Y(Y) \le u\} \qquad \text{since } U=F_Y(Y)$$

$$=\; P\{Y \le F_Y^{-1}(u)\} \qquad \text{since } F_Y(\,) \text{ is a monotonically increasing function}$$

$$=\; F_Y(F_Y^{-1}(u)) \qquad \text{by definition of } F_Y(\,)$$

$$=\; u$$

and the fact that the CDF for a random variable that is Uniform on 0 to 1 is such that $F_U(u) = u$, for $0 \le u \le 1$ (see Equation 3.8 with a=0 and b=1). Recall, the inverse function notation defines the function $F_Y^{-1}(y)$ such that $F_Y^{-1}(F_Y(y))=y$. (Note: $F_Y^{-1}(u)$ can be found by solving for y in the equation $F_Y(y)=u$.)

This theorem allows us to construct probability plots for *any* distribution using arithmetic graph paper. We do it as follows: If we think that a set of data y comes from a CDF, $F_Y(y)$, then we plot the ECDF of the transformed data $u_i =F_Y(y_i)$; in other words evaluate the proposed CDF, $F_Y(y)$, at each of the data points, y_i, to get the transformed data u_i. Next, order the transformed data u_i, to get the ordered values $u_{(i)}$, and plot these ordered values versus the estimate of $P\{U \le u_{(i)}\}=(i-\frac{1}{2})/n=\hat{F}_i$. If $F_Y(y)$ is the appropriate cumulative distribution function for the data, this plot will appear as a straight line like the ECDF plot for the Uniform data in Figure 5.4.

We can accomplish the same result preserving the original scale for our data axis by plotting the original data versus the transformed ECDF, rather than the transformed data versus the untransformed ECDF. In other words, plot the ordered values of the original data $y_{(i)}$ on the horizontal axis versus the transformation of the ECDF values $F_Y^{-1}(\hat{F}_i)=F_Y^{-1}((i-\frac{1}{2})/n)$ on the vertical axis. Next we will show examples of this method of plotting for the Exponential and Normal Distributions.

5.3.2 Exponential Distribution

The CDF for the Exponential Distribution was given in Equation 3.10 as $F(y)=1-e^{-(y/\lambda)}$, for $y \ge 0$. Solving for y in the equation $F(y)=1-e^{-(y/\lambda)} = u$, we find $F^{-1}(u) = -\lambda\ln(1-u)$. Therefore, if we think our data comes from an Exponential Distribution, we will order the data and plot the ordered values, $y_{(i)}$, on the horizontal axis versus the exponential scores, $F^{-1}(\hat{F}_i) = -\ln(1-\hat{F}_i) = -\ln(1-(i-\frac{1}{2})/n)$, on the vertical axis. If the data are from an Exponential Distribution, the points should fall along a straight line with zero intercept and slope equal to $1/\lambda$.

This procedure is demonstrated in Table 5.2 and Figure 5.8. The first column in Table 5.2 is a sample of data that was computer generated from an Exponential Distribution with $\lambda=.5$. The second column is the data in sorted order. The third column is the rank values for the data and the fourth column is the vertical coordinates for the probability plot. The last column is the vertical coordinates for the probability plot, calculated as shown in the previous paragraph. Figure 5.8 shows virtually a straight line of points, as we would expect since the data were generated from an Exponential Distribution. We can see the slope of the plot is about 2, which is $1/\lambda$.

If an exponential plot of real data reveals a straight line of points, we can feel comfortable in assuming that the data represents a sample from an Exponential Distribution. Adding a hand-drawn straight line with zero intercept to the plot (similar to that done in Figure 5.6) helps in judging how close the points fall to a straight line. Also, from the slope of the hand-drawn line we can obtain an estimate for λ, the expected value.

Table 5.2 Data from Exponential Distribution

Data y_i	Ordered Data $y_{(i)}$	Ranks i	ECDF $\hat{F}_i=(i-\frac{1}{2})/15$	F^{-1}(ECDF) $-\ln(1-\text{ECDF})$
0.22068	0.06561	1	0.033333	0.03390
0.49427	0.14048	2	0.100000	0.10536
1.16031	0.15007	3	0.166667	0.18232
0.65707	0.22068	4	0.233333	0.26570
0.68926	0.25176	5	0.300000	0.35667
0.15007	0.27596	6	0.366667	0.45676
1.79750	0.34398	7	0.433333	0.56798
0.44230	0.44230	8	0.500000	0.69315
0.06561	0.49427	9	0.566667	0.83625
0.59037	0.59037	10	0.633333	1.00330
0.34398	0.65707	11	0.700000	1.20397
0.88510	0.68926	12	0.766667	1.45529
0.27596	0.88510	13	0.833333	1.79176
0.25176	1.16031	14	0.900000	2.30258
0.14048	1.79750	15	0.966667	3.40120

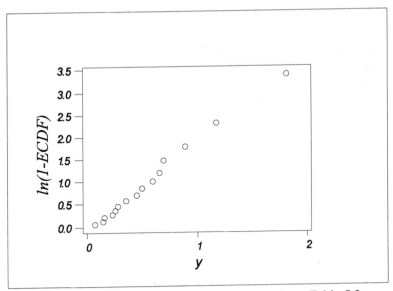

Figure 5.8 Exponential Probability Plot of Data from Table 5.2

5.3.3 Normal Distribution

The CDF for the standard Normal Distribution (i.e., with $\mu=0$ and $\sigma=1$) is given by the integral equation (see Equation 3.12 for the general case):

$$\Phi(y) = \int_{-\infty}^{y} \frac{1}{\sqrt{2\pi}} e^{-y^2/2} dy$$

With this complicated equation it is impossible to solve analytically for y in $\Phi(y)=u$ to get the inverse CDF function $\Phi^{-1}(u)$. However, a numerical approximation to the solution can be found and is available for use in most statistical programs like MINITAB. Some spreadsheet programs such as EXCEL® and Lotus 1-2-3® also have an approximation to the inverse normal CDF function built in (e.g., =NORMSINV( ) in EXCEL). If the program you use does not have an inverse normal CDF, a simple approximation good enough for graphical presentations of data is:

$$z_{(i)} = \Phi^{-1}(\hat{F}_i) = 5.05[\ \hat{F}_i^{.135} - (1-\hat{F}_i)^{.135}] = 5.05[((i-\tfrac{1}{2})/n)^{.135} - (1-(i-\tfrac{1}{2})/n)^{.135}]. \quad \textbf{(5.1)}$$

A Normal Probability plot is constructed on arithmetic paper by ordering the data and plotting the ordered values $y_{(i)}$ on the horizontal axis versus $\Phi^{-1}((i-\frac{1}{2})/n)$ on the vertical axis. If the data is a sample from a Normal Distribution, the points will fall approximately along the straight line $y_{(i)}=\mu+\sigma z_i$, where μ is the mean, σ is the standard deviation, and $z_i = \Phi^{-1}((i-\frac{1}{2})/n)$ are called the Normal scores, or z-scores. For an example, the data from Table 4.1 along with their calculated z-scores are shown below in Table 5.3, and the plot is shown in Figure 5.9. Compare this figure to the Normal plot of the same data in Figure 5.6 that uses the nonlinearly scaled vertical axis. The same pattern in the points can be seen in each figure. The hand-drawn dotted lines added to Figure 5.9 help us to graphically estimate μ, and σ. By projecting the horizontal line from axis at $z_i=0$ to the straight line approximating $y_{(i)}=\mu+\sigma z_i$, and dropping a straight line to the y axis we get an estimate of μ, and the reciprocal of the slope of the line estimates σ.

Table 5.3 Calculation of z-Scores for Data from Table 5.1

Order Number (i)	Data $y_{(i)}$	ECDF $\hat{p}_i=(i-\frac{1}{2})/14$	Z-score $z_{(i)} =\Phi^{-1}((i-\frac{1}{2})/14)$
1	22.4	0.035714	-1.80274
2	24.4	0.107143	-1.24187
3	25.6	0.178571	-0.92082
4	27.1	0.250000	-0.67449
5	28.5	0.321429	-0.46371
6	28.6	0.392857	-0.27188
7	29.8	0.464286	-0.08964
8	30.2	0.535714	0.08964
9	30.6	0.607143	0.27188
10	33.0	0.678571	0.46371
11	36.0	0.750000	0.67449
12	40.6	0.821429	0.92082
13	44.0	0.892857	1.24187
14	53.5	0.964286	1.80274

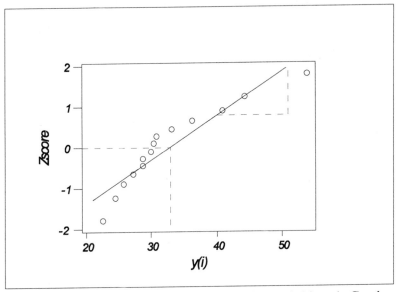

Figure 5.9 Normal Plot of Data from Table 5.1 on Arithmetic Graph Paper

5.4 Interpreting Normal Plots

Normal plots are quite useful and we will use them frequently in the remainder of the book for interpreting data from experimental studies. Two of the main purposes for using Normal plots are to: 1) determine if there are a few outlier data points that don't appear to come from the same distribution as the rest, and 2) determine whether a set of data is approximately Normal (symmetrically distributed) or skewed. If there are outliers in a data set that otherwise appear to be Normally distributed, they will stick out from a straight line of points at the upper right or lower left as shown in Figure 5.10. The points on the upper right are outliers in the upper tail of the distribution, and the points to the lower left are outliers in the lower tail of the distribution. At the bottom of the figure, the distribution is drawn where it should appear, within the points that fall along the straight line. There should not be any points sticking away from the line in the center portion of the plot. If there are, most likely an error was made in constructing the plot. If data from a right-skewed distribution, like an Exponential or Lognormal, are plotted on a Normal Probability Plot, the points fall along a concave curve like that shown in Figure 5.11 rather than a straight line.

Those unfamiliar with interpreting Normal plots often ask how far off the straight line (at the upper right or lower left) should a point be, before it is considered to be an outlier; or how large a departure from linearity is necessary before deciding the data is from a skewed distribution. The answer to these questions comes with experience in interpreting the plots. But, that experience can be gained rapidly by use of a computer. Many statistical programs such as MINITAB® have commands to generate a set of random data drawn from a Normal Distribution. You can quickly

several such plots you will get a feel for how much variation from the straight line you will get with Normal data. When using a spreadsheet that does not have a built-in command to generate random

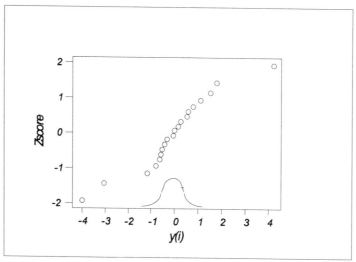

Figure 5.10 Normal Plot with Two Outliers in Left Tail and One in Right Tail

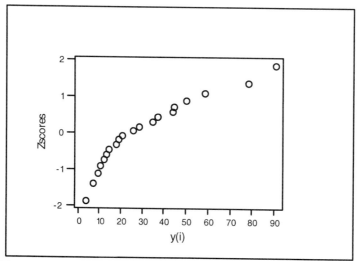

Figure 5.11 Normal Plot of Right Skewed Data

Normal data you can always use the @RAND function to generate a column of random Uniform u_i data between 0 and 1. Next, a column of random Normal data y_i can be obtained by letting $y_i=\Phi^{-1}(u_i)$.

As an example, Figure 5.12 shows Normal plots simultaneously of 6 data sets comprised of 15 values generated by MINITAB to be from Normal Distributions with mean zero and standard deviation one. Figure 5.13 is a similar plot for 6 sets of 31 data values. From these two graphs, you can see that the points fall closer to a straight line with 31 data values than 15. If you examine the individual values at the upper right and lower left of each line of points, you will get a feel for how removed from the line a point can be without being considered an outlier. If you generate and plot several sets of random Normal data like those shown in these two figures you will quickly become an expert in judging outliers and departures from Normality using the Normal plot. The next section gives another alternative for helping to spot outliers.

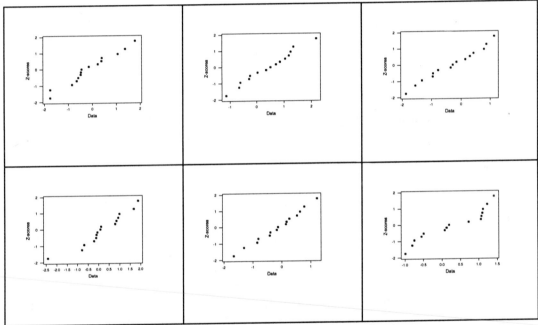

Figure 5.12 Normal Plot of Six Sets of Normal Data (15 observations)

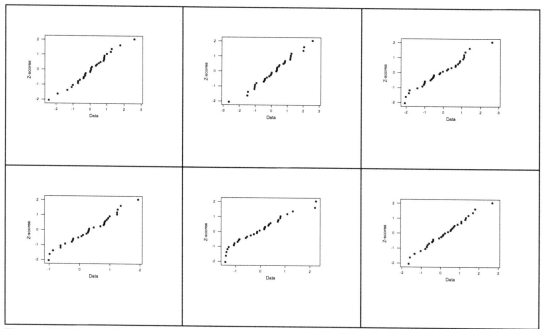

Figure 5.13 Normal Plot of Six Sets of Normal Data (31 observations)

5.5 Half-Normal Plots

In most of the cases presented in Parts III, IV and V of this book, when we examine data for outliers, we can assume the data comes from a distribution with zero mean. It is easier to identify outliers on the probability plot when they only appear at one place rather than at the upper right and the lower left as shown in Figure 5.10. When plotting a set of data that we think comes from a Normal Distribution with zero mean, we can construct a probability plot where the outliers will appear only at the upper right if we plot the absolute value of the data, rather than the original numbers. The rational for constructing a probability plot for the absolute values follows the logic for making full Normal Probability plots. When making a full Normal Probability plot, we plot the ordered data $y_{(i)}$ on the horizontal axis versus $z_i = \Phi^{-1}((i-\frac{1}{2})/n)$ on the vertical axis, because $(i-\frac{1}{2})/n$ is our estimate of $P(Y \le y_{(i)})$ and $\Phi(z)$ is the CDF of a Standard Normal Distribution evaluated at z. In other words, if we plot data, y, on the horizontal axis versus numbers, z_p, on the vertical axis, where $P(Y \le z_p) = p$ if Y follows a Standard Normal Distribution. We will do the analogous thing for absolute values of y.

If we let $W = |Y|$, then $P(W \le w) = P(-w \le Y \le w)$. If we think the data come from a Standard Normal Distribution,

$$P(-w \le Y \le w) = \Phi(w) - \Phi(-w)$$

$$= 2\Phi(w) - 1 \quad \text{since } \Phi(-w) = 1 - \Phi(w)$$

132

Therefore if z_p is such that $P(W \le z_p) = p$ for the absolute values W, then $p = 2\Phi(z_p) - 1$, and $z_p = \Phi^{-1}((p+1)/2)$. Therefore to make a probability plot of the absolute values $|y_i| = w_i$ we can plot the ordered absolute values $w_{(i)}$ on the horizontal axis versus $z_i = \Phi^{-1}\left(\dfrac{\left(\dfrac{i-\frac{1}{2}}{n}\right) + 1}{2}\right)$ on the vertical axis. If the data truly are from a Standard Normal Distribution, the points will fall along the straight line $y_i = z_i$. If the data are from a Normal Distribution with standard deviation σ, the points will fall along the straight line $y_i = \sigma z_i$ or $z_i = (1/\sigma)y_i$.

Figure 5.14 shows a Half-Normal plot of the same data shown in Figure 5.10. In this plot it can be seen that the three outliers are extremely apparent in one area to the right of the line at the upper end. The straight line was added to help judge the outliers. Since the line on this Half-Normal plot should go through the origin, (0,0), it is easier to draw it than it is to draw the appropriate straight line on the full Normal plot, like Figure 5.9.

Table 5.4 shows the calculations necessary for constructing the Half-Normal plot. The first column in the table, labeled y_j, is the raw data that is plotted in Figure 5.10. The second column is the absolute values of the data. The third column is the rank orders of the absolute values. In the fourth column, the ranks are converted to quantiles that are greater than $\frac{1}{2}$ using the formula

$$\hat{q}_i = \frac{[(i-\frac{1}{2})/n + 1]}{2}$$. In the last column are z-scores calculated from the quantiles. The plot is then constructed by plotting the second column on the horizontal axis versus the fifth column on the vertical axis.

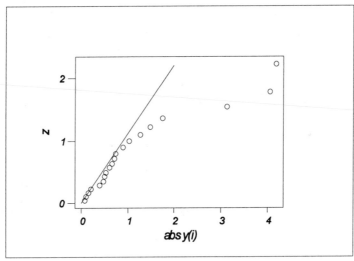

Figure 5.14 Half-Normal Plot of Data from Figure 5.10

133

Table 5.4 Calculation of Coordinates for Half Normal Plot

y_j	$\lvert y_j \rvert$	Rank = i	$\hat{q}_i = \dfrac{\left[\dfrac{(i-\frac{1}{2})}{n} + 1\right]}{2}$	$\Phi^{-1}\left(\dfrac{\left(\dfrac{(i-\frac{1}{2})}{n}\right) + 1}{2}\right)$
-0.60096	0.60096	9	0.7125	0.56070
-0.48390	0.48390	7	0.6625	0.41930
-0.14737	0.14737	3	0.5625	0.15731
-3.08450	3.08450	18	0.9375	1.53412
4.14690	4.14690	20	0.9875	2.24140
-0.88452	0.88452	13	0.8125	0.88715
0.18910	0.18910	4	0.5875	0.22112
1.72298	1.72298	17	0.9125	1.35631
1.47331	1.47331	16	0.8875	1.21334
0.44752	0.44752	6	0.6375	0.35178
-0.06781	0.06781	1	0.5125	0.03134
-1.24090	1.24090	15	0.8625	1.09162
-0.38564	0.38564	5	0.6125	0.28584
-0.65235	0.65235	10	0.7375	0.63566
0.08841	0.08841	2	0.5375	0.09414
0.50211	0.50211	8	0.6875	0.48878
1.01547	1.01547	14	0.8375	0.98423
-4.02221	4.02221	19	0.9625	1.78046
0.72245	0.72245	12	0.7875	0.79778
-0.70280	0.70280	11	0.7625	0.71437

5.6 Summary

In this chapter we have described some graphical techniques for determining what density function, f(y), or CDF, F(y), might best represent a set of data. We have focused our attention on the Normal Distribution which will be used often in later chapters of this book. We have seen that we can make a Normal Probability plot by plotting the ECDF curve on special nonlinearly scaled graph paper that you can purchase, or that by transforming the ECDF probability values the same plot can be made on arithmetic graph paper or using any computer plotting routine (such as those available in spreadsheets or statistical programs). Two of the common departures from Normality that can be easily seen on Normal Probability plots are skewness and outliers. Half Normal plots and some examples were presented to help you learn to judge outliers on the plots. The important terms presented in this chapter were:

5.7 Exercises for Chapter 5

5.1 With the data given below:

1.1055   4.6498   10.0012   8.8353   1.7290   3.3886   1.1712
6.1434   0.4148   0.5427   4.8682   1.7767   4.9621   6.3045
1.1594   2.1804   4.3250   4.2600   0.9796   11.6982

    a) Construct an ECDF plot

    b) Construct a Normal Probability plot

    c) Construct a Exponential Probability plot

    d) Which distribution (Normal, Exponential, or Uniform) fits the data best.

5.2 With the following data set:

0.64741 -1.04835   0.67364 -2.09966 -0.04745 -0.56537 -1.55935
-0.18851   0.75867 -0.32164   1.94487 -3.02096   1.16839   3.21664
3.21668   2.45447   2.31765   0.33819   0.69906 -0.64287   1.17212
1.58019   1.29125   1.29156   3.62930

Construct a Half-Normal Plot.

5.3 Construct Normal Probability plots with each of the four columns of data below. From your plots determine whether the data from each column appears

      1) Normally Distributed
      2) Skewed right
      3) Skewed left
      4) Normal with outliers

1	1.60327	5.54663	3.21760	8.2200
2	3.42026	7.94648	2.20320	7.2674
3	2.89703	6.58496	3.15130	5.6797
4	1.34470	7.60782	4.80087	7.3377
5	2.90145	3.76222	3.47448	5.5763
6	1.34010	6.55311	3.86474	5.9151
7	2.33279	4.59498	3.16708	3.6310
8	0.22202	4.83435	3.80504	4.3702
9	2.11883	1.68404	5.87149	6.4024
10	7.14723	7.74493	1.71194	5.8095
11	0.69078	6.44058	4.23613	6.0493
12	0.85511	4.56899	3.99870	5.5324
13	0.81389	6.01735	3.31773	4.4949
14	1.46992	7.28637	4.40063	5.4703
15	7.38681	6.31690	4.60899	7.4545
16	0.11227	5.99325	3.96832	0.5436
17	2.85990	0.31911	3.30732	4.5539
18	3.34002	7.50001	2.95898	6.1810
19	0.48030	5.51283	3.24880	7.2233
20	3.78336	7.86373	4.47591	3.3858
21	3.17522	7.57471	4.56505	4.1820
22	2.33071	7.35470	5.04033	4.9988
23	2.95426	6.02772	3.69590	12.2346
24	1.32146	6.85340	4.60420	4.1331
25	4.24604	5.74675	3.07348	4.1086

5.4 a) Make a Normal Probability plot of the Test 1 data from Problem 4 in Chapter 4. Draw a line through the data by eye. Does the data appear to be Normally Distributed? Are there any outliers? What is the estimate of the mean and standard deviation of the data using the line you drew through it?

  b) Make a Half-Normal plot of the same data (from Test 1). Remember that you must first subtract the mean of the data from the scores before you can plot them on a Half-Normal plot. The mean of the data is $\bar{y}$ = 80.

5.5 a) Repeat Problem 4(a) using the Test 2 data from Problem 4 in Chapter 4.

b) Repeat Problem 4(b) using the Test 2 data from Problem 4 in Chapter 4.

5.6 a) Make a probability plot of the data from Problem 6 in Chapter 4, assuming that the data follow an Exponential Distribution.

b) Draw a line through the data by eye. Does the plot show that the data is exponentially distributed?

c) What is the estimate of λ based on your line?

d) Are there any anomalous points?

5.7 a) Repeat Problem 4 (a) using the data from Problem 7 in Chapter 4.

b) Repeat Problem 4 (b) using the data from Problem 7 in Chapter 4. (Remember to subtract the mean of the data which is 51.0.)

CHAPTER 6
Inferential Statistics
Prediction and Decision Rules

6.1 Introduction

In Chapters 4 and 5 we presented some descriptive statistics that are useful for summarizing data and getting a graphical picture of what the data tell us. In this chapter we will present the theory behind inferential statistics which consist of prediction (or estimation) and formal decision rules. We will then present several examples of statistical inference for enumerative and analytic studies. In the remainder of the book, we will use statistical inference frequently in analysis of data.

Some examples of prediction and formal decision rules were already given in Chapter 2. The run of points below the center line in the control chart for casting porosity defects allowed the PQI team to predict that the added feeders to the casting form would reduce the porosity defects in the future. The control chart for shank dimensions was an example of a decision rule. Each time the lathe operator plotted a point on the control chart, he was following a decision rule, i.e., if the point was within the limits he would not adjust the machine, if the point was outside the limits he would adjust.

In a formal sense, statistical inference is an exercise in inductive logic where we try to predict or infer some characteristic of a population, or make a decision regarding some aspect of a population based on a sample of data from that population. In that sense it is the opposite of probability theory which is deductive logic. In probability theory we deduce from the known population (represented by a probability mass function or probability density function) the probability of observing a specific result in a sample (or realization of a random variable). We can easily see the difference when thinking about games of chance. If we are playing a game with fair dice (i.e; the probability mass function is known) and the rules are explained, we can use the addition and multiplication rules of probability to calculate the odds of winning the game. This is deductive logic. On the other hand, if we know a game is being played with loaded dice, we can observe the results of multiple dice roles and try to induce or estimate the unknown probabilities of getting a 1, 2, etc. This is inductive logic or statistical inference.

For a practical example of sample and population, again think of the control chart examples in Chapter 2. In the control chart of porosity defects, the population was the frequency of porosity defects in future transmission housings cast with the newly designed casting forms. The characteristics of this population could also be described by a probability mass function which might appear something like Figure 4.1 (with the vertical frequency axes scaled into probabilities). The sample was the data in subgroups 26 - 40 in Figure 2.10. The statistical inference consisted of predicting that future castings, using the new casting form, would have less defects than castings made using the old casting form. In other words they predicted that the probability mass function was shifted to the left.

When making decisions about a population with only a sample of data, we will always be in danger of making an error. Errors cannot be avoided, but we can reduce the chance or probability of making an error by using a statistical decision rule rather than an ad hoc method. The purpose of this chapter is to define a theoretical framework for statistical inference and present some practical techniques that will be used in the remaining part of the text.

6.2 Populations

In statistical inference there are two types of populations we will consider. The first type is a *finite population*, wherein every item can theoretically be enumerated. An example of this might be all the weld joints in a lot of manufactured seat brackets. The second basic type of population is a *conceptual population*. In this type of population, we can't get our hands on and count all the elements in the population. Rather, the elements of this type of population are usually generated by a process or phenomena. An example might be all the future transmission housings that could be cast using the new casting form. This population exists only in our mind. Deming has described the difference in making inference about these two types of populations[1].

A study where we collect a sample of data and draw inference about a finite population was called an *enumerative study* by Deming. Here we are trying to predict some characteristic of the population, or make a decision regarding some aspect of the population based on just a sample of the elements, rather than an exhaustive census of all the elements. The validity of this type of study is guaranteed, if a random sampling plan is utilized. We will describe this further in Section 6.3. Common examples of enumerative studies in industrial work are lot acceptance sampling plans and lot acceptance reliability tests.

Deming defined an *analytic study* as one in which data are collected from a process at one point in time, for the purpose of drawing inference or making decisions about the output of that process at other times. In this kind of study the population is a conceptual population. Most engineering studies tend to be of this type. The statistical methods used in these studies are similar to, or the same as, those used for enumerative studies. However, the justification and interpretation are different. The probability of making an incorrect decision in an enumerative study can be determined, and the methods of determination are justified by the sampling procedure used to obtain the data. In analytic studies, however, we are attempting to make a prediction or decision about a population that doesn't even exist, based on a sample of data from the present point in time. The probability of making an error in this type of study can be calculated using the same formulas as used for an enumerative study, but in this case they are only conditional probabilities[2]. They are conditional on the fact that the data we have collected are representative of the output of the process in the future.

The studies using control charts and the tile formulation experiment described in Chapter 2 were examples of analytic studies. Conclusions from these studies were conditional upon the assumption that the production processes would continue to be affected by the same system of common and special causes that they were affected by during the course of the study. This assumption is reasonable when predicting probabilities in unfair games of chance through observations (e.g., rolling a loaded die or playing computer games that employ random number generators). In games of chance the data-generating process is static, but in physical tests, conditions are never constant and caution must be exercised in interpreting results. Confirmation tests and repeat tests under different environmental conditions are usually advisable when conducting analytic studies, and any conclusions drawn should be qualified by stating the assumptions made (i.e., no future change expected in underlying system of common and special causes).

6.3 Statistical Inference

There are two methods of data analysis used in statistical inference. The first of these techniques is called *estimation*, and the second is called *statistical hypothesis testing*. Estimation, sometimes called prediction, is the act of estimating parameters of the probability function that represents the population we are trying to draw inference about. The estimates or predictions are made by collecting sample data and forming a summary statistic, like a sample average, to estimate the population parameter. It is obvious that an estimate will never be 100% correct. However we would like our estimate to be close, in some sense, to the population values it estimates. Oftentimes an interval estimate is formed which has a high probability of containing the parameter being estimated. The width of this interval, called a *confidence interval*, gives us some idea how accurate an estimate is.

A statistical hypothesis test is a formal way of determining whether a set of data supports or contradicts a statistical hypothesis. A statistical hypothesis is a conjecture or assertion concerning the parameters of a population. For example, when monitoring a process for control, we might hypothesize the mean of the distribution is the same as it has been historically. The test is conducted by collecting data from the process, and calculating a test statistic (e.g., the sample average). If the test statistic calculated from the sample data has a value that is very unlikely to occur if the hypothesis is true, then there is evidence to reject the hypothesis. If not, we would conclude that we don't have enough evidence to reject the hypothesis. Statistical hypothesis testing, like the scientific method in general, is a good way of disproving a hypothesis but not vice versa. Therefore, if the experimenter seeks evidence to support a given contention, he should state the hypothesis in the way that its rejection favors the contention.

In this section we will discuss the statistical methods of estimation and hypothesis testing in more detail. Following sections will show specific examples of these methods in enumerative and analytic studies.

6.3.1 Estimation

Estimation is concerned with finding a statistic that will be in some sense close to the population value that it is used to estimate. A statistic like the sample mean, $\bar{Y}$, or the sample standard deviation, s, is defined to be a quantity calculated from sample data. In estimation, we define a statistic we will use to estimate a population parameter before any data is collected. These statistics, or functions of the sample data, are then called point estimators or simply estimators. For example, if we are trying to estimate the mean, or expected value, of a population we could use the sample mean, $\bar{Y}$, or the sample median as our estimator. For a particular sample of data, any statistic we calculate will be unlikely to assume the exact value of the population mean, so statisticians have devised criterion to judge how close a particular statistic will be in the long run (i.e, over repeated use) to the parameter it is estimating. Comparing various statistics with respect to these criterion allows mathematical statisticians to select appropriate estimators.

One measure of the closeness of a statistic to the parameter it is estimating is called the *bias*. Bias is defined as the difference between the expected value (or long run average) of the statistic and the population parameter. For example, suppose we use the sample mean, $\bar{Y}$, to estimate the mean, μ, of a population. We learned in Chapter 3 that if we repeat the process of taking samples of data and calculating the sample mean, that the resulting sample means will follow approximately a Normal distribution with mean, μ, and variance σ^2/n. Therefore the expected value of the sample mean is μ, and its bias is zero. When an estimator has zero bias we say that it is an *unbiased estimator*. Unbiasedness is a desirable property for an estimator to have, because it means that on the average the estimator will be correct.

Another measure of closeness of an estimator is its mean square error. The mean square error is the expected squared difference between the estimator and the parameter it is estimating. For the sample mean this is $E(\bar{Y}-\mu)^2$. Since the sample mean is unbiased, $E(\bar{Y})=\mu$, its mean square error reduces to its variance $\sigma^2/n = E(\bar{Y}-\mu)^2$. In general the mean square error of unbiased estimators is their variance, and if two unbiased estimators exist for estimating the same parameter, we would prefer the one with the smaller variance. For example, if we know that the population can be represented by a Normal Distribution, then both the median and the sample mean are unbiased estimators of the population mean, μ, but the variance of the median is $(\pi/2)\sigma^2/(n-1)$ which is larger than σ^2/n, the variance of the sample mean, $\bar{Y}$. Therefore the sample mean is preferred as an estimator of the population mean, when the population can be assumed to be Normal. In general if two unbiased estimators exist, the one with the smaller variance will be closer on the average to the parameter being estimated, and it is called more *efficient*.

Even in cases where an efficient and unbiased estimator, like $\bar{Y}$, is available, we know that for a particular sample of data it will not be equal to the population mean. Therefore interval estimates or confidence intervals are sometimes preferred over the simple point estimates previously described. A confidence interval is a pair of statistics determined in a way so that the probability that interval contains the population parameter it is estimating is high (e.g., 90 or 95%). The statistics forming the interval estimates are often functions of the unbiased point estimators previously described. Specific examples of these will be shown in Sections 6.4 and 6.5.

6.3.2 Statistical Hypothesis Tests

A statistical hypothesis test is similar in some ways to a court of law. The hypothesis is assumed true, as the defendant in a court of law is assumed innocent, unless there is contrary evidence beyond a reasonable doubt. To introduce statistical hypothesis tests, we will describe the six-step procedure for completing a statistical hypothesis test that is given in most elementary statistics books. We will explain each step of this outline with brief examples.

(1) State the hypothesis, and the alternative

(2) Choose a significance level

(3) Select the statistic or quantity to be calculated from the data

(4) Calculate the critical region or rejection region

(5) Compute the statistic

(6) Draw a conclusion. If the statistic lies in the critical region reject the hypothesis, otherwise fail to reject.

The first step in the outline is to state the hypothesis, sometimes called the null hypothesis and the alternative. Since populations can be described by a probability distribution, statistical hypotheses are stated in terms of the parameters of the distribution. The null hypothesis is what we are trying to contradict with data beyond a reasonable doubt, and in its place confirm the alternative hypothesis. Since the null hypothesis is assumed true, unless there is adequate evidence to the contrary, the hypothesis is always written as an equality. For example, the hypothesis we implicitly test when we place an average $(\bar{X})$ on a control chart, is that the mean of the population (which can be represented by a Normal Distribution and represents the results of some defined method of measuring all items in the population) has not changed from the historical value it had the time the control chart limits were developed. This is usually stated symbolically as:

$$H_o : \mu = \mu_o$$

where H_o represents the null hypothesis, and μ_o represents the historical mean of the distribution. The alternative hypothesis is what would be assumed if the null hypothesis were proven false beyond a reasonable doubt. It is usually written like:

$$H_a : \mu \neq \mu_o$$

which is to say the mean is not equal to the hypothesized value. This is called a two-sided alternative since it would be true if either $\mu > \mu_o$ or $\mu < \mu_o$.

A one sided alternative is also used in some cases. For example, after adding additional feeders to the casting form for auto transmission housings the null hypothesis would have been $H_o : \lambda = \lambda_o$ and the alternative would have been $H_a : \lambda < \lambda_o$, where λ is the parameter of the Poisson Distribution and represents the average number of porosity defects. The Poisson Distribution would be used to represent the population in this case since the defect counts are discrete random variables. In this case the PQI teams would not be interested in increasing porosity, therefore they would only want to reject H_o if they could show beyond a reasonable doubt that their improved casting form actually decreased defects.

The second step of the hypothesis testing procedure is to choose the significance level. By the significance level we mean the probability of declaring the null hypothesis, H_o, to be false based on the sample data, when in fact the hypothesis is true. In practice there are two errors that can be made when testing a statistical hypothesis. These errors are summarized in Table 6.1 below:

Table 6.1 Possible Errors in Hypothesis Testing

		The True State	
		H_o True	H_o False
Result of Hypothesis Test	Accept H_o	Correct Decision	Type II Error
	Reject H_o	Type I Error	Correct Decision

The probability of an error of the first type, $P(I) = \alpha$, is called the *significance level* or sometimes the α level, and $(1-\alpha)$ is called the *confidence level* of the test. The probability of an error of the second type, $P(II) = \beta$ is called the *operating characteristic (OC)* of the test, and $1-\beta$ is called the *power* of the test. The significance level can be set at whatever level is desired. For example, if it is costly to adjust a process unnecessarily, then we would want to set the significance level very low to avoid unnecessary adjustments. On the other hand if adjustments were inexpensive (in terms of their potential negative effects along with the time or cost of making them) we could set the significance level higher. Typical values for α are 0.05 and 0.01. Choice of α automatically determines the critical region for the hypothesis test.

The probability of an error of the second type, β, will depend upon the value of the actual population parameter and the amount of data we have. If the value of the population parameter is close, but not equal, to the hypothesized value we are more likely to make a type II error and accept H_o when it is false. On the other hand, when the population parameter is far from the hypothesized value and lies in the alternative, we are much less likely to make a type II error. Likewise with very little data we are more likely to make a type II error, while with a lot of data the chances of this type of error are minimized.

The third step in the hypothesis testing procedure is to select the test statistic to be calculated from the data. For an $\bar{X}$ chart, this statistic would be simply $\bar{X}$, the average of subgroup measurements. We will discuss other test statistics later in the chapter.

The fourth step in the hypothesis testing procedure is to determine the critical region or rejection region. This is the set of values of the test statistic that would be very unlikely to occur (with probability α or less) if H_o were actually true. To determine this region, we need to know α, and the probability distribution of the test statistic when H_o is true. For example, the distribution of sample averages, $\bar{X}$ is approximately Normally distributed, due to the Central Limit Theorem. Therefore, the critical region for $\bar{X}$ control charts is determined by the Normal Distribution. If we

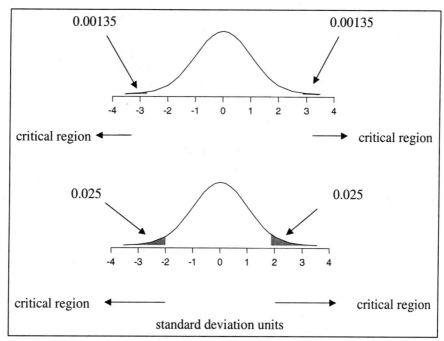

Figure 6.1 Critical Regions for α=0.0027 and α=0.05

set the significance level equal to 0.05, the critical region for a control chart would be the region shown in the bottom half of Figure 6.1, beyond 1.96 standard deviations above or below the mean. The traditional significance level for a control chart is 0.0027, which corresponds to ±3 standard deviations, and results in the usual critical region shown in the top half of Figure 6.1.

The fifth step in the statistical hypothesis testing procedure is to collect the data and compute the value of the test statistic, such as $\bar{X}$.

The sixth and final step of the hypothesis testing procedure is to compare the computed statistic to the rejection region. If the statistic falls in the critical region, a very rare event has occurred if H_o is true, and we should reject the hypothesis. If the test statistic does not fall in the critical or rejection region, we would not have enough evidence to reject the hypothesis.

6.4 Statistical Inference for Enumerative Studies

In this section we will give two examples of statistical hypothesis tests for enumerative studies. The first example concerns an acceptance sample, and the second concerns a reliability test.

6.4.1 Acceptance Sampling

A frequent use of hypothesis tests is in acceptance sampling. Here a manufacturer inspects a sample of incoming raw materials to determine whether they meet a minimum quality standard.

Example 1: A airplane manufacturer receives a shipment of seat brackets. The seat brackets are welded components, and the airline manufacturer requires that the number of brackets with defective weld joints in the lot be acceptably low, or they will return the lot to their supplier. Since a lot is large (>3000 brackets) the manufacturer will not want to inspect every bracket before making a decision. Therefore, only a sample of brackets will be inspected, and a hypothesis test is used to make a decision about all the parts in the lot.

The first step is to define the null and alternative hypothesis of the form H_o: $p = p_o$, and the alternative H_a: $p > p_o$, where p_o represents an acceptably low fraction defective (such as 0.02) which is sometimes called the AQL or acceptable quality level in acceptance sampling language.

The second step is to define the significance level or α. The typical significance level for a test of this type is 0.05, and in acceptance sampling language it is called the *producer's* or *supplier's risk*, since it represents the probability the manufacturer will reject the lot of seat brackets, when in fact the lot has an acceptable quality level of 2% defectives.

The third step is to define the statistic to be calculated from the sample data. In this case the natural statistic to use is Y, the total number of brackets with defective weld joints in the sample. If the sample consists of inspecting 100 brackets, this statistic is the number of failures that occur in 100 inspections.

The fourth step is to define the critical region. This will be a result of the choice of the significance level, α, and the distribution of the statistic, Y. When N, the number of items in the lot is large relative to the sample size (in this case the lot size is 3000 compared to the sample size of 100), the distribution of the number of failures, Y, in a sample of size n is approximated by the Binomial Distribution with parameters n=100 and $p=p_o=0.02$. In other words, the probability that the number of defects in a sample, Y , is equal to some specific value, y, is given by the Binomial

mass function $P(Y=y) = \binom{100}{y}(.02)^y(.98)^{100-y}$. Therefore, the critical region will be

$\{5, ..., 100\}$ because the probability of getting a value of Y greater than 4 in a sample of size 100 is $\alpha \cong 0.05$. The probability is approximate in this case because the Binomial Distribution is discrete, and this is the closest to $\alpha=0.05$ that we can come, i.e.,

$$\sum_{k=5}^{100}\binom{100}{k}(.02)^k(.98)^{(100-k)} \cong 0.05$$

(See Figure 6.3 for the probabilities of values of y from 0 to 10, and Figure 6.2 for a histogram of those values.) Acceptance sampling plans are usually referred to by, n, the number of items drawn in the sample and, c, the largest number of defects that can be tolerated in the sample. This is an example of a plan with n=100 and c=4.

The fifth step is to inspect a sample of brackets and count the number defective. The way the sample is taken in an enumerative study, such as this, is extremely critical. Judgment samples in this situation are unreliable and will frequently lead to faulty conclusions. To ensure the sample of brackets inspected is representative of the lot, we need to select a so-called *random sample* wherein each item in the lot has an equal chance of being in the sample. To do this, we must number

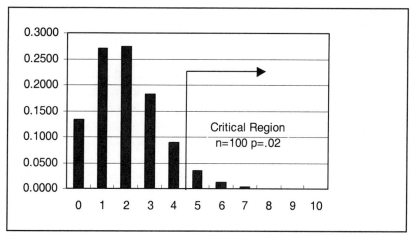

Figure 6.2 Critical Region for Sampling Plan with n=100, $\alpha \cong .05$

each seat bracket in the lot, put these numbers in a hat, mix them well, and draw a sample of 100. Equivalent ways of doing this are to use a table of random numbers or a compute- generated random subset. For example, the table of random digits, A.1 in Appendix A can be used to draw a random sample. Suppose that a sample of 10 from a lot of 1000 is required. First number the items in the lot from 0 to 999. Next start reading 3-digit numbers from A.1, ignoring any repeats. Starting at the top of the table and reading down the first column, the sample would consist of items 167, 178, 543, 444, 188, 308, 827, 642, 469, and 23. Normally start at a random place in the table by closing your eyes and pointing to a spot on the page.

The final step is to compare the number of defective brackets in the sample to the critical region. If five or more defective brackets are found, the lot should be rejected and returned to the supplier, otherwise the lot should be accepted and used in production.

If the airplane manufacturer would like to estimate the proportion of defective seat bracket welds in the lot, in addition to testing the hypothesis H_o: $p = 0.02$, a simple unbiased point estimator is Y/n, the proportion of defective bracket welds in the sample. For example, if one defective bracket were found in the sample, H_o would be accepted and the estimated proportion of defective bracket welds in the lot would be 1/100=0.01. If however, four defective brackets were found in the sample of 100, H_o would be accepted, but the proportion defective in the lot would be estimated to be 0.04. An interval estimate could also be formed, but in this case the limits are not simple functions of the point estimator and will not be presented here.

In acceptance sampling language, the probability of a type II error, β or the operating characteristic of the test, is often called the *customer's risk* because it is the probability of accepting a lot that has a proportion defective above the Acceptable Quality Level (AQL is 0.02 for the example). To compute β we must first define a value of $p > p_o$. Once p is specified, the probability of a type II error, or failing to reject the lot, is given by:

147

$$\beta = \sum_{k=0}^{4} \binom{100}{k} p^k (1-p)^{100-k}$$

which is the probability of having the number of defectives in the sample being in the acceptance region $\{0,1,2,3,4\}$. Table 6.2 lists several values of p in the alternative, and the corresponding values of β for the acceptance sampling test with n=100, c=4. Figures 6.3 and 6.4 show graphically and numerically how the values in Table 6.2 were determined. As can be seen in the Figures, β is equal to the CDF for y=c (4 in this example).

Table 6.2 $\beta = P(II)$ for Various Values of p in H_a

p	β
.04	.629
.06	.277
.08	.090

From this table we can see the chances of making a Type II error are high for values of p close to $p_o = .02$. The only way to reduce β is to increase the sample size. For example, with a sample size of n=200, the critical region becomes $\{8,9,...200\}$, or c=7 since,

$$\sum_{k=8}^{200} \binom{200}{k} p^k (1-p)^{200-k} \cong 0.05$$

With this sample size and critical region we compute β as

$$\beta = \sum_{k=0}^{7} \binom{200}{k} p^k (1-p)^{200-k}$$

Y	Probability	CDF
0	0.1326	0.1326
1	0.2707	0.4033
2	0.2734	0.6767
3	0.1823	0.8590
4	0.0902	0.9492
5	0.0353	0.9845
6	0.0114	0.9959
7	0.0031	0.9991
8	0.0007	0.9998
9	0.0002	1.0000
10	0.0000	1.0000

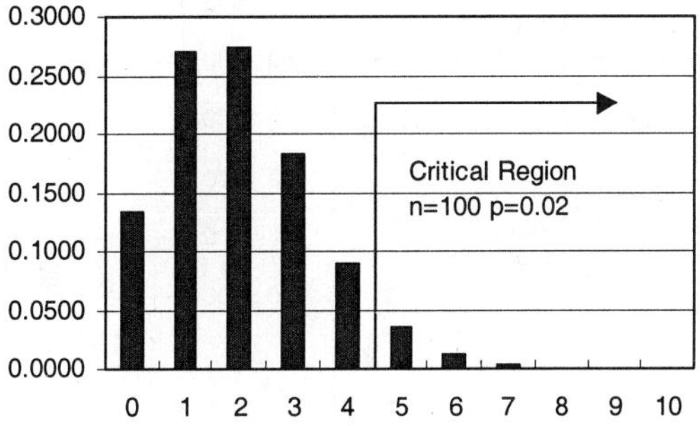

Y	Probability	CDF
0	0.0169	0.0169
1	0.0703	0.0872
2	0.1450	0.2321
3	0.1973	0.4295
4	0.1994	0.6289
5	0.1595	0.7884
6	0.1052	0.8936
7	0.0589	0.9525
8	0.0285	0.9810
9	0.0121	0.9932
10	0.0046	0.9978

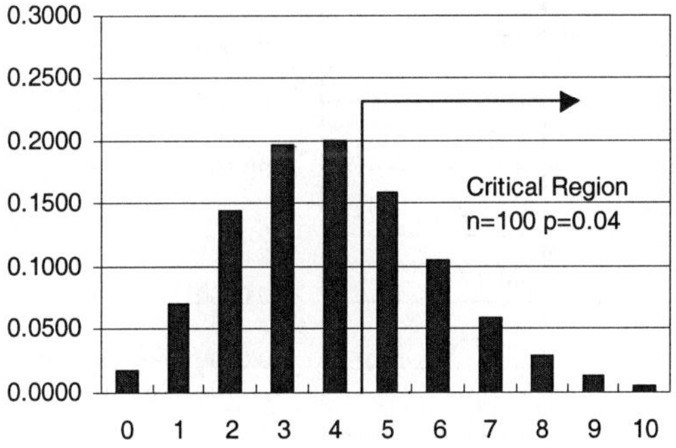

Figure 6.3 Calculation of the Probability of a Type II Error for p=0.02 and p=0.04

Y	Probability	CDF
0	0.0021	0.0021
1	0.0131	0.0152
2	0.0414	0.0566
3	0.0864	0.1430
4	0.1338	0.2768
5	0.1639	0.4407
6	0.1657	0.6064
7	0.1420	0.7483
8	0.1054	0.8537
9	0.0687	0.9225
10	0.0399	0.9624

Y	Probability	CDF
0	0.0002	0.0002
1	0.0021	0.0023
2	0.0090	0.0113
3	0.0254	0.0367
4	0.0536	0.0903
5	0.0895	0.0799
6	0.1233	0.3032
7	0.1440	0.4471
8	0.1455	0.5926
9	0.1293	0.7220
10	0.1024	0.8243

Figure 6.4 Calculation of the Probability of a Type II Error for p=0.06 and p=0.08

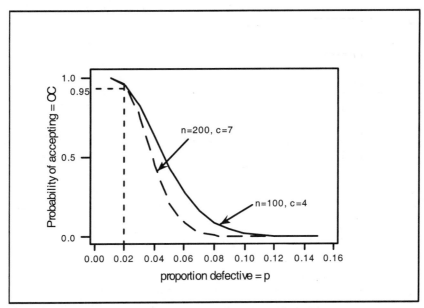

Figure 6.5 OC Curves for Two Acceptance Sampling Plans

Figure 6.5 shows β as a function of p for the two different sampling tests plans (n=100, c=4) and (n=200, c=7). These curves are commonly called the OC curves (which stands for Operating Characteristic curves) for the sampling test, or sampling plan. These curves show the β=customer's risk for all values of p in the alternative H_a. We can see that the risk for the plan with the larger sample size is lower than for the plan with the smaller sample size, illustrating what we said earlier about the probability of a type II error being both a function of the population parameter and the sample size.

When choosing the values for n and c in an acceptance sampling plan, it is often desirable to choose them such that not only will OC = 1-α be large at the AQL (i.e., p_o), but also so that the OC will be small, equal to β, at a specified value of p=p_a in the alternative that is often called the LTPD or lot tolerance percent defective. Computer programs have been developed (for example see Snyder and Storer[3] or King[4]) that will calculate values of n and c that will give as closely as possible, the specified values of α, β, p_o, and p_a.

6.4.2 Reliability Testing

Reliability tests, another form of hypothesis testing, are often used to qualify the reliability of a lot of components or material. For example, a supplier of torsion bars for military vehicles may want to guarantee that the mean time to failure of the parts in a lot is greater than a certain value; or a supplier of plastic tubing used in manufacturing catheters may want to ensure that the median burst strength of the material is greater than a minimum value. In both of these cases, testing would be destructive, and therefore a reliability test must be devised to qualify the lot with only a sample of items being tested.

Example 2: The supplier of plastic tubing wants to ensure the median burst strength of a lot of tubing, $\tilde{m}$, is greater than 60 psi. Therefore, $H_o: \tilde{m} = 60$ and $H_a: \tilde{m} > 60$. Again the null hypothesis is expressed as an equality, and the alternative hypothesis expresses what the supplier would like to show.

The reliability test consists of taking a random sample (as defined in the last section) of lengths of 10 tubes from the lot, then inflating them until they burst. The test results in psi were: 29.5, 482.9, 120.3, 216.1, 484.7, 95.9, 237.7, 218.0, 401.0, 31.5.

The probability distribution of the strength at burst within the lot is known to be approximately Lognormal, since the mode of failure is usually due to propagation of voids or cracks in the plastic, and the parameter, σ, was known to be 1.0 from a previous history of reliability tests. Recall from Chapter 3, that if the distribution of burst strengths, X, is Lognormal, then the distribution of Y=ln(X) is the Normal Distribution. The natural test statistic for this reliability test is the mean, $\bar{Y}$, of the log burst strengths in the sample of 10.

The significance level α is set at 0.05. Knowing that the distribution of the natural log of the burst strengths is Normal, with parameter $\mu = \ln(\tilde{m}) = \ln(60) = 4.0943$ under the null hypothesis, and σ known to be 1.0, the hypothesis and alternative can be stated in an equivalent way as $H_o: \mu = \mu_o = 4.0943$, $H_a: \mu > \mu_o$. If the null hypothesis is true, the distribution of the mean burst strengths, $\bar{Y}$, is Normal with mean 4.0943 and standard deviation $(\sigma / \sqrt{n}) = (1.0 / \sqrt{10}) = 0.31622$. The extreme of the critical region, c, is called the critical value and is such that:

$$P\{\bar{Y} > c \mid H_o \text{ is true}\} = 0.05$$

This expression is read as: "the probability that $\bar{Y}$ is greater than c, given that the null hypothesis is true equals 5%." To find the value, c, we standardize by subtracting the mean and dividing by the standard deviation as in Example 8 of Chapter 3.

$$P\left\{\frac{\bar{Y} - \ln 60}{1.0/\sqrt{10}} > \frac{c - \ln 60}{1.0/\sqrt{10}}\right\} = 0.05$$

or

$$P\left\{Z > \frac{c - 4.0943}{0.31622}\right\} = 0.05$$

and using Table A.3 in Appendix A we see that

152

$$(c-4.0943)/0.31622 \quad = \quad 1.645$$

$$\therefore \ c = 4.6144$$

The logarithms of the test data given above are: 3.38, 6.18, 4.79, 5.38, 6.18, 4.56, 5.47, 5.38, 5.99, 3.45. The mean of the log data is $\bar{y}= 5.077$ which is above the critical limit. Therefore, the manufacturer can feel 95% = 100(1-α) confident in rejecting H_o in favor of what he would like to ensure the median burst strength is greater than 60 psi.

      If the manufacturer would like an estimate of the log median burst strength, they can use the sample mean $\bar{y}= 5.077$ which is an unbiased estimator. In this case, where the point estimator is the sample mean, it is also easy to compute an interval estimate or confidence interval for the population parameter $\mu=\ln(\tilde{m})$. This interval estimate gives us some idea how accurate our estimate is. Since $\bar{Y}$ is Normally distributed with mean $\mu=\ln(\tilde{m})$, and standard deviation $\sigma/\sqrt{n}$, then $P(\mu-Z_{.025}\sigma/\sqrt{n}<\bar{Y} <\mu+Z_{.975}\sigma/\sqrt{n})= .95$, where $Z_{\alpha/2}$ is the normal score such that $P(Z<Z_{\alpha/2})=\alpha/2$, for $\alpha/2=.025$ $Z_{\alpha/2}=1.96$ (from Table A.3). The interval can be written equivalently as:

$$P(\bar{Y}-1.96\,\sigma/\sqrt{n}<\mu < \bar{Y}+1.96\,\sigma/\sqrt{n})= .95$$

and we can use the end points of this interval ($\bar{Y}-1.96\,\sigma/\sqrt{n}$, $\bar{Y}+1.96\,\sigma/\sqrt{n}$) as a 95% confidence interval for μ. This means that, before we collect the data, we are 95% confident that the mean log median burst strength will be between $\bar{Y}-1.96\,\sigma/\sqrt{n}$ and $\bar{Y}+1.96\,\sigma/\sqrt{n}$. For our example where $\bar{y}=5.077$ and σ was known to be 1.0, the interval is computed as:

$$(5.077- 1.96(1/\sqrt{10}), 5.077+1.96(1/\sqrt{10})) = (4.457 ,5.697)$$

      Once the data is collected and the confidence interval is computed, the interpretation changes slightly. The population parameter, μ, is a constant, and it is either in the interval (4.457, 5.697), or it is not. The probability statement is interpreted by saying if we were to repeat this procedure (i.e., take a sample and compute the confidence limits) over and over, 95% of the time our computed interval would contain the population value we are trying to estimate. The width of the confidence interval is a function of the sample size, n, and the confidence coefficient. If we increase our confidence from 95% to 99% the interval will get wider. If we increase our sample size, from 100 to 200, the interval will become narrower.

6.4.3 The t-Distribution

      In the previous example, it was assumed that the standard deviation of the natural logs of the burst strengths was known to be 1. This could be true in practice where reliability tests are performed for each lot, and the standard deviations were reasonably constant from lot to lot with a long term average of 1.0. However, if the standard deviation for the lot was unknown, a different calculation is required to determine the critical region. In the last example, we used the fact that

$$Z = \frac{\bar{Y} - \mu_o}{\sigma/\sqrt{n}}$$

follows the Standard Normal Distribution, in order to calculate a critical value. If σ is unknown, we can replace it by an estimate, s (i.e., the sample standard deviation), and the quantity

$$t_{n-1} = \frac{\bar{Y} - \mu_o}{s/\sqrt{n}} \qquad\qquad (6.1)$$

follows the student's t-distribution with n-1 degrees of freedom, rather than the Standard Normal Distribution. This distribution gets its unique name from the fact that W. Gosset, who first derived and published the t-distribution, wrote under the pen name "Student" because he worked for a brewery that didn't want their competitors to know they were using statistical methods. The statistic in Equation 6.1 is called the *t-statistic*. The Student's t-distribution is a symmetric family of distributions with mean zero, like the Standard Normal Distribution. The parameter of the t-distribution is the degrees of freedom (df). The tails of the t-distribution are slightly heavier than the tails of the Standard Normal Distribution, reflecting the uncertainty added when the estimate s replaces the known quantity σ. For example, Figure 6.6 compares the Standard Normal density function to two different t-distributions.

As can be seen, the tails of the t-distribution are heavier than the tails of the Standard Normal Distribution. But as the sample size n, or degrees of freedom n-1, increases, the t-distribution approaches the Standard Normal Distribution. Table A.5 in Appendix A tabulates critical values for the t-distribution family. The table is indexed on the left by the degrees of freedom, and across the top by the confidence level. The confidence level is 1 minus the significance level, and as shown by the graph at the top of the table it represents the area under the bell curve between -t* and t*, where t* is the critical value for a two-sided test. The act of comparing the test statistic in Equation 6.1 to critical value in Table A.5, is called a *t-test*.

If we assume that σ was unknown and return to the reliability test described above, we can illustrate the t-test. The degrees of freedom for this example are 10-1= 9, and the critical value (from Table A.5) with significance level 0.05 is 1.833. This value was determined by looking on the row with df = 9, and the column with confidence level =90%. The confidence level is 90% rather than 95% in this case because the alternative hypothesis is H_a: $\mu > \mu_o$, and the entire α=0.05 area should be in the upper tail of the figure shown on top of Table A.5.

The sample standard deviation of the log data is 1.026 and the computed statistic is:

$$t_9 = \frac{\bar{y} - \mu_o}{s/\sqrt{n}} = \frac{5.077 - 4.0943}{1.026/\sqrt{10}} = 3.03 > 1.833$$

154

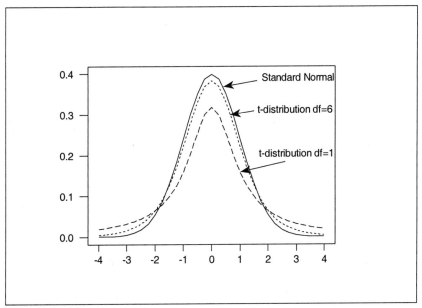

Figure 6.6 Comparison of Standard Normal and t-Distributions

Therefore, the hypothesis $H_o: \tilde{m} = 60$ would again be rejected in favor of $H_a: \tilde{m} > 60$, and the manufacturer of catheters would be at least 95% confident that the median burst strength was greater than 60psi. When σ is unknown, the formula for the $(1-\alpha)\%$ two-sided confidence interval changes from $\bar{Y} \pm Z_{\alpha/2} \sigma/\sqrt{n}$ to $\bar{Y} \pm t_{(n-1),\ \alpha/2} (s/\sqrt{n})$, so the confidence interval would be

$$5.077 \pm 2.262(1.026/\sqrt{10}) \quad \text{or} \quad (4.343, 5.811) \text{ and on the anti-log scale this is } (76.93, 334).$$

6.4.4 Discussion

The accuracy of the critical value determined from the significance level and operating characteristic (OC) for hypothesis tests, and the accuracy of confidence intervals in enumerative studies, are justified by the method of random sampling, and the assumed distribution of the measured value in the entire lot (i.e., finite population). For inference about the population mean, the distribution of the measured characteristic over the finite population does not even have to be known. When the sample size is large (i.e., n>30), the Central Limit Theorem will often justify approximate normality of statistics such as the sample mean, $\bar{Y}$, or the t-statistic given in Equation 6.1.

The example presented above is an illustration of the situation where we are trying to draw an inference about the mean of a single population (which can be represented by a probability density function). Table 6.3 summarizes the statistics used in the situations described in this example.

Table 6.3 Statistics for Inference about a Population Mean

Sample Size	Population Standard Deviation	Test Statistic	$(1-\alpha)\%$ Confidence Interval	Distribution of Test Statistic
Small n<30	Known	$Z = \dfrac{\bar{Y} - \mu_o}{\sigma/\sqrt{n}}$	$\bar{Y} \pm Z_{\alpha/2}(\sigma/\sqrt{n})$	Standard Normal
Small n<30	Unknown	$t = \dfrac{\bar{Y} - \mu_o}{s/\sqrt{n}}$	$\bar{Y} \pm t_{(n-1)\alpha/2}(s/\sqrt{n})$	Students t with (n-1) Degrees of Freedom
Large n≥30	Unknown	$Z = \dfrac{\bar{Y} - \mu_o}{s/\sqrt{n}}$	$\bar{Y} \pm Z_{\alpha/2}(s/\sqrt{n})$	Standard Normal (approximate)

6.5 Statistical Hypothesis Tests in Analytic Studies

In analytic studies we are trying to make a decision or predict the characteristic of the future output of a process, based on a sample of output taken now. To illustrate the difference between analytical and enumerative studies, let's consider an example where the computation of test statistics, confidence interval, critical values, etc. are the same as they were in the last example shown in Section 6.4.

6.5.1 Testing the Effect of a Process Change

In an electric furnace smelting process, production management wanted to know if switching to a less expensive supply of coke (reducing agent) would affect the future furnace efficiency in the plant. The current furnace efficiency was about 99%, and as long as switching to the less expensive supply of coke did not decrease furnace efficiency by more than 1%, it would be beneficial to switch. To answer the question, a plant test was run wherein several rail cars of coke from the alternate supply were used in the process, and the furnace efficiency determined. It was possible that the less expensive coke could actually increase furnace efficiency. Therefore, the hypothesis to be tested with the data resulting from the plant test was

$$H_0: \mu = 99$$

against the alternative

$$H_a: \mu < 99.$$

The data collected after 10 carloads of the alternate coke supply was:

96.219, 97.103, 96.921, 101.355, 98.551, 99.483, 97.186, 94.246, 97.536, 97.645

The significance level was chosen to be $\alpha=0.05$. The population standard deviation was unknown so the appropriate test statistic according to Table 6.3 is the t. The corresponding critical value for the t-statistic is -1.833. This critical value was found in Table A.5 looking at the row with 9 degrees of freedom and the column with confidence level 90%, since the alternative is one-sided, and we want 5% in the left tail which is the critical region. The sample mean and standard deviation computed from the test data are: $\bar{y}= 97.624$, and $s = 1.903$. The t-statistic calculated according to Equation 6.1 is:

$$t_9 = \frac{\bar{y}-\mu_o}{s/\sqrt{n}} = \frac{97.624-99.0}{1.904/\sqrt{10}} = -2.286 < -1.833$$

There is enough evidence to reject H_o and the production management should feel confident (95%) that the new coke supply will reduce furnace efficiency, and a switch would be detrimental. A 95% confidence interval for the mean furnace efficiency with the new coke supply is:

$$\bar{y} \pm t_{(n-1)\alpha/2}(s/\sqrt{n}) = 97.624 \pm 2.262(1.903/\sqrt{10}) = (96.26-98.98)$$

In this case we use the tabled t-value 2.262 from the row with 9 degrees of freedom and the column with 95% confidence level. We use this rather than 1.833 because this is a two sided confidence interval.

For this example, the significance level (or confidence level) does not have the same meaning that it did for an enumerative study. In the enumerative study the significance level was guaranteed if the sample selected was a random sample. By guaranteed, we mean that if the same sample selecting and statistic calculating procedure were done repeatedly with the same finite population or lot, a type I error would only be made 5% of the time, due to sampling variability. However, in the analytic study described above, the sample data we have collected is more like a judgment sample, and past performance is never a guarantee of the future performance. We hope that everything in the furnace now (i.e., ore, temperature, flux, etc.) is the same or representative of what the conditions will be in the future, and that the difference in furnace efficiency, $\bar{y}-\mu_o$, is accurate. But, it is possible that during the test period management was watching over the process, and things (in addition to changing the coke supply) were being done to increase furnace efficiency. In that case the conclusion could be totally faulty. Even when effort is made to assure conditions are representative of what they will be in the future, management can never be sure of this, and conclusions of the study should always be qualified by stating that they are only valid if all other conditions remain the same.

In addition to calculating a t-statistic from the sample data, and comparing it to the critical value, confidence in the conclusions of a study can be increased by more in-depth analysis of the data. For example, if past experience at the plant has shown daily furnace efficiency reports to be approximately Normally Distributed with a process that is in statistical control, then we should expect the same of the test data. If there are unusual trends or patterns in the data, it may be a signal that our sample data is not representative of the process. Figure 6.7 shows a line graph and Normal Probability plot of the furnace efficiency data. Since nothing unusual appears, management should feel more confident in their conclusion. In practical situations, plots like these will be more sensitive for detecting unusual changes if there is more data. If enough data are collected to actually construct a control chart, as will be shown in Chapter 17, and a state of statistical control can be demonstrated in the furnace with both the old and new coke supply, the difference in control chart center values will be a much more reliable prediction of the effect of switching coke supplies.

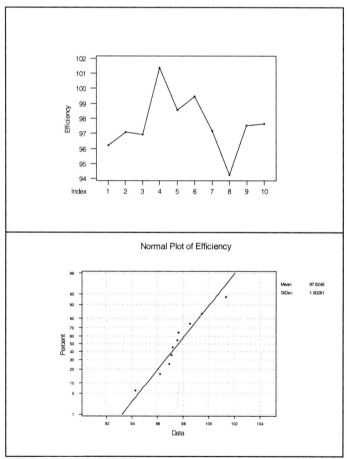

Figure 6.7 Diagnostic Plots of Furnace Data

The results of analytic studies are more reliable when they are repeated, and yield similar results, under different environmental conditions, and when a direct comparative test is run under the same environmental conditions, instead of comparing it to historical results which may have occurred under different conditions. In the next section we will present an example comparative test, and later we will talk about the use of replicates and blocking (i.e., repeating a test under different environmental conditions) as methods to increase the validity of analytical studies.

6.5.2 A Simple Comparative Test for Means (Two-Sample t-Statistic)

A better way to perform an analytic study is to do a comparative test. Through comparative tests, we can reduce some of the risk of faulty conclusions. Consider the following approach to the furnace test discussed in the last section. The plant obtained 10 rail cars of the alternate coke supply. Instead of running these through the furnace consecutively, and risking that something could be different during those runs (in addition to the coke) than it was in the past or will be in the future, it was decided to intersperse the cars among those from the normal coke supply. In this way a side-by-side comparison of the furnace efficiency was obtained using the two different coke supplies. Better than simply alternating the two coke supplies, randomization of the ordering can prevent confusion in the comparison that could be caused by cyclical patterns of change in factors (that may also affect furnace efficiency) like ore supply or furnace temperatures.

Randomization or random ordering of the use of the alternate coke supplies can be facilitated through use of random numbers like those given in Table A.1. The procedure for doing this would be to number the 20 consecutive tests that will be performed, from 0 to 19. Next turn to the random number Table A.1 place your finger on the page and read consecutive two-digit numbers down a column or across a row. The first 10 unique numbers between 0 and 19 will be the tests where alternate coke supply is used and the second 10 unique numbers will be tests that use the normal supply. For illustrative purposes consider the ordering obtained using the first row of two digit numbers from Table A.1. We read the numbers 16, 77, 66, 52, 93, 76, 69, 80, 61, 05, etc. We ignore numbers outside the range 0 to 19 and make the following table of orderings shown in Table 6.4. In general this randomization procedure will avoid any biases due to warm-up time, or learning curves, etc. in comparative tests.

Table 6.4 Orderings for a Comparative Coke Test

Test Number	Random Number	Coke Supply
1	16	Normal
2	05	Alternate
3	03	Alternate
4	17	Normal
5	10	Normal
6	08	Alternate
7	09	Alternate
8	11	Normal
9	18	Normal
10	06	Alternate
11	00	Alternate
12	14	Normal
13	19	Normal
14	13	Normal
15	15	Normal
16	02	Alternate
17	12	Normal
18	01	Alternate
19	04	Alternate
20	07	Alternate

Table 6.5 shows the data from the comparative test, along with summary means and standard deviations for each coke type.

Table 6.5 Comparative Test Results

Normal Coke Supply		Alternate Coke Supply	
Test Number	Efficiency	Test Number	Efficiency
1	98.649	2	99.5626
4	99.794	3	96.6727
5	98.885	6	98.4924
8	98.009	7	97.0059
9	98.270	10	96.7918
12	99.719	11	99.9202
13	96.337	16	96.5977
14	97.346	18	96.2397
15	101.203	19	96.5478
17	100.160	20	98.6172
	$\bar{y}_N = 98.84$		$\bar{y}_A = 97.64$
	$s_N = 1.44$		$s_A = 1.37$

Simple graphs of the data are useful for determining if anything unusual happened during the course of the tests. Figure 6.8 shows comparative boxplots of the efficiency data from each coke supply and Figure 6.9 shows a simple line graph. In this case there does not appear to be anything remarkable like trends or outlier data points. Management wanted to know if the difference in average efficiency between the two coke supplies was real or within the range of variability to be expected. To answer this question, the difference in average efficiencies must be judged by the variability that naturally occurs in furnace efficiency when the coke supply is constant. There are two measures of the variability with the coke held constant (i.e., s_N, and s_A shown in Table 6.5). Using the summary statistics in Table 6.5, a t-statistic can be calculated that will help answer management's question.

The numerator (i.e., signal) for the t-statistic, $\bar{y}_A - \bar{y}_N$, is the difference in the sample averages. But, for the denominator (noise) we need an estimate of the standard deviation of this difference. Recalling the formulas for the variance of an average and the variance of a difference shown in Examples 14 and 15 in Chapter 3, a reasonable estimate of the standard deviation of

161

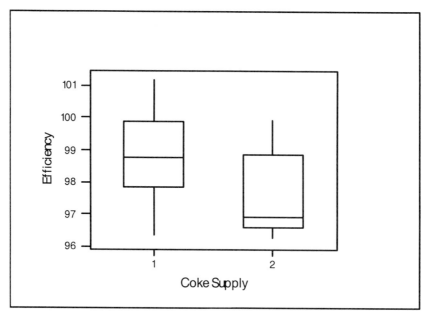

Figure 6.8 Comparison of Test Data

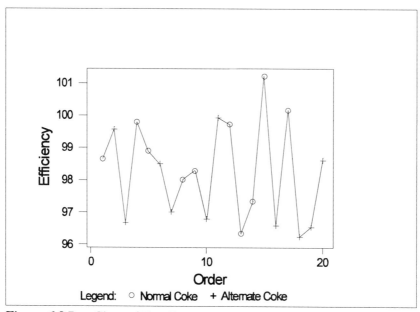

Figure 6.9 Run Chart of Test Data

$\bar{y}_A - \bar{y}_N$, is $\sqrt{s_A^2/n_A + s_N^2/n_N}$. Therefore, the signal-to-noise ratio is:

$$t = \frac{\bar{y}_A - \bar{y}_N}{\sqrt{s_A^2/n_A + s_N^2/n_N}}$$

(6.2)

If we assume that the variances are the same in both groups and that s_A^2 and s_N^2 are independent estimates of the same common variance, we can pool them together by taking a weighted average of the variances (weighted by their degrees of freedom) to get the pooled variance as shown below:

$$= \frac{(n_A - 1)s_A^2 + (n_N - 1)s_N^2}{n_A + n_N - 2}$$

(6.3)

where:

s_p^2 is the pooled sample variance
n_A is number of test data points with alternate coke
n_N is number of test data points will normal coke
s_A^2 is sample variance of the alternate coke data
s_N^2 is sample variance of the normal coke data

Assuming equal variances, the denominator for the t-ratio can be simplified to

$$s_d = \sqrt{s_p^2 \left(\frac{1}{n_A} + \frac{1}{n_N} \right)}$$

where s_d is the standard error of the difference in the two means $\bar{y}_A - \bar{y}_N$. The t-statistic or signal-to-noise ratio for comparing two sample averages, when the variances are assumed equal for the two groups, is then written as:

$$= \frac{\bar{y}_A - \bar{y}_N}{\sqrt{s_p^2 \left(\frac{1}{n_A} + \frac{1}{n_N} \right)}} = \frac{\bar{y}_A - \bar{y}_N}{s_d}$$

(6.4)

and should be compared to the critical t values (t*) in Table A.5 with $n_A + n_B - 2$ degrees of freedom. Equation 6.4 is called the *two-sample t-statistic* for comparing means from two samples of data. By pooling the two variances to form s^2_p, we assume that they both individually estimate a common variance, σ^2. If from a graph like Figure 6.8 this assumption appears unjustified, then Equation 6.2 should be used along with a modified formula for the degrees of freedom[5]. Since the variability in the two groups appears similar in Figure 6.8, we will use Equation 6.4 for the example data from Table 6.5. The pooled variance s^2_p is calculated using Equation 6.3 as:

$$s^2_p = \frac{9(1.44^2) + 9(1.37^2)}{10 + 10 - 2} = 1.96$$

and the two-sample t-statistic or signal-to-noise ratio is:

$$t = \frac{97.64 - 98.84}{\sqrt{1.96\left(\dfrac{1}{10} + \dfrac{1}{10}\right)}} = -1.90$$

This calculated t value is larger in absolute value than the tabled t* value with 90% confidence and 18 degrees of freedom (1.734), but smaller than the tabled values with 95 or 99% confidence (2.101, 2.878). Therefore, for this particular example management could be 95% confident that there is a real decrease in furnace efficiency using the alternate coke, not enough evidence to be 97.5 or 99.5% confident.

An unbiased point estimate of the difference in mean furnace efficiency when using the different coke supplies is the difference in the two sample means, $\bar{y}_A - \bar{y}_N$. A formula for a confidence interval on the difference of means is given by:

$$\bar{y}_A - \bar{y}_N \pm t_{n_A+n_N-2,\alpha/2}(s_d)$$

Using this formula, a 95% confidence interval for the difference in means is 97.64-98.84±2.101(.626) or (-2.51, 0.115).

Generally, in the physical world no two things are identical. This includes means of conceptual populations. Even if, in a particular comparative experiment the signal-t- noise t-ratio is not large enough to say with any confidence (i.e., at least 90%) that there is a difference in means, in reality there is almost certainly some small difference in the population means. Therefore, when a t-ratio is smaller in absolute value than the tabled t* values in Table A.5, the correct conclusion should be that there is not enough evidence to prove that there is a difference in the means, rather than there is no difference in the means.

6.5.3 Statistical and Practical Significance

There is a difference between what we call statistical significance and practical significance. We say that a signal, $\bar{y}_1 - \bar{y}_2$, or a difference in sample averages, is statistically significant whenever it is large with respect to the noise (i.e., s_d). On the other hand, a difference in actual population means ($\delta = \mu_1 - \mu_2$) is of practical importance whenever it is large enough to have operational or economic benefits when considered over a long period of time. In this example, a decrease of 1% in furnace efficiency would cost the plant more than the savings on coke. Therefore, a difference of $\delta = 1\%$ or more has practical significance.

In some cases two conceptual populations may have a practical difference in means, but we would not be able to see that difference in a study such as the one described above, because not enough data were taken and too much variability was seen in the data collected. In other cases, if we collect a lot of data in a study, we may be able to show a statistically significant difference by having a large signal-to-noise t-ratio, when there is no practical significance to the difference in means observed. In Chapter 9, we give some guidelines about how much data to collect or how many experiments to perform in order to make practical and statistical significance coincide.

6.5.4 A Simple Comparative Test of Variances

The objective of some comparative experiments may be to compare variances rather than means. For example, a problem in Chapter 2 described a situation in the Ina tile company of Japan where a reduction in the variability in tile sizes was sought by testing different clay formulas prior to firing in the kiln. In this section, we will present a test statistic that can be used to compare variances obtained in comparative experiments. For testing a hypothesis about variances, the null hypothesis is $H_0: \sigma^2_1 = \sigma^2_2$ and the alternative is $H_a: \sigma^2_1 \neq \sigma^2_2$.

As an example, consider the following situation. An engineer sought to reduce variability in the robot placement of brackets on a rocket engine casing, by attempting to reduce robot gear backlash through a software algorithm. There was no prior information about robot repeatability in the form of a standard deviation s, therefore data was to be collected under two situations: (1) using the gear backlash compensating algorithm and (2) not using the gear backlash compensating algorithm. The robot would be programmed so that the gear backlash compensating algorithm could be turned on and off at will, and it would be decided randomly (e.g., by result of a coin toss) when the algorithm would be used. This would prevent any biases to the comparison caused by factors other than robot repeatability, such as orientation of the engine casing, etc.

Results of the comparative experiment are shown in Table 6.6. And visualized in Figure 6.10. The boxplots seem to indicate a wider spread in the uncompensated values. Sample variances can be computed to quantify the difference in spread. These are shown for each group of data in Table 6.6.

Table 6.6 Comparative Experiment in Bracket Placement
Error in Placement

Without Backlash Compensation	With Backlash Compensation
-2.714	.948
-.632	.971
3.245	-.002
1.643	-1.516
-2.741	-2.237
.342	-1.342
1.213	-.178
-4.472	.587
1.783	.919
$s_1^2 = 6.584$	$s_2^2 = 1.473$

$$F_{8,8} = 6.584 / 1.473 = 4.47$$

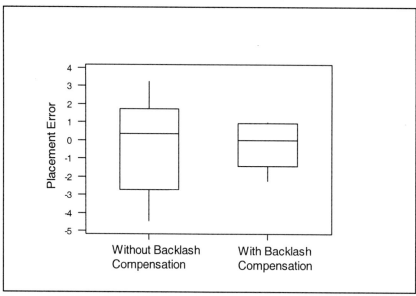

Figure 6.10 Box plots of Bracket Placement Data

Testing the hypothesis $H_o: \sigma^2_1 = \sigma^2_2$ is equivalent to testing the hypothesis $H_o: \sigma^2_1 / \sigma^2_2 = 1$. The test statistic and point estimator for the ratio of population variances is the ratio of the two sample variances (sometimes called the *variance ratio*). For the data in Table 6.6 the ratio of sample variances is $s_1^2/s_2^2 = 6.584/1.473 = 4.47$. If H_o is true, this ratio should be 1, but it will vary from that ideal due to random variability. If the two populations can be represented by Normal Distributions with the same variance, then the ratio of sample variances will follow an F-Distribution. The F-Distribution is a skewed distribution that has two parameters $v_1 = n_1 - 1$, and $v_2 = n_2 - 1$ called the numerator degrees of freedom and the denominator degrees of freedom. Figure 6.11 shows an example of an F-Distribution with $v_1 = 8$ and $v_2 = 8$. Table A.6 in Appendix A lists the upper 5% points of the F-Distribution. For example looking at 5% point with $v_1 = 8$ (indexed by the column of the table) and $v_2 = 8$ (indexed by the row of the table) we find the value 3.44. This means $P(F_{8,8} > 3.44) = 0.05$. We can use this table to determine a critical region for our test statistic.

For the robot repeatability example, $v_1 = 9 - 1 = 8$ and $v_2 = 9 - 1 = 8$ like the example above, so the critical value is 3.44. Since the statistic $s_1^2/s_2^2 = 4.47 > 3.44$ it can be concluded with greater than 95% confidence that the gear backlash algorithm had a beneficial effect in reducing the variability.

When comparing variance ratios to the tabled F-distribution, the larger variance should always be placed in the numerator since the table only contains the upper 95 percentage points.

6.5.5 Comparison of Population Means with Paired or Blocked Observations

In analytic studies, one way to increase confidence in conclusions is to repeat the test over different environmental conditions. For example, suppose a chemist is comparing two different assay methods for determining the iron content of ore samples. He could collect his data by splitting one ore sample into 10 parts and using the standard assay on 5 randomly selected parts and the new assay on the remaining 5 parts. The two-sample t-statistic described in Section 6.5.2 could be used to test the hypothesis that the two methods gave the same mean assay. However, the conclusions for his study would really only be applicable to the one ore sample, unless it could be argued or assumed

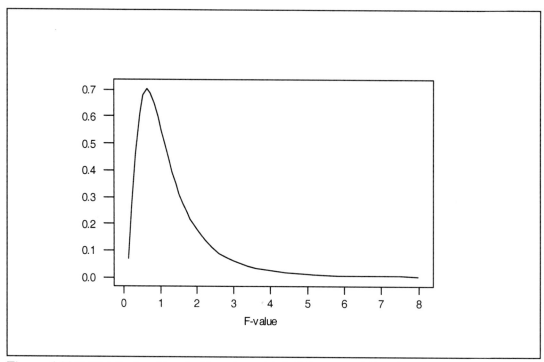

Figure 6.11 F-Distribution Density Function with 8, 8 Degrees of Freedom

that this sample was representative of all other samples.

A better way to collect the data would be to obtain several different ore samples. Table 6.7 shows the results of each assay on 10 different samples of ore, which are assumed to be representative of the types of ore that will be assayed in the future. The hypothesis to be tested is whether the two assay methods are equivalent.

Table 6.7 Comparative Assays

Sample	Method 1	Method 2	Difference
1	37.9247	37.9886	0.0639
2	48.0850	48.1299	0.0449
3	32.8171	32.8854	0.0682
4	36.0239	36.0739	0.0499
5	36.8719	36.9291	0.0571
6	42.3003	42.3562	0.0559
7	40.7925	40.8413	0.0487
8	48.6603	48.7205	0.0601
9	43.0184	43.0603	0.0419
10	35.7591	35.8050	0.0458
Means	40.23	40.28	0.0537
Standard Deviations	5.32	5.31	0.0087

The results from the two assays on each ore sample are very similar, but Method 2 tends to be consistently higher than Method 1. Also it can be seen that there are large differences from one ore sample to another. Figure 6.12 shows line graphs of the data for each method, and it can be seen that the differences between methods is overwhelmed by the differences between samples within each method.

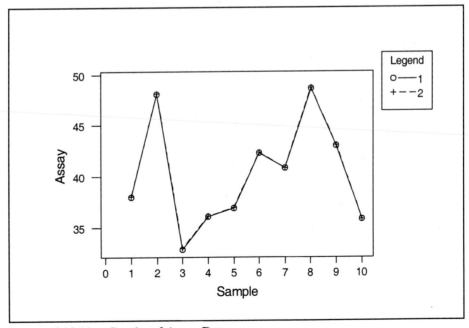

Figure 6.12 Line Graphs of Assay Data

169

If the two-sample t-statistic defined in Equation 6.1 were used to test the hypothesis H_o: $\mu_1=\mu_2$, the large variation in assays within each method (caused by the differences in ore samples) would overshadow the small but consistent differences in the two methods. The t-statistic with 18 degrees of freedom is t=-0.02, which is clearly not significant.

However, when the observations are blocked or paired (by sample) as they are in Table 6.7, we can create a more powerful test of the differences in means that won't be affected by the variation from pair to pair or block to block. This is done by analyzing the differences, rather than the individual assay data. The differences, denoted $D_i = Y_{1i} - Y_{2i}$, between the assay for Method 1 and Method 2 on the ith sample, reflects only the differences between methods and not the differences between samples. The expected value of D_i is $\mu_d = (\mu_1-\mu_2)$, and a test of the hypothesis $H_o:\mu_d =0$ is equivalent to the hypothesis $H_o: \mu_1=\mu_2$. The one-sample t-statistic defined in Equation 6.1 can be used to test this hypothesis by treating each of the differences in Table 6.7 as an individual observation. Used on the differences in this form it is usually called the *paired t-statistic*. The test statistic is calculated from the data in this table as:

$$ t_{n-1} = \frac{\bar{d} - 0}{s_d/\sqrt{n}} = \frac{0.0536 - 0}{0.0087/\sqrt{10}} = 19.17 $$

where n=10 is the number of pairs, $\bar{d}$ is the average difference shown at the bottom of Table 6.7 and s_d is the standard deviation of the differences. This value exceeds the critical limit of 2.262 in Table A.5 with 95% confidence and 9 degrees of freedom. Therefore, by removing the sample-to-sample variation from the denominator of the test statistic, we see that the small but systematic difference in the two assays is significant. Also, the conclusion of this analytic study is more defensible than it would be if only one sample of ore had been used. With one sample of ore it must be assumed that the results of the test are applicable to all ore samples to be analyzed in the future. With the present study design, the same assumption must be made, but it is made with more solid footing, since several ore samples were used in the study. In this sense we say there is a broader basis for our conclusion.

6.5.6 Common Statistics for Inference with Two Populations

Three new test statistics have been described in this section that are used for comparing the means or variances of two populations. Table 6.8 summarizes them. Each statistic can be used for testing a hypothesis in either an enumerative or analytical study, although the justification for the Type I and Type II error rates and their interpretation are different.

Table 6.8 Common Test Statistics for Two Populations

Hypothesis	Test Statistic	Assumptions	Distribution of Test Statistic	Table for Critical Values
$H_o: \mu = \mu_o$	$\dfrac{\bar{d} - 0}{s_d / \sqrt{n}}$	Paired or Blocked Observations	t-distribution with n-1 degrees of freedom	A.5
$H_o: \mu_1 = \mu_2$	$\dfrac{\bar{y}_1 - \bar{y}_2}{s_P \sqrt{\dfrac{1}{n_1} + \dfrac{1}{n_2}}}$	Two Independent sets of Observations and $\sigma_1 = \sigma_2$	t-distribution with $n_1 + n_2 - 2$ degrees of freedom	A.5
$H_o: \sigma_1 = \sigma_2$	$\dfrac{s_1^2}{s_2^2}$	Two Independent sets of Observations and Normal Distribution of Populations	F-distribution with $n_1 - 1$ and $n_2 - 1$ degrees of freedom	A.6

6.6 Summary

In this chapter we have introduced the method of statistical inference through hypothesis testing. Two types of populations have been described along with the concept of enumerative and analytic studies. Random sampling was introduced and given as the justification for the Type I and Type II error probabilities in enumerative studies. Additional judgment is required when drawing inference from analytic studies, since it must be assumed that conditions during collection of the sample or test data are representative of the future population about which inference is to be made. The concepts of repetition, comparative tests, blocking or pairing of observations, randomizing the order of tests, and simple graphs of the test data have been introduced to help broaden the basis for conclusions and justify the assumptions made when drawing inference from analytic studies. For the remainder of the book we will deal exclusively with analytic studies, and although we won't explicitly state our assumption of representative data in all examples presented, it must always be kept in mind.

Important terms presented in this chapter were:

References

1.Deming, W. E., *Out of the Crisis*, (Cambridge, MA,: MIT Press, 1986), p. 132.

2.Deming, W. E. , "On Probability As a Basis for Action," *The American Statistician*, Vol. 29, No. 4, (Nov. 1975).

3.Snyder, D. C. and Storer, R. F., "Single Sampling Plans Given an AQL, LTPD, Producer and Consumer Risks," *Journal of Quality Technology*, Vol 4., No.3, (1972).

4.King, D. W., *Statistical Quality Control Using the SAS System*, SAS Institute, Cary, N.C. (1995), p. 264.

5.Snedecor, G. W. and Cochran W. G., *Statistical Methods 7th Edition*, (Ames IA: The Iowa State University Press, 1980), p. 96.

6.7 Exercises for Chapter 6

6.1 A manufacturer of electric washing machines receives large shipments of ceramic resistors from a supplier. A lot would be acceptable if 1% or less of the resistors in that lot are defective. The manufacturer established the following decision rule. First a random sample of 100 resistors is taken from each lot and each resistor is tested and classified as defective or nondefective. If the number of defective resistors in a sample is three or greater, the lot is rejected.

a) State the null and alternative hypothesis for this acceptance sample test

b) Calculate the probability of a Type I error

c) Is this an enumerative or analytic study ?

6.2 In a continuous chemical process, control charts were kept on the purity of a solvent recycle stream. Historical data showed the process was in a state of statistical control and the solvent purity was approximately Normally distributed with mean $\mu=97\%$ and standard deviation $\sigma=.20$.

The plant had to switch suppliers of filters used in the recycle stream due to delivery problems with the old supplier. Production management was concerned whether use of the new filters would reduce the purity of the recycled solvent. After the change in filters, 10 hourly assays of the solvent purity (note this was the normal frequency of sampling used in the past) resulted in the following data.

Assay	Purity
1	96.908
2	96.724
3	96.978
4	96.503
5	96.995
6	96.661
7	96.579
8	97.041
9	96.870
10	97.134

a) State the null and alternative hypothesis to test the production management's concern.

b) Using a significance level of $\alpha=.05$, define the test statistic and rejection or critical region.

c) Calculate the test statistic and draw a conclusion.

d) Are there any other factors that could have biased the results of this test, and should the conclusions be qualified?

6.3 An investigation was conducted on a lathe to determine if the spindle speed affected the surface finish along the path of the cutting tool. Sixteen material blanks were cut; half at a low spindle speed of 500 RPM and half at a high spindle speed of 1060 RPM. The order of the cutting speed was randomized. The surface finish for each piece cut was measured on a profilometer, and the square roots of the surface finish measurements are shown below:

500 RPM	1060 RPM
21.84	23.99
23.16	24.04
20.46	23.97
21.51	24.58
22.30	24.07
21.42	23.69
20.92	23.74
21.36	23.56

a) Make comparative boxplots of the two sets of data

b) Calculate the means and standard deviations for each set of data

c) Calculate the pooled variance s_p^2

d) Calculate the t-test statistic for comparing the two sample averages.

e) What confidence do you have that RPM affected surface finish?

6.4 M. Murphy[1] recorded the mileages he obtained while commuting to school in his 9-year-old economy car. He kept track of the mileages for 10 different tankfuls of gas involving two different octanes.

87 Octane	90 Octane
26.43	30.57
27.61	30.91
28.71	31.21
28.94	31.77
29.30	32.86

a) Find s_p for this data

b) Conduct a significance test of $H_0 : \mu_{87} = \mu_{90}$ versus the alternative $H_a : \mu_{87} \neq \mu_{90}$. Follow the six-step outline and use significance level $\alpha=.05$

6.5 Suppose the manufacturer of floor tiles described at the end of Chapter 2 had measured the length of the diagonal of tiles coming out of the kiln, rather than just classifying each tile as a defective or not. The data below are measurements of the diagonals of samples of tiles coming out of the kiln. The first column is for tiles made with the normal clay formula, and the second is for tiles made with 5% lime added to the formula.

Normal	+5% Lime
25.2666	27.9590
25.4827	27.8757
29.8007	28.1675
27.7890	28.4736
27.8867	27.6106
27.9696	27.3288
28.8402	27.9597
27.7995	27.7812
30.0379	28.1976
28.1331	28.0433
25.9897	28.3916
27.5365	28.2880

Use the six-step outline to test the hypothesis that the variability in tile dimensions is the same

---

[1]Taken from Exercise 6-29 *Statistics for Engineering Problem Solving*, by Vardeman, S., PSW Kent Publishers, Boston, 1994.

for both clay formulas versus the alternative that adding 5% lime decreases variability. Use $\alpha=.05$ for the significance level.

6.6 Your company assembles floppy disks and it purchased the components. You buy the plastic cases in lots of 10,000. You have several specifications for the cases. One of the specifications is that 99.0% of the cases must survive a drop impact test (i.e. 99% of the cases must not crack when a cylindrical weight with a hemispherical end is dropped on the side of each case from a height of 18 inches). You test 200 cases in this test. (Note: The test is destructive, even the cases that pass the test are dented and must be discarded, so you do not want to test any more cases than you have to.)

(a) How many cases (out of the 200 cases tested) must fail the test for you to reject the lot of 10,000 cases?

(b) What is the power of the test for guarding against accepting a shipment (lot) of plastic cases that has a survival rate of 97.5% rather than the 99.0% that you require?

6.7 You are the same manufacturer of floppy disks described in Problem 1. Another specification you have for the cases you purchase is that 75% of the cases must survive the drop impact test when the weight is dropped from 30 inches.

(a) What is the critical region for rejecting a shipment of cases if 200 cases are randomly selected and tested? HINT: Use the Normal Distribution approximation to the Binomial to determine your answer.

(b) What is the power of the test for guarding against accepting a shipment of plastic cases that has a survival rate of only 70.0% rather than the 75.0% that you require?

6.8 You purchase a liquid plasticizer (in tank truck quantities) that is an additive to the clear plastic parts that you manufacture. The color of the material, on some scale, must be less than 650. You take nine samples from the tank truck, measure the color of each sample, and average the nine reading to determine if the shipment is acceptable. The analytical method has a standard deviation of $\sigma = 24$.

(a) What is the critical region for rejecting the shipment?

(b) What is the power of the test if the actual color of the material is 680?

(c) If the standard deviation of the test was unknown, and it was estimated from the sample results (n=9) to be, s=25, what is the critical region for rejecting the shipment?

6.9   A proposed change in the formulation of an adhesive is being considered because it would be less expensive. However, it is important that the strength of the adhesive does not decrease (by a significant amount). To check on the consequences of the formulation change, the strengths of several samples of each formulation were measured. The results are summarized below (where A denotes the standard, and B denotes the proposed formulation):

	A (Standard)	B (Proposed)
Average strength	563	542
Standard Deviation	15.3	16.8
Number of Samples	12	18

(a) Is the proposed formulation significantly weaker than the standard formulation?
(b) What is the power of the test to detect a drop of 30 if it occurs?

6.10   Two bailing machines measure, bale, and package six ft^3 bags of peat moss. The two machines were tested prior to purchase to determine which one was better, i.e., which one has the smaller standard deviation. Sixteen bales were selected at random from the trial run of Machine A, and the volumes baled were found to have a standard deviation, $s_A = 0.26$; likewise, twelve bales were selected at random from the trial run of Machine B, and the volumes found to have a standard deviation, $s_B = 0.35$. Is Machine B significantly worse than Machine A?

6.11   A flow rate sensor, namely a venturi nozzle, is known to give a standard deviation in the measured flow rate of 0.55%. A change in the machining of the nozzle was made, and it is feared that the standard deviation has deteriorated (gotten bigger). To investigate this possibility, a sample of 25 new nozzles was tested and they were found to have a standard deviation of 0.75%. Are the new nozzles less consistent?

Part III
Good Experiments Make for Good Statistics

Every technical investigation involving experimentation embodies a strategy (either consciously or unconsciously) for deciding what experiments to perform, when to quit and how to interpret the data. There are as many strategies as there are investigators. Some are good, and others are not so good. This part presents several statistically derived strategies that have greater efficiency than other more intuitive approaches. That is, the use of statistical experimental strategies will *generally* get the investigator to his or her goal in the shortest time (i.e., running the fewest experiments), give the greatest degree of reliability to the conclusions, and keep the risk of overlooking something of practical importance to a minimum.

This part of the book is a technical manual which presents methods an engineer can use to plan experimental program in two basic situations: (1)screening which variables are important from a multitude of possible candidates and (2) roughly optimizing with respect to a relatively few important variables to find the combination (of those tested) that gives the best response.

CHAPTER 7
Strategies for Experimentation with Multiple Factors

7.1 Introduction

In Chapter 1, the use of statistics in the scientific method was described. In subsequent chapters we demonstrated the use of statistical data analysis techniques for helping us to see if a given set of data supports or contradicts a theory or hypothesis. It was shown, in fact, that a statistical analysis of any set of data is crucial to objectively determining which conclusions are warranted from that data and which are not. But in the beginning chapters of this book we also explained that statistical methods were just as useful in deciding what data to collect and how that data should be collected. This other area of application is called the statistical design of experiments, or just *design* for short. In practice, using a good strategy for data collection (i.e., a good design) is even more important than a statistical analysis of the results. That is because it is difficult, if not impossible, to remedy the problems inherent in a bad strategy, even with the most sophisticated statistical analysis. As the old adage goes: "You can't make a silk purse out of a sow's ear." On the other hand, it is difficult (but not impossible) to mangle the interpretation of a well-designed set of experiments.

These two tools — statistical design and statistical analysis — are used very fruitfully in finding empirical solutions to industrial problems. The areas of application include research, product design, process design, production troubleshooting, and production optimization. The response or objective in an industrial study is usually a function of many interrelated factors, so solving these problems is typically not straightforward. There are two broad categories of approaches to these problems: 1) finding solutions by invoking known theory or facts, including experience with similar problems or situations, and 2) finding solutions through trial and error or experimentation. Statistical design and analysis methods are very useful for the second approach, and they are much more effective than any traditional, nonstatistical method.

7.2 Classical Versus Statistical Approaches to Experimentation

A typical strategy for an industrial study used by someone unfamiliar with statistical plans is the *one-at-a-time* design. In this approach, a solution is sought by methodically varying one factor at a time (usually performing numerous experiments at many different values of that factor), while all other factors are held constant at some reasonable values. This process is then repeated for each of the other factors, in turn. One-at-a-time is a very time consuming strategy, but it has at least two good points. First, it is simple, which is not a virtue "to be sneezed at." And second, it lends itself easily to a graphical display of results. Since people think best in terms of pictures, this is also a very important benefit.

Unfortunately, one-at-a-time usually results in a less than optimal solution to a problem, despite the extra work and consequent expense. The reason is that one-at-a-time experimentation is only a good strategy under what we consider to be unusual circumstances: (1) It is a good strategy if the response is a complicated function of the factor, X, (perhaps multi-modal) which requires many levels of X to elucidate, and (2) It is a good strategy only if the effect of the factor being studied (and therefore it's optimum value) is not changed by the level of any of the other factors. That is, the one-at-a-time strategy works only if the effects are strictly additive, and there are no interactions. As was just stated, but worth repeating, these circumstances do not typically exist in the real world, and so one-at-a-time ends up being an exceedingly poor approach to problem solving.

A more typical set of circumstances characterize the assumptions made by statistical designs: (1) Over the experimental region, the response is smooth with, at most, some curvature but no sharp kinks or inflection points, and (2) The effect of one factor can depend on the level of one or more of the other factors. In other words, there may be some interactions between the factors. If these two assumptions hold, the classical one-at-a-time approach could do very badly. If the first assumption holds, one-at-a-time would require many more experiments to do the same job, because a smaller number of levels of a factor are needed to fit a smooth response than to fit a complicated one. And if the second assumption is true, one-at-a-time could lead to completely wrong conclusions. For example, if we look at the effect of time and temperature on yield in one-at-a-time fashion, we would get the results shown in Figure 7.1 (top and middle). We would think that the best we can do is about 86% yield. But that is, in fact, very far from the true optimum shown in Figure 7.1 (bottom). The reason for the failure of one-at-

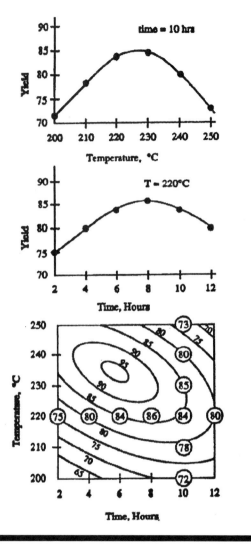

Figure 7.1 -- One-at-a-Time Experimentation Used to Optimize Process Yield

a-time experimentation is the interaction between time and temperature that exists. In other words, the best time depends on the temperature (or vice versa). It can't be repeated too often, that these interactions are the rule, not the exception. A statistical strategy (i.e., a factorial design) would not have led us astray. Incidentally, the potential for failing with a one-at-a-time strategy gets even stronger in the real world where experimental errors cloud the issue further.

But one-at-a-time is still far better than another common strategy used in industrial research, namely sheer guesswork. Using this strategy, researchers assume that they know the solution to a problem, as if they had the appropriate theory or experience to guide them. Then they run a few confirmatory experiments for validation. If the confirmatory experiments produce less than the anticipated results, more guessing and confirmatory experiments are run. The result is a very haphazard approach. These poor strategies for studying many factors simultaneously in industrial research, often deplete all research funds before a satisfactory solution can be reached. This, in turn, leads to less than optimal product designs, inefficient production processes, and the complete abandonment of many promising ideas.

Studying the effects of many factors simultaneously in an efficient way, while at the same time allowing valid conclusions to be drawn, is the purpose of the industrial experimental design techniques described in this text. We will start in the next chapter (Chapter 8) with a focus on the design that will be our cornerstone — the two-level factorial design. It is one of the most basic, and yet most powerful experimental strategies that exists.

7.3 Some Definitions

But before beginning the discussion on the two-level factorial design, it is important to ensure that we are all speaking the same language. So let us define several of the terms that we will use frequently.

- *Experiment* (also called a *Run*) — an action in which the experimenter changes at least one of the factors being studied and then observes the effect of his/her action(s). Note that the passive collection of historical data is not experimentation.
- *Experimental Unit* — the item under study upon which something is changed. In a chemical experiment it could be a batch of material that is made under certain conditions. The conditions would be changed from one unit (batch) to the next. In a mechanical experiment it could be a prototype model or a fabricated part.
- *Factor* (also called an *Independent Variable* and denoted by the symbol, X) — one of the variables under study which is being deliberately controlled at or near some target value during any given experiment. It's target value is being changed in some systematic way from run to run in order to determine what effect it has on the response(s).
- *Background Variable* (also called a *Lurking Variable*) — a variable of which the experimenter is unaware or cannot control, and which could have an effect on the outcome of an experiment. The effects of these lurking variables should be given a chance to "balance out" in the experimental pattern. Later in this book we will discuss techniques (specifically randomization and blocking) to help ensure that goal.
- *Response* (also called a *Dependent Variable* and denoted by the symbol, Y) — a characteristic of the experimental unit which is measured during and/or after each run. The value of the response depends on the settings of the independent variables (X's).
- Experimental *Design* (also called Experimental *Pattern*) — the collection of experiments to be run. We have also been calling this the experimental *strategy*.

● Experimental *Error* — the difference between any given observed response, Y, and the long run average (or "true" value of Y) at those particular experimental conditions. This error is a fact of life. There is variability (or imprecision) in all experiments. The fact that it is called "error" should not be construed as meaning it is due to a blunder or mistake. Experimental errors may be broadly classified into two types: bias errors and random errors. A bias error tends to remain constant or follow a consistent pattern over the course of the experimental design. Random errors, on the other hand, change value from one experiment to the next with an average value of zero. The principal tools for dealing with bias errors are blocking and randomization (of the order of the experimental runs). The tool to deal with random error is replication. All of these will be discussed in detail later.

Some of the terms are illustrated in Figure 7.2. It is a study of a chemical reaction (A + B → P) consisting of nine runs to determine the effects of two factors (time & temperature) on one response (yield). One run (or experiment) consists of setting the temperature, allowing the reaction to go on for the specified time, and then measuring the yield. The design is the collection of all runs that will be (or were) made.

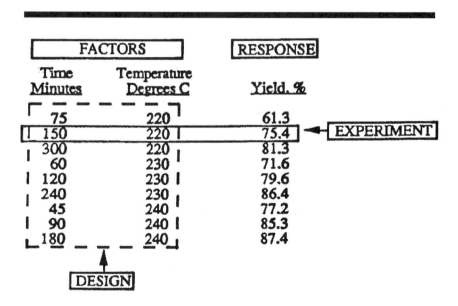

Figure 7.2 — Example of an Experimental Study with Terminology Illustrated

184

7.4 Why Experiment? (Analysis of Historical Data)

Frequently in industrial settings, there is a wealth of data already available about the process of interest in the form of process logs and, more recently, computer databases. It is tempting, therefore, to abandon an (expensive) experimental study in favor of a (cheap) analysis of the historical data on hand. On the surface, this looks very reasonable. After all, there has certainly been variation in all of the factors of interest over the life of the process. And there is certainly much more data available from the logs than could ever be collected in an experimental program, so the precision should be excellent. Why then ever bother with experiments at the plant level?

The reasons are many. All of them have to do with the inadequacies inherent in unplanned data:

- **Correlation ≠ Causation[1]**

  If a significant correlation is found between the response and an independent variable (factor), it may be interesting, but it *does not prove causation*. For example, let us say that someone who had hip surgery years ago noticed that when their hip ached it was a sure sign that it would rain within 12 hours. It may be a perfect correlation, but it does not mean that they should bruise their hip to make it ache if their lawn needs rain. Notice that there may be value in the correlation as a predictor of what will happen. But to use it to try to control the system is very likely folly.

- **Correlated Factors**

  Over time, there may have been quite a bit of variation in the factors of interest, but often they move up and down together. It is then impossible to sort out which of the factors, if any, is having an impact on the response. For example, in the gasification of coal, it may be useful to know which impurities in the coal are deleterious to yield. The amounts of the impurities may change quite a bit, but if they stay in roughly the same proportions, it is impossible to sort out which impurity is the culprit, if any.

- **No Randomization → Bias**

  Since there is no randomization in the variations of the factors with time, the door is wide open for biases to influence the results. For example, if a plant switched from one supplier of raw material to another and noticed a deterioration in yield (i.e., the average yield for the month before the switch was better than the average yield for the month after the switch), one explanation is that the new raw material is worse than the old one. However, another explanation is that normal drifting of the process due to other things changing gave a worse yield after the switch. So the raw material source had nothing to do with the drop in yield. Note: A valid test would be to randomly switch back and forth several times between suppliers.

- **Tight Control of Factors**

  If an independent variable is known (or thought) to be important, then it will be controlled tightly if possible. Therefore, there will not be enough variation in that factor to get a measurable effect on the response. The few data points that may exist in which these factors do vary more widely will have occurred during plant "upsets." That is not the kind of data that one wants to use to draw sound conclusions.

● **Incomplete Data**

Plants are quite different from laboratories and pilot plants, in that as few variables are measured and recorded as possible, consistent with being able to control the process. Therefore, even in the case when all important variables are known, which is unusual, they are not usually recorded. And to make matters worse, in older plants where much of the data was/is recorded by hand, even the variables to be measured and recorded may be missing or suspect.

Therefore, the analysis of historical data should be done with great care. This does not mean that it is a totally worthless exercise. Some interesting correlations may emerge. These can be used for predictive purposes, or they may point to some factors that were previously thought to be unimportant that should be studied in an experimental program. But, this analysis should be undertaken with the knowledge that the chances of gleaning any worthwhile information from the data are quite low (<10%). And it must be kept in mind that an analysis of historical data is NEVER a replacement for an experimental program if one needs to determine causation between the factors and the response(s). "To find out what happens to a system when you interfere with it, you have to interfere with it (not just passively observe it)."[2] For a more complete discussion of this topic see Reference 2, pages 487–498.

7.5 Diagnosing the Experimental Environment

There is no single experimental design that is best in all possible cases. The best design depends very much on the environment in which the experimental program will be carried out[3]. Figure 7.3 on the following page shows different environments that will be discussed in this book. In Chapter 8, the two-level factorial experimental designs for the objective of constrained optimization will be presented. These designs provide the building block which will be used to construct designs for the other situations. Chapter 12 will present experimental designs for the objective of screening, and Chapter 13 will discuss experimental designs applicable for unconstrained optimization (response surface studies). Designs for mechanistic modeling are not discussed in this text, the reason being that they are too specific to the model being used.

The characteristics that define the environment are:

● **Number of Factors**

The single most important characteristic of the experimental environment is how many independent variables are to be studied. If the number is small (say three or less), then a design giving fairly complete information on all of them may be reasonable "right off the bat." However, if there are many variables, it is usually more reasonable to proceed in stages — first sifting out the variables of major importance, and then following up with more effort on them.

● **Prior Knowledge**

The amount of prior knowledge also shapes the experimental program to a very large degree. When the area to be studied is new, there are generally a large number of potential variables that may have an effect on the responses. However, when the area has been studied extensively in the past, the scope of the experiments is generally to further elucidate in detail the effects of a few of the key variables. If theory is available, a mechanistic model may be desirable and experiments can be set up for determining the unknown parameters in the model. The best

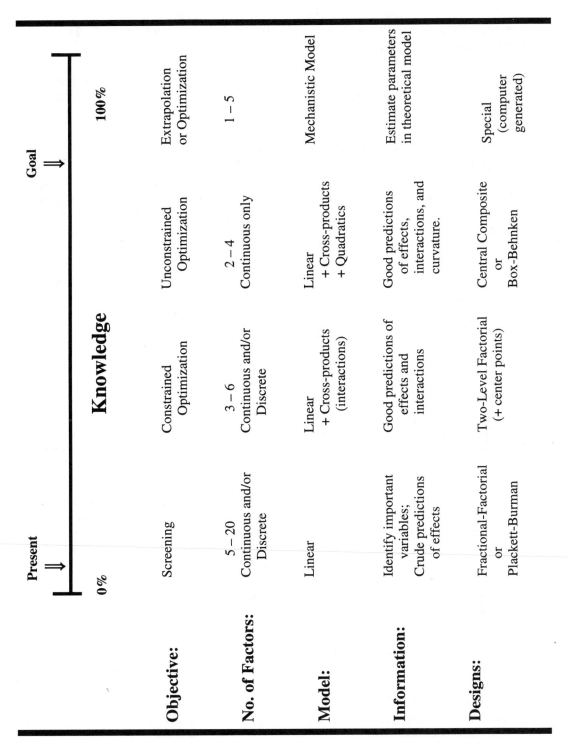

Figure 7.3   The Objective of Experimentation is to Increase Our Knowledge

experiments to run are specific to the model, but often experimental designs used for empirical models are good as a starting point[4].

- **Cost of an Experiment**

    The size of a reasonable experimental program is, of course, dictated by the cost of an experimental run versus the potential benefits. The cheaper an experiment is, the more thoroughly we can study the effects of the independent variables for a reasonable total cost.

- **Precision**

    Generally speaking, the reason for experimenting is to be able to make predictions about what will happen if you make similar actions in the future. For example, the reason for studying the effect of pH on the yield of a chemical reaction is to be able to say that the yield is 4% higher at a pH of 9 than it is at a pH of 7. The more precise you want your predictions to be, and the less precise your individual data are, the greater the number of experiments required.

- **Iteration Possible**

    If the duration of an experiment is relatively short, it is usually reasonable to experiment in small "bite sizes" and iterate toward your final goal. If, on the other hand, the time for an experiment to be completed is long (such as in stability testing), it would be necessary to initially lay out a fairly extensive pattern of experiments. When it is possible to iterate, the first stage of an experimental program should usually be a set of screening experiments. At this stage, all of the factors that could conceivably be important are examined. Since the cost of looking at an extra five or ten variables is relatively low with a screening design, it is much cheaper in the long run to consider some extra variables at this stage than to find out later that you neglected an important variable. The next stage is generally a constrained optimization design in which the major variables are examined for interactions and better estimates are obtained of their linear effects. The minor variables are dropped from consideration after the screening stage (i.e., they are held at their most economical values). If necessary for further optimization, a full unconstrained optimization (response surface) design may now be run to allow for curvature in the effects of the factors on the responses.

    Generally, the constrained optimization design builds upon the screening design, and the unconstrained optimization design just adds more points to the constrained optimization design, so that no points are wasted going from one stage to the next.

    If it is necessary to go further and use a theoretical model, the full, unconstrained optimization design is usually a good starting point for the estimation of parameters in the model. Additional runs would have to be designed via computer. This topic is beyond the scope of this text, and so it will not be discussed any further.

7.6 Example of a Complete Experimental Program

7.6.1 Chemical Process System

An example may be useful at this point to illustrate the typical steps involved in solving a specific industrial problem experimentally. Let us say we have a chemical process under development. It consists of a batch reaction to make a product, and the objective is to find the conditions which give the best yield. The process involves charging a stirred tank reactor with solvent, catalyst, and an expensive reactant (Reactant 1). A second reactant (Reactant 2, which is inexpensive) is added slowly. Reactant 2 is added in excess to ensure that all of Reactant 1 is consumed. Some yield is lost due to the formation of byproducts. After all Reactant 2 is added, the reaction is quenched by adding cold solvent, and then the product is separated by distillation.

Several variables were thought to possibly have an important influence on yield. They are listed in Table 7.1 along with a reasonable range of values for each factor.

Table 7.1   Variables to Be Studied for Chemical Reaction Example

Variable	Definition	Label	Range of Variables	
			Low Level (-1)	High Level (+1)
A	Temperature, °C	X_1	75	85
B	Reactant 2, % excess	X_2	4	8
C	Time to Add Reactant 2, min.	X_3	10	20
D	Agitation, rpm of stirrer	X_4	100	200
E	Solvent : Reactant 1 Ratio	X_5	1 : 1	2 : 1
F	Catalyst Concentration, mg/l	X_6	20	40

7.6.2 Step One: Screening

Since there were six factors to be investigated, the first step in the experimental program was to find out which variables were the most important and focus the most attention on them. It is simply too expensive to study every factor thoroughly, and it is a waste of resources to lavish attention on a variable that has a minor impact. Therefore, we start with screening experiments, which corresponds to the first step in increasing our knowledge. This step is shown in Figure 7.3 semi-graphically as the first column in the figure.

In our example, we used one of the screening designs discussed in detail in Chapter 12, a twelve run Plackett-Burman design, which was copied from Table B.1-1 in Appendix B. The design is shown in Table 7.2 along with the yield data that was collected for each run. Some details in the analysis of the data are shown in the table, but will not be discussed here. The main object of the experiments was to find out which variables were important, which means which variables had the greatest impact on yield when the variables were changed. The measure of that importance is the effect of the variable, which is how much the yield changed (on average) when the variable was changed from its low value to its high value. These effects are shown near the bottom of Table 7.2.

For example, when the reaction temperature, Factor A, was increased from 75 °C to 85 °C, the yield increased by 24.1% on average. When the variability of the measurements was taken into account, only three variables were found to be significant: A, C, and F. They are the reaction temperature, the addition time of Reactant 2, and the catalyst concentration. Furthermore, they all had positive effects on the yield, which means that the higher the variable value, the higher the yield. The best yield obtained in this set of experiments can be seen to be at the high values of the three important factors (Run 10).

Table 7.2 Experiments Run to Determine Which Variables Are Important

Screening Design (Plackett-Burman — Table B.1-1)

Run No.	A X1	B X2	C X3	D X4	E X5	F X6	X7	X8	X9	X10	X11	Yield
							---------	Unassigned	---------			
1	1	1	-1	1	1	1	-1	-1	-1	1	-1	62.7
2	1	-1	1	1	1	-1	-1	-1	1	-1	1	74.9
3	-1	1	1	1	-1	-1	-1	1	-1	1	1	44.9
4	1	1	1	-1	-1	-1	1	-1	1	1	-1	72.1
5	1	1	-1	-1	-1	1	-1	1	1	-1	1	61.3
6	1	-1	-1	-1	1	-1	1	1	-1	1	1	54.1
7	-1	-1	-1	1	-1	1	1	-1	1	1	1	43.2
8	-1	-1	1	-1	1	1	-1	1	1	1	-1	79.8
9	-1	1	-1	1	1	-1	1	1	1	-1	-1	8.6
10	1	-1	1	1	-1	1	1	1	-1	-1	-1	84.2
11	-1	1	1	-1	1	1	1	-1	-1	-1	1	77.4
12	-1	-1	-1	-1	-1	-1	-1	-1	-1	-1	-1	10.8

	A	B	C	D	E	F	X7	X8	X9	X10	X11
Effects	24.1	-3.35	32.08	-6.13	6.85	23.88	0.859	-1.37	0.974	6.6	6.237
t	7.56	-1.05	10.08	-1.93	2.152	7.502			$s_E = 3.183$		

t for Unassigned   0.27 -0.43 0.306 2.073 1.959
t*(5) = 2.571
Important Variables are X1, X3, and X6   (A, C, and F)

The other important conclusion at the end of our screening experiments was that the remaining factors were not of major importance, and therefore they were set to some reasonable values and ignored for the rest of the experimental program. This means that Factor B, the excess of Reactant 2, was set to 4% excess, Factor D, the speed of the agitator, was set to 100 rpm, and Factor E, the solvent to Reactant 1 ratio, was set to 1:1. All were the low values of the factors, which were picked to minimize cost.

7.6.3 Step Two: Crude Optimization

Follow-up experiments were then conducted on the three important factors to determine optimum operating conditions. The strategy used was based upon a constrained optimization design called a *factorial design*, which makes sure we are in the appropriate region. The factorial design is outlined in the second column of Figure 7.3. An important feature of this design is that it can be augmented easily to find an unconstrained optimum, which is our ultimate goal.

Table 7.3 Experiments Run to Optimize Important Variables (Factorial Design)

Factorial Design   (Table B.3-2,   Block 1)

Run No.	A	C	F	X_1	X_3	X_6	X_1X_3	X_1X_6	X_3X_6	Curvature	Yield		
1	80	15	30	-1	-1	-1	1	1	1		70.2		
2	90	15	30	1	-1	-1	-1	-1	1		71.1		
3	80	25	30	-1	1	-1	-1	1	-1		74.6		
4	90	25	30	1	1	-1	1	-1	-1		55.3		
5	80	15	50	-1	-1	1	1	-1	-1		69.5		
6	90	15	50	1	-1	1	-1	1	-1		50.1		
7	80	25	50	-1	1	1	-1	-1	1		75.5		
8	90	25	50	1	1	1	1	1	1	62.0	29.8	$s =$	1.117
9	85	20	40	0	0	0	0	0	0		78.8	$s_E =$	0.79
10	85	20	40	0	0	0	0	0	0		80.3	$s_C =$	0.756
11	85	20	40	0	0	0	0	0	0	80.0	81.0		

| Effect | | | | -20.8 | -6.42 | -11.5 | -11.7 | -11.7 | -0.7 | 18.01 | | | |
| t | | | | -26.4 | -8.13 | -14.6 | -14.8 | -14.8 | -0.89 | 23.82 | | | |

$t^*(2) =$ 4.303

The factorial design used in this study is shown in Table 7.3. It should be noted that the experiments were centered around the conditions found to be best up to that point: temperature (A) of 85 °C, addition time (C) of 20 minutes, and a catalyst concentration (F) of 40 mg/l. The resulting yield data are also given in the table. An analysis of the data, which will be discussed in much more detail in Chapter 8, showed that all the variables continued to be important, and they also had some interactions. However, the main point that was learned from the data was that the response could not be described by a straight line model – a quadratic equation was needed. This can be seen directly from the data; the average yield in the center of the experimental region was 80%, while the average response at the corners was only 62%. This difference is extremely significant, and it means that the yield is a curved function of the three important factors. It also means that we are in the vicinity of the optimum, and can move on to the final optimization phase.

7.6.4 Step Three: Final Optimization

Curved (e.g., quadratic) functions cannot be elucidated by data from a factorial design alone, so the data was augmented by the extra points to form a complete central composite design, which is outlined in the third column of Figure 7.3. These extra points are shown in the bottom half of the design in Table 7.4 (called Block 2) along with the associated yield data. Notice that Block 1 consisted of the factorial design that was already in hand. This composite set of data was used to complete the optimization.

Table 7.4 Experiments Run to Complete Optimization of Important Variables
(Central Composite Design)

Central Composite Design   (Table B.3-2)

Run No.	A	C	F	X_1	X_3	X_6	Block	Yield
1	80	15	30	-1	-1	-1	1	70.2
2	90	15	30	1	-1	-1	1	71.1
3	80	25	30	-1	1	-1	1	74.6
4	90	25	30	1	1	-1	1	55.3
5	80	15	50	-1	-1	1	1	69.5
6	90	15	50	1	-1	1	1	50.1
7	80	25	50	-1	1	1	1	75.5
8	90	25	50	1	1	1	1	29.8
9	85	20	40	0	0	0	1	78.8
10	85	20	40	0	0	0	1	80.3
11	85	20	40	0	0	0	1	81.0
12	76.4	20	40	-1.73	0	0	2	82.6
13	93.7	20	40	1.73	0	0	2	44.6
14	85	11.4	40	0	-1.73	0	2	67.0
15	85	28.7	40	0	1.73	0	2	51.2
16	85	20	22.7	0	0	-1.73	2	77.2
17	85	20	57.3	0	0	1.73	2	58.2
18	85	20	40	0	0	0	2	80.9
19	85	20	40	0	0	0	2	77.7
20	85	20	40	0	0	0	2	83.5

The procedure used was to fit a full quadratic equation to the twenty data points, and then use the equation to predict the best operating conditions. Fitting the equation is done using regression analysis, which is a tool included in spreadsheet programs as well as statistical software. Table 7.5 shows the output from MINITAB®, a common statistical software package.

Table 7.5 MINITAB Regression Analysis of Data from the Central Composite Design

Response Surface Regression

The analysis was done using coded units

Estimated Regression Coefficients for Yield

Term	Coef	StDev	T	P
Constant	80.3670	0.7726	104.02	0.000
Block	0.5927	0.4263	1.39	0.198
A	-10.6708	0.5060	-21.09	0.000
C	-3.7920	0.5060	-7.49	0.000
F	-5.6607	0.5060	-11.19	0.000
A*A	-5.8437	0.4782	-12.22	0.000
C*C	-7.3472	0.4782	-15.37	0.000
F*F	-4.4738	0.4782	-9.36	0.000
A*C	-5.8125	0.6691	-8.69	0.000
A*F	-5.8375	0.6691	-8.72	0.000
C*F	-0.3625	0.6691	-0.54	0.601

S = 1.892   R-Sq = 99.2%   R-Sq(adj) = 98.4%

The quadratic equation can then be used to find the maximum yield analytically, but a more common approach is to plot the equation, since people tend to think pictorially. Plots for this example are shown in Figure 7.4. The optimum conditions can be seen to be a reaction temperature (Variable A) of 80 °C, an addition time for Reactant 2 (Variable C) of 20 minutes, and a catalyst concentration (Variable F) of 40 mg/l.

7.7 Good Design Requirements

Before jumping in and explaining the details of our cornerstone strategy — the two-level factorial design — which is considered to be very good and widely applicable, it is worth discussing what "good" means first. In other words, what would an ideal experimental design do for us? The following are some attributes of a good design:

● **Defined Objectives**

The experimenter should clearly set forth the objectives of the study before deciding on an experimental design and proceeding with the experiments. Generally this takes the form of what model will be fit to the data: a simple linear model, a linear model with interactions, or a full quadratic model. In addition, the desired precision of the conclusions needs to be specified. A good design must meet all of the objectives. Once a design is selected, the experimenter can and should detail (for the sponsor of the work) not only what information will be obtained form the experimental data, but what information will not be learned, so that there are no misunderstandings.

193

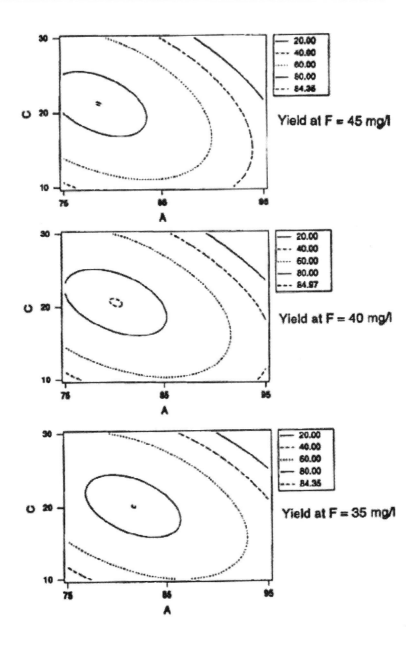

Figure 7.4 Contour Plots Used to Optimize Process Yield

- **Unobscured Effects**

    The effects of each of the factors in the experimental program should not be obscured by other variables insofar as possible. All the designs discussed in this book meet this goal, so it will not be mentioned again.

- **Free of Bias**

    As far as possible, the experimental results should be free of bias, conscious or unconscious. The first step in assuring this is to carefully review the experimental setup and procedure. However, some statistical tools are helpful in this regard:

    ○ BLOCKING (planned grouping) of runs lets us take some lurking variables into account.

    ○ RANDOMIZATION of the run order within each block enables us to minimize the confounding (biasing) of factor effect with background variables.

    ○ REPLICATION aids randomization to do a better job, as well as giving more precision.

- **Variability Estimated**

    In order to be able to decide whether the effects of factors that were found in the experimental program are real or whether they could just be due to the variability in the data, the experimental design should provide for estimating the precision of the results. The only time this is not needed is when there is a well-known history of the process or system being studied, with quantitative estimates of the process standard deviation, σ, from process capability studies. Replication provides the estimate of precision, while randomization assures that the estimate is valid.

- **Design Precision**

    The precision of the total experimental program should be sufficient to meet the objectives. In other words, enough data should be taken so that effects that are large enough to have practical significance will be statistically significant. Greater precision can sometimes be achieved by refinements in experimental technique. Blocking can also sometimes help improve precision a great deal. However, the main tool at our disposal to increase precision is replication. Just how much replication is needed for a desired level of precision is discussed in Section 9.2.

7.8 Summary

In this chapter we discussed what properties are desirable in an experimental strategy. Some terms were defined to help communicate the ideas more clearly. And a few statistical tools were introduced that are used to help us get the best results we can for our efforts The following is a list of the important terms and concepts discussed in this chapter:

References

1. Box, G. E. P., "Use and Abuse of Regression", *Technometrics*, Vol. **8**, (1966), pp. 625–629.

2. Box, Hunter and Hunter,   *Statistics for Experimenters*, John Wiley & Sons, New York, pp. 487-498, (1978).

3. Lucas, J. M., "Discussion of Off-line Quality Control by Kackar," *Journal of Quality Technology*, Vol. **17**, No. 4, (1985), pp.195-197.

4. Kittrell and Erjavec, "Response Surface Methods in Heterogeneous Kinetic Modeling", *Industrial & Engineering Chemistry*, Vol **7**, (1968), p. 321.

7.9 Exercises for Chapter 7

7.1 List five of the major attributes of a good experimental design. Explain briefly why each attribute is important.

7.2 List five of the major characteristics of an experimental environment. Explain briefly how each characteristic is different for each of the major design categories.

7.3 When attempting to optimize plant operations, it is important to collect new data by running experiments on the particular process(es), rather than simply relying on existing data from past plant operations. Give five reasons why this is true, and briefly explain/justify your reasons.

7.4 List the strengths and weaknesses of One-At-A-Time designs.
   How well does this type of design fit the good design requirements listed in the text?
   Is One-At-A-Time ever the best design to use? If so, under what circumstances?

7.6. Imagine that you work for a pharmaceutical company, and you are testing the effectiveness of a new drug on human subjects. The measure of performance is the improvement in the health of the subject given the new drug compared to the improvement in the health of a subject given a placebo (sugar pill with no active ingredients). What precautions would you take to ensure that there was no bias introduced into the design? To answer the question, list at least five of the possible sources of bias, and how you would ensure that each of the sources was addressed by your design.

7.7 (a) Give a specific application for a screening design (Fractional Factorial or Plackett-Burman). List the factors to be studied and the response(s).
   (b) Give a specific application for a constrained optimization design (Factorial Design). List the factors to be studied and the response(s).
   (a) Give a specific application for an unconstrained optimization design (Central Composite or Box-Behnken). List the factors to be studied and the response(s).

CHAPTER 8
Basic Two-Level Factorial Experiments

8.1 Introduction

Our "quest for knowledge" as shown again in Figure 8.1 generally begins with a screening design in order to get the number of factors that we need to examine down to a manageable level. Then we move on to an optimization strategy. So it would seem logical to begin our discussion of experimental designs with screening. However, we will not do that.

Instead we will choose as our first topic for discussion, the two-level factorial design usually used as the *second* step in an investigation (for crude optimization). This second phase of experimentation is highlighted in Figure 8.1. The reason for this apparent jumbling of the order of our discussion is that the two-level factorial design is the building block for almost all of the experimental strategies discussed in this book (as was mentioned in Chapter 7). The typical screening design is formed by taking a piece, or fraction, of the full two-level factorial design. Moreover, the typical comprehensive, unconstrained optimization design is built up by adding a group of experiments to the two-level factorial design. Therefore, it makes good, pedantic sense to begin our discussion "in the middle of the story." Once the two-level factorial design is fully understood, it is much easier to go back and discuss the fractional factorial designs used for screening (Chapter 12) and the composite designs used for unconstrained optimization (Chapter 13).

8.2 Two-Level Factorial Design Geometry

By definition, a factorial design consists of all combinations of the factor levels of two or more factors. For example, let us say that you are interested in checking to see if the accuracy of a voltmeter depends on ambient temperature or warm-up time. Further, you wanted to look at four different ambient temperatures (20, 25, 30, and 35 °C) and three different warm-up times (0.5, 3.0, and 5.5 min.). The full 4×3 factorial design to do the job is shown in Table 8.1, and it can be seen to be all possible combinations of the factor levels.

Since frugality is a major virtue in any practical situation, we would like to run the fewest number of levels of the factors while still learning about their impacts. Since we must change a factor to discover what effect it has on our response, the smallest number of levels of each factor is two. Therefore, our cornerstone design will be the two-level factorial design, in which all possible combinations of two levels for each of the factors are run as experiments. Figure 8.2 shows this design geometry for two factors, and Figure 8.3 shows a two-level factorial design for three factors.

The two levels of each factor are denoted symbolically by a "+" and a "-" to indicate a high and a low level of each particular factor. When the factor is quantitative, it is obvious what this means. For example, if one of the factors is temperature, and it is being studied at 32 and 22°C, the "+" would indicate 32°C, while "-" would indicate 22°C. If the factor is discrete such as voltmeter (or operator, machine, raw material lot, etc.), one of the voltmeters (or operators, machines, lots, etc.) is arbitrarily called "+" and the other is called "-".

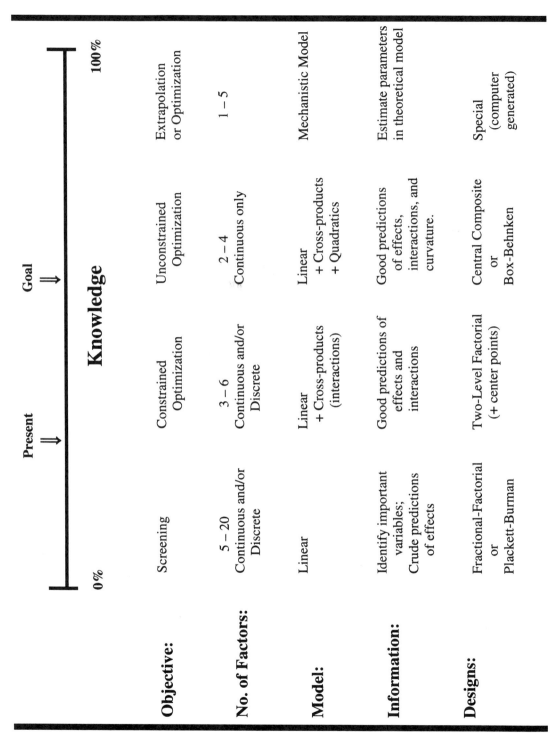

Figure 8.1 Objective of Constrained (or Crude) Optimization

Table 8.1 Full 4×3 Factorial Design to Check the Accuracy of a Voltmeter

Run Number	Ambient Temperature	Warm-up Time	Measured Voltage
1	20.0 °C	0.5 min.	---
2	25.0 °C	0.5 min.	---
3	30.0 °C	0.5 min.	---
4	35.0 °C	0.5 min.	---
5	20.0 °C	3.0 min.	---
6	25.0 °C	3.0 min.	---
7	30.0 °C	3.0 min.	---
8	35.0 °C	3.0 min.	---
9	20.0 °C	5.5 min.	---
10	25.0 °C	5.5 min.	---
11	30.0 °C	5.5 min.	---
12	35.0 °C	5.5 min.	---

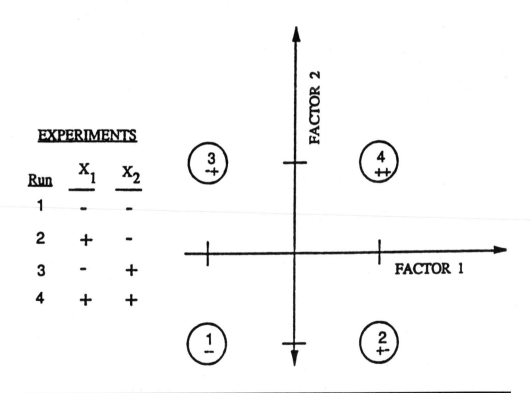

Figure 8.2 Two-Level Factorial Design for Two Factors (2^2 Factorial Design)

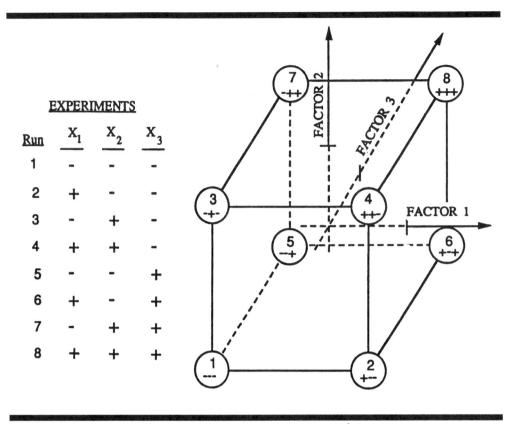

Figure 8.3 Two-Level Factorial Design for Three Factors (2^3 Factorial Design)

Plotting the experimental design points as in Figures 8.2 and 8.3 is useful and should be done whenever possible. This is because one of the basic goals of a good experimental design is to cover the region to be studied as thoroughly as possible with the number of runs to be made. Plotting the experimental design allows us to assess just how well, in fact, we are covering the region. The factorial design, which looks at all the extreme points in the experimental region, is doing this job well. The only things that are missing are some points in the interior of the region. However, in practice some replicated points in the center of the region are usually run. This topic will be discussed in more detail later, and for the time being we will assume only the factorial (corner) points are to be run.

8.3 Main Effect Estimation

Once we have run the experiments and measured a response (or several responses) at each set of conditions, we must analyze the data to determine the impacts of each of the factors under study. To see how this is done, let us look at the 2^2 factorial design shown in Figure 8.4. From this figure we can see that one way of thinking about the design is to view it as two one-at-a-time experiments in X_1: one at a low level of X_2 (runs 1 and 2), and the other at a high level of X_2 (runs 3 and 4). We can calculate the effect of X_1 for both of these pairs of runs. Since we want to summarize the effect of X_1 by one number, called the *main effect* of X_1, the most natural thing to do is to take the average of these two effects.

$$\text{Effect of } X_1 \text{ (at low } X_2) = Y_2 - Y_1$$
$$\text{Effect of } X_1 \text{ (at high } X_2) = Y_4 - Y_3$$

$$\text{Average Effect of } X_1 = [(Y_2 - Y_1) + (Y_4 - Y_3)] / 2$$

An equivalent way of performing the calculation of this main effect is to average the runs at a high level of X_1 and subtract the average of the runs at a low level of X_1. For this reason, the main effect is often called the *average main effect*.

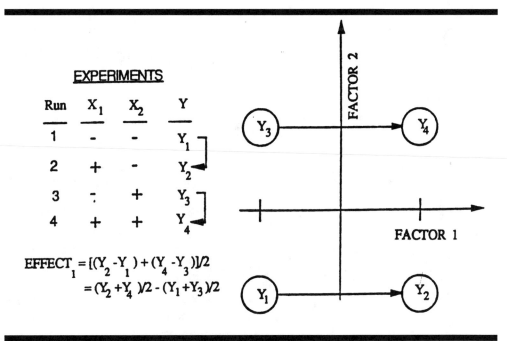

Figure 8.4 Estimating the Main Effect of Factor 1 (X_1) for the 2^2 Factorial Design

The power (efficiency) of two-level factorial designs compared to one-at-a-time designs is due to the fact that these same four experiments can be used to calculate the effect of X_2. However, in this case we are comparing Y_3 to Y_1, and Y_4 to Y_2 (see Figure 8.5). The calculation of the effect of X_2 is identical in principle to that for X_1.

Effect of X_2 (at low X_1) = $Y_3 - Y_1$
Effect of X_2 (at high X_1) = $Y_4 - Y_2$

Average Effect of X_2 = $[(Y_3 - Y_1) + (Y_4 - Y_2)] / 2$

Likewise, this expression can be written as the difference between the average of the Y's at the high X_2 minus the average of the Y's at the low X_2.

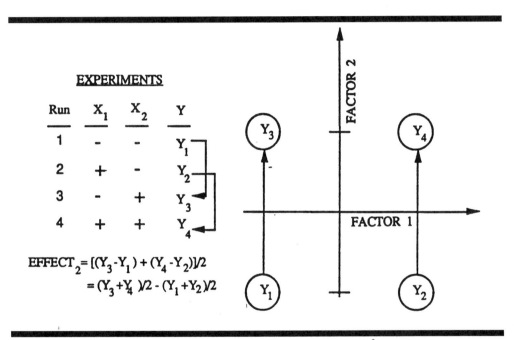

Figure 8.5   Estimating the Main Effect of Factor 2 (X_2) for the 2^2 Factorial Design

These concepts can be readily extended to factorial designs with more than two factors. Figure 8.6 shows the estimation of the main effect of X_1 for a 2^3 factorial design. In this case, we have four pairwise comparisons that give us estimates of the effect of X_1. The main effect of X_1, is simply the average of these four comparisons. Or, equivalently, the effect of X_1 is seen to be the average of the four Y's at the high X_1 minus the average of the four Y's at the low X_1. These same eight experiments can also be used to calculate the effects of X_2 and X_3 in a similar fashion (by subtracting the average of the Y's at the low level of the factor from the average of the Y's at the high level of the factor).

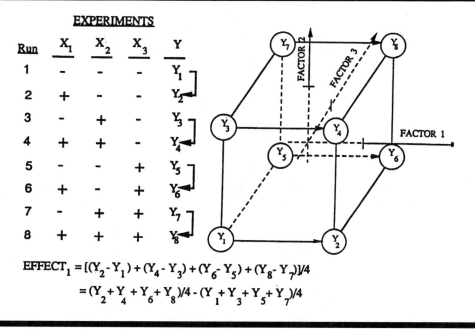

$$\text{EFFECT}_1 = [(Y_2 - Y_1) + (Y_4 - Y_3) + (Y_6 - Y_5) + (Y_8 - Y_7)]/4$$

$$= (Y_2 + Y_4 + Y_6 + Y_8)/4 - (Y_1 + Y_3 + Y_5 + Y_7)/4$$

Figure 8.6 Estimating the Main Effect of Factor 1 (X_1) for the 2^3 Factorial Design

At this juncture, it may be useful to point out two more advantages of factorial designs over one-at-a-time designs (in addition to being space-filling). The simple 2^2 design shown in Figure 8.2 is used for illustration. The simplest one-at-a-time design for calculating the effects of two factors would have three points: a base point (-,-), a point at which only X_1 was changed (+,-), and a point at which only X_2 was changed (-,+). These three points allow the estimation of the effect of X_1 at low X_2 and the effect of X_2 at low X_1.

By the addition of only one more point (+,+), thus making a factorial design, we get: (1) greater accuracy, and (2) a broader basis for our conclusions. The accuracy comes about because of "hidden replication." Although no single pair of conditions is replicated to increase the accuracy of the estimate of the main effect, we measured the effect of each factor at two levels of the other factor. The average of the two estimates is then more precise than either single estimate. The one-at-a-time design would have to have two runs at each of the three conditions (a total of six runs) to achieve the same precision as the factorial design with four runs. This increase in efficiency becomes even more pronounced as the number of factors increases. We are also more confident in our conclusions about the effects of X_1 and X_2 on Y because we measured the effect of each factor at both levels of the other factor (or at all possible combinations of the other factors if there are more than two factors). Thus, our conclusions have a broader basis.

8.4 Interactions

Perhaps the most important advantage of the factorial design is that it allows for estimation of interactions between the factors. Interaction means that the effect of one factor depends on the settings of one or more of the other factors. The simplest type of interaction is called a two-way or two-factor interaction. This interaction is shown graphically for a number of situations in Figure 8.7. If X_1 and X_2 have a two-way interaction, this means that the slope of the linear plot of Y versus X_1 depends on the value of X_2. If there is no interaction, the slope of the line does not depend on X_2, although the position of the line might shift.

The magnitude of the interaction for a 2^2 factorial design is defined as one-half of the difference between the effect of X_1 at "high X_2 and the effect of X_1 at "low" X_2 (see Figure 8.8). This will be shown latter to be equal to the difference between the slopes of the two lines. Equivalently, the X_1X_2 interaction can be seen to be half the difference between the effect of X_2 at "high" X_1 and the effect of X_2 at "low" X_1. It is important that these two ways of calculating the X_1X_2-interaction (half the change in the effect of X_1 or half the change in the effect of X_2) come out the same, otherwise it would matter which variable we chose to be Factor 1 and which variable then became Factor 2. The X_1X_2-interaction definition is easily extended to larger factorials. It is still half the difference between the (average) effect of X_1 at "high" X_2 and the (average) effect of X_1 at "low" X_2.

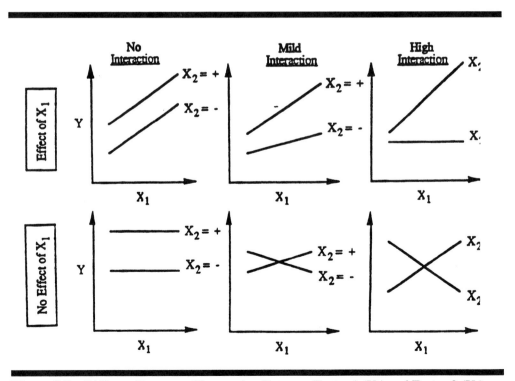

Figure 8.7 Different Degrees of Interaction Between Factor 1 (X_1) and Factor 2 (X_2)

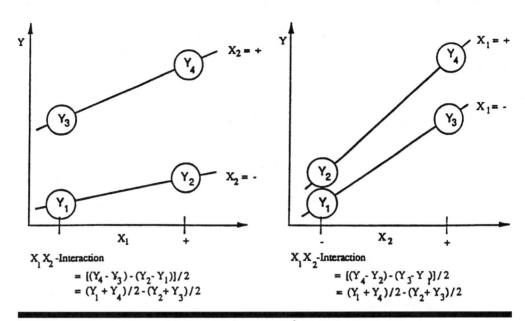

Figure 8.8   Calculation of the Interaction for a 2^2 Factorial Design

For higher order factorials (three or more factors) we can extend the definition of the 2-factor interaction to higher order interactions, although these are harder visualize and are, fortunately, quite rare. So, the $X_1X_2X_3$- interaction can be computed as half the difference between the (average) X_1X_2-interaction at "high" X_3 and the (average) X_1X_2-interaction at "low" X_3. Equivalently, it could be computed as half the change in the X_1X_3-interaction (as X_2 changes), or it could be computed as half the change in the X_2X_3-interaction (as X_1 changes).

8.5 General 2^k Factorial Designs

A factorial pattern of experiments can easily be written down for any number of factors, k. The pattern of experiments is shown in Table 8.2 for up to five factors. This table is reproduced in Appendix B.2, Table B.2-1, for handy reference. The total number of experiments is 2^k and represents all possible combinations of the k factors. The "+" and "-" signs follow an easily remembered pattern which can be readily written down. The first column alternates minuses and pluses, the second column alternates pairs of minuses and pluses, the third column alternates groups of four minuses and four pluses, etc. Remember that the "+" stands for one setting of a factor (the high setting if the factor is quantitative) and the "-" stands for another setting of the factor (the low setting if the factor is quantitative). Each of the rows corresponds to a set of experimental conditions which will be run. Any number of responses can be measured in each experiment. At the end of the experimental program, each response is analyzed separately.

Tables B.2-2 through B.2-5 in Appendix B.2 list the factorial designs for 2 to 5 levels respectively. In addition to the columns of + and - signs for the factors, as shown in Table 8.2, Tables B.2-2 through B.2-5 contain additional columns for interactions. The interaction columns are used to define how the interaction effects are calculated. In Figure 8.8 it was shown that the X_1X_2-interaction in a 2^2 design could be calculated as:

$$X_1X_2\text{-interaction} = (Y_1 + Y_4)/2 - (Y_2 + Y_3)/2$$

It is very important to notice that the interaction can be calculated as the difference between the average of half the responses and the average of the other half of the responses, even though the interaction is defined as something more complicated than that. Remember that this is exactly the way the effects of the factors are calculated. For the factors, we average the responses at the high level (+) and average the responses at the low level (-), and then we take the difference to get the effect. We would like to do the same thing to calculate the interactions if we only knew which responses were at the "high level of the interaction" and which responses were at the "low level of the interaction." That is exactly what the interaction columns in the tables give us. For example, it can be seen that the + and - signs in the X_1X_2 column of Table B.2-2 define exactly how that interaction was calculated by having "+" signs for Runs 1 and 4 and "-" signs for Runs 2 and 3. Therefore, use of the table eliminates the need for going back to the definition (or, equivalently, resorting to a figure like Figure 8.8) to determine which observations need to be averaged and subtracted to calculate interactions. Using the table, the responses (the Y_i's) are simply added up with the appropriate sign and averaged to calculate the interactions (and effects). Therefore, Tables B.2-2 through B.2-5 are called computation tables. Their use is illustrated in Sections 8.7 and 8.9.

These extra interaction columns were created by multiplying together the columns of the individual factors involved in the interaction. Specifically, to create any particular interaction column, each row is examined separately, and the + and - signs for the factors in the interaction are multiplied together to get the + or - sign for the interaction. This procedure is continued for each row (element) of the interaction. For example, in Table B.2-2 on the row for Run 1, the "-" under the X_1 column is multiplied by the "-" under the X_2 column to form the "+" in the X_1X_2 column. This multiplication is repeated for each row in the table.

Table 8.2   The Factorial Pattern of Experiments

	Run	X_1	X_2	X_3	X_4	X_5
	1	-	-	-	-	-
k = 1	2	+	-	-	-	-
	3	-	+	-	-	-
k = 2	4	+	+	-	-	-
	5	-	-	+	-	-
	6	+	-	+	-	-
	7	-	+	+	-	-
k = 3	8	+	+	+	-	-
	9	-	-	-	+	-
	10	+	-	-	+	-
	11	-	+	-	+	-
	12	+	+	-	+	-
	13	-	-	+	+	-
	14	+	-	+	+	-
	15	-	+	+	+	-
k = 4	16	+	+	+	+	-
	17	-	-	-	-	+
	18	+	-	-	-	+
	19	-	+	-	-	+
	20	+	+	-	-	+
	21	-	-	+	-	+
	22	+	-	+	-	+
	23	-	+	+	-	+
	24	+	+	+	-	+
	25	-	-	-	+	+
	26	+	-	-	+	+
	27	-	+	-	+	+
	28	+	+	-	+	+
	29	-	-	+	+	+
	30	+	-	+	+	+
	31	-	+	+	+	+
k = 5	32	+	+	+	+	+

8.6 Randomization

The order of the experiments shown in Table 8.2 is called the Yates order or standard order. This order is convenient for listing experimental results or analyzing them but, the experiments should *not* be run in this order, since the risk of bias would be very great. For example, if you were running a 2^4 factorial design (16 runs) and the response drifted down during the course of the experimental program (say, due to a catalyst aging) or shifted in the middle of the program (e.g., due to a raw material change) this effect would be mistakenly interpreted as an effect of X_4. This is shown in Figure 8.9A. If none of the four factors under study have any real impact, but the first 8 runs were made at the low level of X_4 (the "-" level) and the next 8 runs were made at the high level of X_4 (the "+" level), we would calculate the main effect of X_4 as the difference between the average response of the last eight runs, $\bar{Y}_+$, and the average response of the first eight runs, $\bar{Y}_-$. This difference can be seen to be large simply because of process drift. So even though X_4 has no effect at all, we would think it was very important.

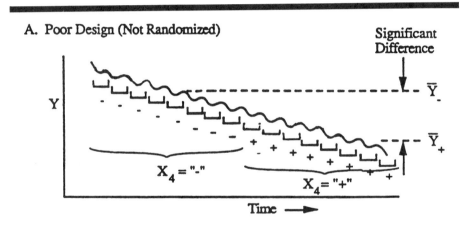

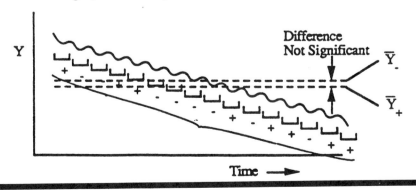

Figure 8.9 Example of Response Data from a Drifting Process

To minimize this type of risk, the experiments should be run in random order. Any method of generating a random run order can be used, like writing the run numbers on pieces of paper, shuffling the pieces of paper and using the shuffled run order. For convenience, a list of random numbers is given in Table A.1 which can be used to randomize run orders. Additionally, typical random orders for 8, 16, and 32 run experimental designs are given in Table 8.3. The impact of randomization is shown in Figure 8.9B. Now when we calculate the main effect of X_4 (or any of the other insignificant factors) the "effect" will usually be small, even when there is drift in the process (or other sources of bias are present). The importance of running experiments in random order cannot be stressed too much. Randomization is usually expensive, but it is worth every penny!

Table 8.3 Possible Random Run Orders for Experiments

8 Runs					16 Runs					32 Runs				
1	5	8	6	5	10	1	11	14	5	29	19	15	3	30
6	7	4	4	3	2	5	9	13	4	32	16	8	6	20
4	4	6	2	4	4	2	4	4	14	18	30	13	24	4
8	1	3	7	1	16	6	16	1	7	2	10	10	9	7
5	6	5	5	2	9	11	13	10	15	17	13	3	28	3
3	3	7	1	7	5	3	12	16	2	15	25	25	26	14
7	2	1	3	8	8	8	14	2	1	13	28	1	2	10
2	8	2	8	6	3	12	6	5	3	20	15	6	29	16
					7	7	10	12	8	14	4	9	31	27
					11	13	5	9	11	6	22	20	1	2
					1	4	2	6	12	22	32	16	15	32
					12	15	15	8	16	16	21	28	12	12
					14	10	8	15	10	1	24	19	23	24
					15	14	7	3	6	25	12	27	17	23
					6	9	1	7	9	12	17	17	18	5
					13	16	3	11	13	30	9	30	27	25
										5	20	2	30	8
										23	2	26	32	28
										19	8	11	25	1
										28	11	5	20	21
										31	5	29	22	26
										8	6	31	7	6
										11	1	4	14	22
										7	31	7	5	17
										24	23	12	11	18
										9	27	22	10	29
										27	3	18	19	11
										4	26	24	8	15
										21	7	32	13	31
										3	29	14	4	19
										26	18	21	16	9
										10	14	23	21	13

8.7 An Example of a 2^3 Factorial Experiment

8.7.1 Background and Design

Students in a university electronics lab often complained about taking a measurement, and then retaking it sometime later and finding the reading had changed. The lab instructor decided to make a small study of the causes for variation in the electronic measurements. He set up a 2^3 factorial experiment. The main objective of the experiment was to examine the effects of three factors on voltmeter readings. A simple circuit (shown in Figure 8.10) was built, and the voltage across the base-to-emitter junction of a transistor was the measured value.

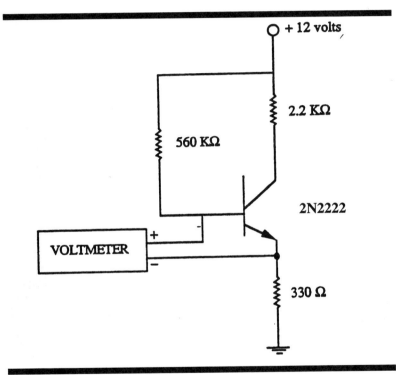

Figure 8.10　Circuit Used for Voltage Measurements

The experimental factors chosen for study were:

X_1 (Factor 1): Ambient temperature, or the temperature of the air surrounding the circuit. Two levels were examined, 22 °C (room temperature), and 32 °C (close to the temperature in some industrial settings). An oven was used, and the circuit was allowed to stabilize for at least five minutes before measurements.

212

X_2 (Factor 2): Voltmeter warm-up time. The voltmeters were allowed to warm up after being turned on for 5 minutes, or for 30 seconds. In the latter case, the meter was shut off for at least 5 minutes prior to turning it on and taking a measurement. Often students are in a hurry to complete a lab, and don't take time to allow the equipment to warm up and stabilize.

X_3 (Factor 3): Circuit warm-up time. The time that the circuit was connected to power was also varied at the levels of 5 minutes and 30 seconds. Again, in both cases, the circuit was allowed to cool more than 5 minutes between each reading.

The experimental plan is shown in Table 8.4. Two replicate measurements were made at each condition and the order of experimentation was randomized by choosing the last column of the 16 run random orders in Table 8.3, which was copied into the run order column of Table 8.4. Using this ordering, the first replicate of Run 4 was completed first by warming the circuit to 32°C, allowing the voltmeter to warm up 5 minutes, then turning on the circuit and measuring the voltage. The first replicate of Run 3 was completed second and eventually all 16 data points were then collected in the appropriate order and the results are also listed in the far right column of Table 8.4.

Table 8.4 2^3 Factorial Design to Investigate Accuracy of Voltmeter

FACTORS AND LEVELS

Factors and Response	Levels	
	-	**+**
X_1 = Ambient temperature, °C	22°	32°
X_2 = Voltmeter warmup time, minutes	0.5	5.0
X_3 = Time that power is connected, minutes	0.5	5.0
Y = measured voltage, millivolts		

DESIGN AND DATA

	Coded Values			Actual Values			Run	Y	
Run	X_1	X_2	X_3	X_1	X_2	X_3	Order	millivolts	
1	-	-	-	22	0.5	0.5	5, 4	705,	680
2	+	-	-	32	0.5	0.5	14, 7	620,	651
3	-	+	-	22	5.0	0.5	15, 2	700,	685
4	+	+	-	32	5.0	0.5	1, 3	629,	635
5	-	-	+	22	0.5	5.0	8, 11	672,	654
6	+	-	+	32	0.5	5.0	12, 16	668,	691
7	-	+	+	22	5.0	5.0	10, 6	715,	672
8	+	+	+	32	5.0	5.0	9, 13	647,	673

The averages of the voltage readings at each set of experimental conditions were calculated and they are given in Table 8.5. The variances were also calculated for each set of experimental conditions as:

$$s^2 = \sum_{i=1}^{r} (Y_i - \bar{Y})^2 / (r - 1) \qquad \textbf{(8.1)}$$

which, for the special case of two observations (r = 2), reduces to:

$$s^2 = (Y_1 - Y_2)^2 / 2 \qquad \textbf{(8.2)}$$

The variances are also listed in Table 8.5. For the time being, this information about the variance is "extra", and will not be used. But, it will become very necessary when we want to calculate the statistical significance of effects and interactions, which is discussed in the next section.

Table 8.5 Summary of Average Voltage Measurements and Variability

Run No.	X_1	X_2	X_3	Y millivolts		Avg Y	Variance	DF
1	-	-	-	705	680	692.5	312.5	1
2	+	-	-	620	651	635.5	480.5	1
3	-	+	-	700	685	692.5	112.5	1
4	+	+	-	629	635	632.0	18.0	1
5	-	-	+	672	654	663.0	162.0	1
6	+	-	+	668	691	679.5	264.5	1
7	-	+	+	715	672	693.5	924.5	1
8	+	+	+	647	673	660.0	338.0	1

8.7.2 Calculation of Effects and Interactions

The data from any 2^3 experiment can be conveniently displayed on a cube which represents the experimental region as in Figure 8.3. For the voltage measurement experiment, this is done in Figure 8.11. In this figure, the direction of the arrows indicates the higher setting of each factor. We can see that X_1 = ambient temperature seems to have a negative effect (the voltage measured at the higher temperature was lower in 3 out of 4 cases).

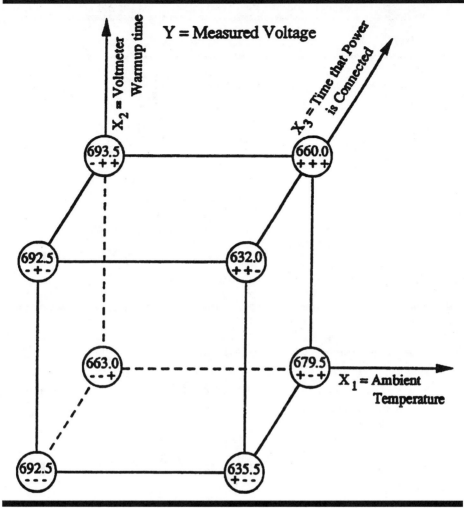

Figure 8.11 Graphical Display of Data from a 2^3 Factorial Design with Voltmeter

The calculation of the average main effects and interactions as defined in Sections 8.3 and 8.4 is accomplished via the use of the computation Table B.2-3 in Appendix B which was described in Section 8.5. The +'s and -'s in the table indicate what sign is associated with responses for each run in the calculation of the various effects. Table B.2-3 was copied into Table 8.6, with the average voltage reading added to the right. Then, to calculate any of the effects, the Y's are simply added up (with the appropriate sign) and averaged by dividing by the number of +'s for the given effect.

Table 8.6 Worksheet for Computation of Effects and Interactions for 2^3 Factorial with Voltmeter

Run	Mean	X_1	X_2	X_3	X_1X_2	X_1X_3	X_2X_3	$X_1X_2X_3$	Avg Y
1	+	−	−	−	+	+	+	−	692.5
2	+	+	−	−	−	−	+	+	635.5
3	+	−	+	−	−	+	−	+	692.5
4	+	+	+	−	+	−	−	−	632.0
5	+	−	−	+	+	−	−	+	663.0
6	+	+	−	+	−	+	−	−	679.5
7	+	−	+	+	−	−	+	−	693.5
8	+	+	+	+	+	+	+	+	660.0
$\Sigma(+)$	5348.5	2607.0	2678.0	2696.0	2647.5	2724.5	2681.5	2651.0	
$\Sigma(-)$	0	2741.5	2670.5	2652.5	2701.0	2624.0	2667.0	2697.5	
$\Sigma(+) + \Sigma(-)$	5348.5	5348.5	5348.5	5348.5	5348.5	5348.5	5348.5	5348.5	
$\Sigma(+) - \Sigma(-)$	5348.5	−134.5	7.5	43.5	−53.5	100.5	14.5	−46.5	
Effects	668.56	−33.6	1.9	10.9	−13.4	25.1	3.6	−11.6	

For example:

Effect of X_1 = Average Response when X_1 is high (+) - Average Response when X_1 is low (−)

$= (Y_2 + Y_4 + Y_6 + Y_8)/4 - (Y_1 + Y_3 + Y_5 + Y_7)/4$ ← Grouping +'s and −'s

$= (- Y_1 + Y_2 - Y_3 + Y_4 - Y_5 + Y_6 - Y_7 + Y_8)/4$ ← Using Computation Column

$= (-692.5 + 635.5 - 692.5 + 632.0 - 663.0 + 679.5 - 693.5 + 660.0)/4$

$= (2607.0 - 2741.5)/4 = -134.5/4 = -33.6$

X_1X_3-Interaction $= (+Y_1 - Y_2 + Y_3 - Y_4 - Y_5 + Y_6 - Y_7 + Y_8)/4$

$= (692.5 - 635.5 + 692.5 - 632.0 - 663.0 + 679.5 - 693.5 + 660.0)/4$

$= (2724.5 - 2624.0)/4 = 100.5/4 = 25.1$

Notice that the responses can be added together with the appropriate sign (which is given in the computation table), or they can be grouped by sign. When the calculations are being done by hand, it is generally better to group the responses by sign. In that case, the Y's with +'s are added first, and the sum of the Y's with −'s are then subtracted. Adding the two sums (denoted by $\Sigma(+)$ and $\Sigma(-)$) gives a check that the arithmetic has been done correctly up to this point. Then, to calculate the effect, the difference between the sums is divided by the number of +'s for the given effect (because the difference between *averages* is what is desired). This is all shown in Table 8.6. If a spreadsheet is used to do the calculations, adding all the Y's together with the appropriate sign is easier and less prone to error. For example, LOTUS 1-2-3 has a SUMPROD function (and other spreadsheets have equivalent functions) that can be used to multiply the elements of a column in the calculation matrix with the corresponding elements of the column of responses and sum the result. No check on the arithmetic is available in that case, but none is really needed. Note: When using the spreadsheet function, the levels of X must be denoted by +1 and −1, not by just + and −.

8.8 Significance of Effects and Interactions

8.8.1 Pooled Variance

In Section 6.5.2 the *pooled standard deviation*, s_p, was calculated as the square root of the weighted average of two variances (calculated from different samples). Remember that the pooling was justified because we assumed that the variation at both sets of conditions was the same. So both sample variances estimate the same common variance, σ^2. If we have more than two estimates of a common variance, σ^2, the general formula for pooling all the estimates into a single number is still a weighted average of the individual estimates, the weighting factor for each estimate being the degrees of freedom for that estimate. If we use $v_i = n_i - 1$ to denote the degrees of freedom for a particular variance estimate, s_i^2, the general formula is:

$$s_p^2 = \sum_{i=1}^{m} v_i \, s_i^2 \Big/ \sum_{i=1}^{m} v_i \tag{8.3}$$

and the degrees of freedom for s_p^2 (denoted by v_p, which is the degrees of freedom for any t-tests) is just the sum of the degrees of freedom of all of the estimates being pooled:

$$v_p = \sum_{i=1}^{m} v_i \tag{8.4}$$

In a replicated factorial experimental design, we have 2^k different sets of replicates — one set at each experimental condition. For example, in the voltmeter experiment we have $2^3 = 8$ pairs of replicates. From each set (pair) of replicates we can compute a variance as shown in Table 8.5. If we assume that the variability of replicates is the same at each experimental condition, we can come up with one *pooled variance* using Equation 8.3. The value of the pooled variance is:

$$s_p^2 = [\,(1)(312.5) + (1)(480.5) + \ldots + (1)(338.0)\,] / (1+1+ \ldots +1) = [2613]/8 = 326.6$$

and the pooled standard deviation is: $\quad s_p = \sqrt{326.6} = 18.07 \quad$ with $\quad v_p = 8$

As already stated above, this pooled value is calculated assuming that the variability is the same at all experimental conditions. This assumption should generally be checked. A simple plot of the variance or standard deviation versus the average at each experimental condition is useful for this purpose. This is often called the *plot of cell standard deviations versus cell means*, where *cell* refers to an experimental condition or run. For example, Figure 8.12 shows this plot for the voltmeter data. In this plot there is no apparent pattern or trend. Therefore, our assumption that each sample standard deviation is estimating the same σ may be reasonable.

If there had been a pattern in this plot, like that shown in Figure 4.17(c), it would indicate that the variability tends to increase as the average response increases. That would contradict our assumption of equal variability at all experimental conditions. In that situation, transformations of the data, that will be explained in Chapter 10, may be required before a valid determination of the significance of effects can be made. But for the voltmeter experiment, we feel somewhat confident that the value of s_p we calculated represents the common variance of the data at all conditions.

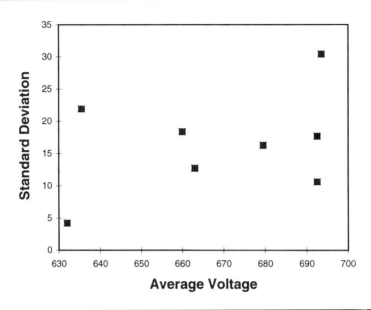

Figure 8.12 Plot of Cell Standard Deviation vs. Cell Mean for Voltmeter Experiment

8.8.2 Statistical Significance of Results

Experimentally measuring the effects the factors have on the response(s) and the interactions of the factors is only part of our job. We must keep in mind that our measurements are subject to experimental error. Even when a particular factor has no real impact on a response, it would be very unusual to measure an effect of exactly zero. Therefore, we have to be careful that we don't draw unwarranted conclusions from our data. As discussed earlier, if a measured effect or interaction can be reasonably explained by experimental variation, we will *assume* it is zero. Only if the effect or interaction is so large in comparison to experimental variation that it cannot reasonably be said to be zero, will we claim that the factor has a statistically significant effect on the response. Often, in this text we might leave out the term "statistically," and just say there is a significant effect of a factor, but that should not be confused with practical significance. We are always referring to statistical significance.

So, how do we determine the significance of our results? We actually already know how. Since each of the average effects (both main effects and interactions) are differences of two averages (i.e., $\bar{Y}_+ - \bar{Y}_-$) the statistical significance of these effects can be judged by computing a signal-to-noise t-ratio exactly as in Section 6.5.2. The numerator of the t-ratio is the effect or interaction (minus the hypothesized value which is presumed to be zero). Recall that the denominator of the signal-to-noise ratio is the standard deviation of the difference in the numerator:

218

$$s_E = \sqrt{s_p^2 \left(\frac{1}{n_+} + \frac{1}{n_-} \right)} \tag{8.5}$$

where s_E = the standard deviation of an effect or interaction (which are the same)

s_p = the pooled estimate of the standard deviation of individual responses

n_+ = the number of observations in the average response at the high level of the effect, $\bar{Y}_+$

and n_- = the number of observations in the average response at the low level of the effect, $\bar{Y}_-$

Since half of the factorial points are at the high level of any factor (or interaction) and the other half of the points are at the low level of that factor, $n_+ = n_- = n_F/2$, where n_F is the number of factorial points (including replication). Taking that relationship into account, the equation for s_E simplifies to:

$$s_E = \sqrt{s_p^2 \left(\frac{1}{n_F/2} + \frac{1}{n_F/2} \right)} = \sqrt{s_p^2 \left(\frac{4}{n_F} \right)} = 2 s_p / \sqrt{n_F} \tag{8.6}$$

The signal-to-noise t-ratios are then called t_E and are given by the formula:

$$t_E = \text{Effect} / s_E = \text{Effect} / (2 s_p / \sqrt{n_F}) \tag{8.7}$$

The statistical significance of each effect and interaction is judged by comparing its t-ratio, t_E, to the critical t-value, denoted by t^*, tabulated in Table A.5 in Appendix A. The t^* value is looked up for the degrees of freedom in s_p and the appropriate confidence level.

8.8.3   Statistical Significance of Results for Voltmeter Example

We will now use our value of the pooled standard deviation to estimate the standard deviation of the effects, s_E. Then we can compute the signal-to-noise t-ratios to assess the statistical significance of the effects and interactions (given in Table 8-6). Using Equation 8.4, and remembering that n_F is the number of factorial points, including replication ($n_F = 16$), we find:

$$s_E = 2 s_p / \sqrt{n_F} = 2(18.07) / \sqrt{16} = 9.036$$

With this value of s_E in hand, the signal-to-noise t-ratio, t_E, was calculated for each effect and interaction using Equation 8.5. These t-ratios are given immediately below the effects in Table 8.7. This table is a repeat of Table 8.6, produced using the EXCEL spreadsheet. The line labeled $\Sigma(+) - \Sigma(-)$ in Table 8.6 is reproduced in Table 8.7 using the SUMPRODUCT function in EXCEL, thus reducing the number of lines of intermediate sums.

Table 8.7 Computation of Significance of Effects and Interactions for 2^3 Factorial

Run	Mean	X_1	X_2	X_3	X_1X_2	X_1X_3	X_2X_3	$X_1X_2X_3$	Avg Y
1	1	-1	-1	-1	1	1	1	-1	692.5
2	1	1	-1	-1	-1	-1	1	1	635.5
3	1	-1	1	-1	-1	1	-1	1	692.5
4	1	1	1	-1	1	-1	-1	-1	632.0
5	1	-1	-1	1	1	-1	-1	1	663.0
6	1	1	-1	1	-1	1	-1	-1	679.5
7	1	-1	1	1	-1	-1	1	-1	693.5
8	1	1	1	1	1	1	1	1	660.0
Sumproduct	5348.5	-134.5	7.5	43.5	-53.5	100.5	14.5	-46.5	
Effect	668.6	-33.63	1.875	10.88	-13.38	25.13	3.625	-11.63	
t_E =		-3.721	0.208	1.204	-1.48	2.781	0.401	-1.287	$t*(8)=2.306$

The statistical significance of the effects is judged by comparing the t_E's to the tabulated critical t-value, t*, given in Table A.5 (in Appendix A). In this case the t* value is looked up with 8 degrees of freedom in s_p, our estimate of σ (see Table 8.5), which then becomes our degrees of freedom for s_E. At the 95% confidence level, t* = 2.306. It can be seen that both the X_1 (ambient temperature) main effect and the X_1X_3-interaction (ambient temperature × circuit warm-up time) are statistically significant. The main effect for circuit warm-up time, X_3, was not significant.

8.8.4 Interpretation of Results for Voltmeter Example

The results of the statistical analysis indicate that the voltmeter warm-up time (X_2) and all of the interactions involving X_2 (i.e., X_1X_2, X_2X_3 and $X_1X_2X_3$) are not significant. So we conclude that X_2 does not affect voltage reading. Remember – this conclusion is really an assumption, like innocent unless proven guilty (beyond a reasonable doubt). All we can really say definitely is that there is not enough evidence to conclude that voltmeter warm-up time does affect the meter reading. Therefore, students in the lab trying to repeat a voltage reading need not be concerned that they let the meter warm up for exactly the same time as they did during a previous reading.

The results did show, with greater than 95% confidence, that ambient temperature had an effect on the voltage readings (the readings made at 32°C being 34 millivolts lower than readings made at 22°C) and there was a significant interaction between X_1 and X_3. This interaction means that the (average) effect of ambient temperature (X_1) does not tell the whole story. More specifically, the interaction means that the temperature effect will be different depending on the circuit warm-up time (X_3). This can best be seen by examining a table of the average voltage readings in the four combinations of temperature and circuit warmup time shown in Table 8.8, or a graph of the results as shown in Figure 8.13.

Table 8.8 Average Voltage Readings (to Show the Interaction of X_1 and X_3)

			X_1 = Ambient Temperature, °C	
			- = 22°	+ = 32°
X_3	Time that power is connected	- = 0.5	692.50	633.75
	to circuit, minutes	+ = 5.0	678.25	669.75

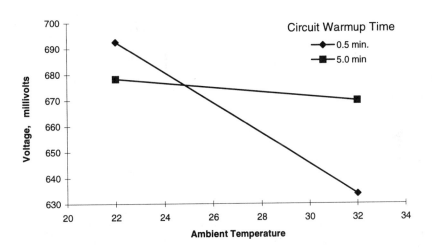

Figure 8.13 Interaction Plot for X_1 and X_3

It can be seen that the effect of ambient temperature is much smaller when the time that power is connected to the circuit (X_3) is 5 minutes (i.e., 669.8 - 678.3 = -8.5) versus when power is only connected to the circuit for 30 seconds (i.e.. 633.8 - 692.5 = -58.7). Therefore, to reduce variability in voltage readings students should be advised to always connect the power to a circuit for at least 5 minutes prior to the reading, and try to hold temperature constant if that is possible.

The pure replication variability (the variability of measurements made under the same conditions as defined by the three factors in the experiment) was estimated by s_p to be 18 millivolts. Therefore, single voltage readings that do not vary by more than $\pm 2\,s_p$ (or $\pm$ 36 millivolts) should be considered to be within the capability of the voltmeter.

In general, results of factorial experiments should be interpreted as in this example. Insignificant effects can be ignored. Significant main effects can be interpreted as the average difference in the response or dependent variable as the factor is changed from low to high level. And any factors having significant interactions must be interpreted by referring to a table of averages like Table 8.8 or a graph like Figure 8.13, because the effects alone are an oversimplification of how the system under study behaves. Ignoring the interactions and pretending that the (average) effect describes the system is like the old joke about the statistician with his feet in the freezer and his head in the oven: on the average he was comfortable!

8.8.5 An Example of Computer Analysis of Data

Many computer programs have been written which can automate the tedious analysis of data as shown in Table 8.7. As an example, Table 8.9 and Figures 8.14 and 8.15 show selected pages of output produced by the program MINITAB (Version 12).

Table 8.9 Sample Output from MINITAB for Voltage Readings

```
Estimated Effects and Coefficients for Y (coded units)
Term          Effect     Coef   StDev Coef       T       P
Constant              668.56         4.518  147.99  0.000
X1            -33.62  -16.81         4.518   -3.72  0.006
X2              1.87    0.94         4.518    0.21  0.841
X3             10.87    5.44         4.518    1.20  0.263
X1*X2         -13.38   -6.69         4.518   -1.48  0.177
X1*X3          25.13   12.56         4.518    2.78  0.024
X2*X3           3.62    1.81         4.518    0.40  0.699
X1*X2*X3      -11.63   -5.81         4.518   -1.29  0.234

Analysis of Variance for Y (coded units)

Source              DF     Seq SS    Adj SS    Adj MS       F       P
Main Effects         3     5009.7    5009.7    1669.9    5.11   0.029
2-Way Interactions   3     3293.2    3293.2    1097.7    3.36   0.076
3-Way Interactions   1      540.6     540.6     540.6    1.66   0.234
Residual Error       8     2612.5    2612.5     326.6
  Pure Error         8     2612.5    2612.5     326.6
Total               15    11455.9
```

Table 8.9 shows the calculated effects (along with some other information) that are the same as those previously calculated in Table 8.7. The next column labeled "Coef" gives the coefficients when using the variables in an equation. These coefficients (which will be discussed fully in Chapters 9 and 10) are simply half of the effects. The next column, "StDev Coef," gives the standard deviation of the coefficients, which is half the standard deviation of the effects, s_E. The column labeled "T" gives the t-values. The column labeled "P" represents the probabilities of obtaining an effect of greater magnitude than those observed (by chance if the real effect is zero). This column saves us the effort of looking up a t* value; any P's less than 0.05 are statistically significant at the 95% confidence level. Therefore, the X_1 main effect and the X_1X_3 interaction are significant (as we determined by hand). The "Constant" is the grand mean of all the data, and its standard deviation is $s_P / \sqrt{16}$ = 18.07) / 4 = 4.518, which agrees with the entry in the table.

Table 8.9 also shows an Analysis of Variance table, which will be explained in detail in Chapter 11. For now, suffice it to say that the table allows checking of the significance of *groups* of effects or interactions. An F-ratio is used to calculate significance rather than the t-value. The other important item of information given in that part of the table is the "Pure Error" adjusted mean square ("Adj MS") which is our estimate of the variance $s_p^2 = 326.6$. Thus $s_p = \sqrt{326.6} = 18.07$, all of which agrees with the calculations on page 217.

In addition to tabular output, computer programs often produce graphs that make it easier to interpret the data and results of the statistical analysis. For example, Figure 8.14 is a cube plot produced by MINITAB, which is equivalent to Figure 8.11. The top row in Figure 8.15 contains a graphs of the main effects of each of the three factors. It can be seen from the graphs exactly how much the voltage changed when each factor was changed from its low to high level, and that X_1 (temperature) was the most important factor. Figure 8.15 also contains a plot of the interactions. As expected, the X_1X_2 interaction is the largest. If this interaction is compared to the one shown in Figure 8.13, they can be seen to be the same.

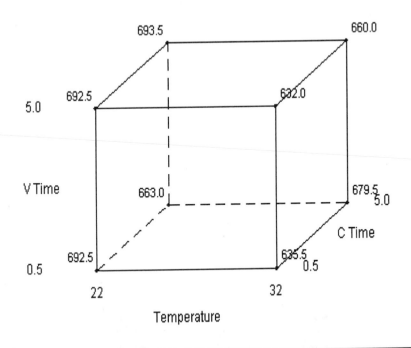

Figure 8.14 Cube Plot of the Voltage Data Produced by MINITAB (Ver 12)

Main Effects Plot (data means) for Voltage

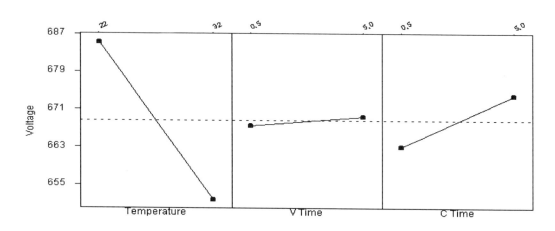

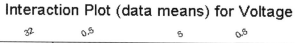

Interaction Plot (data means) for Voltage

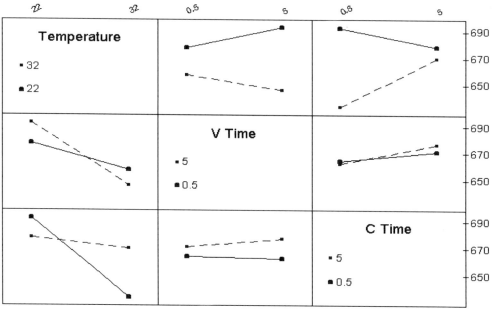

Figure 8.15 Plot of Main Effects and Interactions Produced by MINITAB (Version 12) for Voltage Readings Example

8.9 An Example of an Unreplicated 2^4 Design (Stack Gas Treatment)

In this section we present another example of a 2^k experiment. This example will again illustrate the interpretation of results from a factorial experiment and introduce additional techniques for analysis.

8.9.1   Background and Design

This example is taken (a bit loosely) from a study done at the University of North Dakota Energy and Environmental Research Center[1]. The overall research goal was to control SO_2 emissions during coal combustion in power plants via the injection of dry sodium carbonate into the flue gas upstream of the baghouse (used to control particulate emissions). The method worked well in the lab, but when implemented on a plant scale some NO_2 occasionally was formed resulting in a brown plume from the stack. To investigate the conditions under which NO_2 would be produced, a factorial experimental design was run in the lab.

The response measured was the amount of NO_2 formed (ppm) and the four factors studied, along with the levels of each, are given in Table 8.10. The factors were SO_2 concentration (X_1), temperature (X_2), oxygen concentration (X_3), and the amount of moisture (X_4). A full factorial design was planned. The design, in both coded and uncoded X's, is given in Table 8.11.

There was a small "wrinkle" in this study that should be pointed out. It was known that the variability in the measured response, NO_2 concentration, was Normally distributed but it did not have a constant variance. Rather, the standard deviation of the error in the measurements was a constant percentage, namely 25%. Since our analysis assumes that all the data have the same precision (constant variance), the response had to be transformed (more will be said about this in Chapter 10) to make that assumption valid. The log transformation accomplishes this goal; constant percentages become constant amounts on a log scale. Therefore, $Y = \log(NO_2)$ becomes the actual response to be analyzed. The standard deviation was known to be 25% of the measurement (as stated already), so on the logarithmic scale the standard deviation is the difference of the logs, i.e., $\sigma = \log(1.25) - \log(1.0) = 0.10$.

Once the design and response were settled, the next step was to run the experiments. Of course, they could not be run in the order listed in Table 8.11, or a large risk of biases invalidating the results would have been incurred. The tool that was absolutely necessary to use was randomization. In this study the factorial points were run in completely random order; a random number generator on a calculator was used to obtain the sequence. The 16 data points were then collected (i.e., each experiment was run and the NO_2 concentration was measured) in the appropriate order. The results are also given in Table 8.11.

Table 8.10 Stack Gas Treatment Example of 2^4 Factorial Design

	Levels	
Factors:		+
$X_1 = SO_2$ concentration, ppm	0	3000
$X_2 = $ Temperature, °F	150	350
$X_3 = O_2$ concentration, %	0	6
$X_4 = $ Moisture, %	0	20

Response: Concentration of NO_2 formed, ppb
$Y = \log(NO_2)$

Variability: Experimental variability (in NO_2 determination) has a
Standard error of 25% (well-known)

Y has a standard deviation of log (1.25), $\sigma_Y = 0.10$

Table 8.11 Stack Gas Treatment Example of 2^4 Factorial Design with Runs and Data

	Coded X Values				Actual Factor Levels				Run	NO_2	
Run	X_1	X_2	X_3	X_4	SO_2	Temp	O_2	Moist	Order	(ppb)	$Y=\log(NO_2)$
1	-	-	-	-	0	150	0	0	3	130	2.11
2	+	-	-	-	3000	150	0	0	13	150	2.18
3	-	+	-	-	0	350	0	0	16	210	2.32
4	+	+	-	-	3000	350	0	0	4	200	2.30
5	-	-	+	-	0	150	6	0	15	110	2.04
6	+	-	+	-	3000	150	6	0	1	18,000	4.26
7	-	+	+	-	0	350	6	0	8	110	2.04
8	+	+	+	-	3000	350	6	0	5	150,000	5.18
9	-	-	-	+	0	150	0	20	7	130	2.11
10	+	-	-	+	3000	150	0	20	2	150	2.18
11	-	+	-	+	0	350	0	20	10	250	2.40
12	+	+	-	+	3000	350	0	20	6	140	2.15
13	-	-	+	+	0	150	6	20	14	140	2.15
14	+	-	+	+	3000	150	6	20	9	22,000	4.34
15	-	+	+	+	0	350	6	20	12	150	2.18
16	+	+	+	+	3000	350	6	20	11	170,000	5.23

8.9.2 Graphical Analysis

Before mechanically cranking out a numerical analysis of a design, it is a good idea to get as much insight into the situation as possible. The first step in doing this is to make simple plots of the data. Figure 8.16 is a line graph of the data in the order the experiments were conducted. In this figure we see no patterns or trends. If there was an increasing or decreasing trend, or a consistent cycling pattern in the plot, it would lead us to suspect that changes in some lurking variable might have added variability to our experimental results. In that case, we might make additional plots of the data versus any other recorded variables that were known to have changed during the course of the experiment. If we could find a graphical relationship between the response and an uncontrolled variable, we might be able to adjust for this variable and reduce the standard error of the effects using regression analysis, that will be explained in Chapter 10. Recall that we can never adjust for biases in effects caused by changes in lurking variables if we failed to randomize our experiments. Regression analysis will only allow us to correct the effect of lurking variables on the standard error of an effect.

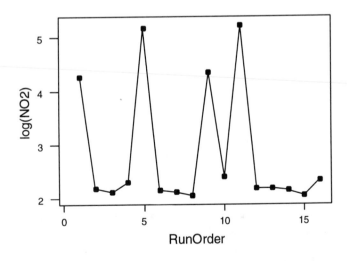

Figure 8.16 Plot of Data vs. Random Run Order from Volt Meter Experiment

Simple Boxplots of the data separated by levels of each factor can give us a general feeling about the importance of the factor effects. Figure 8.17 shows Boxplots separated by levels of each of the factors. In this figure, it can be seen that SO_2 and O_2 seem to have the largest effects on $Y=\log(NO_2)$, while moisture seems to have a negligible effect. The wider spread in the response at the high levels of SO_2 and O_2 may indicate that these factors interact with others. If a Boxplot program is not available, a simple scatter plot of the response on the y-axis and factor levels on the x-axis will give essentially the same information as Figure 8.17.

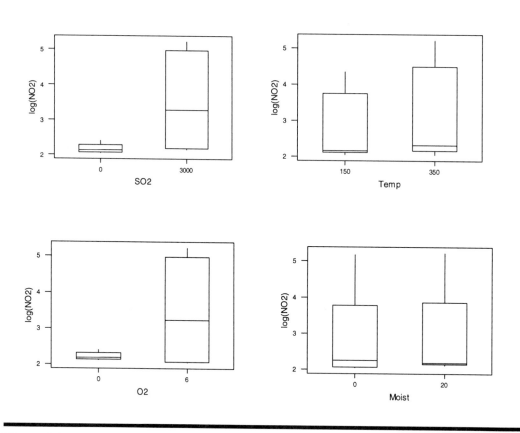

Figure 8.17   Boxplots of Data vs. Factor Levels from Gas Treatment Experiment

A good way to begin visualizing potential interactions is to plot the data on the cube which represents the experimental region. For our gas treatment experiment, this is shown in Figure 8.18 (a) and (b).

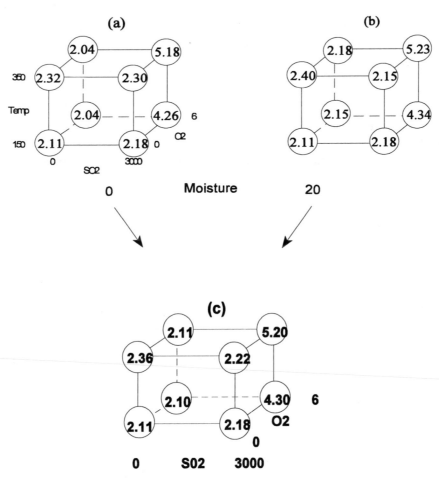

Figure 8.18 Cube Plots of Data from Gas Treatment Example

From a comparison of the cubes for low moisture and for high moisture, it can be judged that moisture had very little impact, if any. Therefore, the two moisture levels were averaged and the averages shown in Figure 8.18 (c). From this figure, it seems like all three remaining variables had positive effects. It also seems that a $X_1X_2X_3$-interaction exists, since the interaction of X_1X_2 (i.e., temperature and SO_2) is negligible at low $X_3=O_2$ (the front of the cube), but appears quite large at high $X_3=O_2$ (at the back of the cube). By the interaction of temperature and SO_2 we mean the effect of temperature is different at high SO_2 than at low SO_2. The negligible interaction, or difference in temperature effects, can be seen at the front of the cube as the $Log(NO_2)$ increases from 2.11 to 2.36 as temperature increases (with low SO_2) and again increases from 2.18 to 2.225 as temperature increases with high SO_2. On the other hand at the back of the cube (i.e., high O_2), the $Log(NO_2)$ increases from 2.095 to 2.110 as temperature increases (with low SO_2), but increases from 4.30 to 5.205 (a much larger increase) as temperature increases with high SO_2.

8.9.3 Numerical Analysis

In order to determine the exact average effects and interaction, and even more importantly, whether or not they are statistically significant, a numerical analysis is necessary.

The calculation of effects and interactions is accomplished via the use of the computation table for a 2^4 factorial design listed in Appendix B.2-4. The table was copied into a spreadsheet and the results are shown in Table 8.12. The response, $Y = \log(NO_2)$, was appended to the right for handy reference. Then, to calculate each effect, the Y's are simply added up (with the appropriate sign) and averaged by dividing by the number of observations at the high level (or low level) of each factor (or interaction). The calculation of effects and interactions was shown in the example in Section 8.7. As a review, let us calculate one of the interactions.

$$X_1X_2 = [Y_1 - Y_2 - Y_3 + Y_4 + Y_5 - Y_6 - Y_7 + Y_8 + Y_9 - Y_{10} - Y_{11} + Y_{12} + Y_{13} - Y_{14} - Y_{15} + Y_{16}] / 8$$
or $\quad X_1X_2 = [\ 2.11 - 2.18 - 2.32 + 2.30 + ... -2.18 + 5.23\ ] / 8 = 0.171$

The simultaneous calculation of all the effects was accomplished using the SUMPRODUCT function in the spreadsheet. The results of all the computations are shown in Table 8.12.

Run	Mean	X_1	X_2	X_3	X_4	X_1X_2	X_1X_3	X_1X_4	X_2X_3	X_2X_4	X_3X_4	$X_1X_2X_3$	$X_1X_3X_4$	$X_2X_3X_4$	$X_1X_2X_3X_4$	Y
1	1	-1	-1	-1	-1	1	1	1	1	1	1	-1	-1	-1	1	2.11
2	1	1	-1	-1	-1	-1	-1	-1	1	1	1	1	1	-1	-1	2.18
3	1	-1	1	-1	-1	-1	1	1	-1	-1	1	1	-1	1	-1	2.32
4	1	1	1	-1	-1	1	-1	-1	-1	-1	1	-1	1	1	1	2.3
5	1	-1	-1	1	-1	1	-1	1	-1	1	-1	1	1	1	-1	2.04
6	1	1	-1	1	-1	-1	1	-1	-1	1	-1	-1	-1	1	1	4.26
7	1	-1	1	1	-1	-1	-1	1	1	-1	-1	-1	1	-1	1	2.04
8	1	1	1	1	-1	1	1	-1	1	-1	-1	1	-1	-1	-1	5.18
9	1	-1	-1	-1	1	1	1	-1	1	-1	-1	-1	1	1	-1	2.11
10	1	1	-1	-1	1	-1	-1	1	1	-1	-1	1	-1	1	1	2.18
11	1	-1	1	-1	1	-1	1	-1	-1	1	-1	1	1	-1	1	2.4
12	1	1	1	-1	1	1	-1	1	-1	1	-1	-1	-1	-1	-1	2.15
13	1	-1	-1	1	1	1	-1	-1	-1	-1	1	1	-1	-1	1	2.15
14	1	1	-1	1	1	-1	1	1	-1	-1	1	-1	1	-1	-1	4.34
15	1	-1	1	1	1	-1	-1	-1	1	1	1	-1	-1	1	-1	2.18
16	1	1	1	1	1	1	1	1	1	1	1	1	1	1	1	5.23
SUMPROD	45.17	10.47	2.43	9.67	0.31	1.37	10.73	-0.35	1.25	-0.07	0.45	2.19	0.11	0.07	0.17	
Effects		1.31	0.30	1.21	0.04	0.17	1.34	-0.04	0.16	-0.01	0.06	0.27	0.01	0.01	0.02	
t(E)		26.18	6.08	24.18	0.78	3.43	26.83	-0.88	3.13	-0.18	1.13	5.48	0.28	0.18	0.43	

Table 8.12 Calculation of Effects and Interactions for Stack Gas Treatment Example

The next question that should come to mind is: Are any of these calculated effects real, or could they just be due to the random variability of the data? To help us with this decision we divide each of the effects by their standard error as described in Section 8.8. In other words, we calculate the signal-to-noise ratios, denoted by t_E (also given in Table 8.12).

There is a difference between the calculation of the standard error of effects for the gas treatment experiment and the voltmeter experiment described in Section 8.8.2. Recall that in the voltmeter experiment, a measure of variability called the pooled standard deviation, s_p, was calculated from the replicates in the experiment, and then the standard error of the effects or interactions was computed using the formula $s_E = 2\, s_p / \sqrt{n_F}$. In the gas treatment example, there were no replicates and thus no estimate of the variability could be made from the data. However, there was a known constant percentage error that allowed us to determine that the standard deviation of $Y=\log(NO_2)$ was $\sigma=\log(1.25)=0.10$. Therefore, the standard error of an effect or interaction was computed below as:

$$\sigma_E = 2\, \sigma / \sqrt{n_F} = 2\,(0.10) / \sqrt{16} = 0.050$$

and the signal-to-noise ratios in Table 8.12 were computed as:

$$t_E = \text{Effect} / \sigma_E$$

These were compared to the t* values in Table A.5. The degrees of freedom associated with the t* statistic is always the degrees of freedom associated with the estimate of variability. Recall that in the voltmeter experiment we used 8 degrees of freedom, which were the degrees of freedom for s_p. In the gas treatment experiment, the measure of variability $\sigma = \log(1.25) = 0.10$ was known, so the degrees of freedom are essentially infinite. The last row of Table A.5, with degrees of freedom = ∞, is equivalent to the Standard Normal Table A.3, and our signal-to-noise ratios, t_E, are actually the same as the z-statistics used in the reliability test in Section 6.4.2. The critical value with 95% confidence is t* = 1.96, and it can be seen in Table 8.12 that the significant effects were, $X_1, X_2, X_3,$ $X_1X_2, X_1X_3, X_2X_3,$ and $X_1X_2X_3$. The moisture content of the stack gas (X_4) is not important, nor does it have any significant interactions with other factors.

8.9.4 Interpretation of Results

The results of the experiment show that there is not enough evidence to indicate that the moisture content of the stack gas had any effect on NO_2 production. This result was a big surprise, and it meant that wet scrubbing could be used if desired. All three of the other variables ($X_1 = SO_2$ concentration, X_2 = temperature, and $X_3 = O_2$ concentration) do effect the amount of NO_2 formed. However, the highest order interaction that is significant, namely $X_1X_2X_3$, must take precedence in interpreting the effects. The three-way interaction indicated there is a three-way dependence in the way $X_1, X_2,$ and X_3 affect NO_2 formation, and neither the average main effects nor even the two-way interactions should be interpreted alone. In most circumstances it will be extremely rare to discover significant interactions of order higher than two. But in unusual cases like this one where they are significant, the simple cube plots like Figure 8.17(c) can help to make the interpretation straightforward. This figure shows that NO_2 levels (or to be exact, log (NO_2)) increase only when both SO_2 and O_2 concentrations increase, and NO_2 increases most when all three variables (SO_2 and O_2 concentrations and temperature) simultaneously increase. In other words, log (NO_2) is essentially constant except when there is high level of SO_2 and a high level of O_2, and in that case increasing temperature also increases NO_2 production. This would most likely be the condition causing a brown plume of NO_2 to be emitted from the stack. Notice that this result would never have been discovered using the classical one-at-a-time approach to experimentation, wherein each variable would have been separately varied between low and high conditions while the other variables were held constant at their low level.

In presenting the results, the fact that interactions are significant make life more difficult. It means that more is going on than can adequately be described by the average effect of each variable. Therefore, our final results should be summarized and displayed graphically as was done in Figure 8.13 for the voltmeter experiment and Figure 8.18(c) for the gas treatment experiment, or through use of a prediction equation that will be described in Chapter 9.

8.10 Judging Significance of Effects When There Are No Replicates

In the gas treatment experiment presented in Section 8.9, there were no replicate experiments, but there was a known estimate of σ that was used as a yardstick to judge the significance of effects. This was a special case, however. Usually there is no known estimate of σ, and in experiments without replicate experiments the signal-to-noise ratios cannot be computed to judge the significance of effects. We might seem to be at an impasse, but there are still two options available to us if enough runs were made (≥ 16).

The first option is to examine the interactions of third order or higher. These are extremely unlikely to be real, and hence they can be used to estimate s_E directly. If we have q of these interactions, then the formula for estimating s_E is:

$$s_E = \sqrt{\sum_{i=1}^{q} I_i^2 / q} \qquad (8.8)$$

Once we have s_E, we can compute the signal-to-noise ratios (i.e., t-statistics), and then judge the significance of the effects and two-factor interactions.

Another alternative that gets away from computing signal-to-noise ratios entirely is to use graphical methods to judge the significance of effects. The approach is as follows. If none of the effects are actually significant, all of the observations would be random replicates. The factorial main effects and interactions would be differences of averages of random data, and by the Central Limit Theorem discussed in Section 3.6, they would be approximately Normally distributed with a mean of zero. Therefore, a Normal probability plot of the effects would appear as a straight line of points, like that descibed for Normal data in Section 5.3. If any effects are significant, they would stick out from the line of points as an outlier in a sample of Normal data. This method works quite well as long as the majority of effects are unimportant and fall along a straight line.

To illustrate the idea, we will use the effects from the gas treatment experiment and assume no known estimate of σ was available. Table 8.13 shows the ranked effects (from Table 8.12) along with their calculated z-scores (similar to Table 5.3) and Figure 8.19 is a Normal plot of the effects. From the graph, the effects that really stick out like outliers (namely X_1, X_2, X_3, X_1X_2, X_1X_3, X_2X_3, $X_1X_2X_3$.) Are the same effects that were found to be significant using the signal-to-noise t-ratios in Table 8.12. Drawing a straight line through the first eight pairs ($E_{(i)}$, z_i) in Figure 8.19 helps in the determination.

Table 8.13 Worksheet for Calculation of Z-Scores of Effects Gas Treatment Example

Order Number	Ordered Effect $E_{(i)}$	Effect Label	Percentile $p_i = (i-.5)/15$	Z-score $\Phi^{-1}((i-.5)/15)$
1	-0.044	X_1X_4	.033	-1.84
2	-0.036	$X_1X_2X_4$	.100	-1.28
3	-0.009	X_2X_4	.167	-0.97
4	0.009	$X_2X_3X_4$	.233	-0.73
5	0.014	$X_1X_3X_4$	.300	-0.53
6	0.021	$X_1X_2X_3X_4$	.367	-0.34
7	0.039	X_4	.433	-0.17
8	0.056	X_3X_4	.500	0.00
9	0.156	X_2X_3	.566	0.17
10	0.171	X_1X_2	.633	0.34
11	0.274	$X_1X_2X_3$	.700	0.53
12	0.304	X_2	.766	0.73
13	1.209	X_3	.833	0.97
14	1.309	X_1	.900	1.28
15	1.341	X_1X_3	.967	1.84

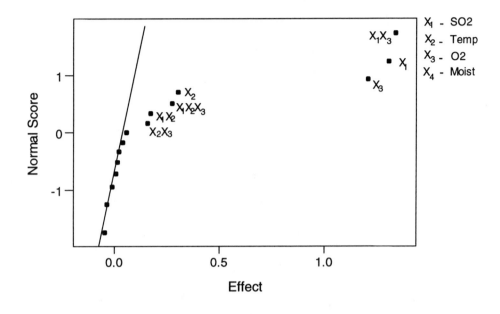

Figure 8.19 Normal Probability Plot of Effects from Stack Gas Treatment Example

The reciprocal of the slope, the straight line drawn on the Normal plot, is a direct estimate of the standard error of an effect (see Section 5.4) $s_E = 1/17.27 = .058$. This is actually very close to the $\sigma_E = .05$ that was determined from the known standard deviation. Thus when no replicates are present in an experiment, we may still be able to get a graphical estimate of σ_E. (Note: Since we have already determined which effects and interactions are significant from the graph, we are obtaining s_E for the sake of curiosity only — we are not going to use it for anything. We could however, if we were interested, use it to estimate σ from the relationship: $\sigma = (\frac{1}{2})\sigma_E \sqrt{n_F}$)

An even simpler graphical approach to determining the significant effects is to make a Pareto Diagram of the absolute value of the effects, as shown below in Figure 8.20. In this plot we can separate the vital few important effects from the trivial many unimportant effects. Again we can see that the most important effects are X_1, X_2, X_3, X_1X_2, X_1X_3, X_2X_3, $X_1X_2X_3$.

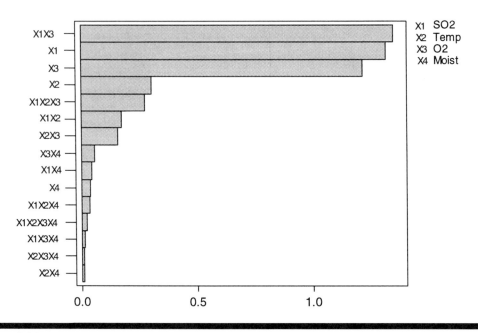

Figure 8.20   Pareto Chart of Absolute Effects from Gas Treatment Experiment

8.11 Summary of Analysis of 2^k Experiments

In this chapter we described the structure of 2^k experiments and showed two examples of analysis and interpretation of results. In this section we present a simple flow diagram to help in remembering the various methods of analysis (both graphical and numerical), as well as when and where to use each. In Figure 8.21 we see the three steps that are usually followed in the analysis of a 2^k experiment. First is the exploratory stage where we examine the data graphically to see roughly what effects and interactions may be important. In the exploratory step we also check the data to see whether the assumption of equal variance at all experimental conditions appears reasonable. The second step in the analysis is the data summary step. Here we calculate the effects and interactions and variability summaries. We test the significance of the effects using either the signal-to-noise ratios if there are replications in the data, or the Normal plot of effects (or Pareto Diagram of absolute effects) if there are no replicates. The final step of the analysis is the interpretation and presentation of results. In this step, verbal descriptions of the effects and interactions are made. These verbal descriptions are usually improved if simple graphical displays like main effect plots and interaction plots are made and described.

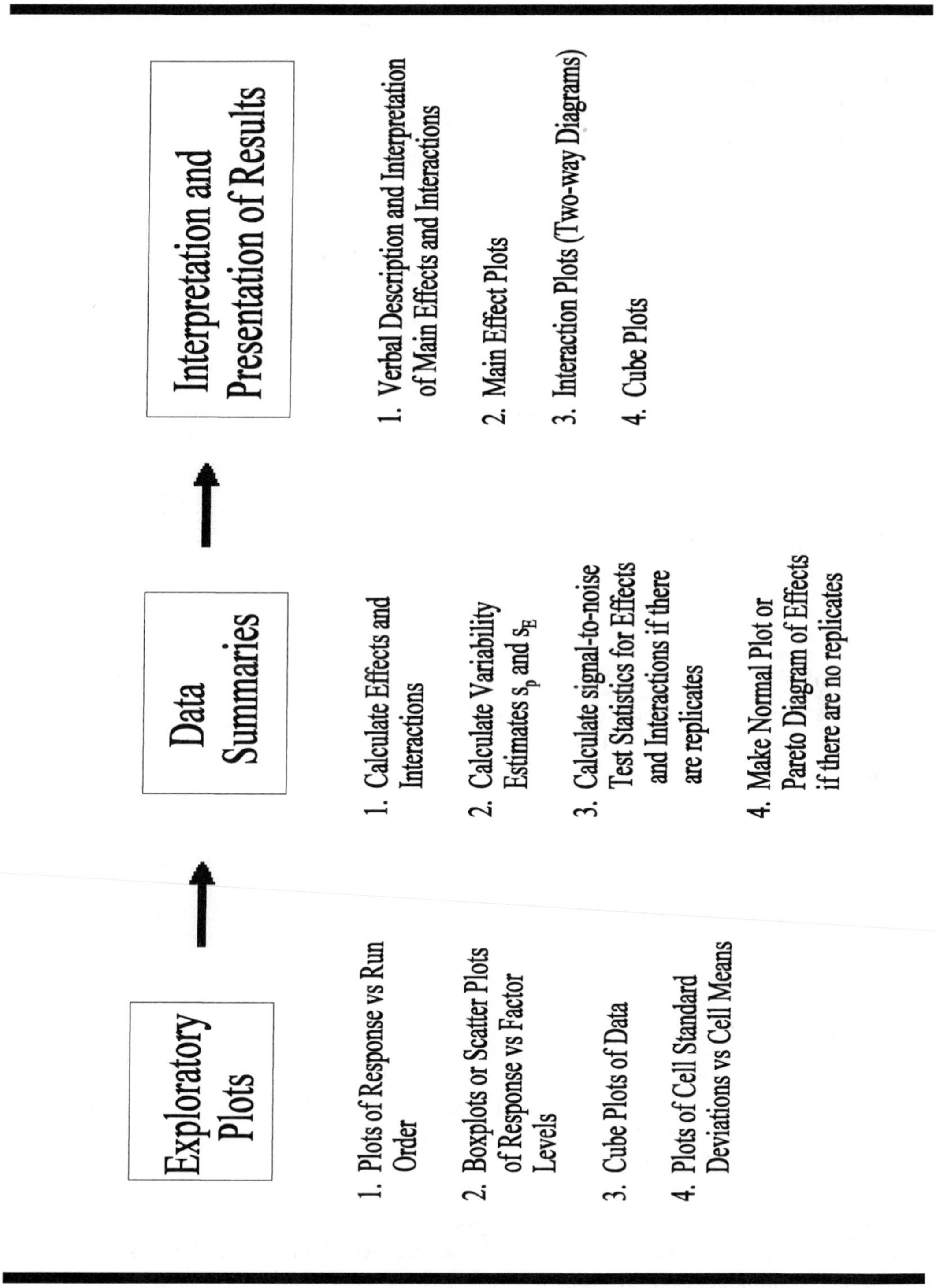

Figure 8.2 Flow Diagram for the Analysis of 2^k Experiments

Exploratory Plots

1. Plots of Response vs Run Order

2. Boxplots or Scatter Plots of Response vs Factor Levels

3. Cube Plots of Data

4. Plots of Cell Standard Deviations vs Cell Means

Data Summaries

1. Calculate Effects and Interactions

2. Calculate Variability Estimates s_p and s_E

3. Calculate signal-to-noise Test Statistics for Effects and Interactions if there are replicates

4. Make Normal Plot or Pareto Diagram of Effects if there are no replicates

Interpretation and Presentation of Results

1. Verbal Description and Interpretation of Main Effects and Interactions

2. Main Effect Plots

3. Interaction Plots (Two-way Diagrams)

4. Cube Plots

8.12 Summary

8.12.1 Important Equations

Effect (or Interaction) $= \overline{Y}_+ - \overline{Y}_-$

$s_E = 2\, s_P / \sqrt{n_F}$

$t_E = \text{Effect} / s_E$

$$s_p^2 = \sum_{i=1}^{m} v_i\, s_i^2 \Big/ \sum_{i=1}^{m} v_i$$

8.12.2 Important Terms and Concepts

The following is a list of the important terms and concepts covered in this chapter.

References

[1.]   G. F. Weber, et. al., "Simultaneous SO_x / NO_x, Control," Final Technical Report for Period 4/1/86 – 3/31/87, prepared for U.S. Dept. of Energy, Office of Fossil Energy, Pittsburgh (Agreement No. DE - FC21 - 86MC10637).

[2.]   Lawson, J. "Computation Strategies in Applied Statistics courses for Engineers," *American Statistical Association Proceedings of the Section on Statistical Education*, American Statistical Association, Washington, D.C. (1990).

8.13 Appendix 8.A Yates Algorithm

When the average response data from a factorial design are written in the standard (Yates) order shown in Table 8.2 (for more examples see Tables 8.5 and 8.11), there is a simple algorithm that can be used and/or programmed to calculate the effects and interactions. This algorithm is called *Yates Algorithm* and works as follows: The response observations, in the standard order, are grouped into successive pairs (for example, see Table 8.A.1 which uses the data from Table 8.7).The pairs are summed to form the first half of the entries in Column 1.

For example, $692.5 + 635.5 = 1328.0$, $692.5 + 632.0 = 1324.5$, etc.

Next the same pairs are differenced (second observation minus first observation) to form the entries in the second half of Column 1.

For example, $635.5 - 692.5 = - 57.0$, $632.0 - 692.5 = - 60.5$, etc.

This same procedure is repeated using the entries in Column 1 (rather than the response data) to form the entries in Column 2, and is repeated again to form more columns until the first entry in the last column is the sum of all the observations (k columns for a 2^k design). The rest of the entries in the last column correspond to the difference of sums (i.e. $\Sigma(+) - \Sigma(-)$) which were used in Table 8.6 to calculate the effects and interactions. The order that the differences appear in the last column in Table 8.A.1 is called the Yates order of effects. The next step would be to divide Column 3 by 4 (i.e., 2^{k-1} in general) to get the effects.

Table 8.A.1 Yates Algorithm Applied to Voltage Measurement Example

Standard	Design			Paired Observations	Yates Algorithm Column			
Order	X_1	X_2	X_3	Y	1	2	3	Label
1	-	-	-	692.5	1328.0	2652.5	5348.5	Mean
2	+	-	-	635.5	1324.5	2696.0	- 134.5	X_1
3	-	+	-	692.5	1342.5	- 117.5	7.5	X_2
4	+	+	-	632.0	1353.5	- 17.0	- 53.5	$X_1 X_2$
5	-	-	+	663.0	- 57.0	- 3.5	43.5	X_3
6	+	-	+	679.5	- 60.5	11.0	100.5	$X_1 X_3$
7	-	+	+	693.5	16.5	- 3.5	14.5	$X_2 X_3$
8	+	+	+	660.0	- 33.5	- 50.0	- 46.5	$X_1 X_2 X_3$

Yates algorithm can easily be utilized in a spreadsheet program like EXCEL or Quatro Pro[2]. This saves creating large worksheets like Table 8.12 to calculate effects. Table 8.A.2 shows an example of the cell formulas needed to reproduce Table 8.A.1. Column A is the response data in standard order. The formulas in Column B create the sums of pairs and differences of pairs in Column 1 of Table 8.A.1. When Column B in the spreadsheet is highlighted and copied and pasted to Column C, the formulas are automatically updated to the sums and differences that form Column 2 of Table 8.A.1. Finally the formulas in Column C of the spreadsheet are copied and pasted into Column D to get the results shown in Column 3 of Table 8.A.1.

Table 8.A.2 Spreadsheet Cell Formulas to Perform Yates Algorithm

	A	B	C	D
1	692.5	+A2+A1	+B2+B1	+C2+C1
2	635.5	+A4+A3	+B4+B3	+C4+C3
3	692.5	+A6+A5	+B6+B5	+C6+C5
4	632.0	+A8+A7	+B8+B7	+C8+C7
5	663.0	+A2-A1	+B2-B1	+C2-C1
6	679.5	+A4-A3	+B4-B3	+C4-C3
7	693.5	+A6-A5	+B6-B5	+C6-C5
8	660.0	+A8-A7	+B8-B7	+C8-C7

8.14 Exercises for Chapter 8

8.1   An experiment is to be performed to determine the effects of Fill Pressure (20psi or 60psi), Barrel Temperature (180° or 200°) and Mold Temperature (180° or 200°) upon the percent shrinkage of injection molded parts.

   a) Set up the list of experiments for a 2^3 factorial in the standard order

   b) Choose a list of random run orders for the experiments.

8.2   Connors, Anderson, Perkins and Cedeno performed an experiment to determine the effects of Spindle Speed (- = 960RPM, + =1800RPM) , Feed Rate (- = 7.2IPM, + = 16IPM) and Depth of Cut (- = 0.005 in., + =0.030in.) upon the surface finish of milled aluminum. The data is listed below; smaller values indicate a smoother finish.

X_1 =SS	X_2 =FR	$X_3 = DC$	Surface Finish
-	-	-	35 42 28
+	-	-	76 124 124
-	+	-	41 18 38
+	+	-	60 65 68
-	-	+	33 34 34
+	-	+	137 144 167
-	+	+	28 28 26
+	+	+	64 52 66

a) Calculate the mean and variance, s^2 , of the replicate responses for each run, and calculate the pooled standard deviation, s_p.

b) Make a plot of cell standard deviations versus cell means to see if the equal variance assumption is justified.

c) Calculate the main effects, two-way interaction effects, and the three-way interaction effect.

d) Make boxplots of the response at each level of the three factors similar to Figure 8.17, and comment on what you see.

e) Calculate the standard error of an effect, s_E , and determine which effects and interactions are significant at the 95% confidence level.

f) Graph any significant interactions and write a sentence or two interpreting each significant effect or interaction.

8.3   The following data (with coded values for factors) were taken from a study dealing with a solar water heating system.[1] The factors were X_1, the total daily insolation, X_2, the tank storage capacity, X_3, the water mass flow rate through the absorber, and X_4, the intermittency of the input from solar radiation. The responses were Y_1, the collection efficiency, and Y_2, the energy delivery efficiency. A computer model was developed from theory, but was too complicated for ready appreciation. Therefore, the model was used to predict the responses at the sixteen conditions defined below according to a 2^4 design. This would allow for a simpler linear approximation model to be determined.

Run	X_1	X_2	X_3	X_4	Y_1	Y_2
1	-	-	-	-	43.5	82.0
2	+	-	-	-	41.3	82.0
3	-	+	-	-	44.9	82.1
4	+	+	-	-	43.0	82.2
5	-	-	+	-	35.0	61.7
6	+	-	+	-	37.5	66.0
7	-	+	+	-	39.7	67.7
8	+	+	+	-	39.9	68.6
9	-	-	-	+	51.3	83.7
10	+	-	-	+	50.2	86.3
11	-	+	-	+	52.4	84.1
12	+	+	-	+	51.9	89.8
13	-	-	+	+	38.4	100.0
14	+	-	+	+	39.2	100.0
15	-	+	+	+	41.3	100.0
16	+	+	+	+	41.6	100.0

a) Calculate the effects for each response and construct a Normal or half-Nnormal plot of the effects to determine which effects and interactions appear to be significant.

b) Make a Pareto Diagram of the absolute effects to determine which effects and interactions appear to be significant

c) For each response, separately construct boxplots at each level of the four factors similar to Figure 8.17 and comment on what you see.

d) Graph any significant interactions and write a sentence interpreting each significant main effect and/or interaction.

---

[1]Close, D. J. " A Design Approach for Solar Processes," Solar Energy, Vol. 11, p 112 (1967).

8.4   Lorenc, Jones, McKee, and Runyan performed a 2^3 experiment to determine the effects of machine settings on Boy 15s injection molding machine upon the weight of injection molded parts. The machine settings or factors they studied were Holding Time (- = 10 seconds, + = 20 seconds), Holding Pressure (- = 4 psi, + = 28 psi) and Injection Speed (- = 4, + = 80). The data is listed below.

X_1 = HT	X_2 = HP	X_3 = IS	Part Weight
-	-	-	13.62 13.60 13.59 13.64
+	-	-	13.61 13.67 13.60 13.61
-	+	-	13.63 13.72 13.74 13.70
+	+	-	13.71 13.72 13.70 13.61
-	-	+	13.81 13.80 13.82 13.81
+	-	+	13.78 13.79 13.79 13.76
-	+	+	14.24 14.32 14.32 14.26
+	+	+	14.24 14.11 14.24 14.56

a) Calculate the mean and variance, s^2, of the replicate responses for each run, and calculate the pooled standard deviation s_p.

b) Make a plot of cell standard deviations versus cell means to see if the equal variance assumption is justified.

c) Calculate the main effects, two-way interaction effects and the three-way interaction effects.

d) Make boxplots of the response at each level of the three factors similar to Figure 8.17, and comment on what you see.

e) Calculate the standard error of an effect, s_E, and determine which effects and interactions are significant at the 95% confidence level.

f) Graph any significant interactions and write a sentence or two interpreting each significant effect or interaction.

8.5   The cutting of small, square, metal plates (used to make square nuts) is done using a shearing process. The variables in the process include shearing pressure, shearing blade width, and the temperature of the metal to be cut. The response is the crispness of the cut, rated on a scale of 1 to 10. The measured (reported) crispness is an average crispness of 10 sheared plates. A process engineer decided to study the system to find the settings of the variables that would maximize crispness. The data are listed below.

Run Order	X_1 = Pressure - = 200 + = 400	X_2 = Width - = 0.2" + = 1.0"	X_3 = Temp. - = 300 °F + = 500 °F	Crispness
7, 4	-	-	-	3.1 4.2
1, 14	+	-	-	8.4 8.8
5, 12	-	+	-	3.8 4.6
10, 15	+	+	-	8.9 9.1
2, 11	-	-	+	3.2 4.3
3, 16	+	-	+	6.8 7.3
6, 13	-	+	+	8.0 8.4
8, 9	+	+	+	9.4 9.9

a) Plot the response versus run and comment on what you see.

b) Calculate the mean and variance, s^2 , of the replicate responses for each run, and calculate the pooled standard deviation s_p.

c) Make a plot of cell standard deviations versus cell means to see if the equal variance assumption is justified.

d) Calculate the main effects, two-way interaction effects and the three-way interaction effects.

e) Make boxplots of the response at each level of the three factors similar to Figure 8.17, and comment on what you see.

f) Calculate the standard error of an effect, s_E , and determine which effects and interactions are significant at the 95% confidence level.

g) Graph any significant interactions and write a sentence or two interpreting each significant effect or interaction.

h) What conditions would maximize crispness?

8.6   Huber, Hendricks, Nichol, and Peterson performed a 2^3 experiment to determine the effects of the CO concentration (- = 0.1, + = 0.2), H_2 concentration(- = 0.2 , + = 0.4) and volumetric flowrate (- = 100 cc/min., + = 200 cc/min.) upon the output of a gas chromatograph used to measure the concentration of H_2 . The data is listed below.

X_1 =CO	X_2 =H_2	X_3 = Flowrate	Peak Area × 10^{-5}
-	-	-	3.082 2.880 4.944
+	-	-	3.954 3.081 2.837
-	+	-	6.612 5.150 5.006
+	+	-	5.013 5.406 4.965
-	-	+	3.249 3.153 3.005
+	-	+	2.971 2.975 3.049
-	+	+	5.595 5.343 5.436
+	+	+	5.388 5.368 5.494

a) Calculate the mean and variance, s^2 , of the replicate responses for each run, and calculate the pooled standard deviation s_p.

b) Make a plot of cell standard deviations versus cell means to see if the equal variance assumption is justified.

c) Calculate the main effects, two-way interaction effects, and the three-way interaction effects.

d) Make boxplots of the response at each level of the three factors similar to Figure 8.17, and comment on what you see.

e) Calculate the standard error of an effect, s_E , and determine which effects and interactions are significant at the 95% confidence level.

f) Graph any significant interactions and write a sentence or two interpreting each significant effect or interaction.

8.7   A quality engineer in a chemical production facility was assigned the task of maximizing the yield of a chemical reaction taking place in a continuous stirred tank reactor (CSTR). The four variables she decided to examine first were temperature, concentration of catalyst, agitator speed, and reactor pressure. A full factorial design was used, and the data are listed below.

Run	X_1 = Temp. - = 80 °C + = 100 °C	X_2 = Cat. - = 4 g/l + = 5 g/l	X_3 = Agit. - = 400 rpm + = 500 rpm	X_4 = Press. - = 700 psi + = 1000 psi	Yield
1	-	-	-	-	79
2	+	-	-	-	81
3	-	+	-	-	82
4	+	+	-	-	84
5	-	-	+	-	85
6	+	-	+	-	84
7	-	+	+	-	85
8	+	+	+	-	86
9	-	-	-	+	87
10	+	-	-	+	88
11	-	+	-	+	89
12	+	+	-	+	91
13	-	-	+	+	93
14	+	-	+	+	94
15	-	+	+	+	95
16	+	+	+	+	97

a) Calculate the main effects and all interactions regardless of order.

b) Estimate the standard error of an effect, s_E , from the high order interactions. Then determine which effects and interactions are significant at the 95% confidence level.

c) Make a normal plot of the effects, and a Pareto Diagram of the absolute effects and judge graphically which effects and interactions are significant. How does your result here compare to what you found in b) ?

d) Graph any significant interactions and write a sentence or two interpreting each significant effect or interaction.

e) What conditions would maximize yield?

8.8   Wagner, Stockett, Swensen, and Wright studied the factors that affect the force to dissociate the femoral head from the bipolar cup in a hip replacement socket. In order to perform hip replacement, doctors need to cut through a membrane surrounding the hip joint. This membrane holds the hip tightly in place, and does not recover after surgery. Therefore a hip replacement is more susceptible to dislocation than a healthy hip joint. When a hip is dislocated, the doctor tries to move the hip back into its correct position without surgery. This sometimes results in a dissociation of the bipolar cup and necessitates surgery to replace the cup.

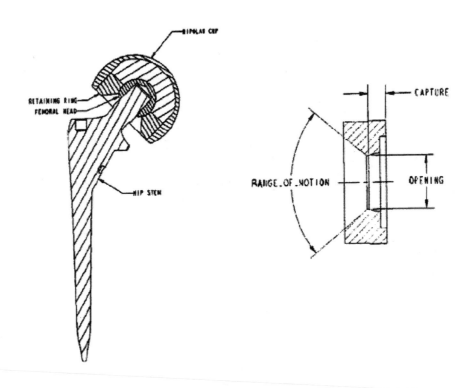

Four factors were studied to find their effects upon the force necessary to dissociate the bipolar cup. These factors are listed below:

	(-)	(+)
Factor	Low Level	High Level
Capture Height (CH)	3mm	8mm
Retaining Ring Opening (OP)	Narrow	Wide
Femoral Head Offset (OFF)	+0mm	+5mm
Ring Size (RS)	42mm	58mm

The following table lists the experimental settings and results obtained by making and testing 16 prototype bipolar cups. The objective was to maximize the force to dissociate while retaining at least 49 degree range of motion.

Run Order	CH	OP	OFF	RS	Force to dissociate	Range of Motion
10	3mm	Narrow	0mm	42mm	58.5	52.5
2	8mm	Narrow	0mm	42mm	88	52.5
4	3mm	Wide	0mm	42mm	39	58.69
16	8mm	Wide	0mm	42mm	46.5	58.6
9	3mm	Narrow	5mm	42mm	49.5	39.5
5	8mm	Narrow	5mm	42mm	66.5	39.5
8	3mm	Wide	5mm	42mm	32	47.2
3	8mm	Wide	5mm	42mm	46.5	47.2
7	3mm	Narrow	0mm	58mm	49.5	54.1
11	8mm	Narrow	0mm	58mm	102	54.1
1	3mm	Wide	0mm	58mm	32.5	59.5
12	8mm	Wide	0mm	58mm	58.5	59.5
14	3mm	Narrow	5mm	58mm	36	41.1
15	8mm	Narrow	5mm	58mm	81	41.1
6	3mm	Wide	5mm	58mm	29	48.3
13	8mm	Wide	5mm	58mm	56.5	48.3

a) Plot the response versus run and comment on what you see.

b) Calculate the effects for each response, and construct a Normal or Half-Normal plot of the effects to determine which effects and interactions appear to be significant.

c) For each response, separately construct boxplots at each level of the four factors similar to Figure 8.17 and comment on what you see.

d) Graph any significant interactions and write a sentence interpreting each significant main effect and/or interaction.

CHAPTER 9
Additional Tools for Design and Analysis of Two-Level Factorials

9.1 Introduction

The last chapter dealt with the basics of two-level factorial designs, including design geometry, main effect and interaction estimation, significance testing, and checking for significance if there are no repeat runs. On the surface, that might seem to be all you need to know in order to charge ahead and run one of these designs. But, in reality, there are a number of issues that must still be addressed in almost any practical application of two-level factorial designs. Some of these issues are: (a) Will the number of runs from the full 2^k design give the accuracy needed? (b) How do you obtain a linear prediction equation from the effects and interactions that were calculated, and how do you check for the adequacy of that model? (c) What do you do with known background or nuisance variables that might influence the results but are not really of interest, like batch of raw material? (d) What can be done when the order of experimentation is difficult or impossible to randomize? These very important questions will be dealt with in this chapter.

9.2 Calculating the Number of Replicates for Needed Precision

Although factorial designs give estimates of the effects of factors which are more precise than any other experimentation strategy, it may still be necessary to replicate the design to obtain the precision that is desired.

Remember our basic assumption when we examine the results of our experiments: If a measured effect could reasonably have occurred as a result of the variability in our measurements, then we say the variable had no effect at all (or, more precisely, no statistically significant effect). We don't want this to happen if the effect is important. In other words, we don't want to end up ignoring any factor if its effect is *economically* significant as described in more detail in Section 6.5.3. So, to make sure that the risk of that oversight happening is small, we use one of the tools in our statistics arsenal to increase the precision of our results — replication.

To facilitate our discussion, let us use the symbol, δ, to denote the size of the smallest effect that we do not want to overlook. (Of course, any effects that are bigger are easier to "see.") And if s_p, as usual, is the estimate of the standard deviation of a single observation, then the number of factorial runs needed to "do the job," is given by

$$n_F = (8 s_p / \delta)^2 \tag{9.1}$$

The number of runs, n_F, should be rounded up to the nearest multiple of 2^k ($n_F = r \times 2^k$, where r is the number of replicates). This will guarantee that the risk of overlooking an effect of size δ is less than 10% (usually closer to 5%), and will allow for testing effects at the 95% confidence or significance level.

Equation 9.1 is given without justification for the time being. So it is important to examine it for reasonableness. It should first be noted that as s_P increases, it requires more experiments for a given accuracy. That makes sense. It should next be noted that it requires more and more experiments to pick out smaller and smaller δ's; this is also as one would expect. So, at least the direction of the impacts of s_P and δ are correct.

When δ is about the same size as s_P, 64 experiments are required. Generally speaking, precision greater than that is impractical (unless experiments are quite inexpensive). One should also be aware that it doesn't matter whether the experiments come from a large factorial or a smaller, replicated factorial, the precision will be the same. So if 32 experiments are required for adequate precision, and the number of factors to be studied is k = 4, then the 16 run (= 2^4) full factorial must be repeated. Alternatively, another factor can be added to the study, and a 32 run (= 2^5) full factorial could be run. The precision will be the same for the two designs: it depends only on the total number of factorial runs carried out, not on how many factors are studied.

Equation 9.1 is *extremely* important, and it should *always* be used before beginning an experimental program of any reasonable size. In the form above, it gives the "price tag," n_F, associated with any particular precision, δ, desired. But in practice it is often used (very appropriately) "backwards," as given in Equation 9.2:

$$\delta = 8 s_P / \sqrt{n_F} \tag{9.2}$$

In this form, the equation gives the precision you can "buy" for a given budget, n_F. If the precision is not good enough, a decision must then be made to either increase the budget (n_F), or to scrap the whole experimental program (or to look for a way to decrease σ, which is not a statistical problem). Scrapping the program is not a desirable thing to do, but it is better than spending the money (running the experiments) and not being able to draw any conclusions from the data, because the variability was too great.

Often students ask, "Where does δ come from?" The answer is that it is decided on a case-by-case basis from the economics of the situation. Although you would always love to have infinite precision (i.e., $\delta = 0$) so that any factor with a nonzero effect on the response would be found to be statistically significant, this is not practical. So δ is then ultimately decided by the sponsor of the experiments (whoever is paying the bills). They must decide how much accuracy is affordable by the use of Equation 9.2.

Before leaving this topic, let us discuss briefly how Equation 9.1 was derived. In order to understand it, we must remember what we are going to do with the effects that we calculate — namely, we are *always* going to test them for statistical significance. There is some value, denoted by $E^* = t^* \times s_E$, which we feel is the largest an experimentally determined effect can be, just due to chance (when the true effect is zero). That critical value, E^*, depends of course on the variability in our measured effects, s_E. Therefore, any effects that we measure to be greater than E^*, we will call real. In other words, we will say that the factors are *significant*. (We should really say that the factors are *statistically significant*.) We must keep in mind that we may be wrong. The true effect of the factor may be zero, but the *random errors* could give us an *apparent* effect. The chance of being wrong is known (in statistical jargon) as the α-risk, which was discussed in more detail in

250

Section 6.3. The amount of α-risk we take is totally under our control, since we decide how big E*
is. The bigger an effect has to be before we are willing to call it real (i.e., the larger the value of E*
we select), the smaller our α-risk. This is shown in Figure 9.1 (A).

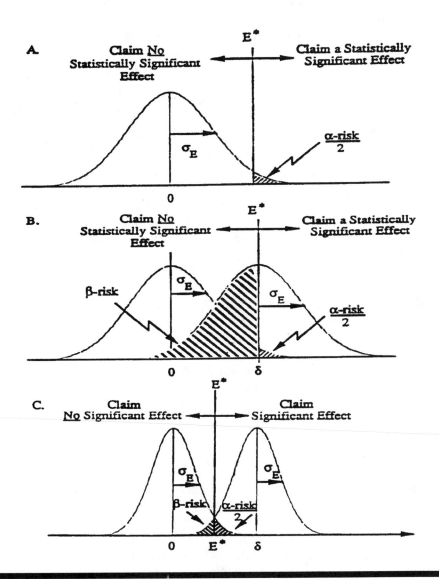

Figure 9.1 Replication Needed for Adequate Accuracy:
(A) Control α-risk by selection of E*, (B) δ at E* gives β-risk of 50%,
(C) δ well beyond E* gives acceptable β-risk

How big should δ be relative to E*? The first response most people would make is that E* and δ should be the same. This seems to make a lot of sense since δ is the size of an effect we do not want to overlook, and when we measure an effect of size E* we will say it is real. The problem is that we "never" measure the true effect of a factor. It is always clouded by experimental error. So sometimes what we measure will be less than δ and sometimes it will be more, as shown in Figure 9.1 (B). The chance of it being less than E* (in this case, less than δ), is the risk we run of calling the factor unimportant. We should say "the effect is not statistically different from zero." That risk is known, in statistical jargon, as β-risk (which was also discussed in more detail in Section 6.3). To repeat, β-risk is the risk of missing an important effect (like a doctor failing to detect that a person is sick, and telling them they are fine). When δ = E*, the β-risk is 50%, which is usually considered much too high.

In order to get the β-risk down to an acceptable level, δ must be a good bit bigger than E*; just how much bigger depends on exactly what level of β-risk is acceptable (see Figure 9.1 (C)). It should be remembered that even though we are talking in terms of making δ bigger than E*, δ is "fixed" by the economics of the situation. So what we actually need to do is make E* a good bit smaller than δ. We can do this because we are able to make the standard deviation of an effect, s_E, as small as we want by taking enough data; the more data, the smaller s_E will be.

If α-risk is selected to be 5%, and we have a Normal Distribution for the measured effects, then $E^* = 1.96\sigma_E$. If β-risk is also set to 5%, then $\delta = E^* + 1.64\sigma_E$. (It is different from the α-risk distance because only one side of the distribution is relevant.) Combining the two relationships gives us:

$$\delta = 1.96\sigma_E + 1.64\sigma_E = 3.60\sigma_E \tag{9.3}$$

In Chapter 8 we developed the relationship between the standard deviation of an effect and the number of factorial points (Equation 8.4), which is repeated here for convenience:

$$\sigma_E = 2\sigma / \sqrt{n_F} \tag{9.4}$$

As previously mentioned, σ is the experimental variability of a single data point. By inserting Equation 9.4 into Equation 9.3, we can express δ as a function of n_F.

$$\delta = 3.60\sigma_E = 3.6 (2\sigma / \sqrt{n_F}) = 7.2\sigma / \sqrt{n_F} \tag{9.5}$$

Rearranging Equation 9.5 gives us the result we were looking for.

$$n_F = (7.2\,\sigma / \delta)^2 \quad \text{for a Normal Distribution.} \tag{9.6}$$

Since we do not usually have a Normal Distribution (because we do not usually know σ), but rather we typically have a t-Distribution, the constant, 7.2, must increase a bit. To be precise about how much it should be increased would require taking into account how well we will know σ as estimated by s_P (i.e., how many "degrees of freedom" we will have in our estimate, s_P). A value of 8.0 rather than 7.2 is recommended by Wheeler[1] as being reasonable in most instances. Therefore,

$$n_F = (8\,s_P / \delta)^2 \quad \text{for a t-Distribution.} \tag{9.7}$$

252

To illustrate how this equation is used, we will present two examples. In the first example an experiment is being planned to study the purification process for a crystalline product. In the second example we will revisit the stack gas treatment problem from Section 8.10.

Example 1: In the purification process for a crystalline product, the product is first dissolved in a solvent, next a second liquid is added to the solution and stirred in, finally the purified crystalline product is precipitated out by steam distillation. In the plans for experimentation, it was decided to vary the temperature of the dissolution solvent, the type of liquid added at the second stage, the stirring time and the steam distillation time. Two levels were chosen for each of the four factors, and a 2^4 factorial experiment was planned. Based on the dollar value of the purified product, it was determined that it would be important to detect any main effect or interaction that was 0.05 or more. Therefore, in this example the size of an important effect $\delta = 0.05$. From past history of the purification process, the standard deviation in final crystalline purity was $s_P = 0.02$. This was an estimate based on run-to-run variability with the factor levels constant.

The number of factorial points needed was calculated, using Equation 9.1, to be:

$$n_F = [8 \, (0.02)/(0.05)]^2 = 10.24$$

Since $10.24 < 2^4 = 16$, the number of replicates, $r = 1$.

In this case, with one replicate of the factorial experiment, n_F is really greater than the requirement. So, we might ask the reverse question, "How small an effect can we reasonably expect to detect with our budget of $n_F = 16$ experiments?" Using Equation 9.2, we calculate:

$$\delta = 8 \, s_P / \sqrt{n_F} = 8 \, (0.02) / \sqrt{16} = 0.04$$

which is even better than required.

Example 2: Recall that in the stack gas treatment experiment from Section 8.10, the variability in measured response (NO_2) was expressed as a constant percentage rather than a constant. Therefore the log transformation was used before the data were analyzed. In order to determine how much data are needed for this situation, we must agree on what precision is desired on the percent scale, then use the log transformation before applying Equations 9.1 or 9.2.

In this example, it was decided that any factors that affected the NO_2 concentration by 50% or more were important to detect. Therefore, $\delta = \log(1.5) = 0.18$. The standard deviation, on the percentage scale, was known to be 25%, so $\sigma = \log(1.25) = 0.10$. Thus, the number of factorial points needed is $n_F = [7.2 \, (0.10)/(0.18)]^2 = 16$. Equation 9.7 (that is a constant of 7.2 instead of 8) was used to calculate n_F because we knew the value of σ, which is fairly unusual.

If we were not confident that we knew σ, then we ran fewer runs than necessary. The δ to be expected with 16 runs (assuming that we would use the t-distribution in our data analysis), is:

$$\delta = 8 \, s_P / \sqrt{n_F} = 8 \, (0.10) / \sqrt{16} = 0.20$$

which corresponds to an error in NO_2 concentration of $\log^{-1}(0.20) = 10^{0.20} = 1.59$, or an increase in NO_2 concentration of 59%. This is not greatly different from the desired detectable increase of 50%.

9.3 Results in Equation Form

In order to use data from a factorial experiment to predict the response at combinations of factor settings that have not been run in the experiment, it is necessary to develop a mathematical model or prediction equation for the response. The mathematical model which is implicit in the 2^k factorial design is:

$$
\begin{aligned}
\hat{Y} = \ & b_0 && \text{-constant term} && \text{(9.8)}\\
& + b_1 X_1 + b_2 X_2 + ... + b_k X_k && \text{-linear terms}\\
& + b_{12} X_1 X_2 + b_{13} X_1 X_3 + ... + b_{k-1,k} X_{k-1} X_k && \text{-cross product terms}\\
& + b_{123} X_1 X_2 X_3 + ... && \text{-higher order terms}
\end{aligned}
$$

where

$\hat{Y} =$ predicted value of Y

$X_i =$ coded value for factor i

$$
= \left(\frac{\text{Factor Value} - \text{Center}}{\text{High Factor Value} - \text{Center}} \right) \text{ and}
$$

$$
\text{Center} = \left(\frac{\text{High Factor Value} + \text{Low Factor Value}}{2} \right)
$$

The unknown constants in the equation are readily obtainable from the factorial effects that we have already calculated.

$$
b_0 = \bar{Y} \tag{9.9}
$$

$$
b_i = \tfrac{1}{2}[\text{Effect for Factor } X_i] \tag{9.10}
$$

$$
b_{ij} = \tfrac{1}{2}[\text{Interaction for } X_i X_j] \tag{9.11}
$$

$$
b_{ijk} = \tfrac{1}{2}[\text{3-Factor Interaction for } X_i X_j X_k] \tag{9.12}
$$

The coefficients are half the value of the effects because coefficients represent the slope or change in Y for a one-unit change in X (e.g., from $X = 0$ to $X = 1$), whereas the effects represent the change in Y for a two-unit change in X (from -1 to +1).

Normally when writing the prediction equation for a factorial design, we only include terms in the model for effects and interactions that were found to be statistically significant. By doing this we average or smooth away noise in predictions.

To see an example of this consider the voltmeter experiment presented in Chapter 8. The prediction equation for this experiment can be written as:

$$\hat{Y} = 668.56 - \left(\frac{33.63}{2}\right)X_1 + \left(\frac{25.13}{2}\right)X_1X_3$$

$$= 668.56 - 16.81X_1 + 12.56X_1X_3$$

Where, 668.6 is the mean of the data, and -33.6 and 25.1 are the values of the significant main effect and interaction. When an interaction such as X_1X_3 is included in the model, many data analysts would also prefer to include both main effects that contribute to the interaction in the model, even though they may not both be statistically significant. For the voltmeter experiment, that would mean including the X_3 as shown below:

$$\hat{Y} = 668.56 - 16.81X_1 + 5.44X_3 + 12.56X_1X_3$$

$$\text{Since } X_1 = \left(\frac{\text{Temperature} - 27}{5}\right), \text{ and } X_3 = \left(\frac{\text{Circuit Warm-up Time} - 2.75 \text{ min.}}{2.25 \text{ min.}}\right)$$

it would be even more meaningful to write the equation in the form:

$$\hat{Y} = 668.56 - 16.81\left(\frac{T - 27}{5}\right) + 5.44\left(\frac{C \text{ Time} - 2.75}{2.25}\right) + 12.56\left(\frac{T-27}{5}\right)\left(\frac{C \text{ Time} - 2.75}{2.25}\right)$$

In this way predictions of the voltage at any combination of T (Temperature) and C Time (Circuit Warm-up Time) can be made by substituting the values directly into the equation. For example the predicted voltage at T=30 Deg and C Time = 2.0 minutes would be:

$$\hat{Y} = 668.56 - 16.81\left(\frac{30 - 27}{5}\right) + 5.44\left(\frac{2.00 - 2.75}{2.25}\right) + 12.56\left(\frac{30 - 27}{5}\right)\left(\frac{2.00 - 2.75}{2.25}\right)$$

$$= 654.2$$

Another way of writing the prediction equation would be to multiply through the brackets and collect the constants and multipliers to form:

$$\hat{Y} = 835.62 \ -6.433 \ T \ - \ 27.73 \ C \ Time \ + \ 1.117 \ T \times C \ Time$$

However, in this form the constants in the model lose their direct interpretation as the grand average and half effects for the experiment. For this reason, the first form of the model is generally preferred.

The prediction equation for the voltmeter experiment is shown graphically in Figure 9.2. Here we can see the actual data from the experiment (represented as black dots), and the predictions (represented as the wireframe surface). This 3-D graph gives us a good feeling for how well the prediction model actually predicts the data. A prediction, made at a certain combination of Temperature and Circuit Warm-up Time, will represent our estimate of the long run average voltage at those conditions. We can expect that individual voltage readings at the same conditions to vary from our prediction, just as the experimental data points in Figure 9.2 differ from the wireframe prediction surface. Since the pooled standard deviation, s_p, is a measure of the variability of individual response values from their long run average, we can use it to get a rough idea how far a single voltage reading will vary from a predicted value. Assuming a Normal Distribution of individual responses at the same conditions, we could roughly say that we are 95% sure that a single measurement should be within $\hat{Y} \pm 2s_p$. More exact formulas for confidence intervals on predictions will be given in Chapter 10.

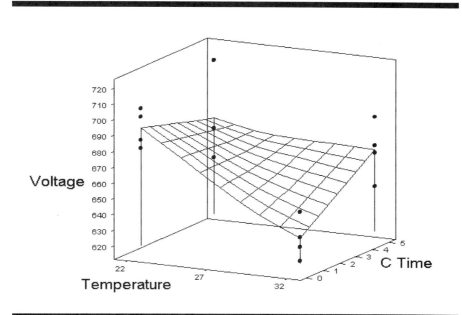

Figure 9.2 3D graph of Prediction Equation for Voltmeter Experiment

To verify the validity of a mathematical model and predictions derived from it, we should always check the assumptions upon which the model is based . The assumptions made for the prediction equation derived from a factorial experiment are: (1) linearity of the underlying relationship, (2) equal variance of response throughout the experimental region, and (3) a Normal Distribution of the responses around the true relationship. A rough check of all of these assumptions can be made by looking at the relative position of the data points to the prediction surface in a picture like Figure 9.2. But, this is a simple situation with only two factors. With more than two factors in the equation, it is impossible to make a figure like 9.2. Instead, various plots of residuals must be made. The residuals are the differences between the actual observed response data and their predicted values from the equation. Chapter 10 will show examples of calculating the residuals and plotting them to check the validity of a prediction equation. Another (more accurate) way of checking the linearity of the underlying relationship will be described in Section 9.4.

A graphical display of the prediction equation like Figure 9.2, can only be used when there are only two factors. Another common way of displaying the prediction equation is a contour plot. Contour plots are similar to a topographic map. In a contour plot, two factors are shown on the X and Y axes and different contour lines represent the level of the predicted response. If there are more than two factors in the prediction equation, the most important factors are put on the axis and multiple plots are made at different levels of the less important factors. Figure 9.3 is an example of a contour plot for the voltmeter prediction equation.

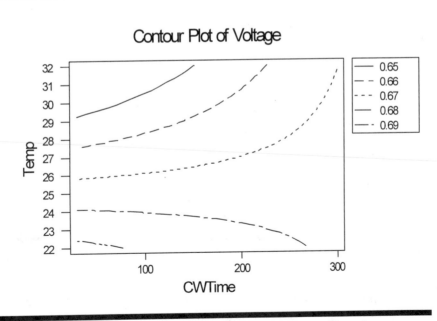

Figure 9.3 Contour Plot of Prediction Equation for Voltmeter Experiment

257

9.4 Testing for Curvature in the Model by Including Center Points in the Design

Replicate experiments at a center point (as shown in Figure 9.4 for one, two, and three factors) are a valuable adjunct to a two-level factorial design for two reasons: (1) they provide additional replicates from which an estimate of error can be calculated, and (2) they provide a check of the adequacy of the linear prediction model. Both of these are extremely important benefits. Since there is so much to be gained from the center points, many factorial designs used in practice do incorporate center points in the total experimental design. Let's talk about both benefits now in a bit more detail.

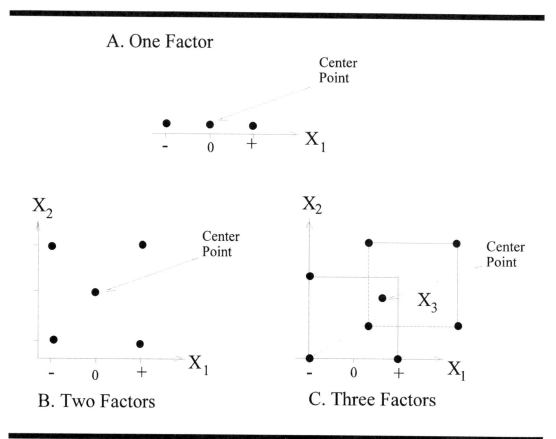

Figure 9.4 Center Points in Two-Level Factorial Designs

The center points provide us with a "pure" estimate of error, since they are replicated points. Without center points (or replication at all the design points), the only way an estimate of error could be obtained would be by assuming some higher order interactions are zero. If the interactions are not zero, then our estimate of error is inflated. This, in turn, inflates our estimate of s_E, which causes our calculated t-values to be too small, which could result in some of our significant effects or interactions being labeled insignificant. Thus, without the estimate of error we get from the replicated design points, we could easily overlook some important effects.

The analysis of data from a factorial design also assumes that a linear model adequately describes the changes in the response resulting from changes in the independent variables. With only two levels of each factor, there is no way to check the validity of this assumption. You cannot, after all, fit anything more complicated than a straight line through two points! If there are nonlinearities (curvature), the factorial design will yield poor predictions except near the actual design points (corners of the cube). Yet, you would like to make predictions about the response throughout the experimental region. How can you tell if you are justified in doing that? Center points come to our rescue. Center points allow us to check for curvature by seeing if there is a big difference between the actual response at the center point and the value that we expect (or predict) at those middle conditions if a linear model was adequate. This is shown graphically in Figure 9.5 for one factor. It should be noted that we are demonstrating curvature with one factor because it is the easiest situation to visualize, although it is not of great importance in practice. However, the concept generalizes readily to more factors. For example, curvature is also shown in Figure 9.6 for two factors.

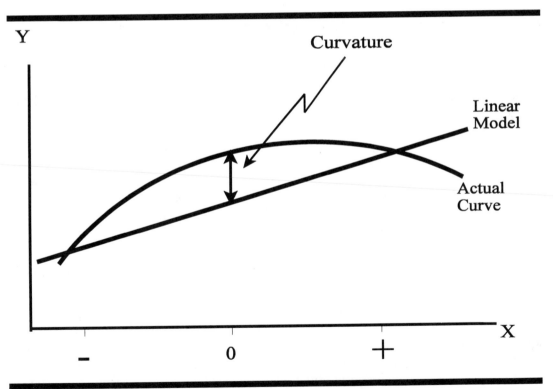

Figure 9.5 Curvature in Two-Level factorial Design with One Factor

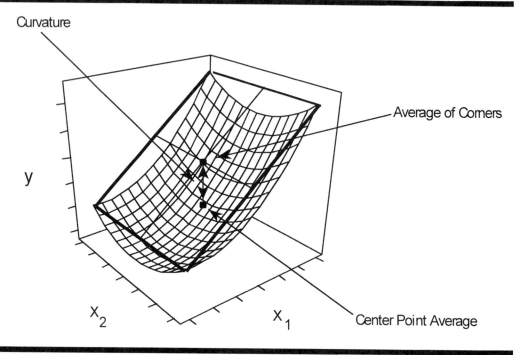

Figure 9.6 Curvature in Two-Level Factorial Design with Two Factors

To simplify our discussion, let us use the term, *curvature*, to denote specifically the difference between the actual response in the middle and the linear prediction, and let us give it the symbol, C. Thus:

C = Actual Response at Center - Predicted Response at Center (via linear model) **(9.13)**

The actual response at the center is estimated by averaging the center points. The prediction at the center using a straight line model through the factorial points is obtained very easily by just averaging the factorial (corner) points. That this procedure works is obvious in the case of only one factor, but it is also true for any number of factors. So the computation of C becomes:

$$C = \overline{Y}_{Center} - \overline{Y}_{Factorial}$$ **(9.14)**

When we calculate C, which we will call curvature for simplicity, it will "never" be exactly zero (due to random error), even when a linear model is perfect. So before we can say that we do or do not have curvature, we must test for the statistical significance of the calculated curvature. In order to do that, we must know how much variability there is in the curvature estimate, which is quantified by its standard deviation, s_C. This s_C is easily determined via the same procedure as we used for s_E in Section 8.8. Since C is the difference between two averages, its standard deviation is given by:

$$s_C = \sqrt{s_p^2\left(\frac{1}{n_C} + \frac{1}{n_F}\right)} \tag{9.15}$$

where n_C = the number of observations in the average response at the center, $\overline{Y}_C$
and n_F = the number of observations in the average response at the factorial points, $\overline{Y}_F$. It must
be remembered that n_F includes replication; it is not just the number of corner conditions.

To test the statistical significance of the measured curvature, C, we could calculate the t-statistic (signal-to-noise ratio),

$$t_C = C / s_C \tag{9.16}$$

and see if it is larger than could reasonably be explained by chance alone (as compared to the t* values in Table A.5 of Appendix A). To illustrate the test for curvature, consider the following example.

Example 3: In a chemical process one reactant is held in a vessel for cooling. When it reaches the desired temperature, a second reactant is quickly charged in an aqueous/organic mixture. Experiments were performed on this process to optimize the yield. Two factors were varied in the experiments. The first was the temperature in the vessel, and the second was the aqueous to organic ratio of the charge. The response was the yield in percent. A 2^2 factorial experiment was performed with two replicates at each of the four experimental conditions, and since both factors were quantitative, four additional replicates were made at a center point.

The coded factor levels were $X_1 = \left(\dfrac{Temp-113}{9}\right)$ and $X_2 = \left(\dfrac{Ratio-.55}{.45}\right)$
and a worksheet with the results are shown in Table 9.1 below.

Table 9.1 Yields from Chemical Experiment and Summary Statistics

Run	Temp X_1	Aq/Org X_2	Observed Yields (Y)		Means $\overline{Y}$	Variances s^2	df
1	-	-	68.5	68.3	68.40	0.020	1
2	+	-	72.0	72.0	72.00	0.000	1
3	-	+	73.0	70.0	71.50	4.500	1
4	+	+	72.4	73.0	72.70	0.180	1
5	0	0	74.0	75.0	74.125	0.396	4
			74.0	73.5			

Figure 9.7 below shows the average response at each of the five experimental conditions. Since the average response at the center point (74.125) appears to be higher than any of the factorial points, a linear prediction equation probably would not be appropriate for this data.

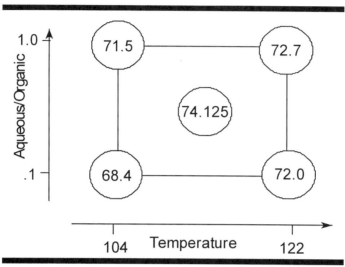

Figure 9.7 Average Yields Chemical Reaction Experiment

To perform a formal test of the curvature effect we start by calculating the pooled variance. In this case we pool (or average) the variance from each of the four replicated factorial points and the replicated center point using Equation 8.3 (given in Section 8.8.1). Notice that s_p^2 is a weighted average of the five estimates of the variance, and the degrees of freedom are different at the center point than at the factorial points. The degrees of freedom for this pooled variance is the sum of the degrees of freedom for the individual variances, or 7.

$$s_p^2 = \frac{1\times(0.020) + 1\times(0.000) + 1\times(4.500) + 1\times(0.180) + 3\times(0.396)}{1 + 1 + 1 + 1 + 3} = 0.8411$$

$$s_p = \sqrt{0.8411} = 0.9171$$

The curvature effect is:

$$C = \text{Center Point Average - Factorial Average}$$

$$= 74.125 - \frac{68.4 + 72.0 + 71.5 + 72.7}{4} = 74.125 - 71.15 = 2.975$$

The standard deviation or standard error of the curvature effect is:

$$s_C = s_p \sqrt{\frac{1}{n_C} + \frac{1}{n_F}} = (0.9171) \sqrt{\left(\frac{1}{4} + \frac{1}{8} \right)} = 0.5616, \quad \text{and the signal-to-noise t-ratio is:}$$

$$t_C = \text{Curvature} / s_C = 2.975 / 0.5616 = 5.297.$$

When this calculated value is compared to the critical value of the Student's t- statistic, t*, from Table A.5 with 95% confidence for 7 degrees of freedom, which is t* = 2.365, we find that there is significant curvature.

When significant curvature is found, as in Example 3, the linear prediction equations developed in Section 9.3 will not be valid for making predictions within the experimental region. To come up with a valid interpolation equation in this case, more experiments are needed as well as techniques for fitting a quadratic model (which will be described in Chapters 13 and 14).

9.5 Blocking Factorial Experiments

In cases where a known background variable can change and affect the results of an experiment, we should take that variable into account to avoid biasing the effects of the factors we are studying. We can take known background variables into account by grouping the experiments (in a factorial design) into sets of runs which have common levels of this variable. This is called *blocking*. For example, Table 9.2 compares a blocked and unblocked method for running two replicates of a simple 2^2 design for studying the effects of temperature and time on the yield of a chemical reaction.

Table 9.2 Unblocked and Blocked Factorial Designs to Study the Effects of Time and Temperature on Yield of a Chemical Reaction

Temp X_1	Time X_2	Run Order	Day	Yield	Temp X_1	Time X_2	Block Day	Run Order	Yield
			Unblocked Design				Blocked Design		
-	-	5	2	Y_1	-	-	1	1	Y_1
-	-	7	2	Y_2	+	-	1	3	Y_2
+	-	4	1	Y_3	-	+	1	2	Y_3
+	-	1	1	Y_4	+	+	1	4	Y_4
-	+	6	2	Y_5	-	-	2	2	Y_5
-	+	3	1	Y_6	+	-	2	3	Y_6
+	+	2	1	Y_7	-	+	2	1	Y_7
+	+	8	2	Y_8	+	+	2	4	Y_8

In the unblocked design on the left, the order of experiments, or runs, is completely randomized and run in two days (runs 1– 4 on Day 1 and runs 5 – 8 on Day 2). In the blocked design on the right, one complete replication of the design is run on each day or block, and the order is randomized within each block. The presumption is that uncontrolled conditions on a given day are more alike (consistent) than conditions from one day to the next.

There are three purposes for blocking:

(1) First of all, we want to prevent the effects of known background variables from biasing or confusing the treatment effects of interest.

(2) The second reason is to broaden the basis for conclusions.

(3) And the third is to increase the precision of the treatment effects.

These concepts will be illustrated with the simple chemical reaction experiment described in right side of Table 9.2.

First, blocking prevents bias caused by known background variables. Let us say that eight runs are needed in total, but only four can be completed in one day. If we use the random run orders listed in the unblocked design, it would cause three of the four experiments at the high level of Factor 1 (temperature) to be run on Day 1, and three of the four for low temperature to be run on Day 2. If there was any change in a background or lurking variable between Day 1 and Day 2 of experimentation, the effect for temperature could be seriously biased. This bias can be completely prevented if the experiments are blocked as in the right side of Table 9.2. The block can be treated just like another factor in the experiment. The systematic randomization in the blocked design makes only the block factor confused or biased by any lurking variables that change from Day 1 to Day 2. The average effects of X_1 and X_2 are completely unconfused with blocks and any lurking variables that change from Day 1 to Day 2.

The second reason for blocking is to broaden the base for conclusions. If only the first block of experiments (Day 1) was completed, and all runs used the same raw materials (i.e., reagents from one batch), then the conclusions regarding the effects of time and temperature are only valid with respect to that particular day or batch of raw materials. However, if the second block of experiments was also completed, and used a different batch of raw materials, then the conclusions can be broadened. If the two batches of raw materials (represented by blocks or days in Table 9.2) are randomly selected from several batches available, then the conclusions regarding the effects of time and temperature are more likely to hold true for all the batches of materials that were available.

The final purpose of blocking is to increase precision. The purposes of blocking are best achieved by grouping homogeneous sets of experimental units together in blocks, and letting the characteristic of the units vary widely between blocks. By doing this, treatment effects are compared to the variability of the homogeneous experimental units within the blocks and the precision is increased. In the example in Table 9.2 the effects of time, temperature and their interaction would be compared to the variability of batches within a day (where the same batch of raw materials were used).

The blocked experiment on the right side of Table 9.2 looks like a 2^3 experiment with the levels of the third factor (block) run in sequence rather than in a randomized order. For this reason many beginning users of experimental designs make an unknowing mistake in conducting their experiments. If they think their list of factorial experiments would be difficult or impossible to run in a completely random order, because it is difficult to change back and forth between high and low levels of one factor, then they include the hard to vary factor as a block (like the example shown in Table 9.2). However, when a factor is included as a block with no randomized sequence of changes between high and low levels, its effect will be biased by other background variables that change during the course of experiments.

If we are really not interested in a block variable (like the Day in Table 9.2) the bias will not bother us. The purpose of blocking is to give us greater precision, by removing the variability caused by the blocking variable from s_E , and to give our conclusions broader base by repeating experiments at different levels of the blocking variable. But the purpose of blocking is not to estimate the effect of the blocking variable! If it is difficult to run experiments in completely random order, because levels of one or more factors are hard to change, a split plot type experiment (to be explained in Section 9.6) should be used, not a blocked experiment.

Originally blocked designs were used in agricultural experimentation where the experimental units were plots of ground. Grouping similar and adjacent plots of ground together in blocks, could prevent (or block out) the effects of soil fertility, etc., from biasing treatment effects or inflating the estimate of s_E. The effects of the different soil plots was not of interest in agricultural experiments, but by choosing blocks or sets of plots from widely different areas and soil types, conclusions from agricultural experiments could be more generally applied.

9.5.1 Calculating the Error of an Effect (s_E) in Blocked 2^k Experiments

Calculation of effects in blocked factorials is no different than calculating effects in the randomized factorials described in Chapter 8. In a blocked design like the right side of Table 9.2, the main effects and interactions of X_1, and X_2 are calculated exactly as was shown in Sections 8.3 and 8.4. But, the addition of the block factor (Day) has changed this from a 2^2 experiment with replicates to a 2^3 design without replicates. Without replicates in a blocked design, the standard error of the effects cannot be computed by Equation 8.4 (in Section 8.8). We can still judge the significance of effects using graphical methods (see Section 8.10), or a more exact method using t-statistics, that will be explained next.

To employ t-statistics, we still need an estimate of the standard error of an effect, s_E . We take advantage of the premise that, in a blocked design, there should be no interactions of the block effects with the other factors in the design (otherwise we couldn't generalize a conclusion about a consistent effect over all blocks). Therefore, the estimates of the block interaction terms are a direct indication of the variability that can manifest itself in the effects. In other words, the interactions of the block factors should all be zero, but the actual measured interactions will exhibit scatter around zero with a standard deviation equal to the standard deviation of the other effects and interactions, s_E. Therefore, s_E can be estimated by calculating the standard deviation of the block interactions around zero (i.e. the square root of the average of the squared block interaction terms). To be more specific consider the following: The usual formulas for estimating the variance of a random variable, Y, are

$$s^2 = \sum (Y - \mu)^2 / n \qquad \text{if we know } \mu, \text{ the mean of the distribution,} \qquad \textbf{(9.17)}$$

or $\qquad s^2 = \sum (Y - \bar{Y})^2 / (n - 1) \qquad \text{if we don't know } \mu \text{ and estimate it by } \bar{Y}. \qquad \textbf{(9.18)}$

In this case, we are not dealing with individual observations, but rather with effects (and interactions) directly. So we will not be calculating s^2 with our formula, but rather we will be calculating s_E^2. Also note that this is an example of the fairly unusual case in which we happen to know the mean of the distribution, μ, which is zero. So, we will use Equation 9.17, with blocking variable interactions, BI's, instead of the Y's.

$$s_E^2 = \sum (BI_i - 0)^2 / m \qquad \textbf{(9.19)}$$

or $\qquad s_E = \sqrt{\sum BI_i^2 / m} \qquad \textbf{(9.20)}$

where BI_i are the block interactions, and m is the number of block interactions. Therefore, m is also the degrees of freedom our estimate of s_E. Once we have s_E, and its degrees of freedom the t-statistics, E/s_E, can be calculated as usual, and used to test the significance of our measured effects.

9.5.2 A Two Block Example

Figure 9.8 shows pictorially the settings of a laboratory experiment conducted to study two factors that may effect the half-life of a chemical pesticide in soil. The experimental unit was a container of soil treated with the pesticide. Each container was stored in a growth chamber at constant temperature and moisture. The moisture level of each container was monitored over time, and water was added periodically to maintain a nearly constant moisture level. The half-life of the pesticide was determined for each container by sampling the soil weekly and analyzing for the concentration of the pesticide. Typical exponential decay curves resulted, as shown in Figure 9.9, that allowed estimation of the half-life.

The temperature level was varied by having two identical growth chambers maintained at different temperatures. Half of the soil containers in each chamber were maintained at a high moisture level, and half were maintained at a low moisture level. Positions of the high and low moisture containers were randomly assigned to shelf positions in the chambers in order to avoid bias caused by any positional effects. The entire set of four experiments was duplicated simultaneously using two different soil types (1 = sandy-loam, 2 = silt-clay), so that the conclusions could be extended to more than one soil type. This created a blocked design.

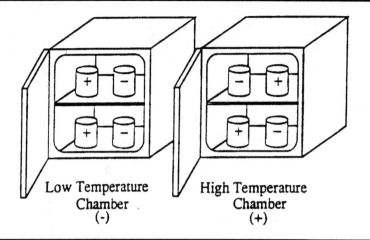

Figure 9.8 Growth Chambers for Pesticide Degradation Experiment
(signs on containers represent moisture level)

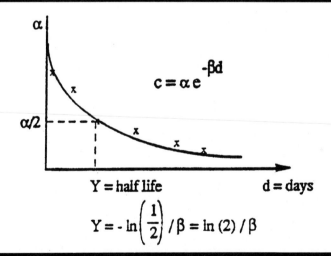

$$c = \alpha e^{-\beta d}$$

$$Y = -\ln\left(\frac{1}{2}\right) / \beta = \ln(2) / \beta$$

Figure 9.9 Typical Degradation Cure Used to Determine Half-Life

The experiments and results are listed in Table 9.3. The four containers of soil from each soil type were randomly assigned (with the help of Table A.1) to receive one of the temperature and moisture combinations, and the container numbers are shown in the eighth column of the table. Whenever the response (half-life for this experiment) ranges over an order of magnitude or more (in this case 31.5 - 886.3) it is common to perform the analysis on the logarithm. (Incidently, it makes no difference whether base 10 or base e is used for the logarithms.) The transformation is often necessary to ensure the validity of the two major assumptions which are invoked in the analysis of the data from a factorial design: 1) a linear model and 2) a constant variance. See Section 8.10 for another example of the use of logarithms. The half-life data are listed in the ninth column of the table, and the logarithms are listed in the tenth column.

A graphical presentation of $Y = \ln(\text{half-life})$ is shown in Figure 9.10. It can be seen that increasing temperature and increasing moisture tend to reduce the half-life (i.e. they have negative effects). To confirm what can be seen graphically, the effects were calculated (using Yates Algorithm —see Appendix 8.A), and are displayed compactly in the last column of Table 9.3.

To judge the significance of the effects we calculate the variance of the effects, s_E^2, from the block interactions ($X_1 \times B$, $X_2 \times B$, and $X_1 \times X_2 \times B$) which are assumed to be zero. Using Equation 9.20, s_E^2 is estimated as the average of the block interactions squared. The standard deviation of an effect, s_E, is then the square root of the result, as shown below:

$$s_E = \sqrt{\sum BI_i^2 / m} = \sqrt{\frac{(-0.106)^2 + (0.232)^2 + (-0.410)^2}{3}} = \sqrt{\frac{0.233}{3}} = 0.279$$

The degrees of freedom for this estimate is 3 (the number of block interactions).

The signal-to-noise t-ratios are calculated for the main effects X_1 and X_2 and their interactions, by dividing the effects by $s_E = 0.279$, as:

$$t_{X_1} = -1.166/0.279 = -4.179, \quad t_{X_2} = -0.630/0.279 = 2.258, \quad \text{and} \quad t_{X_1X_2} = 0.0813/0.279 = 0.291$$

The critical t* from Table A.5 is 3.182 with 95% confidence level and 2.353 with 90% confidence. Therefore, there is 95% confidence that temperature had a significant effect, but less than 90% confidence that moisture had an effect.

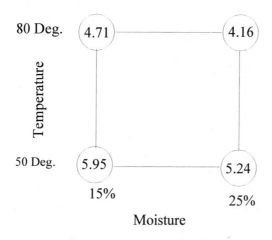

Figure 9.10 Graphical Analysis of Pesticide Half-Life Results

Table 9.3 Blocked Factorial Design to Study Pesticide Degradation in Soil

| Run No. | Coded Levels | | | Uncoded Levels | | | Random Soil Container | Half Life | Y ln(half life) | Computed Effects |
	X_1	X_2	Block	Temp	%Moist	Soil Type				
1	-	-	1	50°F	15%	SL	2	886.3	6.787	5.0156 = Mean
2	+	-	1	80°F	15%	SL	3	187.9	5.236	-1.166 = X_1
3	-	+	1	50°F	25%	SL	1	229.1	5.434	-0.630 = X_2
4	+	+	1	80°F	25%	SL	4	129.8	4.866	0.081 = $X_1 \times X_2$
5	-	-	2	50°F	15%	SC	3	167.5	5.121	-1.130 = B
6	+	-	2	80°F	15%	SC	4	65.2	4.178	-0.106 = $X_1 \times B$
7	-	+	2	50°F	25%	SC	1	156.3	5.052	0.232 = $X_2 \times B$
8	+	+	2	80°F	25%	SC	2	31.5	3.451	-0.410 = $X_1 \times X_2 \times B$

9.5.3 Designs for Blocked Factorials

Factor effects and interactions for 2^k factorials run in blocks can always be computed using the worksheet method or Yates algorithm described in Chapter 8. However, the block effects and block interaction effects cannot be computed by these methods unless the number of blocks in the design is a power of two (i.e., $2^1 = 2$, $2^2 = 4$, etc.). For the more general blocked factorial designs, where the number of blocks is arbitrary, the analysis and significance tests must be completed using the Analysis of Variance (ANOVA) that will be described in Chapter 11. However, there are still many useful blocked designs where the number of blocks is a power of two. Table 9.4 shows how the computation tables in Appendix B.2 can be used to create and analyze various blocked designs with this property.

To demonstrate the use of this table, consider extending the pesticide half-life experiment presented in the last section to four blocks. According to Table 9.4 a 2^2 in four blocks can be created through use of Table B.2-4, by using the X_3 and X_4 terms to define the blocks. A (-,-) combination for X_3 and X_4 would be Block 1, a (-,+) combination would be Block 2, a (+,-) would be Block 3, and a (+,+) would be Block 4. Table 9.5 shows the results of the extended experiment. The analysis would proceed by calculating the effects with use of Table B.2-4 and the standard error of an effect by the square root of the average of the squared effects of the block interactions: X_1X_3, X_1X_4, X_1X_3, $X_1X_3X_4$, X_2X_3, X_2X_4, $X_2X_3X_4$, $X_1X_2X_3$, $X_1X_2X_4$, and $X_1X_2X_3X_4$. The details will be left for an exercise.

Table 9.4 Blocked Full Factorial Designs

Factors	Design	Number of Blocks	Appendix Table	Terms used to define Blocks	Terms used for Calculating Block Effects	Block Interactions used to Calculate Error
2	2^2	2	B.2-3	X_3	X_3	$X_1X_3, X_2X_3,$ $X_1X_2X_3$
2	2^2	4	B.2-4	X_3, X_4	X_3, X_4, X_3X_4	$X_1X_3, X_2X_3, X_1X_4,$ $X_2X_4, X_1X_2X_3,$ $X_1X_2X_4, X_1X_3X_4,$ $X_2X_3X_4, X_1X_2X_3X_4$
2	2^2	8	B.2-5	X_3, X_4, X_5	$X_3, X_4, X_5,$ $X_3X_4, X_3X_5,$ $X_4X_5, X_3X_4X_5$	$X_1X_3, X_2X_3, X_1X_4,$ $X_1X_5, X_1X_3X_4,$ $X_1X_3X_5, X_1X_4X_5,$ $X_1X_3X_4X_5, X_2X_3,$ $X_2X_4, X_2X_5, X_2X_3X_4,$ $X_2X_3X_5, X_2X_4X_5,$ $X_2X_3X_4X_5, X_1X_2X_3,$ $X_1X_2X_4, X_1X_2X_5,$ $X_1X_2X_3X_4, X_1X_2X_3X_5,$ $X_1X_2X_4X_5,$ $X_1X_2X_3X_4X_5$
3	2^3	2	B.2-4	X_4	X_4	$X_1X_4, X_2X_4, X_3X_4,$ $X_1X_2X_4, X_1X_3X_4,$ $X_2X_3X_4, X_1X_2X_3X_4$
3	2^3	4	B.2-5	X_4, X_5	X_4, X_5, X_4X_5	$X_1X_4, X_1X_5, X_1X_4X_5,$ $X_2X_4, X_2X_5, X_2X_4X_5,$ $X_3X_4, X_3X_5, X_3X_4X_5,$ $X_1X_2X_4, X_1X_2X_5,$ $X_1X_2X_4X_5, X_1X_3X_4,$ $X_1X_3X_5, X_1X_3X_4X_5,$ $X_2X_3X_4, X_2X_3X_5,$ $X_2X_3X_4X_5,$ $X_1X_2X_3X_4, X_1X_2X_3X_5,$ $X_1X_2X_3X_4X_5$
4	2^4	2	B.2-5	X_5	X_5	$X_1X_5, X_2X_5, X_3X_5,$ $X_4X_5, X_1X_2X_5,$ $X_1X_3X_5, X_1X_4X_5,$ $X_2X_3X_5, X_2X_4X_5,$ $X_3X_4X_5, X_1X_2X_3X_5,$ $X_1X_2X_4X_5, X_1X_3X_4X_5,$ $X_2X_3X_4X_5,$ $X_1X_2X_3X_4X_5$

Table 9.5 Example of a 2^2 Design × Four Blocks (pesticide degradation)

Run No.	Temp X_1	Moisture X_2	Block	X_3	X_4	Random Soil Container	Half Life
1	-	-	1	-	-	2	886.3
2	+	-	1	-	-	3	187.9
3	-	+	1	-	-	1	229.1
4	+	+	1	-	-	4	129.8
5	-	-	2	+	-	3	167.5
6	+	-	2	+	-	4	65.2
7	-	+	2	+	-	1	156.3
8	+	+	2	+	-	2	31.5
9	-	-	3	-	+	1	253.9
10	+	-	3	-	+	3	65.8
11	-	+	3	-	+	4	188.3
12	+	+	3	-	+	2	54.7
13	-	-	4	+	+	2	240.7
14	+	-	4	+	+	4	75.5
15	-	+	4	+	+	3	166.2
16	+	+	4	+	+	1	45.2

9.6 Split Plot Designs

In many multi-factorial industrial experiments, the experimental unit is different for one factor than it is for another. One of the most common causes for this difference is the inability to completely randomize the order of experiments. In some experimental situations, the levels of one or more factors may be easy to change or manipulate throughout the course of experimentation, while other factor levels may be more difficult or expensive to manipulate. This fact restricts the normal practice of randomization, and in these cases a block of experiments at the low level of the difficult to vary factor are run first, followed by a block of experiments at the high level (or vice versa). When this happens, the effect of the difficult to vary factor could be seriously biased by any background variables that change during the course of experimentation. One way to reduce the chance of bias is to repeat the two blocks of experiments, while possibly reversing the order of the levels of the difficult to vary factor. When a set of experiments is performed in this way, the experimental unit for the difficult to vary factor is a block of runs, but the experimental unit for the easy to vary factors is an individual run.

Whenever the experimental unit is different for one factor than it is for another, we call the experimental design a *split plot experiment*. The name comes from its agricultural origin, where some factors (such as row spacing) were difficult to vary (from narrow to wide) within small plots of ground, but other factors such as seed variety or fertilizer rate could be easily varied within a small plot. The levels of a factor that are difficult to change are called *whole plot factors* because their levels were assigned at random to larger blocks (or plots in the agricultural setting). On the other hand, factors whose levels were easy to change were randomized to smaller experimental units, within the larger blocks, and were called the *split plot factors*.

In a split plot experiment, we must calculate a different standard error of an effect for the whole plot factors than we will calculate for the split plot factors. Unless the list of experiments is repeated, so that there are at least two blocks of experiments for each level of the whole plot factor, it will not be possible to calculate the standard error of the whole plot effect. A frequent mistake in the conduct of experiments is failure to recognize a split plot situation, and failure to replicate the list of experiments. In this case the whole plot factor effect is biased by changes in background variables, and no significance test can be performed on it since there is no estimate of its standard error.

In experiments with restricted randomization, the standard error for the whole plot effects often can be quite different than the standard error for the split plot factors. Errors can easily be made in judging the significance of effects if the two separate standard errors are not properly calculated and used in making the signal-to-noise t-ratios.

Split plot experiments occur frequently when experiments are conducted in processes that are comprised of many steps. In these situations changing the levels of some factors involve changes at one processing step, while changes in the levels of other factors may involve changes at other processing steps. This usually results in different experimental unit sizes for different factors. The following example illustrates an example in a multi-step process. In this example, we will illustrate the proper method of replicating the whole plot and calculating the correct standard errors for the whole and split plot factors.

9.6.1 A Split Plot Example

In the integrated circuit industry, individual circuits, called die, are formed on a silicon wafer. Contact window forming is one of the more critical processing steps. A window is a hole of about 3 μm diameter etched through an oxide layer of about 2 μm. The purpose of the windows is to permit interconnections to the micro-circuit. For this reason the windows are called *contact windows*. The contact windows are formed by a photolithography process, as diagramed in Figure 9.11. In this process the photoresist is applied to spinning oxide coated silicon wafers in order to leave a uniform layer. Next lots of wafers are baked to dry the photoresist. The third step is to expose the photoresist coated wafers to ultraviolet light through a mask. The light passes through the mask in the window areas and causes the photoresist in those areas to become soluble in an appropriate solvent. Next the wafers are dipped in a developer which removes the photoresist in the window areas and exposes the oxide. In a high vacuum chamber, a plasma is established that etches the oxide faster than the remaining photoresist layers. At the end of the etching process, windows are left through the oxide to the underlying silicon.

273

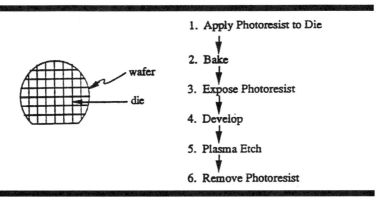

1. Apply Photoresist to Die

2. Bake

3. Expose Photoresist

4. Develop

5. Plasma Etch

6. Remove Photoresist

Figure 9.11 Contact Window Formation

A critical characteristic of contact windows is uniformity of size, and in the manufacturing process it is important to produce windows sizes very near the target dimension. Windows that are not open, or are two small result in loss of contact of the device. Windows that are too large will result in shorted device features.

In order to study the effects of the photoresist baking time, exposure time and developer time on the uniformity of contact windows, an experiment was set up. Test mask patterns were made in the upper left corner of each wafer. In the test patterns, soluble lines of photoresist were imprinted that were later developed and etched. The width of the lines could be measured, and the uniformity of these measurements (as indicated by the variance, s^2) was the response or dependent variable of interest.

In conducting the experiment, the photoresist exposure and development steps were applied individually to each wafer, and it was easy to change the exposure or development time between individual wafers. Thus the experimental unit for these two split plot factors was the wafer. The photoresist baking, however, was done with groups or "lots" of wafers as pictured in Figure 9.12. It would have been very time consuming, tedious, and possibly not representative of the true process if one wafer were baked at a time. Thus the experimental unit for baking time was a lot of wafers, and baking time is the whole plot factor.

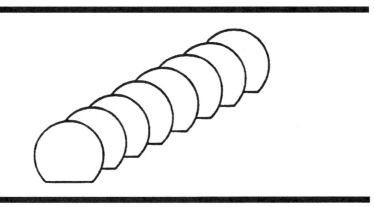

Figure 9.12 A "Lot" of Wafers (the experimental unit for baking)

274

Table 9.6 Experimental Factors and Levels for Contact Window Formation Example

Factor	Levels	
	-	+
X_1 = Development Time (Split Plot Factor)	30 minutes	60 minutes
X_2 = Exposure Time (Split Plot Factor)	20% under normal	20% over normal
X_3 = Photoresist Bake Time (Whole Plot Factor)	20 minutes	40 minutes

The factors and levels for the experiment are shown in Table 9.6. Each run of the experiment was conducted in the following way:

1.   Photoresist was applied to a lot of wafers.

2.   A baking time (20 or 40 min.) was selected at random and all wafers in the lot were baked for the chosen time.

3.   Four wafers from the baked lot were selected at random, and each assigned to one of the four development-exposure time conditions (i.e., [30, -20%], [60, -20%], [30,+20%] or [60, +20%])

4.   The four test wafers were exposed and developed according to the individual conditions they were assigned to, and then they were etched. Each wafer represented a run.

5.   Five measurements of the line width were made at different locations on the test pattern for each test wafer.

The whole process was repeated with four lots of wafers, two lots on each of two days called reps (short for replicates). Table 9.7 shows the coded and uncoded factor levels. The rep factor (X_4) is treated like a blocking factor, with respect to the whole plot factor (X_3 = bake time), and the whole plot standard error of an effect is calculated using the block×whole plot effect interaction. All the runs in Rep 1 were completed first. Within the rep, a lot was baked for 20 minutes followed by a second lot that was baked 40 minutes. In the second rep, the first lot was baked 40 minutes followed by a second lot that was baked for 20 minutes. The random wafer column, at the right side of Table 9.7, identifies the development and exposure conditions that were applied to the four randomly drawn wafers from each lot. The random run order, shown in the far right column of the table, resulted from a two-stage randomization: First the order of the baking times within each rep was randomized, and second the order of exposure-development times was randomized within each bake time and rep.

Table 9.7 Split Plot Design with Random Ordering (contact window example)

Run (Wafer)	Coded Factor Levels				Actual Factor Levels					Random Order
	X_1	X_2	X_3	X_4	X_1 Develop Time	X_2 Exposure Time	Random Wafer #	X_3 Bake Time	X_4 Rep	
1	-	-	-	-	30 min	-20%	2	20 min	1	2
2	+	-	-	-	60 min	-20%	4	20 min	1	4
3	-	+	-	-	30 min	+20%	1	20 min	1	1
4	+	+	-	-	60 min	+20%	3	20 min	1	3
5	-	-	+	-	30 min	-20%	2	40 min	1	6
6	+	-	+	-	60 min	-20%	3	40 min	1	7
7	-	+	+	-	30 min	+20%	4	40 min	1	8
8	+	+	+	-	60 min	+20%	1	40 min	1	5
9	-	-	-	+	30 min	-20%	2	20 min	2	14
10	+	-	-	+	60 min	-20%	3	20 min	2	15
11	-	+	-	+	30 min	+20%	1	20 min	2	13
12	+	+	-	+	60 min	+20%	4	20 min	2	16
13	-	-	+	+	30 min	-20%	4	40 min	2	12
14	+	-	+	+	60 min	-20%	1	40 min	2	9
15	-	+	+	+	30 min	+20%	3	40 min	2	11
16	+	+	+	+	60 min	+20%	2	40 min	2	10

Table 9.8 shows the results of the experiment, the response variable $Y=\log_e(s^2)$, and the calculated effects, in the far right column. The effects are labeled below with abbreviations (1) for X_1, etc.

Since the whole plot factor was randomized to lots within reps, the whole plot standard error of an effect, s_W, is calculated from the rep×bake time (i.e., X_3X_4 interaction). This standard error has one degree of freedom. The standard error for the split plot factors, s_E, is calculated using the other interactions with the rep factor (namely X_1X_4, X_2X_4, $X_1X_2X_4$, $X_1X_3X_4$, $X_2X_3X_4$, and $X_1X_2X_3X_4$). This standard error has 6 degrees of freedom. The calculation of these standard errors is shown at the bottom of the table.

276

Table 9.8 Experimental Results for Contact Window Example

Run	X_1	X_2	X_3	X_4	Replicate Measures of Line Width					s^2	Y $\log_e(s^2)$	Effects
1	-	-	-	-	2.6320	2.6620	2.6210	2.5605	2.4568	0.00661	-5.0185	-4.330 (Mean)
2	+	-	-	-	2.4450	2.6380	2.4680	2.6760	2.5794	0.01040	-4.5660	0.639 (1)
3	-	+	-	-	2.5760	2.5590	2.5964	2.4475	2.6441	0.00530	-5.2398	-0.196 (2)
4	+	+	-	-	2.5520	2.5470	2.5114	2.6750	2.688	0.00655	-5.0287	0.219 (12)
5	-	-	+	-	2.4850	2.4060	2.4735	2.5576	2.657	0.00912	-4.6973	0.923 (3)
6	+	-	+	-	2.7206	2.4750	2.6070	2.6450	2.6994	0.00945	-4.6617	0.084 (13)
7	-	+	+	-	2.4946	2.5586	2.6330	2.5394	2.6559	0.00448	-5.4092	0.180 (23)
8	+	+	+	-	2.5290	2.6445	2.4032	2.3810	2.3785	0.01366	-4.2931	0.168 (123)
9	-	-	-	+	2.4420	2.4630	2.5490	2.6855	2.5353	0.00917	-4.6917	1.069 (4)
10	+	-	-	+	2.4830	2.6290	2.6040	2.7910	2.4863	0.01599	-4.1360	0.185 (14)
11	-	+	-	+	2.6600	2.5790	2.6285	2.4915	2.526	0.00486	-5.3259	0.061 (24)
12	+	+	-	+	2.6930	2.6517	2.5960	2.5970	2.3935	0.01326	-4.3230	0.009 (124)
13	-	-	+	+	2.5273	2.4960	2.4880	2.3305	2.0803	0.03475	-3.3596	0.725 (34)
14	+	-	+	+	2.7836	2.8280	2.3250	2.8140	2.3592	0.06567	-2.7231	-0.038 (134)
15	-	+	+	+	2.3820	2.5190	2.3250	2.6705	2.2109	0.03168	-3.4520	0.095 (234)
16	+	+	+	+	2.3000	2.2060	2.5160	2.7600	2.9445	0.09540	-2.3496	-0.163 (1234)

$$s_W = \sqrt{E_{34}^2/1} = .725, \qquad s_E = \sqrt{\frac{E_{14}^2 + E_{24}^2 + E_{124}^2 + E_{134}^2 + E_{234}^2 + E_{1234}^2}{6}} = \sqrt{\frac{0.185^2 + 0.061^2 + 0.009^2 + 0.038^2 + 0.095^2 + 0.163^2}{6}} = 0.112$$

The t-ratios for all the effects are shown below. Notice that the t-ratio for bake time (X_3) is computed using the whole plot standard error of an effect.

$$t_1 = \frac{0.639}{0.112} = 5.71$$

$$t_2 = \frac{-0.196}{0.112} = -1.75$$

$$t_3 = \frac{0.923}{0.725} = 1.27 \quad \leftarrow \text{ This is the t-ratio for the whole plot effect}$$

$$t_{12} = \frac{0.219}{0.112} = 1.95$$

$$t_{13} = \frac{0.084}{0.112} = 0.75$$

$$t_{23} = \frac{0.180}{0.112} = 1.60$$

$$t_{123} = \frac{0.168}{0.112} = 1.50$$

The significance of the effects is judged by comparing the computed t_E to the tabled t^* with the appropriate degrees of freedom (in this case 1 degree of freedom ($t^*=12.706$) for the bake time (X_3) effect and 6 degrees of freedom ($t^*=2.447$) for all the others.

As can be seen, the only effect that was significant was X_1=Development Time, which had a positive effect. That means that increasing development time increases $Y=\log_e(s^2)$, which means it reduces the uniformity of the contact window size. Therefore the most uniform contact windows are formed with a short (30 minute) development time. Notice that the bake time (X_3) effect would be falsely thought to be significant if the wrong standard error had been used in calculating its t-ratio.

A similar experiment on the contact window formation process was conducted at the Bell Laboratories Integrated Circuits Design Capability Laboratory at Murray Hill (MH ICDCL) in 1980[2]. In three short months of work, these experiments identified improved process parameter settings that accounted for a fourfold reduction in the variance of contact window size, and a threefold reduction in defect density (due to unopened windows). These improvements along with the improved stability and robustness of the process, at the new process parameter settings, gave process engineers added confidence to eliminate a number of in-process checks resulting in a halving of the total throughput time of wafers in the window photolithography process.

9.6.2 Designs for Split Plot 2^k Experiments

The factor effects and interactions for 2^k factorials run in split plot designs can always be computed using the worksheet method or Yates algorithm shown in Chapter 8. However, the rep effects and rep interaction effects that are used to compute the whole plot standard error of an effect (s_W) and the standard error for the split plot factors (s_E) cannot be computed using these methods unless the number of replicates is a power of two (i.e., $2^1 = 2$, $2^2 = 4$, etc.). For the more general split plot 2^k factorial designs, where the number of reps is arbitrary, the analysis and significance tests must be completed using the Analysis of Variance (ANOVA) that will be described in Chapter 11.

There are still many useful 2^k split plot designs where the number of reps is equal to two. Table 9.9 shows how the computation tables in Appendix B.2 can be used to create and analyze various 2^k split plot designs with this property. The replicate term as well as the terms that should be used to calculate the whole plot and split plot errors are shown in this table.

Table 9.9 Two Rep Split Plot 2^k Designs (with breakdown of error terms)

Number of Whole Plot Factors	Number of Split Plot Factors	Appendix Table	Whole Plot Factors	Slit Plot Factors	Rep Factor	Whole Plot Error Factors	Split Plot Error Factors
1	1	B.2-3	X_2	X_1	X_3	X_2X_3	X_1X_3, $X_1X_2X_3$,
1	2	B.2-4	X_3	X_1, X_2	X_4	X_3X_4	X_1X_4, X_2X_4, $X_1X_2X_4$, $X_1X_3X_4$, $X_1X_2X_3X_4$
2	1	B.2-4	X_2, X_3	X_1	X_4	X_2X_4, X_3X_4, $X_2X_3X_4$	X_1X_4, $X_1X_2X_4$, $X_1X_3X_4$, $X_1X_2X_3X_4$
1	3	B.2-5	X_4	X_1, X_2, X_3	X_5	X_4X_5	X_1X_5, X_2X_5, X_3X_5, $X_1X_2X_5$, $X_1X_3X_5$, $X_2X_3X_5$, $X_1X_2X_3X_5$, $X_1X_4X_5$, $X_3X_4X_5$, $X_1X_2X_4X_5$, $X_1X_3X_4X_5$, $X_2X_3X_4X_5$, $X_1X_2X_3X_4X_5$
2	2	B.2-5	X_3, X_4	X_1, X_2	X_5	X_3X_5, X_4X_5, $X_3X_4X_5$	X_1X_5, X_2X_5, $X_1X_2X_5$, $X_1X_3X_5$, $X_1X_4X_5$, $X_2X_3X_5$, $X_2X_4X_5$, $X_1X_2X_3X_5$, $X_1X_2X_4X_5$, $X_1X_3X_4X_5$, $X_2X_3X_4X_5$, $X_1X_2X_3X_4X_5$
3	1	B.2-5	X_2, X_3, X_4	X_1	X_5	X_2X_5, X_3X_5, X_4X_5, $X_2X_3X_5$, $X_2X_4X_5$, $X_3X_4X_5$, $X_2X_3X_4X_5$	X_1X_5, $X_1X_2X_5$, $X_1X_3X_5$, $X_1X_4X_5$, $X_1X_2X_3X_5$, $X_1X_2X_4X_5$, $X_1X_3X_4X_5$, $X_1X_2X_3X_4X_5$

9.7 Summary

In this chapter we have discussed three additional choices that must be made in planning 2^k type factorial experiments and five additional methods for analyzing data from 2^k experiments. The three choices needed for planning 2^k experiments are: 1) choice of the number of replicates 2) choice of whether or not to run a blocked experiment, and 3) choice of whether or not to run a split plot experiment. Choice of the number of replicates depends upon the smallest effect that we don't want to overlook (δ) and an estimate of the standard deviation of a single observation (σ). Choice of whether to run a blocked experiment depends on the existence of known background variables. Finally, choice of whether to run a split plot design usually depends on whether the runs or experiments can be completely randomized to experimental units, or whether there are difficult to vary factors that prevent complete randomization. Figure 9.13 is a flow diagram that illustrates the choices that must be made.

Additional methods of analysis discussed in this chapter were: 1) expressing the results of an experiment in the form of a linear prediction equation that can be used to predict the response at combination of factor levels not yet tested, 2) including center points in the 2^k design and doing a curvature test to check the adequacy of the linear prediction equation, 3) using contour plots to represent the prediction equation graphically, 4) calculating the standard error of the effect (s_E) from block interaction terms in Blocked experiments, and 5) analysis of split plot experiments that involve calculating the standard error of split and whole plot factors. In Chapter 10 we will expand on the method of fitting prediction equations to data and present additional methods of checking the adequacy of the models.

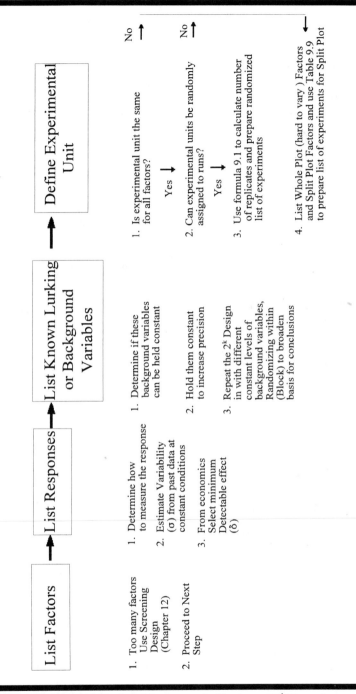

Figure 9.13 Flow Diagram of Choices Necessary in Design of 2^k Experiment

9.7.1 Important Equations

$$n_F = (8\,\sigma\,/\,\delta)^2$$

$$\delta = 8\,\sigma\,/\,\sqrt{n_F}$$

$$b_i = \tfrac{1}{2}[\text{Effect for Factor } X_i\,]$$

$$b_{ij} = \tfrac{1}{2}[\text{Interaction for } X_iX_j\,]$$

$$s_E = \sqrt{\sum_i BI_i^{\,2}/m}$$

9.7.2 Important Terms and Concepts

The following is a list of the important terms and concepts discussed in this chapter.

1.Wheeler, R. E., "Portable Power", *Technometrics,* Vol. 16, (1974) , p. 193.

2. Phadke, M. S., Kackar, R. N., Speeney, D. V., and Grieco, M. J., "Off-Line Quality Control in Integrated Circuit Fabrication using Experimental Design," *The Bell System Technical Journal*, May-June 1983, Vol 61, No 5, p.1273-1309.

9.8 Exercises for Chapter 9

9.1   Suppose that in the experiment described in Problem 8.1, it was desired to detect any main effect or interaction that would effect the percent shrinkage of injection molded parts by 10% or more. Factors with effects this large could then be controlled to reduce shrinkage and save considerable expenses in assembly.   If the historical standard deviation in percent shrinkage from replicate parts molded under the same conditions was $\sigma = 4\%$, how many replicates of the 2^3 design you created for Problem 8.1 are required?

9.2   Doty, Gillian, LeBaron, and Sawaya studied the effects of the number of Automatic Guided Vehicles, AGV's (1-3), Speed of AGV's (120/100 - 180/150 ft./min.) and the pick up and drop off times of AGV's upon the total throughput time in a castings factory. A 2^3 design with replicated center points was used, and the data was generated by the ProModel Simulation program. The data is listed below:

$X_1 =$ Number of AGV's	$X_2 =$ Speed of AGV's	$X_3 =$ Pick/Drop time	Throughput
-	-	-	180 184 181
+	-	-	202 204 207
-	+	-	199 194 195
+	+	-	206 213 210
-	-	+	190 186 184
+	-	+	208 208 209
-	+	+	196 203 204
+	+	+	215 210 195
0	0	0	194 201 195 204 197 203

a) Calculate the mean and variance, s^2, of the replicate responses for each run, and calculate the pooled standard deviation s_p.

b) Do a graphical analysis by plotting the mean responses at each run on a cube.

c) Calculate the main effects, two-way interactions, the three-way interaction, and the curvature effect.

d) Calculate the standard error of an effect, s_E, the standard error of the curvature effect, s_C,

e) Determine which effects and interactions are statistically significant. Is curvature significant?

9.3 The following problems are based on the experiment on milling aluminum in Problem 8.2.

a) Assuming the historical standard deviation of the surface finish of milled aluminum was $\sigma = 12.5$, what is the smallest effect that you could detect with 95% confidence, using three replicates of the full design shown in Problem 8.2.

b) Using the data from Problem 8.2, develop a prediction equation using the significant effects and or interactions.

c) Using the prediction equation you developed in Part (b), predict the value of surface finish with Spindle Speed = 1200 RPM, Feed Rate = 10 IPM, and Depth of Cut = 0.020.

9.4 List the experiments in randomized order for

a) A 2^3 design

b) A 2^3 design run in two blocks

c) A 2^3 design run in four blocks

9.5 Router bits are used to smooth the edges of printed circuit boards. When a bit wears out, dust is created that clings to the edge surface of the circuit boards. This requires an extra cleaning step, or creates difficulty while sliding the boards into slots during assembly. An experiment is to be designed to test the effect of two factors at two levels upon the useable life of router bits. The factors are X_1 = Router RPM, and X_2 = feed rate. An experiment will consist of setting the RPM and feed rate, then routing with one bit until dust begins to form indicating the bit is worn. Router bits have four positions, and thus four experiments can be made with each bit. It is known that useable life differs between bits (i.e., a known lurking variable).

a) Set up a 16 run blocked design to test the effects and interactions of the two factors. Show the run order of the experiments, and explain how you would calculate the error of an effect.

b) Set up a 32 run blocked design to test the effects and interactions of the two factors. Show the run order of the experiments, and explain how you would calculate the error of an effect.

9.6   In the study dealing with a solar water heating system that was given in Problem 8.3, one additional run of the computer model was made at the center point, where the coded factors levels were $X_1 = 0$ (the total daily insolation), $X_2 = 0$ (the tank storage capacity), $X_3 = 0$ (the water mass flow rate through the absorber), and $X_4 = 0$ (the intermittency of the input from solar radiation). The responses at this condition were $Y_1 = 41.6$ (the collection efficiency) and $Y_2 = 100.0$ (the energy delivery efficiency). This additional data would allow for checking the adequacy of a simpler linear approximation model to be determined.

a) Write a simple linear prediction model for each response using the effects that appear important on the plots you constructed in Problem 8.3.

b) Determine an estimate of s_E from the Normal plot of effects you made in Problem 8.3.

c) Calculate the constant that the ratio s_C/s_E would equal for this experiment.

d) With reference to the constant calculated in Part (c), does curvature appear to play an important role?

9.7   Four different fabrics defined by the % rayon and weave style were compared on a Martindale wear tester for suitability for automobile seats. The wear tester can compare four samples of cloth in a single run (block). The weight loss in milligrams from four runs were measured and the following results recorded:

Run Order or Tester Pos.	Block	%Rayon	Weave Style	Weight loss
1	1	-	-	36
4	1	+	-	36
3	1	-	+	38
2	1	+	+	30
3	2	-	-	17
2	2	+	-	26
4	2	-	+	18
1	2	+	+	17
3	3	-	-	30
1	3	+	-	41
4	3	-	+	39
2	3	+	+	34
2	4	-	-	30
3	4	+	-	38
1	4	-	+	40
4	4	+	+	33

a) Determine if there are any significant effects of % rayon, weave style, or their interaction.

b) Was the blocking of experiments useful?

9.8   A design engineer wished to determine the relationship between the type of fastener, thread type, hole size, size of the hole relative to its tolerance, and the clamp load of a fastener constant torque (response). Each experiment consisted of installing a fastener and tightening it to the specified torque then measuring the clamp load between the two sheets being fastened. The factors and levels are listed below:

Factor	-	+
1. Type Fastener	Standard	Self Threading
2. Thread	Rolled	Tapped
3. Hole Size	1/4	3/8
4. Hole Relative to Tolerance	Low end of Tolerance	High end of Tolerance

a) List the experiments in a worksheet form and indicate the order that they should be performed, in order to examine the effects.

b) Differences in clamp load of 10 psi are important (because reduction of 10 psi may cause failure of the final product). From previous control charts produced using one type of fastener, thread type, and hole size the standard deviation of clamp load was estimated to be $\sigma = 6.07$. How many replicates do you recommend for this experiment?

c)   If hole relative to tolerance was hard to vary from experiment to experiment, and it would be much more convenient to run the experiments in groups with constant hole relative to tolerance, how would you change the experimental design? List a new worksheet and show the order in which experiments are conducted.

9.9   For the scenario given in Problem 8.2 ($n_F = 3 \times 2^3 = 24$, s = 12.7):

a) Were an appropriate number of experiments performed if $\delta = 15$? If not, what would be an appropriate number of experiments?

b) For the design that was used ($n_F = 24$) what δ were you fairly sure (95%) of detecting? (Note: For simplicity, just use Equation 9.2.)

c) Summarize the result of your analysis of Problem 8.2 (conducted for Chapter 8 homework) in equation form. Reminder: include only significant effects and interactions.

9.10   Sausage casings are extruded from a gel batch made from collagen (the underside of cow hides). The gel batch was made by soaking the collagen in extremely hot water for a prolonged period.   Next, the gel is extruded into a tube and fixed by inflating it with ammonia.   It was desirable to make the diameter of the extruded tube as consistent as possible. Therefore an experiment was set up to test factors in both the gel-making stage as well as the extrusion stage to see their effect on diameter consistency. Consistency was measured as the log of the variance of several diameter measurements taken along the length of the tube. Below is a list of the factors and levels:

Factor	-	+
1. Gel batch soaking time, X_1	48 hrs	72 hrs
2. Extrusion pump pressure, X_2	30 psi	45 psi
3. Ammonia concentration, X_3	low	high

Since Factors 2 and 3 could be varied within a gel batch, but Factor 1 could only be varied between gel batches, four different gel batches were made in order to replicate Factor 1. The first two gel batches were labeled Replicate 1 and the second two as Replicate 2. The data from the experiment are shown below:

Run	Rep	Gel	X_1	X_2	X_3	$\log(s^2)$
1	1	1	-	-	-	28.2359
2	1	1	-	+	-	26.5620
3	1	1	-	-	+	15.5834
4	1	1	-	+	+	26.4054
5	1	2	+	-	-	37.0237
6	1	2	+	+	-	58.9393
7	1	2	+	-	+	30.8361
8	1	2	+	+	+	51.4071
9	2	3	-	-	-	32.5547
10	2	3	-	+	-	42.3253
11	2	3	-	-	+	20.4696
12	2	3	-	+	+	28.4434
13	2	4	+	-	-	58.6681
14	2	4	+	+	-	73.3822
15	2	4	+	-	+	57.0995
16	2	4	+	+	+	71.9670

a) Perform analysis of this data to determine if there are significant effects of factors X_1 - X_3 or any of their two-way or three-way interactions.

b) Write a sentence or two interpreting each significant effect or interaction you find.

c) Make a recommendation of the conditions that will result in the most consistent casing.

9.11   For the scenario given in Problem 8.3 ($n_F = 3 \times 2^3 = 24$, s = 4.5):

a) Were an appropriate number of experiments performed if $\delta = 10$? If not, what would be an appropriate number of experiments?

b) For the design that was used ($n_F = 24$) what δ were you fairly sure (95%) of detecting? (Note: For simplicity, just use Equation 9.2.)

c) Summarize the result of your analysis of Problem 8.3 (conducted for Chapter 8 homework) in equation form. Reminder: include only significant effects and interactions.

9.12   A study was performed to investigate the effects of iron deficiency in diets of rats on their activity. The measure of activity, Y, was the time spent in vertical movement (seconds per 15-minute interval). The response was measured automatically and continuously using a rat cage with infrared beams connected to a computer monitoring system. The response recorded was the average of 40 15-minute readings taken during 10 hours of light or 10 hours of darkness (with 2 hours in between). The factors studied were:

X_1:   Iron in diet   - = Deficient   + = Adequate
X_2:   Age of rat   - = 3 weeks old   + = 6 weeks old
X_3:   Time of day   - = day (light)   + = night (dark)

Iron in Diet X_1	Age, weeks X_2	Day or Night X_3	Response, Y Litter of Rats			
			1	2	3	4
-1	-1	-1	4.1	9.0	3.7	5.2
1	-1	-1	2.6	8.4	4.3	7.5
-1	1	-1	4.9	7.2	3.8	3.3
1	1	-1	3.1	8.8	4.8	4.9
-1	-1	1	10.3	14.2	12.7	12.8
1	-1	1	10.9	18.5	12.8	15.0
-1	1	1	5.6	12.1	4.8	6.3
1	1	1	10.2	15.9	12.8	10.3

Four rats were used for each set of conditions, each from a different litter of rats. The data collected are given above.

a) Treating the rats as pure replicates, analyze the data to determine which factors and interactions are statistically significant. Summarize your results in equation form.

b) Treating the rat litters as blocks, reanalyze the data to determine which factors and interactions are statistically significant. Note: Use the interactions of blocking variables with the factors studied to estimate s_E. Summarize your results in equation form.

c) What was the impact of treating the litters as blocks on your analysis?

9.13   Let us pretend theat while Phil Jackson was coaching Michael Jordon for the Chicago Bulls (of the National Basketball Association), he was interested in figuring out what factors influenced Michael's performance during a basketball game. He (Phil Jackson) suspected that the number of McJordon burgers Michael consumed the afternoon before the game, the hours of sleep Michael got the night before the game, and the number of glasses of Jordonade Michael drank the evening of the game, all influenced how many points Michael scored during the game. Phil heard, via the grapevine, that you were a statistical whiz, and he called you to design an experiment for the study. You recommended a tried and true factorial design plus center points, and Phil managed to collect the following data:

X_1 = Burgers - = 1 0 = 2 + = 3	X_2 = Sleep - = 6 hours 0 = 7 hours + = 8 hours	X_3 = Jordonade - = 2 glasses 0 = 3 glasses + = 4 glasses	Y = Performance (Points scored in the basketball game)
-	-	-	25
+	-	-	31
-	+	-	49
+	+	-	53
-	-	+	37
+	-	+	14
-	+	+	29
+	+	+	38
0	0	0	30, 27, 31

Do any of the factors have an impact on Michael's performance? Back up your conclusions with the appropriate calculations.

9.14   The following is a continuation of Problem 8.6, which is based on the performance of a gas chromatograph.

(a) Develop a prediction equation based on the significant effects.

(b) Using the prediction equation from Part (a), determine the conditions (within the range of the experimental data) which maximize the peak area response. (Hint: Use a spreadsheet solver function.)

9.15   The following is a continuation of Problem 8.8, which studied prototype replacement hip joints.

(a) Develop a prediction equation based on the significant effects for both force to dissociate, and range of motion.

(b) Using your two prediction equations from Part (a), try to locate conditions (within the range of the experimental data) which will maximize force to dissociate, while maintaining a range of motion that is at least 49 degrees.

CHAPTER 10

Regression Analysis

10.1 Introduction

The goal of the vast majority of experimentation is (or ought to be) to develop a model which adequately describes the system being studied. The model can then be used for whatever the intended objective is: optimization, troubleshooting, control, etc. The model is essentially a concise summary of all the data that was taken. The model smooths out the noise (variability) in the data, and it elucidates the underlying relationships between the factors and the response(s).

As a rule, the *form* of the model is known (or specified) before any experiments are run. It could be a simple straight line, a complicated polynomial, or anything in between. It could even be a mechanistic model, but those are, in general, beyond the scope of this text. But, even though the form is known, the model has constants in it whose values are NOT known. One way to think of the goal of experimentation is that the goal is to allow the estimation of those constants (which we sometimes also call by other names like coefficients or parameters). The main thrust of this chapter describes the process we use to distill the values of those constants from our data.

We will begin this chapter with a description of the method, which is called *Least Squares*. We will then apply the method to a simple problem of fitting a straight line (Section 10.3). After that, the more useful situation of fitting a model with several factors in it will be covered (Section 10.4). We will look at describing how well the model fits the data in Section 10.5, and how good any assumptions are that we had to make along the way (Section 10.6). Included at the end of this chapter is an appendix on matrix algebra. It is intended as a very brief review of matrix notation and manipulation, and it is included because the equations that are used for regression analysis can be expressed very succinctly in matrix form.

10.2 The Method of Least Squares

As stated in the introduction, our goal is to find the best values for the constants in a model (i.e., the values of the constants that give the best fit to a set of data). Before we can do anything we must agree upon what we mean by "best." As the basis of our discussion, let us say we are trying to find the best straight line through the origin and the two data points shown in Figure 10.1.

The centuries-old method of fitting the line is the eyeball method. This involves nothing more than putting a straight-edge on the graph where it looks best by eye and then drawing the line. Although this is still a frequently used method when it comes to drawing a straight line, it has some shortcomings. First of all it is not objective; two people would not necessarily agree on the best straight line through a given set of data. We would also usually like to be able to draw some inferences about the model (like putting error limits on the coefficients and/or the model predictions)

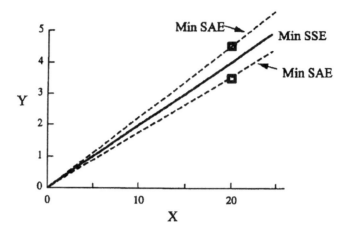

Figure 10.1 Possible Lines Through Two Data Points

which cannot be done with eyeball lines. And last, but not least, the eyeball method can only be used with one X variable. Drawing multi-dimensional surfaces (even linear ones) by eye is virtually impossible. So we need a quantitative method.

A good fit of the line to the data means that the line should be as close to all the data points as possible. Therefore, one possible quantitative measure of goodness is the sum of the absolute values of the vertical distances from the data points to the line, or the SAE (Sum of the Absolute Errors) for short. In equation form:

$$SAE = \sum_{i=1}^{n} | Y_i - \hat{Y}_i | \qquad (10.1)$$

where Y_i is the value of the response for the ith data point, and $\hat{Y}_i$ is the predicted value of the response for the ith data point given by the line. Although this is intuitively a very reasonable measure, it is not necessarily unique. In Figure 10.1, both of the two outside lines and all lines in between them would have the same value of SAE. Therefore, any line in that interval would be judged by the SAE criterion to be equally good. This is not what we would like. We would like the line down the middle, centered between the two data points, to be best. So SAE is not such a wonderful criterion after all. And, as a practical matter, it often turns out to be difficult to find a minimum (i.e., the best model), so a better criterion is needed.

The measure that is almost universally used is the sum of the squared vertical distances from the data points to the line, or the SSE (Sum of the Squared Errors) for short. In equation form:

$$SSE = \sum_{i=1}^{n} (Y_i - \hat{Y}_i)^2 \qquad (10.2)$$

292

Using this criterion, there is only one best line, and it goes down the center of the data as we would expect it to. Although, as has already been indicated, there are other measures of goodness, the criterion used in this text will be SSE exclusively. When using this criterion, values for the coefficients in our model that minimize SSE are sought. Since we are making the sum of the squared distances as small as possible, the whole procedure is often called the *method of least squares*.

To illustrate the method for a more realistic but still simple case, consider the same straight line through the origin as discussed in relation to Figure 10.1, but applied to fitting a few more data points. The physical situation is that we want to predict the tensile strength of a standard alloy of steel as a function of its Brinell hardness number. The response, Y, is therefore the tensile strength, and the independent variable, X, is the Brinell hardness. The mathematical model is:

$$\hat{Y} = bX \qquad\qquad (10.3)$$

Samples of five steel alloys were taken and their Brinell hardness numbers were recorded in Table 10.1. The tensile strength values were determined experimentally and also recorded in that table.

Table 10.1 Data and Calculations for Predicting Tensile Strength of Steel from Brinell Hardness

Steel Sample	(X) Brinell Hardness	(Y) Tensile Strength	X^2	Y^2	XY
1	500	256	250,000	65,536	128,000
2	431	212	185,761	44,944	91,372
3	370	189	136,900	35,721	69,930
4	321	155	103,041	24,025	49,755
5	285	138	81,225	19,044	39,330
Sums:	1907	950	756,927	189,270	378,387

Now that we have data, the method of least squares can be used to estimate the unknown coefficient, b, in Equation 10.3. In other words, we would like to find the best value of b, and we now understand "best" to mean the value (of b) which minimizes SSE. If we put our model into the equation for SSE, we get

$$\text{SSE} = \sum_{i=1}^{n} (Y_i - \hat{Y}_i)^2 = \sum_{i=1}^{n} (Y_i - bX_i)^2 \tag{10.4}$$

Since the Y_i's and the X_i's and known quantities, the only unknown in the equation for SSE (Equation 10.4) is b. Therefore, we could use any minimization procedure that we choose to find the best value for b.

One possible method is "brute force." That is, we can try numerous values b, calculate the SSE for each one, plot the SSE values on a graph versus b, and pick the value that gives the best fit (i.e., the smallest value of SSE). If this were done for our data set, the plot shown in Figure 10.2 would be obtained from which the best value for b can be seen to be 0.50.

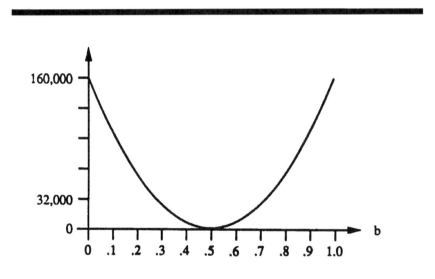

Figure 10.2   SSE As a Function of the Slope of the Line, b

Although brute force will certainly get us to the best value of b, we can be much more efficient. It should be noticed that the graph of SSE versus b looks very much like a quadratic curve, and, in fact, it is. This can be seen if we expand Equation 10.4.

$$\text{SSE} = \sum_{i=1}^{n} (Y_i^2 - 2bX_i Y_i + b^2 X_i^2) \tag{10.5}$$

294

or, upon rearranging,

$$\text{SSE} = \left(\sum_{i=1}^{n} X_i^2\right)b^2 - 2\left(\sum_{i=1}^{n} X_i Y_i\right)b + \left(\sum_{i=1}^{n} Y_i^2\right) \tag{10.6}$$

For the data in Table 10.1, $\sum X_i^2 = 756{,}927$, $\sum X_i Y_i = 378{,}387$, and $\sum Y_i^2 = 189{,}270$, so that Equation 10.6 becomes

$$\text{SSE} = 756{,}927 b^2 - 756{,}774 b + 189{,}270 \tag{10.7}$$

which is a quadratic equation in b. Using elementary calculus, the minimum is located at the point where the first derivative of SSE with respect to b is zero. If we take the first derivative we get

$$\frac{d(\text{SSE})}{db} = 2\left(\sum_{i=1}^{n} X_i^2\right)b - 2\left(\sum_{i=1}^{n} X_i Y_i\right) = 0 \tag{10.8}$$

or, for our example,

$$\frac{d(\text{SSE})}{db} = 2(756{,}927)b - 2(378{,}387) = 0 \tag{10.9}$$

Solving Equation 10.8 (or Equation 10.9) for b, we get

$$\hat{b} = \sum_{i=1}^{n} X_i Y_i / \sum_{i=1}^{n} X_i^2 = 378{,}387 / 756{,}927 = 0.4999 \tag{10.10}$$

This is the same answer that we got by brute force, but we got it directly. Note that the "^" (called a hat) on b indicates that the value is not the true value (which is unknown) but rather our best guess based on the data we have and the method of least squares.

As a check that Equation 10.8 gives us a minimum (although in this case it is obvious from the graph of SSE versus b) we can check the second derivative to make sure it is positive.

$$\frac{d^2(\text{SSE})}{db^2} = 2\left(\sum_{i=1}^{n} X_i^2\right) > 0 \tag{10.11}$$

Since the second derivative is the sum of the X_i's squared, it will always be positive, and so this procedure will always give a minimum in SSE. This continues to be true for larger models also.

The data for this example is shown graphically in Figure 10.3 along with the prediction equation, $\hat{Y} = 0.500 X$, fit by using least squares. The equation can be seen to go through the data nicely.

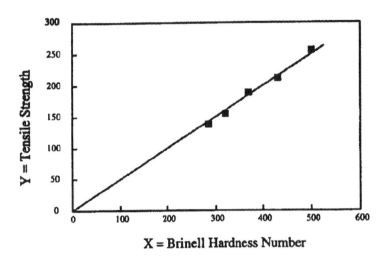

Figure 10.3 Brinell Hardness Data and Fitted Line

10.3 Linear Regression

10.3.1 Estimation of Coefficients

Let us now look a situation that is just a bit more complex — the case where the regression equation is a straight line (as before) but it may have a nonzero intercept. The least squares estimates of the slope and intercept are still found in the same manner. The model is

$$\hat{Y} = a + bX \qquad (10.12)$$

and the SSE (the sum of the squared differences between actual data and predicted values from the equation) can be written as a quadratic function of a as well as b. This is shown in Equations 10.13 and 10.14.

$$SSE = \sum_{i=1}^{n} (Y_i - \hat{Y}_i)^2 = \sum_{i=1}^{n} (Y_i - [a + bX_i])^2 \qquad (10.13)$$

or, upon expanding the squared quantity,

$$SSE = \sum Y_i^2 - (2\sum Y_i)a - (2\sum X_i Y_i)b \\ + na^2 + (\sum X_i^2)b^2 + (2\sum X_i^2)ab \qquad (10.14)$$

The values of a and b which give a minimum SSE can be found, as before, by taking the derivatives with respect to a and b and setting them equal to zero.

296

$$\frac{\partial \text{SSE}}{\partial a} = -2\sum Y_i^2 + (2n)a + (2\sum X_i)b = 0$$

$$\frac{\partial \text{SSE}}{\partial b} = -2\sum X_i Y_i + (2\sum X_i)a + (2\sum X_i^2)b = 0$$

(10.15)

This is a pair of simultaneous linear equations in a and b whose solutions are

$$\hat{b} = \frac{n\sum X_i Y_i - (\sum X_i)(\sum Y_i)}{n\sum X_i^2 - (\sum X_i)^2}$$

(10.16)

$$\hat{a} = (\sum Y_i - \hat{b}\sum X_i)/n$$

For the data in Table 10.1, the calculated coefficients are

$$\hat{b} = \frac{5(378,837) - (1907)(950)}{5(756,927) - (1907)^2} = 0.5425$$

(10.17)

$$\hat{a} = [950 - (0.5425)(1907)]/5 = -16.92$$

The model (straight line) with the least squares estimates for a and b is shown in Figure 10.4 along with the data. It can be seen to be a closer fit to the data than the straight line through the origin. This is to be expected, of course, since the bigger model is more flexible.

The fact that the coefficients of the least squares regression line are so easily calculated has made the method very popular. Even many hand-held calculators can do the calculations automatically, not to mention popular computer software packages like spreadsheets.

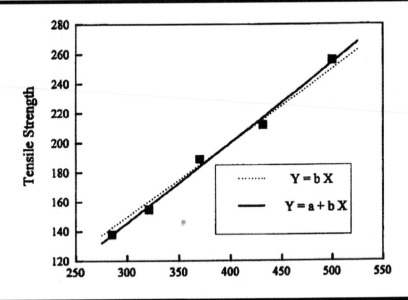

Figure 10.4 Brinell Hardness Data with Model Y= a+bX

10.3.2 Statistical Significance of Coefficients

As early as the nineteenth century, least squares was a commonly used method of fitting regression lines to astronomical data. Karl Gauss sought to determine the conditions under which the line fitted with least squares was the best possible estimate of the true line (*best* in the sense of being unbiased with minimum variance as discussed in Section 6.3.1). His work led to the discovery of the Normal (or Gaussian) Distribution. He also found what he was after — that the least squares estimates are best under the following conditions: (1) the errors, ϵ_i, are Normally distributed with a mean of zero, (2) all the errors have the same variance (i.e., all data have the same precision), and (3) the errors are independent from observation to observation (i.e., knowing the value of one of them does not help you predict any others). Remember that the error, ϵ_i, is what is added to the true response for the ith observation to give the observed response.

$$Y_i = a + bX_i + \epsilon_i \qquad (10.18)$$

Figure 10.5 illustrates this situation. At any specific X value, the long run average or expected value of Y is a + bX, but a particular observation of Y is one random selection from the bell-shaped Normal Distribution centered on the line, $Y = a + bX$. It should also be noticed that the Normal Distributions all have the same variance, σ^2.

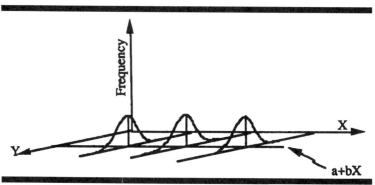

Figure 10.5 Assumptions Made During Regression Analysis

When the distribution of the errors is Normal, not only do we know that the least squares estimates of a and b are best, but we are also able to use statistical theory to perform tests of significance and/or determine confidence intervals. It would be a rare situation where that would not be just as important as determining the best line.

In this section, we will focus on tests of significance. Specifically, we want to determine if the coefficients (a and b) are significantly different from zero. This is crucial if one wants to justify the assertion that X influences Y. Since b quantifies just how much Y changes when X is changed, if b could reasonably be zero, then it is plausible that X has no impact on Y.

Let us consider the case where we have data at only two values of X: x_1 and x_2. Then, as can be seen in Figure 10.6, comparing the slope of the regression line, b, to zero is equivalent to comparing the mean of the Normal Distribution centered at $a + bx_1$ (denoted by μ_1) to the mean

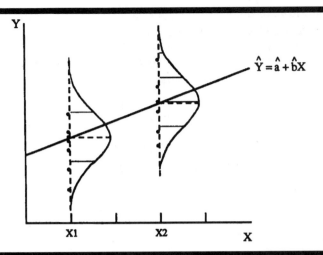

Figure 10.6 Comparing the Slope, b, to Zero Is Like Comparing the Means of Two Distributions

of the Normal Distribution centered at a + bx$_2$ (denoted by μ_2). Recall from Chapter 6, that the signal-to-noise ratio (or two-sample t-statistic) for comparing the means of two Normal Distributions is:

$$t_{(n_1 + n_2 - 2)} = \frac{\bar{Y}_1 - \bar{Y}_2}{s_p \sqrt{1/n_1 + 1/n_2}}$$

(10.19)

If this statistic is larger than the upper tail of the tabulated student's t-distribution, it indicates that $\mu_1 - \mu_2 > 0$, or the mean of the first population is greater than the mean of the second population. If this statistic is smaller than the lower tail of the tabulated student's t-distribution it indicates that $\mu_1 - \mu_2 < 0$, or the mean of the second population is larger. In either case, if the two means are different, the slope, b, is not zero. The student should notice that this is basically the same situation we are in when we analyze data from a two-level factorial design (with or without center points).

In the general case where there are more than two values of X, the t-statistic for testing whether b = 0 is, as expected, the ratio of $\hat{b}$ to its standard deviation:

$$t_{(n-2)} = \hat{b} / s_{\hat{b}}$$

where: $s_{\hat{b}} = s / \sqrt{\Sigma (X_i - \bar{X})^2}$

(10.20)

In this equation, s is the standard deviation of the errors in the individual responses, and $\bar{X}$ is the average of the X values ($\bar{X} = \Sigma X_i / n$). Although the equation for the standard deviation of $\hat{b}$ was given without any derivation (or perhaps especially because it was not derived), it should be examined to see if it makes sense. Equation 10.20 says that the variability in the estimate of the slope is proportional to the variability of the individual observations, s, which is reasonable. The only other thing that affects the precision of $\hat{b}$ is $\Sigma (X_i - \bar{X})^2$; the bigger the sum the more precise the estimate of the slope. The sum gets larger if we have more data and/or the X values are further away from the middle. Both of these impacts make good intuitive sense — more data and more spread in the values of X seem like they would result in a better estimate of the slope.

In order to calculate the standard deviation of $\hat{b}$, we first need to know s, the standard deviation of the errors in the Y values. When we had a group of replicate data points taken at the same conditions, the formula used in Chapter 4 (as well as elsewhere) was:

$$s^2 = \sum_{i=1}^{n} (Y_i - \bar{Y}_i)^2 / (n-1) \tag{10.21}$$

Now, however, we do not (necessarily) have any replicate data points. But, that does not mean that we have no estimate of the variability. The scatter of the data around the line is essentially the same thing, so our equation to calculate the variance becomes:

$$s^2 = \sum_{i=1}^{n} (Y_i - \hat{Y}_i)^2 / (n-c) = SSE / (n-c) \tag{10.22}$$

It should be noticed that the denominator changed from (n-1) to the more general (n-c), where c is the number of constants in the model to be estimated from the data. Before, c was one (just an average), and now c is two (a slope and an intercept).

For our Brinell Hardness example, the quantities needed to calculate the t-statistic are given in Table 10.2. Using them gives:

$$s_{\hat{b}} = 4.43 / \sqrt{29,597.2} = 0.0257$$

So $t_3 = 0.5425 / 0.0257 = 21.08$. As was the case for other t-statistics used in previous sections, a value larger than the tabled t-distribution with the appropriate degrees of freedom indicates that $b \neq 0$. This is far, far greater than the tabled value of 3.182, so we conclude that there is certainly (for all practical purposes) a relationship between Brinell Hardness and tensile strength.

Table 10.2 Worksheet for Calculating the Significance of the Linear Model for Brinell Hardness

X	Y	$\hat{Y} = $ $\hat{a} + \hat{b}X$	$\epsilon_i = $ $Y - \hat{Y}$	ϵ_i^2	$(X - \bar{X})^2$	X^2
500	256	254.34	1.66	2.747	14,066.0	250,000
431	212	216.91	-4.91	24.097	2,460.2	185,761
370	189	183.82	5.18	26.881	130.0	136,900
321	155	157.23	-2.23	4.982	3,648.2	103,041
285	138	137.70	0.30	0.089	9,293.0	81,225
SUM 1907	950			58.796	29,597.2	756,927
AVG 381.4						

$$s^2 = 19.60$$
$$s = 4.43$$

In addition to checking the statistical significance of the slope of the line, we can also check whether the intercept, a, is non-zero. The t-test can be thought of as subtracting $Y = 0 + bX$ from each Y value, then performing a one-sample t-test to determine if the differences have zero mean. The general formula for this t-statistic is

$$t_{(n-2)} = \hat{a} / s_{\hat{a}}$$
$$\text{where:} \quad s_{\hat{a}} = s \sqrt{\Sigma X_i^2 / n \Sigma (X_i - \bar{X})^2}$$

(10.23)

For the Brinell Hardness data,

$$s_{\hat{a}} = 4.43 \sqrt{(756,927)/(5)(29,597.2)} = 10.01$$

Therefore, $t_3 = -16.92 / 10.01 = -1.69$. Since this is not as large as the tabled value, we conclude that the intercept could reasonably be zero. Under the philosophy, "The simpler, the better!" the intercept term in the model should be dropped, and the model of a straight line through the origin is found to be more appropriate.

Although the calculations of least squares estimates for the slope and intercept of straight lines as well as the standard deviations of those estimates can be done by hand (presumably using a calculator at minimum), computer packages now do most or all of the work for us. A sample of a typical spreadsheet regression analysis is given in Figure 10.7. The coefficients and their standard errors are calculated. All that remains for the analyst to do is to compute the t-statistics and check to see if they are significant.

(a) Regression Output:		(b) Regression Output:	
Constant	0	Constant	-16.9
Std Err of Y Est	5.356	Std Err of Y Est	4.427
R Squared	0.987	R Squared	0.993
No. of Observations	5	No. of Observations	5
Degrees f Freedom	4	Degrees f Freedom	3
X Coefficient(s)	0.4999	X Coefficient(s)	0.5425
Std Err of Coef.	0.0062	Std Err of Coef.	0.0257

Figure 10.7 Regression Output from LOTUS 1-2-3 for Windows (Release 4.0): (a) for Zero Intercept, and (b) for Non-Zero Intercept

A sample of output from a typical statistics package is given in Figure 10.8. All calculations are done for you, including assessing the statistical significance of the coefficients (denoted by p in the table rather than our designation of α).

```
The regression equation is
C2 = 0.500 C1

Predictor      Coef       Stdev      t-ratio     p
Noconstant
C1         0.499899    0.006156    81.20     0.000
s = 5.356

The regression equation is
C2 = - 16.9 + 0.543 C1

Predictor      Coef       Stdev      t-ratio     p
Constant     -16.92      10.01       -1.69     0.190
C1          0.54252    0.02573     21.08     0.000
s = 4.427
```

Figure 10.8   Regression Output for MINITAB for Windows (Release 10)

10.3.3 Confidence Intervals for Coefficients

Another very common and useful way of expressing the variability in an estimate of a coefficient in a model is with a *confidence interval.* It is typically used when we are sure that the parameter we are estimating cannot be zero. So we are more interested in putting reasonable error limits on our estimate than we are of testing the null hypothesis. A confidence interval is defined as an interval which has a specified probability of containing the true parameter that we are estimating. The confidence interval is based on our parameter estimate, which is our best guess of the true parameter (obtained from our data), but it recognizes the variability in the estimate.

Mathematically we can state the confidence interval as:

Probability (Lower Limit < true parameter) = confidence, or
Probability (true parameter < Upper Limit) = confidence, or
Probability (Lower Limit < true parameter < Upper Limit) = confidence.

We have three expressions, because we may want only a lower limit, only an upper limit, or both a lower and an upper limit. The lower limit and/or upper limit are calculated from our statistic (estimate of our parameter) and the probability distribution that describes the variability in that statistic. In the case of regression analysis, the appropriate probability density is the t-distribution.

The concept is essentially the same as it was for our significance test — the measured value of our parameter is a random variable which is described by a t-distribution centered on the true parameter value. The true parameter value is assumed to be zero for a test of significance. Then we check to see if we could reasonably get the value we actually did get from our data (under those assumptions). If not, then the assumption must be bad, so our true parameter is not zero (or more properly, it is statistically significantly different from zero). When we calculate a confidence

302

interval, we move the center of the distribution around until it just becomes unreasonable to get the result that we did. How far we can move the center of the distribution to either side becomes our confidence interval.

As an example, let us calculate a confidence interval for the slope of our straight line which was fit to the Brinell Hardness data when our model contains an intercept, a, and a slope, b. From our significance test, it was clear that the slope could not possibly be zero. So our attention has turned to establishing what a reasonable range of values for the slope could be. Our estimate of the slope is $\hat{b} = 0.5425$. If we want a lower limit for what the slope could be, we drag the t-distribution to the left, until it starts to get unreasonable to actually observe our result. This is shown graphically in Figure 10.9 (A). Before we can get a specific value for the lower limit, we need to quantify "unreasonable," which is the same as our Type I error, α, discussed in Chapter 6. If we arbitrarily pick α to be 0.05 (or 5%), the confidence level is 95% (= 1 - α). We then use the critical t-value, $t^* = 2.353$, from Table A.5 with 3 degrees of freedom (the degrees of freedom for s) and the standard deviation of $\hat{b}$, $s_{\hat{b}} = 0.0257$, to determine our limit:

Lower 95% Confidence Limit (one-sided) = $\hat{b} - t^* s_{\hat{b}}$ = 0.5425 - 2.353(0.0257) = 0.4820.

If we wanted an upper 95% confidence limit, we would drag the t-distribution to the right until only 5% of it was to the left of our parameter estimate, as shown in Figure 10.9 (B). The upper limit for our example is:

Upper 95% Confidence Limit (one-sided) = $\hat{b} + t^* s_{\hat{b}}$ = 0.5425 + 2.353(0.0257) = 0.6030.

Finally, if we wanted a two-sided 95% confidence interval, we would drag the t-distribution to the left until only 2½% of it was to the right of our parameter estimate, and we would drag the t-distribution to the right until only 2½% of it was to the left of our parameter estimate. This is shown in Figure 10.9(C). The interval for our example is:

95% Confidence Interval (two-sided) = $\hat{b} \pm t^* s_{\hat{b}}$ = 0.5425 ± 3.182(0.0257) = 0.4607 to 0.6243.

As mentioned above, the confidence level used is strictly a matter of choice. If a confidence level less than 95% were used, the interval would become smaller, since the smaller interval would have less probability of containing the true parameter value. Conversely, if a confidence level greater than 95% were used, the interval would become larger, since the larger interval would have more of a chance of containing the true parameter value.

It should also be stated that confidence intervals can also be used to judge significance. If the confidence interval for the true parameter includes zero, then zero is a reasonable value of the parameter. In that event, the parameter estimate would not be statistically significant, and the null hypothesis, H_o: parameter = 0, would be accepted.

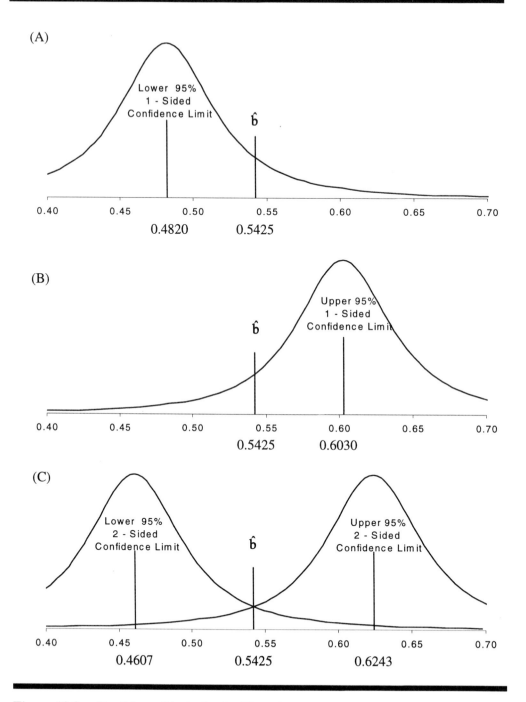

Figure 10.9   Confidence Limits for the Slope, b

10.3.4 Precision in Predictions

In addition to obtaining the coefficients of the best fitting straight line for a set of data and checking their statistical significance, reporting the precision in predicted values obtained from the line is often important. The predicted value of Y when $X = x_0$ is the estimator of the expected value of Y when $X = x_0$ (i.e., $E[Y|X=x_0] = a + bx_0$ as shown in Figure 10.5).

The variance of a predicted value from the equation, $\hat{Y} = \hat{a} + \hat{b}X$, is given by the formula:

$$\text{Var}(\hat{Y} \mid X = x_0) = \frac{\sigma^2}{n} + \frac{(x_0 - \bar{X})^2 \sigma^2}{\Sigma(X_i - \bar{X})^2} \tag{10.24}$$

where: $\text{Var}(\hat{Y} \mid X = x_0)$ denotes the variance of the predicted value, $\hat{Y}$, when the independent variable, $X = x_0$. As usual, n is the number of data points, and $\bar{X}$ is the average of the independent variable values. The estimate of the standard error of a predicted value is then the square root of the right hand side of Equation 10.24, with the true variance, σ^2, replaced by its estimate, s^2.

$$s_{\hat{Y}} = s \sqrt{\frac{1}{n} + \frac{(x_0 - \bar{X})^2}{\Sigma(X_i - \bar{X})^2}} \tag{10.25}$$

For the data in Table 10.2, $s = 4.43$, $n = 5$, x_0 is any value of X (hopefully in the range in the data), $\bar{X} = 381.4$, and the large denominator, $\Sigma(X_i - \bar{X})^2 = 29,597.2$.

The standard deviations of $\hat{Y}$ for a range of values of x_0 are shown in Table 10.3. It should be noticed that the standard error of a predicted value depends upon the point, x_0, where it is calculated. This is obvious from Equation 10.25, but runs counter to many people's intuition. When x_0 is equal to the average of the X values for the data, the standard error is a minimum. As the value of x_0 moves toward the boundary of the data (in either direction), the standard error increases. This is shown graphically in Figure 10.10, which shows the two lines: $\hat{Y} + 2 s_{\hat{Y}}$ and $\hat{Y} - 2 s_{\hat{Y}}$.

Table 10.3 -- Calculation of Error Limits on a Straight Line Model

X_0	$\hat{Y}$	$s_{\hat{Y}}$	$\hat{Y} - 2 s_{\hat{Y}}$	$\hat{Y} + 2 s_{\hat{Y}}$
275	132.28	3.38	125.52	139.03
300	145.84	2.88	140.07	151.60
325	159.40	2.45	154.49	164.31
350	172.96	2.14	168.69	177.24
375	186.53	1.99	182.55	190.50
400	200.09	2.04	196.02	204.16
425	213.65	2.28	209.10	218.21
450	227.22	2.65	221.91	232.52
475	240.78	3.12	234.54	247.02
500	254.34	3.64	247.07	261.62
525	267.91	4.19	259.52	276.29

Although the calculations presented here may seem tedious, it should be mentioned that these standard errors are typically produced quite conveniently by regression computer programs.

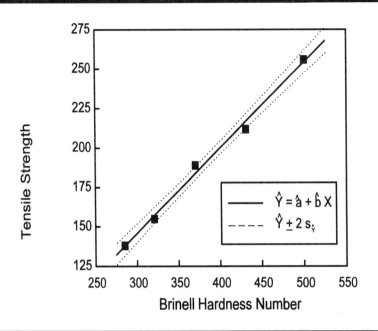

Figure 10.10 Error Limits on a Straight Line Model

10.4 Multiple Regression

In the last section, the ideas of regression analysis were presented in terms of the simple linear model

$$Y = b_0 + b_1 X \tag{10.26}$$

However, the ideas presented extend very easily to the case where there are multiple independent variables, X_i's , and the equation is of the form:

$$Y = b_0 + b_1 X_1 + b_2 X_2 + ... + b_p X_p \tag{10.27}$$

Once we know how to fit a *linear* model with any number of independent variables, it may look like we will still need to learn more in order to fit a more complex polynomial model. This is not the case. For example, if we want to fit a full factorial model in three X variables to some data, the equation we would use is:

$$\hat{Y} = b_0 + b_1 X_1 + b_2 X_2 + b_3 X_3 + b_{12} X_1 X_2 + b_{13} X_1 X_3 + b_{23} X_2 X_3 + b_{123} X_1 X_2 X_3 \qquad (10.28)$$

This is really no different than the equation:

$$\hat{Y} = b_0 + b_1 X_1 + b_2 X_2 + b_3 X_3 + b_4 X_4 + b_5 X_5 + b_6 X_6 + b_7 X_7 \qquad (10.29)$$

where X_1, X_2, and X_3 are the three independent variables, X_4 is really $X_1 X_2$, X_5 is really $X_1 X_3$, X_6 is really $X_2 X_3$, and X_7 is really $X_1 X_2 X_3$. Therefore, in this example, X_4, X_5, X_6, and X_7 are not independently set values for any experiment, but rather they are calculated from X_1, X_2, and X_3. However, our regression analysis does not know (or care) where the values of the variables came from; it only cares that there are now seven *independent* variables (in addition to the constant). So as you can see, the term, independent, does not mean that one variable cannot be calculated from another. It does mean that two (or more) columns may not look the same, or even be a linear function of each other, so we may have to be careful that our *data* supports the calculation of a bigger model. But, once we can fit a linear model in several factors, we will be set, in terms of methodology, to fit any factorial or polynomial model as well.

In order to simplify the formulas used in multiple regression analysis, we will use matrix terminology. For those unfamiliar with matrices, Appendix 10.A should provide the necessary background.

10.4.1 Estimation of Coefficients

Matrix notation can be used as a very concise way of writing the formulas that are used in the various phases of multiple regression analysis. As a basis for discussion, let us continue to use the simple factorial model with three independent variables, X_1, X_2, and X_3, given in Equation 10.28, or its equivalent, Equation 10.29. Let us apply it, for example, to the data in Table 8.5, which was from a 2^3 factorial design replicated twice for a total of sixteen experiments.

First, let us recognize that our responses, y_i's, from our experiments, taken collectively can be thought of as an $n \times 1$ matrix (or column vector), Y, where n is the number of runs (n=16 in this example). And, likewise, the settings of our independent variables are seen to be an $n \times k$ matrix, where k is the number of independent variables (k=3 in this example). If that matrix is expanded to include a column for every coefficient in our model, then we end up with an $n \times (p+1)$ matrix which we will denote by X. In this case, p, the number of coefficients in the model exclusive of the constant term, is seven, so the matrix, B, contains eight coefficients in all. The X and Y matrices are given below, generically in terms of the number of data points, but for our specific number of constants. To reiterate, we had to add columns for the three two-factor interaction terms and the single three-factor interaction term and also a column of 1's (because of the b_0 term) to form the total X matrix. (Remember that we are defining X_4, X_5, X_6, and X_7 to be $X_1 X_2$, $X_1 X_3$, $X_2 X_3$, and $X_1 X_2 X_3$ respectively.) Let us further define two more matrices: the $(p+1) \times 1$ matrix, B, of coefficients in our model, and the $n \times 1$ matrix of predictions from our model, $\hat{Y}$

$$\mathbf{Y} = \begin{bmatrix} y_1 \\ y_2 \\ y_3 \\ \vdots \\ y_n \end{bmatrix} \qquad \mathbf{X} = \begin{bmatrix} 1 & x_{11} & x_{12} & x_{13} & x_{14} & x_{15} & x_{16} & x_{17} \\ 1 & x_{21} & x_{22} & x_{23} & x_{24} & x_{25} & x_{26} & x_{27} \\ 1 & x_{31} & x_{32} & x_{33} & x_{34} & x_{35} & x_{36} & x_{37} \\ \vdots & \vdots & \vdots & \vdots & \vdots & \vdots & \vdots & \vdots \\ 1 & x_{n1} & x_{n2} & x_{n3} & x_{n4} & x_{n5} & x_{n6} & x_{n7} \end{bmatrix} \qquad \mathbf{B} = \begin{bmatrix} b_o \\ b_1 \\ b_2 \\ b_3 \\ b_4 \\ b_5 \\ b_6 \\ b_7 \end{bmatrix} \qquad \hat{\mathbf{Y}} = \begin{bmatrix} \hat{y}_1 \\ \hat{y}_2 \\ \hat{y}_3 \\ \vdots \\ \hat{y}_n \end{bmatrix} \qquad \textbf{(10.30)}$$

We are now in a position to determine the least squares estimates for our coefficients, **B**. The measure we are minimizing is still SSE, the sum of the squared errors. The error for experiment I, ε_i, is (as always) the difference between the actual response, y_i, and the response predicted by our equation, $\hat{y}_i$. The collection of all the errors is an $n \times 1$ matrix, ε, which is:

$$\varepsilon = \mathbf{Y} - \hat{\mathbf{Y}} \qquad \textbf{(10.31)}$$

The sum of the squared errors can be computed very simply once we have ε:

$$SSE = \varepsilon^T \varepsilon = (\mathbf{Y} - \hat{\mathbf{Y}})^T (\mathbf{Y} - \hat{\mathbf{Y}}) \qquad \textbf{(10.32)}$$

where the superscript T on a matrix indicates the transpose of that matrix. The $\hat{\mathbf{Y}}$ in Equation 10.32 can be expressed in terms of **B** as:

$$\hat{\mathbf{Y}} = \mathbf{X}\mathbf{B} \qquad \textbf{(10.33)}$$

so that our equation for SSE becomes:

$$SSE = (\mathbf{Y} - \mathbf{X}\mathbf{B})^T (\mathbf{Y} - \mathbf{X}\mathbf{B}) = \mathbf{Y}^T \mathbf{Y} - 2\mathbf{B}^T \mathbf{X}^T \mathbf{Y} + \mathbf{B}^T \mathbf{X}^T \mathbf{X}\mathbf{B} \qquad \textbf{(10.34)}$$

To find the minimum in SSE we proceed, just as we did for the simple linear model, by taking the derivative of SSE with respect to each element of **B** and setting the derivatives equal to zero. We get a set of simultaneous linear equations which can be written again in matrix form as:

$$\frac{d\,SSE}{d\mathbf{B}} = -2\mathbf{X}^T\mathbf{Y} + 2\mathbf{X}^T\mathbf{X}\mathbf{B} = \mathbf{0} \qquad \textbf{(10.35)}$$

or

$$(\mathbf{X}^T\mathbf{X})\hat{\mathbf{B}} = \mathbf{X}^T\mathbf{Y} \qquad \textbf{(10.36)}$$

The hat on **B** indicates it is the best (least squares) estimates of the values of the coefficients in our model. The equations implied by Equation 10.36 are called the *least squares normal equations*. Since $(\mathbf{X}^T\mathbf{X})$ is a full rank square matrix, it has an inverse defined. Therefore, the solution of Equation 10.36 can be obtained by multiplying both sides of the equation by $(\mathbf{X}^T\mathbf{X})^{-1}$ to get

$$\hat{\mathbf{B}} = (\mathbf{X}^T\mathbf{X})^{-1}\mathbf{X}^T\mathbf{Y} \qquad \textbf{(10.37)}$$

$\wedge$ – estimate

Equation 10.37 is a simple but very powerful matrix equation which defines the least squares estimates for *all* the regression coefficients simultaneously. To see how it works, we will calculate the coefficients in the factorial model for the voltmeter example in Chapter 8. The data are given in Table 8.4. Using the coded values of the X's, we get the following $\mathbf{X}$ and $\mathbf{Y}$ matrices:

$$
\mathbf{X} =
\begin{array}{c}
\text{Int.}\quad X_1\quad X_2\quad X_3\quad X_1X_2\quad X_1X_3\quad X_2X_3\quad X_1X_2X_3 \\
\left[
\begin{array}{rrrrrrrr}
1 & -1 & -1 & -1 & 1 & 1 & 1 & -1 \\
1 & -1 & -1 & -1 & 1 & 1 & 1 & -1 \\
1 & 1 & -1 & -1 & -1 & -1 & 1 & 1 \\
1 & 1 & -1 & -1 & -1 & -1 & 1 & 1 \\
1 & -1 & 1 & -1 & -1 & 1 & -1 & 1 \\
1 & -1 & 1 & -1 & -1 & 1 & -1 & 1 \\
1 & 1 & 1 & -1 & 1 & -1 & -1 & -1 \\
1 & 1 & 1 & -1 & 1 & -1 & -1 & -1 \\
1 & -1 & -1 & 1 & 1 & -1 & -1 & 1 \\
1 & -1 & -1 & 1 & 1 & -1 & -1 & 1 \\
1 & 1 & -1 & 1 & -1 & 1 & -1 & -1 \\
1 & 1 & -1 & 1 & -1 & 1 & -1 & -1 \\
1 & -1 & 1 & 1 & -1 & -1 & 1 & -1 \\
1 & -1 & 1 & 1 & -1 & -1 & 1 & -1 \\
1 & 1 & 1 & 1 & 1 & 1 & 1 & 1 \\
1 & 1 & 1 & 1 & 1 & 1 & 1 & 1
\end{array}
\right]
\end{array}
\qquad
\mathbf{Y} =
\begin{array}{c}
\text{Voltages} \\
\left[
\begin{array}{r}
705 \\ 680 \\ 620 \\ 651 \\ 700 \\ 685 \\ 629 \\ 635 \\ 672 \\ 654 \\ 668 \\ 691 \\ 715 \\ 672 \\ 647 \\ 673
\end{array}
\right]
\end{array}
$$

The least squares normal equation (Equation 10.36) for this example is

$$
\underset{\mathbf{X^TX}}{
\begin{bmatrix}
16 & 0 & 0 & 0 & 0 & 0 & 0 & 0 \\
0 & 16 & 0 & 0 & 0 & 0 & 0 & 0 \\
0 & 0 & 16 & 0 & 0 & 0 & 0 & 0 \\
0 & 0 & 0 & 16 & 0 & 0 & 0 & 0 \\
0 & 0 & 0 & 0 & 16 & 0 & 0 & 0 \\
0 & 0 & 0 & 0 & 0 & 16 & 0 & 0 \\
0 & 0 & 0 & 0 & 0 & 0 & 16 & 0 \\
0 & 0 & 0 & 0 & 0 & 0 & 0 & 16
\end{bmatrix}}
\underset{\mathbf{\hat{B}}}{
\begin{bmatrix}
\hat{b}_0 \\ \hat{b}_1 \\ \hat{b}_2 \\ \hat{b}_3 \\ \hat{b}_4 \\ \hat{b}_5 \\ \hat{b}_6 \\ \hat{b}_7
\end{bmatrix}}
=
\underset{\mathbf{X^TY}}{
\begin{bmatrix}
10{,}697 \\ -269 \\ 15 \\ 87 \\ -107 \\ 201 \\ 29 \\ -93
\end{bmatrix}}
$$

309

and the solution (Equation 10.37) is

$$
\mathbf{\hat{B}} \qquad\qquad (\mathbf{X^TX})^{-1} \qquad\qquad \mathbf{X^TY}
$$

$$
\begin{bmatrix} \hat{b}_0 \\ \hat{b}_1 \\ \hat{b}_2 \\ \hat{b}_3 \\ \hat{b}_4 \\ \hat{b}_5 \\ \hat{b}_6 \\ \hat{b}_7 \end{bmatrix}
=
\begin{bmatrix}
\frac{1}{16} & 0 & 0 & 0 & 0 & 0 & 0 & 0 \\
0 & \frac{1}{16} & 0 & 0 & 0 & 0 & 0 & 0 \\
0 & 0 & \frac{1}{16} & 0 & 0 & 0 & 0 & 0 \\
0 & 0 & 0 & \frac{1}{16} & 0 & 0 & 0 & 0 \\
0 & 0 & 0 & 0 & \frac{1}{16} & 0 & 0 & 0 \\
0 & 0 & 0 & 0 & 0 & \frac{1}{16} & 0 & 0 \\
0 & 0 & 0 & 0 & 0 & 0 & \frac{1}{16} & 0 \\
0 & 0 & 0 & 0 & 0 & 0 & 0 & \frac{1}{16}
\end{bmatrix}
\begin{bmatrix}
10,697 \\
-269 \\
15 \\
87 \\
-107 \\
201 \\
29 \\
-93
\end{bmatrix}
=
\begin{bmatrix}
668.56 \\
-16.81 \\
0.94 \\
5.44 \\
-6.69 \\
12.56 \\
1.81 \\
-5.81
\end{bmatrix}
$$

These values can be compared to the mean, effects, and interactions given in Table 8.6, keeping in mind that coefficients are one-half of effects (except for the mean). Or they can be compared to the coefficients in Table 8.9 directly. Naturally, they agree. This shows that the worksheet method, or Yates algorithm, used for calculating the factorial effects shown in Chapter 8, was simply a step-by-step method for obtaining the least squares estimates.

10.4.2 Statistical Significance of Coefficients

If we group the $\mathbf{X}$ matrices in Equation 10.37, the least squares coefficients in our model can be seen to be calculated as a sum of the y's (each y_i being multiplied by a constant).

$$
\mathbf{\hat{B}} = [(\mathbf{X^TX})^{-1}\mathbf{X^T}]\mathbf{Y} \equiv \mathbf{CY} \tag{10.38}
$$

Since the values of the X variables are assumed to be known without error, the matrix, $\mathbf{C}$, is a matrix of constants that can be calculated exactly. Therefore, any variability in $\mathbf{\hat{B}}$ comes from the variability in the y_i's. When this sort of relationship exists between two matrices (as between $\mathbf{\hat{B}}$ and $\mathbf{Y}$), then it has been shown by others[1] that the variance of the calculated matrix is given by:

$$
\mathrm{Var}(\mathbf{\hat{B}}) = \mathbf{C}\,\mathrm{Var}(\mathbf{Y})\,\mathbf{C^T}
$$
$$
\text{or} \quad \mathrm{Var}(\mathbf{\hat{B}}) = [(\mathbf{X^TX})^{-1}\mathbf{X^T}]\,\mathrm{Var}(\mathbf{Y})\,[(\mathbf{X^TX})^{-1}\mathbf{X^T}]^T \tag{10.39}
$$

If we further assume, as we did in Section 10.3.2, that the errors in $\mathbf{Y}$ are independently distributed like a Normal Distribution with a mean of zero and all have the same variance, σ^2, then the variance of $\mathbf{Y}$ can be written as $\mathbf{I}_n\sigma^2$. $\mathbf{I}_n$ is the nxn identity matrix, which has ones down the diagonal and has zeros everywhere else. Putting this result into Equation 10.39 we get:

$$\text{Var}(\hat{\mathbf{B}}) = [(\mathbf{X^TX})^{-1}\mathbf{X^T}]\,\mathbf{I}_n\sigma^2\,[(\mathbf{X^TX})^{-1}\mathbf{X^T}]^T$$

$$\text{or} \quad \text{Var}(\hat{\mathbf{B}}) = (\mathbf{X^TX})^{-1}\mathbf{X^T}\mathbf{I}_n[\mathbf{X^T}]^T[(\mathbf{X^TX})^{-1}]^T\sigma^2 \tag{10.40}$$

Equation 10.40 can be simplified greatly if we recognize the following: (1) multiplication by the identity matrix does not change anything (much like multiplying an algebraic expression by "1"), and so it can be ignored, (2) the transpose of the transpose of a matrix is the original matrix, and (3) the transpose of a square symmetric matrix (like $\mathbf{X^TX}$ or $(\mathbf{X^TX})^{-1}$) is equal to the matrix itself. Using all these rules we get:

$$\text{Var}(\hat{\mathbf{B}}) = (\mathbf{X^TX})^{-1}\mathbf{X^TX}(\mathbf{X^TX})^{-1}\sigma^2 \tag{10.41}$$

Finally, since multiplying a matrix like $\mathbf{X^TX}$ by its inverse results in an Identity matrix which can be ignored, Equation 10.41 becomes:

$$\text{Var}(\hat{\mathbf{B}}) = (\mathbf{X^TX})^{-1}\sigma^2 \tag{10.42}$$

This equation, like Equation 10.37, is a very succinct yet very powerful formula. It gives the variability in our coefficients for any experimental design, $\mathbf{X}$. The result is a square symmetric matrix whose diagonal terms give the variances of the coefficients, and the off-diagonal terms give the covariances of pairs of the coefficients. It is the diagonal terms we will use when we test the statistical significance of our coefficients.

There is one hitch to using Equation 10.42 — we rarely know σ^2. That does not stop us, however. We can replace it with s^2, its estimate, and proceed. The only change is that we must then use the t-distribution (with the degrees of freedom in s^2) to test for statistical significance, rather than the Normal Distribution.

As a demonstration of the use of Equation 10.42, let us continue with the voltmeter example in Chapter 8. The estimate of s^2 is determined from SSE, which we can calculate since we have already found $\hat{\mathbf{B}}$. The calculation of SSE is detailed in Table 10.4. Once we have SSE, we use Equation 10.22 to evaluate s^2. In this example, the number of constants in our model that we are evaluating from our data, c, is 8 ($= p + 1$) including the intercept. Therefore our degrees of freedom ($= n - c$) is 16-8, or 8. So our estimate of s^2 is SSE/(n-c) = 2612.5/(16-8) = 326.56, and our estimate of s is $\sqrt{326.56}$ = 18.07. This value of s can be compared to the value obtained by using the pooled standard deviation (which was computed in Table 8.5). It is seen to be identical to that value and it has the same degrees of freedom. The reason they agree is that the full factorial model predicts the corner point average at each corner. The pooled standard deviation also computed deviations from the corner point averages. So they were both, in fact, computed in exactly the same way.

Table 10.4   Calculation of s^2 for the Voltmeter Example

Y	$\hat{Y} = X \hat{B}$	$\epsilon = Y - \hat{Y}$	ϵ^2	
705	692.5	12.5	156.25	$s^2 = SSE/[n-c]$
680	692.5	-12.5	156.25	$s^2 = SSE/[n-(p+1)]$
620	635.5	-15.5	240.25	$s^2 = 2612.50 / 8$
651	635.5	15.5	240.25	$s^2 = 326.56$
700	692.5	7.5	56.25	$s = 18.07$
685	692.5	-7.5	56.25	
629	632.0	-3.0	9.00	
635	632.0	3.0	9.00	
672	663.0	9.0	81.00	
654	663.0	-9.0	81.00	
668	679.5	-11.5	132.25	
691	679.5	11.5	132.25	
715	693.5	21.5	462.25	
672	693.5	-21.5	462.25	
647	660.0	-13.0	169.00	
673	660.0	13.0	169.00	
		SSE =	2612.50	

Since we already know $(\mathbf{X}^T\mathbf{X})^{-1}$, we can now use Equation 10.42 to estimate Var $(\hat{\mathbf{B}})$:

$$\text{Var}(\hat{\mathbf{B}}) = \begin{bmatrix} 1/16 & 0 & 0 & 0 & 0 & 0 & 0 & 0 \\ 0 & 1/16 & 0 & 0 & 0 & 0 & 0 & 0 \\ 0 & 0 & 1/16 & 0 & 0 & 0 & 0 & 0 \\ 0 & 0 & 0 & 1/16 & 0 & 0 & 0 & 0 \\ 0 & 0 & 0 & 0 & 1/16 & 0 & 0 & 0 \\ 0 & 0 & 0 & 0 & 0 & 1/16 & 0 & 0 \\ 0 & 0 & 0 & 0 & 0 & 0 & 1/16 & 0 \\ 0 & 0 & 0 & 0 & 0 & 0 & 0 & 1/16 \end{bmatrix} \times 326.56$$

As mentioned earlier in this section, the diagonal terms in this matrix give us the variances of each of the coefficients in our model. Usually, the diagonal elements in the matrix will be different, but in the case of a 2^k-factorial design (with no center points), they are all they same, namely the pooled variance divided by the number of factorial points. So we have found, by using Equation 10.42, that the variance of each of the coefficients in the factorial model is:

$$Var(\hat{b}_i) = s_p^2/n_F \tag{10.43}$$

Since we actually want the standard deviation, we must take the square root of that:

$$Standard\ Deviation(\hat{b}_i),\ s_{b_i} = s_p/\sqrt{n_F} \tag{10.44}$$

By comparing Equation 10.44 to Equation 8.4 (which is, $s_E = 2\ s_p / \sqrt{n_F}$), you can see that the standard deviation of a coefficient is half of the standard deviation of an effect. This is as expected, since the coefficient is half of an effect. So the standard deviation of each $\hat{b}_i$ in this example is $18.07 /\sqrt{16} = 4.52$. This can also be compared to the standard deviation of the coefficients column ("StDev Coeff") in Table 8.9, and again it is the same as expected.

Once we have all of the standard deviations of the coefficients, all that remains is to calculate the t-statistics,

$$t_i = \hat{b}_i/s_{b_i} \tag{10.45}$$

and compare them to the critical (tabled) value with the appropriate degrees of freedom (8 for this example). The t-statistics for our example were calculated using Equation 10.45 in the sample output from MINITAB in Chapter 8, Table 8.9 (as well as in Table 8.7), and so they were not calculated again. The effect of X_1 and the X_1X_3-interaction were found to be statistically significant.

10.4.3 Confidence Intervals for Coefficients

Confidence intervals were discussed in Section 10.3.3 for the coefficients in a straight line model. The confidence intervals for the coefficients in a multiple regression model are exactly the same. Once we have the coefficients in our model, $\hat{b}_i$ (from Section 10.4.1), and the standard deviations of those coefficients, s_{b_i} (from Section 10.4.2), we can express the precision of any of the coefficients as a confidence interval:

Lower $100(1-\alpha)\%$ Confidence Limit for $b_i = \hat{b}_i - t^*_\alpha\ s_{b_i}$, or
Upper $100(1-\alpha)\%$ Confidence Limit for $b_i = \hat{b}_i + t^*_\alpha\ s_{b_i}$, or
Two-sided $100(1-\alpha)\%$ Confidence Interval for $b_i = \hat{b}_i \pm t^*_{\alpha/2}\ s_{b_i}$

For example, let us use two-sided confidence intervals to express the precision of the coefficients in the model for our voltmeter example. The equation is:

$$\hat{Y} = 668.6 - 16.8\ X_1 - 5.4\ X_2 + 12.6\ X_1X_3.$$

The standard deviations of all the coefficients are the same, $s_{b_i} = 4.52$ with 8 degrees of freedom as determined in the previous section. The critical t-value for 95% Confidence is $t^* = 2.306$ (two-sided). So the confidence interval for each of the coefficients is:

$$95\% \text{ Confidence Interval} = \hat{b}_i \pm (2.306)(4.52) = \hat{b}_i \pm 10.4$$

10.4.4 Precision of Predictions

It is often important to report the precision of the predictions calculated using our model. Our predictions at any sets of conditions (we will denote the conditions by $\mathbf{X}_o$) are, of course, calculated using our least squares estimates of $\mathbf{B}$ via the Equation 10.33:

$$\hat{\mathbf{Y}} = \mathbf{X}_o \hat{\mathbf{B}} \qquad (10.46)$$

This equation is totally analogous to Equation 10.38 which said that our estimate of $\mathbf{B}$ is calculated as the product of two matrices — one that is precisely known, and one that has variability. Therefore, we can use the relationship given in Equation 10.39 (the first line) to calculate the variance of $\hat{\mathbf{Y}}$.

$$\text{Var}(\hat{\mathbf{Y}}) = \mathbf{X}_o \text{Var}(\hat{\mathbf{B}}) \mathbf{X}_o^{\mathbf{T}} = \mathbf{X}_o [(\mathbf{X}^{\mathbf{T}}\mathbf{X})^{-1}\sigma^2] \mathbf{X}_o^{\mathbf{T}} \qquad (10.47)$$

$$\text{or} \quad \text{Var}(\hat{\mathbf{Y}}) = \mathbf{X}_o (\mathbf{X}^{\mathbf{T}}\mathbf{X})^{-1}\mathbf{X}_o^{\mathbf{T}}\sigma^2 \qquad (10.48)$$

Remember that $\mathbf{X}$ is the n sets of conditions at which experiments were run, and thus it does not change. On the other hand, $\mathbf{X}_o$ is the collection of points at which we would like to calculate the variance of $\hat{\mathbf{Y}}$. That can be as little as only a single point or a very large number of points, and the matrix changes accordingly.

Continuing with our voltmeter example, let's say we are interested in the line of measured voltage, $\hat{\mathbf{Y}}$ versus ambient temperature, when the voltmeter warm-up time is 5 min. ($X_2 = +1$) and the time that power is connected is 5 min. ($X_3 = +1$). We can calculate predictions (the line) using Equation 10.46 and the variability in those predictions using Equation 10.48. The results of those calculations are shown numerically in Table 10.5 and graphically in Figure 10.11.

It should be noted that the model used (in Table 10.5 and Figure 10.11) had all the non-significant terms deleted, with the exception of the X_3 term. The X_3 term was retained because the X_1X_3-interaction was important, and it did not feel right to delete a first order term in a factor and keep a second order term for that same factor. This is a matter of personal preference. But it is definitely not a good idea to keep all the nonsignificant terms in the model, because the variability in the predictions increases, and there is no offsetting benefit. This concept of trimming insignificant terms from a model is discussed in much more detail in Chapter 14.

To make sure that the use of Equation 10.48 is clear, let us calculate the variability of $\hat{\mathbf{Y}}$ at an ambient temperature of 32 °C and having the power connected for 5 minutes. The voltmeter warm-up time was not found to significantly affect the results, so it will be set to 5 minutes (the same as the time the power is connected) for convenience. Remembering our coding:

Table 10.5 Calculation of Error Limits for Voltmeter Example
$\hat{Y} = 668.6 + 16.8\,X_1 + 5.4\,X_3 + 12.6\,X_1 X_3$ (with $X_2 = X_3 = +1$)

Ambient Temperature	X_1	$s_{\hat{Y}}$	$\hat{Y}$	$\hat{Y} - 2\,s_{\hat{Y}}$	$\hat{Y} + 2$
22.0	-1.0	9.2	678.3	659.8	696.7
24.5	-0.5	7.3	676.1	661.5	690.7
27.0	0	6.5	674.0	660.9	687.1
29.5	0.5	7.3	671.9	657.3	686.5
32.0	1.0	9.2	669.8	651.3	688.2

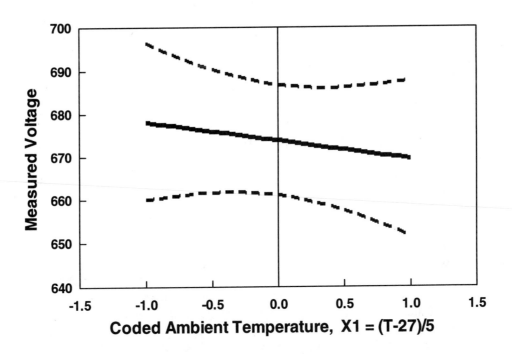

Figure 10.11   Error Limits on Voltage Measurement Model When $X_2 = 1$ and $X_3 = 1$
(Reduced Model: $\hat{Y} = 668.6 - 16.8\,X_1 + 5.4\,X_3 + 12.6\,X_1 X_3$)

315

$$X_1 = [\text{ambient temperature-27}] / 5 = (32\text{-}27) / 5 = 1$$
$$X_2 = [\text{voltmeter warm-up time-2.75}] / 2.25 = (5\text{-}2.75) / 2.25 = 1$$
and $X_3 = [\text{power connect time-2.75}] / 2.25 = (5\text{-}2.75) / 2.25 = 1$

The $\mathbf{X}_o$ matrix (vector in this case) for the reduced model, including a "1" for the constant term, then becomes:

$$\begin{matrix} & \text{Int.} & X_1 & X_3 & X_1X_3 \\ \mathbf{X}_o = [& 1 & 1 & 1 & 1 \ ] \end{matrix}$$

When we use this $\mathbf{X}_o$ matrix in Equation 10.48, and substitute s^2 for σ^2, we get:

$$\text{Var}(\hat{\mathbf{Y}}) = \begin{bmatrix} 1 & 1 & 1 & 1 \end{bmatrix} \begin{bmatrix} 1/16 & 0 & 0 & 0 \\ 0 & 1/16 & 0 & 0 \\ 0 & 0 & 1/16 & 0 \\ 0 & 0 & 0 & 1/16 \end{bmatrix} \begin{bmatrix} 1 \\ 1 \\ 1 \\ 1 \end{bmatrix} (326.6)$$

or $\quad \text{Var}(\hat{\mathbf{Y}}) = \begin{bmatrix} 4/16 \end{bmatrix}(326.6) = \begin{bmatrix} 81.65 \end{bmatrix}$

Notice that the $(\mathbf{X}^T\mathbf{X})^{-1}$ matrix also corresponds to the reduced model, and therefore only has four rows and four columns.

The error limits on the equation values are typically reported as plus or minus twice the standard deviation of $\hat{\mathbf{Y}}$, which are approximate 95% confidence limits on the curve. For these values of X_1 and X_3 the error limits are $\pm 2\sqrt{81.65} = \pm 18.07$. The other error limits in Table 10.5 were calculated in a similar fashion, changing $\mathbf{X}_o$ appropriately.

10.5 Quantifying Model Closeness (R²)

Once a least squares model, $\hat{\mathbf{Y}}$, has been fit to a set of data, generally the first question of interest is: "How well does the equation fit?" Of course, the minimized sum of squared errors,

$$\text{SSE} = \sum (Y_i - \hat{Y}_i)^2 \tag{10.49}$$

is a direct measure of how well the model fits. However, it is highly dependent on the scale of the data, Y_i (not to mention the number of data points). For example, if we are modeling the gross national product, the sum of squares will be very large regardless of how good our model is. On the other hand, if we are modeling the concentration of ozone in air, we will get a very small sum of squares regardless of how poor our model is. Therefore, comparing the sums of squares for two different problems is definitely a situation where we would be comparing apples and oranges. What is needed is some sort of normalization. The normalization that is most commonly used is accomplished by comparing SSE to the total variability in the data, SST (which stands for sum of squares, total). SST is defined as the sum of the squared deviations of the Y data from their average.

In equation form,

$$SST = \sum (Y_i - \bar{Y})^2 \qquad\qquad (10.50)$$

The ratio of SSE to SST represents the fraction of the total variability in the data that *is not* explained by the model. We could use this statistic to compare the models for two different situations, because the ratio is unitless (i.e., it does not depend on the scale of the data, and it essentially makes no difference how much data was used in the analysis).

People actually use the complementary statistic — the fraction of the total variability in the data that *is* explained by the model — which is given the name, R^2.

$$R^2 = 1 - SSE/SST = (SST - SSE)/SST \qquad\qquad (10.51)$$

R^2 is generally greater than 0, and it cannot exceed 1. The bigger it is, the better it is. An R^2 value of 1 means that SSE is zero, and therefore the model fits the data perfectly. At the other end of the scale, an R^2 value of zero means that SSE = SST, or the model is no better than just using the average of the data. As long as the model has a constant term, it is not possible for R^2 to be less than zero.

Table 10.6 Detailed Calculation of R^2 for Voltage Measurement Example

Y_i	$\bar{Y}$	$Y_i - \bar{Y}$	$(Y_i - \bar{Y})^2$	$\hat{Y}$	$Y_i - \hat{Y}$	$(Y_i - \hat{Y})^2$
705	668.56	36.44	1,327.7	692.50	12.50	156.3
680	668.56	11.44	130.8	692.50	-12.50	156.3
620	668.56	-48.56	2,358.3	633.75	-13.75	189.1
651	668.56	-17.56	308.4	633.75	17.25	297.6
700	668.56	31.44	988.3	692.50	7.50	56.3
685	668.56	16.44	270.2	692.50	-7.50	56.3
629	668.56	-39.56	1,565.2	633.75	-4.75	22.6
635	668.56	-33.56	1,126.4	633.75	1.25	1.6
672	668.56	3.44	11.8	678.25	-6.25	39.1
654	668.56	-14.56	212.1	678.25	-24.25	588.1
668	668.56	-0.56	0.3	669.75	-1.75	3.1
691	668.56	22.44	503.4	669.75	21.25	451.6
715	668.56	46.44	2,156.4	678.25	36.75	1,350.6
672	668.56	3.44	11.8	678.25	-6.25	39.1
647	668.56	-21.56	464.9	669.75	-22.75	517.6
673	668.56	4.44	19.7	669.75	3.25	10.6
		SST =	11,455.9		SSE =	3,935.3

$$R^2 = (SST-SSE)/SST = (11,455.9 - 3,935.3) / 11,455.9 = 0.656$$

The detailed calculation of R^2 for the voltage measuring example is shown in Table 10.6, with the voltages given in milli-volts. The value for R^2 is 0.650, which is not particularly high (close to 1.0). This means that there is still quite a bit of variation in the data (namely 35% of it) that is not explained by the model.

Figure 10.12 (a) shows graphically how the data and the model compare for the voltage measurement example by plotting the data, Y (on the y-axis) vs. the model predictions, $\hat{Y}$, (on the x-axis). The 45 degree line on the graph represents the perfect fit (where $Y = \hat{Y}$). As expected with a low R^2 value, the points are not close to the 45 degree line. The three other graphs in Figure 10.12 show the kind of scatter that would be seen for other values of R^2. As R^2 increases from 0.65 to 0.95, the scatter around the 45 degree line gets less and less pronounced. And when R^2 decreases to 0.50, the scatter around the 45 degree line increases.

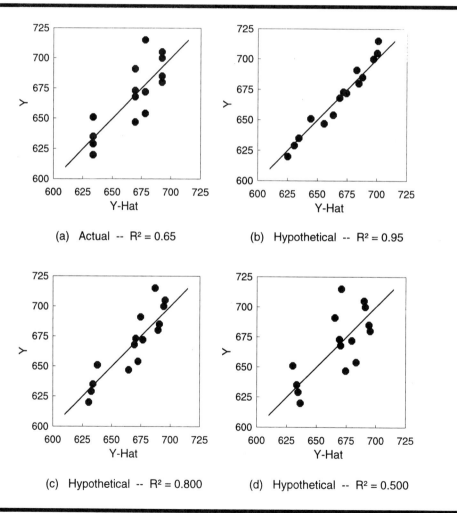

(a) Actual -- R^2 = 0.65

(b) Hypothetical -- R^2 = 0.95

(c) Hypothetical -- R^2 = 0.800

(d) Hypothetical -- R^2 = 0.500

Figure 10.12 Data Values vs. Model Predictions for the Voltage Measurement Example

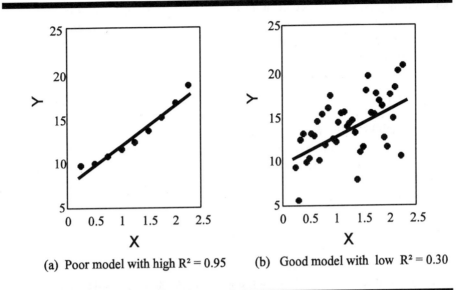

(a) Poor model with high $R^2 = 0.95$ (b) Good model with low $R^2 = 0.30$

Figure 10.13   Models with Deceptive R^2 Values

Although R^2 is a useful indicator of how closely the regression equation matches the data, it should not be used as the only criterion for judging models. The reason is that R^2 is not a definitive measure of how good a model is (i.e., a high R^2 does not necessarily mean that the model is good, nor does a low R^2 necessarily mean the model is poor). For example, if we have the simple Y vs. X data shown in Figure 10.13 (a), a straight line model would have a high R^2 value ($R^2 = 0.95$). But it is clear from the plot that a straight line is a poor model and a quadratic is needed to describe the data. On the other hand, if we have the Y vs. X data shown in Figure 10.13 (b), the data definitely show a trend of increasing Y with increasing X. A straight line is statistically significant and all that is warranted. Therefore the model is good even though the R^2 value is low ($R^2 = 0.30$). The low R^2 value is an indication that there is high random variability in the data, not that the model is poor.

In the case where there are multiple independent X variables, checking for systematic versus random deviations is more difficult, and other statistics are required. These statistics are discussed in Chapter 14.

10.6 Checking Model Assumptions (Residual Plots)

We fit a model to some data because we would like to use the model for some purpose (like optimizing a process). Before we invest too much time looking at what the model has to say about our system, and perhaps get too excited about our great/poor results, we should check our modeling assumptions. In other words, we should make sure that our model is worth looking at.

Our assumptions, which were mentioned in Section 10.3.2 already, are:
(1) The errors are Normally distributed.
(2) The errors have the same variance.
(3) The errors are independent (or the experiment was randomized).
(4) The errors have a mean of zero (i.e., the model is adequate).

If one or more of our assumptions turns out to be incorrect, we must reanalyze our data taking this into account. We could then end up with a different (more correct) model. The new model is the one on which we should be basing our conclusions.

Notice that all four of our assumptions deal directly with the errors in our data. Since our estimates of the errors are the residuals, $\varepsilon_i = Y_i - \hat{Y}_i$, it makes sense that the residuals need to be examined closely. The residuals for our voltage measurement example were given in Table 10.6 and are given again in Table 10.7.

10.6.1 Checking for Normal Distribution of Errors

The first assumption we made was that the errors in our data are Normally distributed. Therefore, if we plot the residuals, they should look like a random sample from that bell-shaped distribution. The obvious plot to make would be a dot plot, in which the values are plotted as dots on the x-axis. This was done for our voltage measurement example in Figure 10.14 (a). The plot seems to look much like a Normal Distribution, but it is actually difficult to tell.

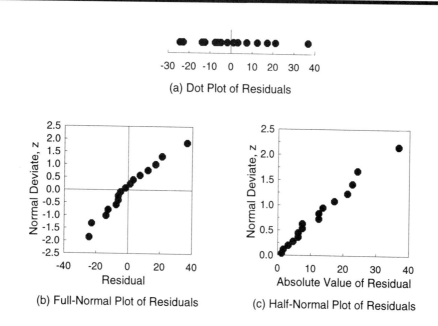

(a) Dot Plot of Residuals

(b) Full-Normal Plot of Residuals

(c) Half-Normal Plot of Residuals

Figure 10.14   Plots of Residuals to Check Normality of Errors for the Voltmeter Example

320

Table 10.7 Detailed Calculations Required for the Full-Normal or Half-Normal Plot for the Voltage Measurement Example

Run	Y_i	$\hat{Y}_i$	ε_i	$\varepsilon_{(i)}$	$\dfrac{[(i) - \frac{1}{2}]}{n}$	z	$\|\varepsilon_i\|$	$\|\varepsilon_{(i)}\|$	$\dfrac{0.5 + [(i) - \frac{1}{2}]}{2n}$	z
1	705	692.50	12.50	-24.25	0.031	-1.86	12.50	1.25	0.516	0.04
2	680	692.50	-12.50	-22.75	0.094	-1.32	12.50	1.75	0.547	0.12
3	620	633.75	-13.75	-13.75	0.156	-1.01	13.75	3.25	0.578	0.20
4	651	633.75	17.25	-12.50	0.219	-0.78	17.25	4.75	0.609	0.28
5	700	692.50	7.50	-7.50	0.281	-0.58	7.50	6.25	0.641	0.36
6	685	692.50	-7.50	-6.25	0.344	-0.40	7.50	6.25	0.672	0.45
7	629	633.75	-4.75	-6.25	0.406	-0.24	4.75	7.50	0.703	0.53
8	635	633.75	1.25	-4.75	0.469	-0.08	1.25	7.50	0.734	0.63
9	672	678.25	-6.25	-1.75	0.531	0.08	6.25	12.50	0.766	0.72
10	654	678.25	-24.25	1.25	0.594	0.24	24.25	12.50	0.797	0.83
11	668	669.75	-1.75	3.25	0.656	0.40	1.75	13.75	0.828	0.95
12	691	669.75	21.25	7.50	0.719	0.58	21.25	17.25	0.859	1.08
13	715	678.25	36.75	12.50	0.781	0.78	36.75	21.25	0.891	1.23
14	672	678.25	-6.25	17.25	0.844	1.01	6.25	22.75	0.922	1.42
15	647	669.75	-22.75	21.25	0.906	1.32	22.75	24.25	0.953	1.68
16	673	669.75	3.25	36.75	0.969	1.86	3.25	36.75	0.984	2.15

The header groups: "For Full-Normal Plot" spans ε_i, $\varepsilon_{(i)}$, $\frac{[(i)-\frac{1}{2}]}{n}$, z. "For Half-Normal Plot" spans $\|\varepsilon_i\|$, $\|\varepsilon_{(i)}\|$, $\frac{0.5 + [(i)-\frac{1}{2}]}{2n}$, z.

A more sensitive tool is the Normal plot (or Half-Normal plot) discussed in Chapter 5. The information needed to construct a Full-Normal (or Half-Normal) plot is also given in Table 10.7 as a review of the procedure. The plots are shown in Figures 10.14 (b) and (c). Remember that if the points come from a Normal Distribution, the plot will be a straight line. Again, the plots look good, indicating that our assumption is fine. In general, if an outlier is detected, the point should be removed and the analysis should be done again. This was not needed for this example.

A word of caution — dropping a data point should not be done lightly. A reason for the bad result should be diligently sought in order to more fully justify deleting the point and in order to help prevent more bad data in the future. Also, the experiment should be repeated if at all possible so that the integrity of the statistical design is maintained.

10.6.2 Checking for Constant Variance

An important assumption in regression analysis is that each of the data points has the same precision, and therefore each of the data points should be given equal weight. However, it is possible that some of the data have greater precision than others (i.e., the errors in those more precise data have a smaller variance than the others). The most common case of this is the situation in which the variance gets bigger as the response gets bigger. Therefore, to check the assumption, the residuals should be plotted against $\hat{Y}$, as shown in Figure 10.15 for the voltage measurement example. In this example it may appear as if the scatter is bigger at higher values of $\hat{Y}$, but there are more data at higher values, and so there would naturally be more scatter. The other plots of residuals shown in Section 10.6.4 should also be checked for evidence of nonconstant variance, as well as for trends (which is the main goal of that section).

If the assumption of constant variance is seen to be poor, two main remedies are available. The first would be to transform the responses to new ones like taking the logarithm if the responses had a constant percentage error. A second solution would be to use a weighted least squares analysis, in which each data point would be given a weight inversely proportional to its variance. A discussion of these procedures is beyond the scope of this text, and the reader is referred to a book on regression analysis (e.g., Draper and Smith[1]).

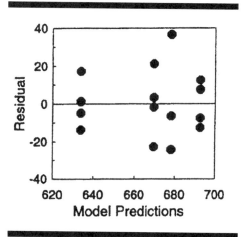

Figure 10.15 Plots of Residuals Model Predictions for the Voltage Measurement Example

10.6.3 Checking for Independence of Errors

Although there are some checks for independence of errors (such as the calculation of an auto-correlation coefficient), correcting for correlations and biases in data is difficult at best, and impossible at worst. The best solution to this problem is to nip it in the bud by the diligent use of *randomization*. The randomization of run order (within blocks) is critical to any experimental program, and virtually guarantees valid inferences from the final model.

10.6.4 Checking for a Mean of Zero for the Errors

If our model is adequate, the values predicted by it will be the true responses (within experimental error). It follows that, if our model predicts the true responses without any biases, the residuals will be scattered around a mean of zero. In other words, the assumption that the errors have a mean of zero is the same as the assumption that we have an adequate model. To check this assumption, we look for trends in the residuals. No trends should exist if our model is adequate. The residuals should have no information contained in them other than how big the random variability is for our system. The other information in our data was presumably all accounted for (and hence removed) by our model. Any nonrandomness is information, and our model could be expanded to take the specific trend into account, thereby making it better (more precise).

The most likely modeling problems are failure to take an independent variable into account correctly, or to overlook an important factor. Therefore, the residuals should be plotted against each independent variable in the experimental design, and they should be plotted against any suspected factors that were not part of the design but inadvertently changed from one experiment to the next. They should also be plotted against run order (to look for a time trend), block, and anything else that might show a trend. The suggested plots were made for the voltage measurement example, and the plots are shown in Figure 10.16. None of the plots show any trends, and so the model appears to be adequate in this regard.

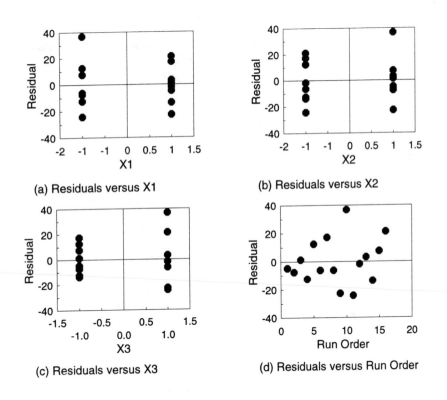

Figure 10.16 Plots of Residuals vs. Independent Variables and Run Order to Check Model Assumptions for the Voltage Measurement Example

10.7 Data Transformation for Linearity

In Chapters 9 and 10 we have been fitting linear models to data. But, if a two-level factorial design (without center points) is used to collect data, there is no statistical test we can perform to determine if the linear model is adequate. With only two levels of a factor, one at $X = -$, and the other at $X = +$, a straight line joining the mean response at each level would seem to be the best we can do. However, there are clues in the data and residual plots that can make us aware of nonlinearities in the underlying relationship.

When the response data spans two or more orders of magnitude (i.e., 1, 10, 100) it is a clue that there is a nonlinear relationship between the response and factors. Another related clue that usually occurs in these situations is a tendency for the variance to increase as the response value increases. Figure 10.17 illustrates this situation with an exponential relationship. Here we see an exponentially increasing relationship between the response, Y, and the factor, X. At the same time we see that there is more variability in the large response values than there is in the small response values.

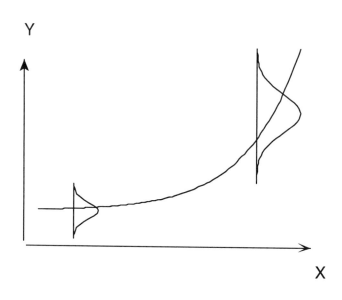

Figure 10.17   An Exponential Relationship with Increasing Variance of Response

In a situation like this the response data will normally range over orders of magnitude, and a plot of the residuals versus the predicted values will show a megaphone shape with a wider spread in the residuals at the right than on the left of the graph. The accuracy of a model fit by least squares can be improved in situations like this by transforming the response data prior to fitting the model. For example, if the true relationship shown in Figure 10.16 is $Y = ae^{bX}$, then there is a linear relationship $\ln(Y) = b_o + bX$, where $b_o = \ln(a)$. Consider the following example:

Beecher, Gomm, and Horn conducted a 2^3 factorial experiment to study the effect of $X_1 =$ Laser Current, $X_2 =$ Monochronometer Exit Slit Width, and $X_3 =$ Photomultiplier Voltage upon the Detector Response, Y (in voltage) to determine the sensitivity of a fluorescence instrument used in spectroscopic determination and identification of chemical compounds. Table 10.8 shows the factor levels and actual response data.

Table 10.8 Data from Fluorescence Spectroscopic Experiment

Run	X_1 (- = 15A + = 20A)	X_2 (- = 25 μm + = 50 μm)	X_3 (- = 1000 V + = 1100 V)	Y (Detector Response, Volts)
1	-	-	-	0.9, 0.9, 0.7
2	+	-	-	2.1, 2.1, 2.1
3	-	+	-	1.7, 1.5, 1.7
4	+	+	-	3.9, 4.1, 3.9
5	-	-	+	2.1, 2.3, 2.1
6	+	-	+	5.3, 5.3, 5.3
7	-	+	+	4.3, 3.9, 4.1
8	+	+	+	10.4 ,10.1,9.3

We can see that the response values range over an order of magnitude from 0.7–10.4. The least squares linear model fit to the data was:

$$Y = 3.754 + 3.142X_1 + 2.308X_2 + 3.242X_3 + 0.942X_1X_2 + 1.342X_1X_3 + 0.975X_2X_3 + 0.408X_1X_2X_3$$

where X_i are the coded factor levels. All coefficients in the model were significant at the 99% confidence level, but the plot of residuals versus predicted values, shown in Figure 10.18, and the Normal Probability plot of residuals, Figure 10.19 illustrates the characteristic pattern of a nonlinearity and nonconstant variance. The plot of residuals versus predicted values shows the megaphone pattern with wider spread at the right than on the left. The Normal plot of residuals exhibits a non-Normal pattern characteristic of distributions with long tails. This would result when fitting a linear model to an exponential relationship like that shown in Figure 10.17, because some residuals would come from the narrow distribution at the left while others would come from the wide distribution at the right.

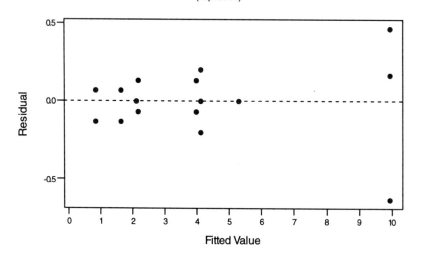

Figure 10.18 Megaphone Pattern for Fluorescence Data

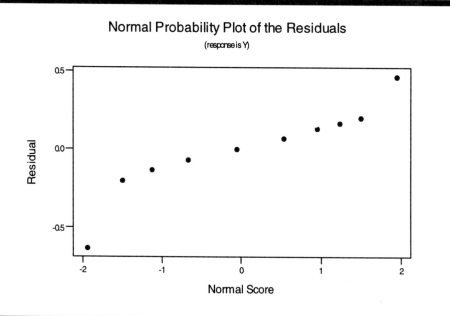

Figure 10.19 Non-Normal (Long-Tailed) Pattern in Fluorescence Residuals

326

The model can be substantially improved if we take a natural log transformation of the response data. Doing this, the natural logs of the response data was computed, and then the least squares method was used fit the model. The results were:

$$\ln(Y) = 1.071 + 0.900\, X_1 + 0.645\, X_2 + 0.931\, X_3$$

The interaction terms are not included in this model this time, because they were not significant. On the original scale of the data, interactions were significant in the model because they tried to account for the nonlinearity. When a transformation is used that linearizes the relationship, the interactions are no longer important in the model. In addition to the removal of interactions from the model, the transformation also changed the plot of residuals versus predicted values. Making this plot will be left for an exercise, where it will be seen that it does not take the characteristic megaphone shape shown in Figure 10.18. Also the Normal Plot of Residuals also will appear more linear.

As shown in this example, nonconstant variance, non-Normality of residuals, and nonlinearity of the model generally go hand in hand. A simple transformation, like the logarithm, can often linearize the model and provide a better fit by justifying the assumptions that make the least squares estimates best. Box and Cox proposed a whole family of transformations that can be used to linearize a model and simultaneously make the variance in residuals constant. Their transformations are of the form Y^λ, where in the special case of $\lambda = 0$ we use $\ln(Y)$. Table 10.9 shows some typical values of λ that are useful.

Table 10.9 Typical Box-Cox Transformations

λ	Transformation
2	Y^2
1.5	$Y^{1.5}$
1	Y (Original Scale, No Transformation)
0.5	$Y^{.5} = \sqrt{Y}$
0	ln (Y)
-0.5	$Y^{-.5} = 1/\sqrt{Y}$
-1	$Y^{-1} = 1/Y$

To choose an appropriate transformation from the family, first fit the linear model in the original scale of the data and make a plot of the residuals versus predicted values. If the range of the residuals increases as a function of the predicted values (like the megaphone pattern) choose a value of $\lambda<1$. On the other hand, if the range of the residuals tends to decrease as a function of the predicted value, then choose a value of $\lambda>1$. After applying the transformation and refitting the data, again look at the plot of residuals versus predicted values and the Normal plot of residuals to see if the problem has been solved. If not choose a different value of λ.

A more formal way of choosing the appropriate value of λ, is to try several values, and examine the R^2 statistic for each fitted equation. For example, if we had tried the values $\lambda = -0.50$, -0.25, 0.00, 0.25, 0.50, and 1.00 for the data in Table 10.8, the resulting R^2 are shown in Figure 10.20. Here it can be seen that a value of $\lambda = 0.5$ ($\sqrt{Y}$) appears to give the best fit to this data, even better than the natural log shown earlier. The plot of residuals verses predicted values and Normal plot of residuals also looks best when this square root transformation is used. Construction of these residual plots will be left for an exercise.

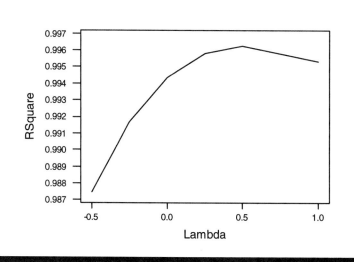

Figure 10.20   Plot of R^2 vs. λ for Fluorescence Data

10.8 Summary

10.8.1 Important Equations

Estimation of Coefficients, $\hat{\mathbf{B}}$

$$\hat{\mathbf{B}} = (\mathbf{X}^T\mathbf{X})^{-1}\mathbf{X}^T\mathbf{Y} \qquad (10.37)$$

Variability of Coefficients, $\hat{\mathbf{B}}$

$$\text{Var}(\hat{\mathbf{B}}) = (\mathbf{X}^T\mathbf{X})^{-1}\sigma^2 \qquad (10.42)$$

Confidence Intervals for Coefficients,

Two-sided $100(1-\alpha)\%$ Confidence Interval for $b_i = \hat{b}_i \pm t^*_{\alpha/2}\, s_{\hat{b}_i}$.

Variability of Model Predictions, $\hat{\mathbf{Y}}$

$$\text{Var}(\hat{\mathbf{Y}}) = \mathbf{X}_o(\mathbf{X}^T\mathbf{X})^{-1}\mathbf{X}_o^T\sigma^2 \qquad (10.48)$$

10.8.2 Important Terms and Concepts

The following is a list of the important terms and concepts covered in this chapter.

References

[1] N. R. Draper & H. Smith, "Applied Regression Analysis," John Wiley & Sons, New York, (1966).

10.9 Exercises for Chapter 10

10.1 A study of the relationship between rough weight (X) and finished weight (Y) of castings was made. A sample of 12 casings was examined and the data presented below:

X Rough weight	Y Finished Weight
3.715	3.055
3.685	3.020
3.680	3.050
3.665	3.015
3.660	3.010
3.655	3.015
3.645	3.005
3.630	3.010
3.625	2.990
3.620	3.010
3.610	3.005
3.595	2.985

a) Make a scatter plot to see the relationship between X and Y.

b) Use the method of least squares to calculate the coefficients in the simple linear regression model, $Y = a + bX$.

c) Calculate the standard errors of the estimated coefficients and determine if they are significant at the 95% confidence level.

10.2 Use the method of least squares fit the equation:

$$\log(NO_2) = Y = b_0 + b_1 X_1 + b_2 X_2 + b_3 X_3 + b_{12} X_1 X_2 + b_{13} X_1 X_3 + b_{23} X_2 X_3 + b_{123} X_1 X_2 X_3$$

to the data in Table 8.11 (excluding the center points) and verify that the coefficients obtained are consistent with the effects given in Table 8.12.

10.3   In a project examining the corrosion resistance of steel plates, a 10% solution of hydrochloric acid (HCl) was run over coated steel plates for different times and temperatures and the weight loss measured. The data is recorded below.

Sample	Time X_1	Temp X_2	Weight Loss Y
1	4	160	0.00068
2	4	160	0.00760
3	4	160	0.00810
4	4	180	0.00960
5	4	180	0.00920
6	4	180	0.00910
7	4	200	0.01150
8	4	200	0.01330
9	4	200	0.01240
10	6	160	0.00900
11	6	160	0.02090
12	6	160	0.03870
13	6	180	0.01000
14	6	180	0.01060
15	6	180	0.07640
16	6	200	0.01480
17	6	200	0.03940
18	6	200	0.01300
19	8	160	0.00760
20	8	160	0.00770
21	8	160	0.00830

a) Fit the equation $Y = b_o + b_1X_1 + b_2X_2$ to the data using the method of least squares and determine if any of the coefficients are significant at the 95% confidence level.

b) Calculate the predicted values and residuals from the model you fit in a) and plot residuals versus X_1, residuals versus X_2, residuals versus predicted values, and a Normal or Half-Normal plot of residuals as shown in Section 10.6.

c) Fit the equation $Y = b_o + b_1X_1 + b_2X_2 + b_{11}X_1^2$ to the data using the method of least squares and determine if any of the coefficients are significant at the 95% confidence level.

d) Calculate the residuals and predicted values from the model you fit in Part (c) and make the same plots as in Part (b).

e) Fit the equation $\log(Y) = b_o + b_1X_1 + b_2X_2 + b_{11}X_1^2$ to the data using the method of least squares and determine if any of the coefficients are significant at the 95% confidence level.

f) Calculate the residuals and predicted values from the model you fit in Part (e) and make the same plots as in Part (d).

g) Which model do you prefer and why?

10.4 Since humidity influences evaporation, a knowledge of the relationship will allow a painter, when applying water-based paints, to adjust his/her spray gun to account for humidity. The following data were obtained (based on "Evaporation During Spray out of a Typical Water-Reducible Paint at Various Humidities", *Journal of Coating Technology*, Vol 65, 1983):

Run	X Relative Humidity	Y Solvent Evaporation	Run	X Relative Humidity	Y Solvent Evaporation
1	35.3	11.0	14	39.1	9.6
2	29.7	11.1	15	46.8	10.9
3	30.8	12.5	16	48.5	9.6
4	58.8	8.4	17	59.3	10.1
5	61.4	9.3	18	70.0	8.1
6	71.3	8.7	19	70.0	6.8
7	74.4	6.4	20	74.4	8.9
8	76.7	8.5	21	72.1	7.7
9	70.7	7.8	22	58.1	8.5
10	57.5	9.1	23	44.6	8.9
11	46.4	8.2	24	33.4	10.4
12	28.9	12.2	25	28.6	11.1
13	28.1	11.9			

a) Estimate the coefficients in the model, $Y = a + bX$, using least squares.

b) Estimate the standard deviation of the errors, s, by calculating the predicted value (using the coefficients from Part a)) at each data point, and then calculating the standard deviation of the residuals (the differences between the data points and the predictions). How many degrees of freedom does the estimate, s, have?

c) Calculate the 95% confidence intervals for the intercept, a, and the slope, b.
Hint: The 95% confidence intervals are $\hat{a} \pm t s_a$ and $\hat{b} \pm t s_b$.

d) Calculate error limits on for the predictions of the equation at X = 25, 50, and 75. The error limits are usually just reported as $\hat{y} \pm 2 s_{\hat{y}}$ for simplicity.

10.5 Verify that if Equation 10.42 ($V(\hat{B}) = X^TX^{-1} \sigma^2$) is used for the case where the model is $Y = a + bX$, the standard deviation of $\hat{b}$ (i.e. s_b) is the same as given in Equation 10.20.

10.6   An equation is to be developed from which we can predict the gasoline mileage of an automobile based on its weight and the temperature at the time of operation. The model being estimated is:

$$Y = b_0 + b_1X_1 + b_2X_2$$

The following data are available:

Car Number	X_1 Weight tons	X_2 Temperature °F	Y Miles per Gallon
1	1.35	90	17.9
2	1.90	30	16.5
3	1.70	80	16.4
4	1.80	40	16.8
5	1.30	35	18.8
6	2.05	45	15.5
7	1.60	50	17.5
8	1.80	60	16.4
9	1.85	65	15.9
10	1.40	30	18.3

a) Estimate the coefficients in the model using least squares. Use Equation 10.37. Note that all spreadsheet programs have functions for matrix multiplication, matrix transposition, and matrix inversion.   Verify your answers using the regression function in a spreadsheet program.

b) Estimate the standard deviation of the errors, s, by calculating the predicted value (using the coefficients from Part (a)) at each data point, and then calculating the standard deviation of the residuals (the differences between the data points and the predictions).   How many degrees of freedom does the estimate, s, have? Verify your answers using the spreadsheet regression function output from Part (a).

c) Calculate the statistical significance of the coefficients, b_1 and b_2.

d) Calculate the predicted mileage (mpg) at $X_1 = 1.8$ tons and $X_2 = 70$ °F. Also calculate the error limits on that prediction.   The error limits are usually just reported as $\hat{y} \pm 2 s_{\hat{Y}}$ for simplicity.

e) Calculate R^2 for the model. Verify your answers using the spreadsheet regression function output from Part (a).

f) Check your model assumptions by using plots of residuals. These should include a Half-Normal plot and plots versus X_1, X_2, and $\hat{Y}$. Make sure you comment on what you learn/verify from each graph.

10.7 You have the following four X-Y pairs (data points):

X	Y
5	3
6	3
8	5
9	5

a)   What are the coefficients for the best-fitting straight line?

b)   What is your estimate of the standard deviation of the data, s?

c)   Are the intercept and the slope statistically significant?

10.8   You purchase a chemical raw material that has trace amounts of impurity that is detrimental to your process (namely it poisons the catalyst). You have a specification on the maximum amount of impurity that your supplier is allowed to ship to you, but you do not want to rely on your supplier to only ship you good material. So you test each batch of raw material when it comes in. The bad news is that the test is expensive. The good news is that you think you found a cheap replacement for the test, namely measuring the absorbence at a specific frequency. You took the data below to test your proposed new analytical method.

X	Y
Absorbence	Concentration, ppm
0.0335	3.91
0.0489	7.81
0.0571	15.63
0.0488	31.25
0.0827	62.5
0.1662	125
0.3174	250
0.5927	500
0.8877	750
1.1705	1000

(a)   What is the best straight line calibration line based on your data?

(b)   Is a straight line an adequate model?

(c)   How good is the new method? In other words, what is the error in a prediction (particularly at low concentrations / absorbences)?

10.A APPENDIX: Matrix Algebra

A matrix is defined to be a rectangular array of numbers (elements) such as,

$$\begin{bmatrix} 2 & 3 & 4 & 1 \\ 7 & 2 & 1 & 6 \\ 3 & 5 & 1 & 4 \\ 4 & 8 & 9 & 2 \\ 7 & 9 & 6 & 2 \end{bmatrix} \quad \begin{matrix} 5 \times 4 \\ \nwarrow \quad \nwarrow \\ \text{rows} \quad \text{columns} \end{matrix}$$

which has 5 rows and 4 columns. This is referred to as a 5×4 matrix (or 5 by 4 matrix). Any two matrices are said to be equal if they have the same number of rows and columns and exactly the same elements.

For matrices, as with numbers, the operations of addition, subtraction, and multiplication have been defined. For two matrices to be added together or subtracted from one another, they must be of the same dimensions. For example,

$$\begin{bmatrix} 3 & 2 & 1 \\ 5 & 7 & 6 \\ 2 & 1 & 2 \end{bmatrix} + \begin{bmatrix} 7 & 3 & 5 \\ 4 & 2 & 1 \\ 8 & 9 & 1 \end{bmatrix} = \begin{bmatrix} 3+7 & 2+3 & 1+5 \\ 5+4 & 7+2 & 6+1 \\ 2+8 & 1+9 & 2+1 \end{bmatrix} = \begin{bmatrix} 10 & 5 & 6 \\ 9 & 9 & 7 \\ 10 & 10 & 3 \end{bmatrix}$$

In other words the sum or difference of two matrices is just the matrix of sums or differences of individual elements. In shorthand notation if we let

$$\mathbf{A} = \begin{bmatrix} 3 & 2 & 1 \\ 5 & 7 & 6 \\ 2 & 1 & 2 \end{bmatrix} \quad \text{and} \quad \mathbf{B} = \begin{bmatrix} 7 & 3 & 5 \\ 4 & 2 & 1 \\ 8 & 9 & 1 \end{bmatrix}$$

we can write the sum as **A+B**, or the difference as **A-B**. In symbols, capital letters are usually chosen to represent matrices and small subscripted letters to represent individual elements, i.e.,

$$\mathbf{A} = \begin{bmatrix} a_{11} & a_{12} & a_{13} \\ a_{21} & a_{22} & a_{23} \\ a_{31} & a_{32} & a_{33} \end{bmatrix}$$

335

Multiplication is defined only for pairs of matrices where the number of columns in the first matrix equals the number of rows in the second matrix. (Note: This will always be true for square matrices of the same size.) The ij^{th} element of the product of two matrices is then defined to be the sum of products of the ith row of the first matrix times the jth column of the second matrix. For example, if

$$A = \begin{bmatrix} a_{11} & a_{12} & a_{13} \\ a_{21} & a_{22} & a_{23} \end{bmatrix} \quad \text{and} \quad B = \begin{bmatrix} b_{11} & b_{12} & b_{13} \\ b_{21} & b_{22} & b_{23} \\ b_{31} & b_{32} & b_{33} \end{bmatrix} \quad \text{then}$$

$$AB = \begin{bmatrix} (a_{11}b_{11} + a_{12}b_{21} + a_{13}b_{31}) & (a_{11}b_{12} + a_{12}b_{22} + a_{13}b_{32}) & (a_{11}b_{13} + a_{12}b_{23} + a_{13}b_{33}) \\ (a_{21}b_{11} + a_{22}b_{21} + a_{23}b_{31}) & (a_{21}b_{12} + a_{22}b_{22} + a_{23}b_{32}) & (a_{21}b_{13} + a_{22}b_{23} + a_{23}b_{33}) \end{bmatrix}$$

but notice that BA is not defined since the number of columns of B is not equal to the number of rows of A. In general if

$$A = \begin{bmatrix} a_{11} & a_{12} & \cdots & a_{1p} \\ a_{21} & a_{22} & \cdots & a_{2p} \\ \vdots & \vdots & & \vdots \\ a_{r1} & a_{r2} & \cdots & a_{rp} \end{bmatrix} \quad \text{and} \quad B = \begin{bmatrix} b_{11} & b_{12} & \cdots & b_{1q} \\ b_{21} & b_{22} & \cdots & b_{2q} \\ \vdots & \vdots & & \vdots \\ b_{p1} & b_{p2} & \cdots & b_{pq} \end{bmatrix}$$

then the ij^{th} term of AB is $\displaystyle\sum_{k=1}^{p} a_{ik}b_{kj}$

In matrix multiplication, the product of a matrix A and B denoted by AB will have the same number of rows as A and the same number of columns as B. More complicated matrix multiplications such as $(A+B)(C+D) = AC + AD + BC + BD$ can easily be written just as with individual numbers.

There is also an identity element defined for matrix multiplication, just as for multiplication of numbers. For numbers, the identity element is 1, since, for any number, $c \times 1 = c$. For matrix multiplication, we define the identity matrix, I, to be a square matrix with 1's down the diagonal and zeros everywhere else. For the 3×3 case I is defined as:

$$I = \begin{bmatrix} 1 & 0 & 0 \\ 0 & 1 & 0 \\ 0 & 0 & 1 \end{bmatrix}$$

and $\mathbf{AI} = \mathbf{A}$ for any matrix $\mathbf{A}$ that has 3 columns. Likewise $\mathbf{IB} = \mathbf{B}$ for any matrix $\mathbf{B}$ that has 3 rows. This can be verified by actually doing the multiplication:

$$\mathbf{AI} = \begin{bmatrix} (a_{11} \times 1 + a_{12} \times 0 + a_{13} \times 0) & (a_{11} \times 0 + a_{12} \times 1 + a_{13} \times 0) & (a_{11} \times 0 + a_{12} \times 0 + a_{13} \times 1) \\ (a_{21} \times 1 + a_{22} \times 0 + a_{23} \times 0) & (a_{21} \times 0 + a_{22} \times 1 + a_{23} \times 0) & (a_{21} \times 0 + a_{22} \times 0 + a_{23} \times 1) \\ (a_{31} \times 1 + a_{32} \times 0 + a_{33} \times 0) & (a_{31} \times 0 + a_{32} \times 1 + a_{33} \times 0) & (a_{31} \times 0 + a_{32} \times 0 + a_{33} \times 1) \end{bmatrix}, \text{ or}$$

$$\mathbf{AI} = \begin{bmatrix} a_{11} & a_{12} & a_{13} \\ a_{21} & a_{22} & a_{23} \\ a_{31} & a_{32} & a_{33} \end{bmatrix} = \mathbf{A}$$

Also in matrix multiplication, as in numerical multiplication, there is an *inverse* element. In numerical multiplication, the inverse of a number, c, is defined as the quantity that gives 1 as the result when multiplying c. Therefore, $(1/c)$ or c^{-1} is the inverse of c, since $(1/c)$ times c gives 1. Analogously, in matrix multiplication, the inverse of a matrix, $\mathbf{A}$, is defined as the matrix that multiplies $\mathbf{A}$ to give $\mathbf{I}$, and it is denoted by $\mathbf{A}^{-1}$. In equation form, this is written as

$$\mathbf{A}\mathbf{A}^{-1} = \mathbf{A}^{-1}\mathbf{A} = \mathbf{I}.$$

Notice that this inverse can only be defined for square matrices, otherwise $\mathbf{A}\mathbf{A}^{-1}$ would not be the same size as $\mathbf{A}^{-1}\mathbf{A}$.

Also, very importantly, $\mathbf{A}^{-1}$ exists only when $\mathbf{A}$ has a nonzero *determinant*. The determinant is a scalar, and it is a numerical function of the elements of a matrix. It is denoted by $|\mathbf{A}|$ or "det $(\mathbf{A})$." The determinant of a 1×1 matrix is the value of its sole element. The determinant of a 2×2 matrix is

$$|\mathbf{A}| = \begin{vmatrix} a_{11} & a_{12} \\ a_{21} & a_{22} \end{vmatrix} = a_{11}a_{22} - a_{12}a_{21}$$

The determinant of a 3×3 matrix is

$$|\mathbf{A}| = \begin{vmatrix} a_{11} & a_{12} & a_{13} \\ a_{21} & a_{22} & a_{23} \\ a_{31} & a_{32} & a_{33} \end{vmatrix} = \begin{array}{l} (a_{11}a_{22}a_{33} + a_{12}a_{23}a_{31} + a_{13}a_{32}a_{21}) \\ - (a_{31}a_{22}a_{13} + a_{32}a_{23}a_{11} + a_{33}a_{12}a_{21}) \end{array}$$

For larger matrices, the task of computing a determinant by hand is very tedious. Because of this, a computer is almost always used, and that is what we will do as well.

A square matrix is said to be *singular* when its determinant is zero, and *nonsingular* when its determinant is not zero. Singularity is therefore a property of square matrices only, and it is only nonsingular matrices that have inverses.

For a 2×2 matrix, the inverse can be computed relatively easily by hand:

$$\text{if}\quad \mathbf{A} = \begin{bmatrix} a_{11} & a_{12} \\ a_{21} & a_{22} \end{bmatrix} \quad \text{then}\quad \mathbf{A}^{-1} = \frac{1}{|\mathbf{A}|} \begin{bmatrix} a_{22} & -a_{12} \\ -a_{21} & a_{11} \end{bmatrix}$$

For 3×3 and larger matrices, there is no simple and general way of computing inverse matrices. People now depend upon computer programs to compute inverses, and we will concentrate on their use in Least Squares.

The *transpose* of a matrix $\mathbf{A}$, denoted by $\mathbf{A}^T$, is defined to be the matrix whose rows are the columns of $\mathbf{A}$. For example:

$$\text{if}\quad \mathbf{A} = \begin{bmatrix} 1 & 2 & 3 \\ 7 & 5 & 2 \end{bmatrix} \quad \text{then}\quad \mathbf{A}^T = \begin{bmatrix} 1 & +7 \\ 2 & 5 \\ 3 & 2 \end{bmatrix}$$

With this definition, we can always multiply a matrix by its transpose (such as $\mathbf{A}^T\mathbf{A}$), because the number of columns in $\mathbf{A}^T$ must equal the number of rows in $\mathbf{A}$ (it is just $\mathbf{A}$ flipped around). The transpose of a sum of matrices is defined to be the sum of the transposes. That is, $(\mathbf{A}+\mathbf{B})^T = \mathbf{A}^T+\mathbf{B}^T$. The transpose of a product is the product of the transposes in reverse order, i.e., $(\mathbf{AB})^T = \mathbf{B}^T\mathbf{A}^T$. It is important to recognize that the order in matrix multiplication is critical. For the matrix $\mathbf{A}$ above,

$$\mathbf{A}\,\mathbf{A}^T = \begin{bmatrix} 14 & 23 \\ 23 & 78 \end{bmatrix} \quad \text{while}\quad \mathbf{A}^T\mathbf{A} = \begin{bmatrix} 50 & 37 & 17 \\ 37 & 29 & 16 \\ 17 & 16 & 13 \end{bmatrix}$$

A *symmetric* matrix is one whose rows are equal to its corresponding columns. For example,

$$\mathbf{A} = \begin{bmatrix} a_{11} & a_{12} \\ a_{12} & a_{22} \end{bmatrix}$$

is a symmetric matrix, since its first row is equal to its first column, and its second row is equal to its second column. Notice that if a matrix is symmetric, then its transpose is equal to itself ($\mathbf{A}^T = \mathbf{A}$). Also, for any matrix, $\mathbf{A}$, the product of its transpose and itself, $\mathbf{A}^T\mathbf{A}$, is a symmetric matrix since $(\mathbf{A}^T\mathbf{A})^T = \mathbf{A}^T(\mathbf{A}^T)^T = \mathbf{A}^T\mathbf{A}$.

338

The inverse of a symmetric matrix is also symmetric. For example, if

$$\mathbf{B} = \begin{bmatrix} 2 & 3 \\ 3 & -2 \end{bmatrix}, \quad \text{then} \quad \mathbf{B}^{-1} = \frac{1}{-13} \begin{bmatrix} -2 & -3 \\ -3 & 2 \end{bmatrix} = \begin{bmatrix} 2/13 & 3/13 \\ 3/13 & -2/13 \end{bmatrix}$$

CHAPTER 11
Multiple Level Factorial Experiments

11.1 Introduction

In Chapters 8 and 9 we presented the simplest of factorial experiments, namely those with only two levels for each of the factors under study. Examples of two-level factorials were shown when the factors were quantitative such as moisture %, or qualitative such as thread type (rolled or tapped). It makes sense to use two levels for quantitative factors if we are interested in linear relationships. However, two levels are only appropriate for qualitative factors when there are just two alternatives. In many practical situations there may be more than two alternatives for qualitative factors, such as, component type A, B, C, D in a prototype, or material types 1, 2, or 3. Also, it may be desirable to study nonlinear relationships for quantitative factors. Part IV (Chapters 13 - 15 of this book) presents experimental designs specifically for studying nonlinear relationships when only quantitative factors are present. In this chapter we study an extension of the factorial experiments that we have already presented in Chapters 8 and 9. The factorial experiments we present in this chapter will accommodate more than two levels for each factor. This will allow us to study quadratic relationships for quantitative factors and use more than two alternatives for qualitative factors. Quantitative and qualitative factors can be used simultaneously in these designs.

We will introduce an alternate form of the mathematical model for representing the data and making predictions in multilevel factorials, and we will present a new technique called the Analysis of Variance (ANOVA) for determining which factors and interactions are significant. Orthogonal contrasts will be introduced as a tool for studying specific differences among factor levels, and we will show that the calculation worksheet used in Chapter 8 was a special case of orthogonal contrasts. Finally in this chapter we will show how blocking and split plots can be used in multiple level factorials, and how to analyze the data from these experiments.

11.2 Multiple Level Factorial Designs

Multiple level factorial designs consist of experiments run at all combinations of levels of the factors involved. In Chapter 8, when a three-factor two-level design was presented, the list of experiments consisted of all possible $2\times2\times2 = 8$ treatment combinations. Likewise a factorial design with three factors that have 2, 2, and 3 levels respectively would consist of all $2\times2\times3 = 12$ possible combinations of the three factor's levels. The list of experiments can be written down in standard order by varying the levels of the first factor fastest, the levels of the second factor second fastest and levels of the third factor slowest; for example, if we designate the levels of the factors by the integer codes 1,2, etc. The twelve runs of a $2\times3\times2$ factorial experiment and the twelve runs of a 3×4 factorial are both shown in Table 11.1. In this table we include a column for the symbolic response.

Table 11.1 A 2×3×2 Factorial and a 3×4 Factorial

Run No.	2×3×2 Factorial				3×4 Factorial		
	A	B	C	Y	A	B	Y
1	1	1	1	Y_{111}	1	1	Y_{11}
2	2	1	1	Y_{211}	2	1	Y_{21}
3	1	2	1	Y_{121}	3	1	Y_{31}
4	2	2	1	Y_{221}	1	2	Y_{12}
5	1	3	1	Y_{131}	2	2	Y_{22}
6	2	3	1	Y_{231}	3	2	Y_{32}
7	1	1	2	Y_{112}	1	3	Y_{13}
8	2	1	2	Y_{212}	2	3	Y_{23}
9	1	2	2	Y_{122}	3	3	Y_{33}
10	2	2	2	Y_{222}	1	4	Y_{14}
11	1	3	2	Y_{132}	2	4	Y_{24}
12	2	3	2	Y_{232}	3	4	Y_{34}

These designs can be represented geometrically as shown in Figure 11.1. The effects of changing factor levels can be visualized by plotting the average response in each level of the factor and connecting the points with lines. Figure 11.2 is an example for a four-level factor. It is called a *main effect plot*. In the figure the symbol $\bar{Y}_i$ for i= 1, 2, 3, 4 represents the average response in the ith level of Factor A. The horizontal line drawn in this figure is at the average of all the data. For a factor that had no effects, all the points in the graph would be close to the horizontal line, while a factor that had large effects would differ from the line substantially at two or more points.

Two-factor interactions can be visualized in the same way as in the two-level designs. Figure 11.3 is an example of an interaction plot for the interaction between a four-level Factor A and a two-level Factor B. Again $\bar{Y}_{ij}$ for i=1, 2, 3, 4 and j=1, 2 represents the average response in the ith level of Factor A and the jth level of Factor B. If Factor B had more than two levels, there would be more than two segmented lines on the graph. If there was no interaction between Factors A and B the two segmented lines would be parallel within each segment. Since the lines in Figure 11.3 are not parallel (in fact crossing) it indicates that there is an interaction between Factor A and Factor B.

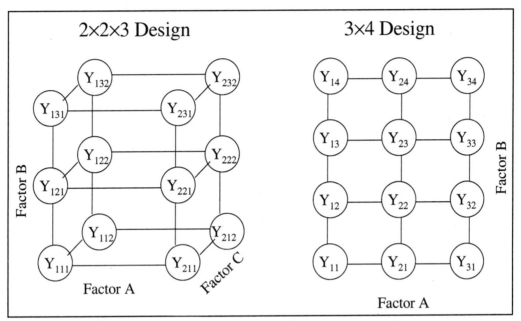

Figure 11.1 Geometric Representation of a 2×3×2 Design and a 3×4 Design

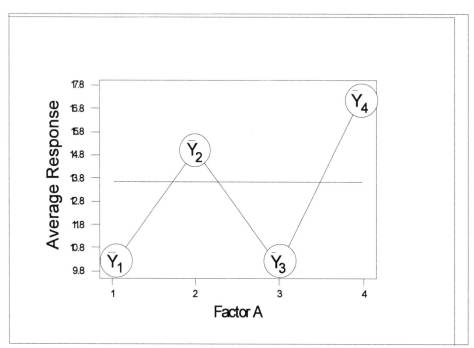

Figure 11.2 Main Effect Plot for Four-Level Factor

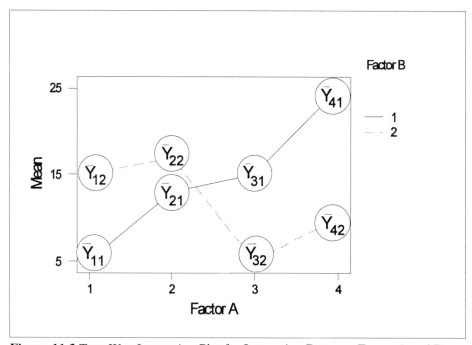

Figure 11.3 Two-Way Interaction Plot for Interaction Between Factor A and B

11.3 The Mathematical Model for Multilevel Factorials

In Chapter 9, a mathematical model for a three-factor factorial was of the form:

$$\hat{Y} = b_o + b_1X_1 + b_2X_2 + b_3X_3 + b_{12}X_1X_2 + b_{13}X_1X_3 + b_{23}X_2X_3 + b_{123}X_1X_2X_3 \qquad (11.1)$$

where b_i was half the calculated effect for a factor (or interaction), and X_i was the coded factor level (i.e., -1 or +1). This model was useful for reporting experimental results or for calculating predicted values. The effects that were used to calculate the b coefficients in this model were the differences, $\bar{Y}_+ - \bar{Y}_-$, in average response between the high (+) and the low (-) levels for each factor or interaction column in the worksheet. Using this model we can make predictions for any of the eight combinations of factor levels in the 2^3 experiment by setting (X_1, X_2, X_3) to the appropriate combinations of +1 and -1. Thus the particular combination of factor levels is identified in model (11.1) by the combination of +1 and -1 used for (X_1, X_2, X_3). However, the mathematical model in Equation (11.1) is inappropriate for multilevel factorials since for some factors there are more than two levels. Therefore, we must find another way of identifying a combination of factor levels, and a new definition for the effects and mathematical model.

The notation we use to identify a particular combination of factor levels in a multilevel factorial is the set of subscripts on the symbolic response. For example, in the 2×2×3 factorial in Table 11.1, we can see that Y_{111} is the response in the first level of Factor A, first level of Factor B, and first level of Factor C. Likewise Y_{123} is the response in the first level of Factor A, second level of Factor B, and third level of Factor C. Thus the set of subscripts, ijk, on the symbolic response Y_{ijk} identifies it to be the response from the ith level of Factor A, jth level of Factor B, and kth level of Factor C.

With this notation for a combination of treatment factors in mind, we will define K estimated effects for a factor, F, that has K levels. The first effect, $\hat{F}_1$, is defined to be the difference in the average response in the first level of Factor F minus the overall average response. In general, we define the kth effect, $\hat{F}_1$, for k=1, ..., K to be the difference of the average response at the kth level of factor F minus the overall average. If F is the third factor in a three-factor design and the first two factors have I and J levels, then we write $\hat{F}_k = \bar{Y}_{..k} - \bar{Y}_{...}$, where the dots replacing subscripts indicate we have averaged the response over all possible values of the subscript in that position, i.e.,

$$\bar{Y}_{..k} = (\sum_{i=1}^{I} \sum_{j=1}^{J} Y_{ijk})/I \times J$$

We will illustrate this definition of effects using the following example data: Consider the third Factor C.

Subscripts=Factor Levels   Response

A	B	C	Y
1	1	1	11
2	1	1	12
1	2	1	22
2	2	1	7
1	1	2	13
2	1	2	7
1	2	2	18
2	2	2	8
1	1	3	8
2	1	3	4
1	2	3	24
2	2	3	13

$\bar{Y}_{..1} = (11+12+22+7)/4 = 13.0.$

$\bar{Y}_{..2} = (13+7+18+8)/4 = 11.5$

$\bar{Y}_{..3} = (8+4+24+13)/4 = 12.25$

$\bar{Y}_{...} = (\bar{Y}_{..1} + \bar{Y}_{..2} + \bar{Y}_{..3})/3 = (13 + 11.5 + 12.25) = 12.25$, and the estimated effects are:

$\hat{C}_1 = 13.0 - 12.25 = 0.75$

$\hat{C}_2 = 11.5 - 12.25 = -0.75$

$\hat{C}_3 = 12.25 - 12.25 = 0.0$

These effects can be visualized graphically on the main effect plot in Figure 11.4.

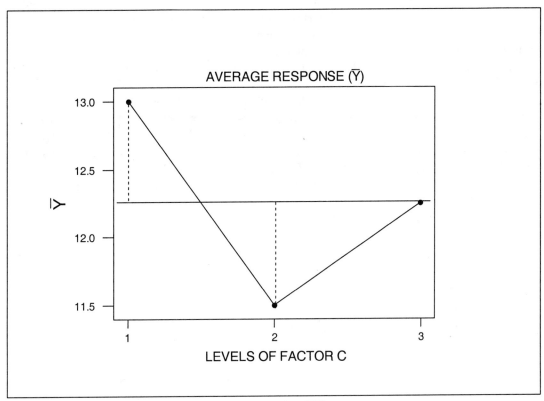

Figure 11.4 Main Effect Plot for Factor C Illustrating Effects

Recall that the main effects in two-level factorials are interpreted as the change in average response between the high and low levels of the factors. The effects in the multilevel factorial have a similar interpretation. They are defined as the difference or offset between the overall average response and the average response in the kth level of a factor, as can be seen in Figure 11.4.

The mathematical model for the multilevel factorial (assuming no interaction effects) can be written in terms of our newly defined effects as follows:

$$\hat{Y}_{ijk} = \hat{\mu} + \hat{A}_i + \hat{B}_j + \hat{C}_k \tag{11.2}$$

where $\hat{Y}_{ijk}$ is the predicted value for the ijkth combination of levels of Factors A, B, and C, and $\hat{\mu} = \bar{Y}_{...}$, the overall average, and $\hat{A}_i$, $\hat{B}_j$, and $\hat{C}_k$ are the effects or offsets from the overall average that result from being in the ith level for Factor A, the jth level of Factor B, and the kth level of Factor C. This model is often called the *additive model*, because predictions can be made from this model by simply adding the estimated main effects for each factor to the overall average. It will

347

yield accurate predictions, if the effects of Factor A were the same for each combination of levels for Factors B and C; the effects of Factor B were the same for each combination of levels of A and C; and the effects of Factor C were the same for each combination of levels of Factors A and B. The additive model will not produce accurate predictions if there are interactions between factors. A better model can be obtained by adding interaction effects to the additive model (11.2), just as they were added to model (11.1).

Two-way interaction effects, for multilevel factorials, are defined and interpreted similar to the main effects. There are I×J effects defined for a two-way interaction between a Factor A, with I levels, and another Factor B, with J levels. The ijth effect is defined to be the average response in the ij combination of levels for Factors A and B minus the overall average response minus the ith effect for Factor A minus the jth effect for Factor B. For example, the $\hat{AB}_{12} = \bar{Y}_{12.} - \bar{Y}_{...} - \hat{A}_1 - \hat{B}_2$. Thus the I×J two-way interaction effects be can be thought of as the difference between the average response in the ijth cell and the predicted value from the mathematical model that includes only main effects. Below we illustrate calculations of the BC interaction effects using the example data from above.

Subscripts=Factor Levels   Response

A	B	C	Y	
1	1	1	11	
2	1	1	12	$\bar{Y}_{.11} = (11+12)/2 = 11.5$
1	2	1	22	
2	2	1	7	$\bar{Y}_{.21} = (22+7)/2 = 14.5$
1	1	2	13	
2	1	2	7	$\bar{Y}_{.12} = (13+7)/2 = 10.0$
1	2	2	18	
2	2	2	8	$\bar{Y}_{.22} = (18+8)/2 = 13.0$
1	1	3	8	
2	1	3	4	$\bar{Y}_{.13} = (8+4)/2 = 6.0$
1	2	3	24	
2	2	3	13	$\bar{Y}_{.23} = (24+13) = 18.5$

$$\hat{BC}_{11} = \bar{Y}_{.11} - \bar{Y}_{...} - \hat{B}_1 - \hat{C}_1 = 11.5 - (12.25 - 3.0833 + 0.75) = 11.5 - 9.9166 = 1.5833$$

$$\hat{BC}_{21} = \bar{Y}_{.21} - \bar{Y}_{...} - \hat{B}_2 - \hat{C}_1 = 14.5 - (12.25 + 3.8033 + 0.75) = 14.5 - 16.0833 = -.1.5833$$

$$\hat{BC}_{12} = \bar{Y}_{.12} - \bar{Y}_{...} - \hat{B}_1 - \hat{C}_2 = 10.0 - (12.25 - 3.0833 - 0.75) = 10.0 - 8.4166 = 1.5833$$

$$\hat{BC}_{22} = \bar{Y}_{.22} - \bar{Y}_{...} - \hat{B}_2 - \hat{C}_2 = 13.0 - (12.25 + 3.0833 - 0.75) = 13.0 - 14.5833 = -1.5833$$

$$\hat{BC}_{13} = \bar{Y}_{.13} - \bar{Y}_{...} - \hat{B}_1 - \hat{C}_3 = 6.0 - (12.25 - 3.0833 + 0.0) = 6 - 9.1667 = -3.1667$$

$$\hat{BC}_{23} = \bar{Y}_{.23} - \bar{Y}_{...} - \hat{B}_2 - \hat{C}_3 = 18.5 - (12.25 + 3.0833 + 0.0) = 18.5 - 15.333 = 3.1667$$

These effects can be visualized graphically in Figure 11.5. The heavy lines in this graph connect the average response values. The thin, dotted lines connect the predicted values from the additive model with no interaction effects. Notice that the thin, dotted lines are parallel because they represent predictions from a model that includes no interactions. The solid vertical lines represent the effects, or differences in the average response and the predicted values from the *additive model*.

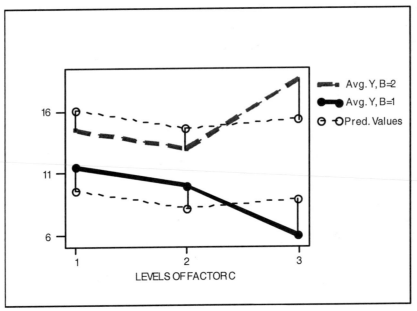

Figure 11.5--Interaction Effects for BC Interaction

Interaction effects for AB and AC are defined similarly, and a mathematical model that includes all two-factor interactions would be written as follows:

$$\hat{Y}_{ijk} = \hat{\mu} + \hat{A}_i + \hat{B}_j + \hat{C}_k + \hat{AB}_{ij} + \hat{AC}_{ik} + \hat{BC}_{jk} \qquad (11.3)$$

If there are replicate experiments at each combination of levels of the three factors, A, B, and C, then it is possible to define three-way interaction effects similar to the way we defined two-way interactions effects. We define the three-way interaction effects to be the difference between the average response at the ijk combination of factor levels and the predicted value form the mathematical model that includes all main effects and two way interactions i.e., $\hat{ABC}_{ijk} = \bar{Y}_{ijk} - \hat{Y}_{ijk}$. This definition of effects can be extended to factorials involving any number of factors, and we can add four -way interaction effects, five-way interaction effects, etc. to the mathematical model.

The estimated effects such as $\hat{A}_i$, $\hat{B}_j$, and $\hat{ABC}_{ijk}$ are really random variables since their values will differ if we repeat the entire experiment and calculate the effects with new data. If we could repeat the experiment a large number of times, and compute the long run average, or expected value, for each of the effects we would represent them as A_i, B_j, and ABC_{ijk} without the hats. These are constants rather than random variables and they represent the real effects that we will never actually observe. Writing the mathematical model in terms of these real effects,

$$Y_{ijk} = \mu + A_i + B_j + C_k + AB_{ij} + AC_{ik} + BC_{jk} + ABC_{ijk} + \varepsilon_{ijk} \qquad (11.4)$$

we must add an error term, ε_{ijk}, to represent random differences between the observed value, Y_{ijk} in the ith level of Factor A, jth level of Factor B, and kth level of Factor C in our experiment, and the long run or expected value in that combination of factor levels. We are specifically interested in the variance of these error terms, ε_{ijk}, because they will become the yardstick for judging whether estimated effects from an experiment are significant. Replicate experiments allow estimation of the variance of, ε_{ijk}, and we will discuss how this is done in Section 11.5.

350

11.4 Example of a 2×2×3 Design

In practice, the coded levels of the factors (i.e., 1, 2, etc.) would be replaced by the actual factor levels to be tested and the order of the experiments would be randomized. For example, consider an experiment to study vibration in an electric motor. Three factors were studied, with levels as shown in Table 11.2. In this experiment there are two two-level quantitative factors and one three-level qualitative factor. The levels of Factor B are reported in coded levels. Since Factor C has three alternative levels, the standard 2^3 design of Chapter 8 cannot be used.

Table 11.2 Factors and Levels for Motor Noise Experiment

	Levels		
Factors	1	2	3
A. Rotor Slot Skew	1.0	1.5	-
B. Air Gap	g	$g + \Delta g$	-
C. Cover Type	Cast Iron	Aluminum	Cast Iron with thick section

Translating the levels of the 2×2×3 factorial from Table 11.1 into actual factor levels would result in the data collection worksheet for the vibration experiment shown in Table 11.3. Each row of the worksheet represents one prototype electric motor. In order to judge the significance of effects, replicate experiments must be made. Therefore, two assemblies were made for each prototype and tested in the random order prescribed by the second column in the worksheet. The resulting vibration, in microns, was measured for each test and recorded in the far right column of the worksheet.

Table 11.3 Data Collection Worksheet

| | | Factors | | | Response Y |
| | | A | B | C | Vibration |
Run No.	Random Order	Rotor Slot Skew	Air Gap	Cover	(microns)
1	19, 16	1.0	g	Cast Iron	11, 9
2	10, 13	1.5	g	Cast Iron	12 ,13
3	15, 4	1.0	g + Δg	Cast Iron	22, 14
4	22, 21	1.5	g + Δg	Cast Iron	7, 10
5	24, 12	1.0	g	Aluminum	13, 14
6	17, 9	1.5	g	Aluminum	7, 9
7	20, 2	1.0	g + Δg	Aluminum	18, 16
8	8, 11	1.5	g + Δg	Aluminum	8, 10
9	5, 6	1.0	g	thick Cast Iron	8, 9
10	1, 23	1.5	g	thick Cast Iron	4, 6
11	3, 7	1.0	g + Δg	thick Cast Iron	24, 15
12	18, 14	1.5	g + Δg	thick Cast Iron	13, 15

Below we show the mean and variance of the vibration for each experiment, and we calculate the pooled standard deviation in the same way we did for the worksheets in Chapter 8.

Run Number	$\bar{Y}$	Variance
1	10.0	2.00
2	12.5	0.50
3	18.0	32.00
4	8.5	4.50
5	13.5	0.50
6	8.0	2.0
7	17.0	2.0
8	9.0	2.0
9	8.5	0.50
10	5.0	2.0
11	19.5	40.50
12	14.0	2.00

Pooled variance:

$$s^2_p = (2.0 + 0.5 + \ldots + 2.0)/12 = 7.5416$$

This pooled variance is an estimate of the variance of the random terms, ε_{ijk}, in model (11.4).

The means for each experiment can be represented geometrically as shown in Figure 11.6 . Here we can see that the lowest vibration (5.0) occurs at the back lower right side where cover type is Thick Cast Iron, Air Gap is the minimum (g) and Rotor Slot Skew is the maximum (1.5). From this figure it appears that there are interactions between the factors. For example, increasing Air Gap from the low to high levels sometimes causes an increase in vibration and sometimes causes a decrease in vibration depending on the levels of Rotor Slot Skew and Cover Type.

Table 11.4 presents the calculated means and effects for the factors and interactions. This

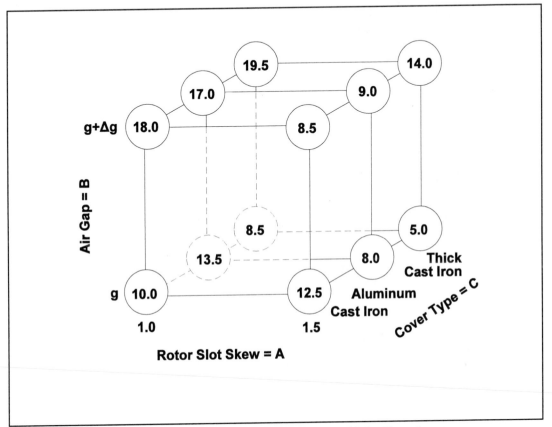

Figure 11.6 Geometric Representation of Vibration in Motor Noise Experiment

table shows that at the high level of Factor A (i.e., Rotor Slot Skew =1.5) there is less vibration since its mean (9.33) is lower and its effect (-2.58333) is negative. This confirms what we can see in Figure 11.6, where lower vibration averages are seen on the right side of the cube. The table also shows us that the low level of Factor B (i.e., Air Gap =g) on the average results in less vibration. This again confirms what we can see in Figure 11.6 where the lower vibration averages are on the bottom of the cube. The absolute values of the effects indicate that Factors A and B have the largest effects. It is difficult to interpret the interaction effects in the tabular form. Interaction plots as

Table 11.4 Means and Effects for Vibration

Term	Count	Mean	Std.Error	Effect
ALL	24	11.91667	11.91667	
Factor A				
1	12	14.5	.8122329	2.583333
2	12	9.333333	.8122329	-2.583334
Factor B				
1	12	9.666667	.8122329	-2.25
2	12	14.16667	.8122329	2.25
Factor C				
1	8	12.375	.994778	.458333
2	8	11.875	.994778	-4.16E-02
3	8	11.5	.994778	-.416667
Interaction AB				
1,1	6	10.83333	1.148671	-1.416667
1,2	6	18.16667	1.148671	1.416666
2,1	6	8.5	1.148671	1.416667
2,2	6	10.16667	1.148671	-1.416666
Interaction AC				
1,1	4	14.25	1.406829	-.708333
1,2	4	15.25	1.406829	.791667
1,3	4	14	1.406829	-8.33E-02
2,1	4	10.5	1.406829	.708334
2,2	4	8.5	1.406829	-.791666
2,3	4	9	1.406829	8.333E-02
Interaction BC				
1,1	4	11.5	1.406829	1.375
1,2	4	10.75	1.406829	1.125
1,3	4	6.75	1.406829	-2.5
2,1	4	13.25	1.406829	-1.375
2,2	4	13	1.406829	-1.125
2,3	4	16.25	1.406829	2.5
Interaction ABC				
1,1,1	2	10.5	1.989556	-1.458333
1,1,2	2	13.5	1.989556	.791667
1,1,3	2	8.5	1.989556	.666667
1,2,1	2	18	1.989556	1.458334
1,2,2	2	17	1.989556	-.791666
1,2,3	2	19.5	1.989556	-.666666
2,1,1	2	12.5	1.989556	1.458333
2,1,2	2	8	1.989556	-.791667
2,1,3	2	5	1.989556	-.666667
2,2,1	2	8.5	1.989556	-1.458334
2,2,2	2	9	1.989556	.791666
2,2,3	2	13	1.989556	.666666

shown in Figures 11.7 and 11.8 make it easier to see how the effects of one factor depend on the levels of the other factors. Figure 11.7 looks like the interaction plots of Chapter 8 since it involves

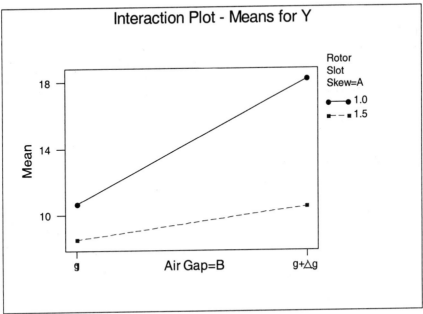

Figure 11.7 Interaction Plot for Factors A and B

two-level factors. It shows that the effect of Factor B (Air Gap) is much larger when the Factor A (Rotor Slot Skew) is at its low level. Figure 11.8 is an interaction plot between a two-level factor (Air Gap) and a three-level factor (C= Cover). In this figure it can be seen that the difference in average vibration between Cast Iron Covers and Aluminum Covers are virtually the same when using the smaller Air Gap or larger Air Gap. The parallel line segments between Cast Iron and Aluminum, make this clear visually. However, the difference between Thick Cast Iron Covers and Cast Iron (or Aluminum) Covers depends very much on whether the small or larger Air Gap is used. With the larger Air Gap, upper dotted line, Thick Cast Iron Covers seem to cause more vibration than Aluminum or Cast Iron. With a smaller Air Gap, bottom solid line, Thick Cast Iron Covers seem to cause less vibration.

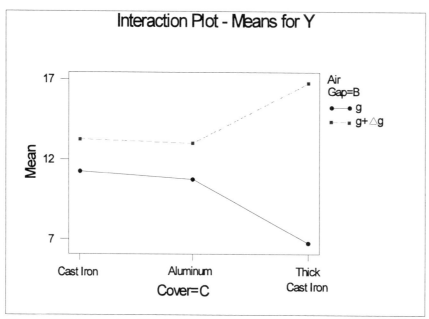

Figure 11.8 Interaction Plot for Factors B and C

In general, there may be many two-way, three-way and higher order interactions to consider in the analysis of a factorial design. We need to have a yardstick for determining which factors and interactions are significant and worth plotting. In the next section we will describe the Analysis of Variance. It is the method used to determine which factors and interactions are significant in multilevel factorials.

11.5 Judging the Significance of Effects in Multilevel Factorials

One of the main purposes of experimentation is to determine whether or not factors have effects that are significantly different from zero. Recall that we judged the significance of effects in two-level factorials using the signal-to-noise t-ratio.

$$t_E = \frac{\bar{Y}_+ - \bar{Y}_-}{2s_p/\sqrt{n_f}}$$

This worked because the effects were simply the differences in average response from two separate (determined by the levels of a factor) classifications of data.

We can't use the t-ratio for testing the significance of effects in multilevel factorials, since the effects, $\bar{Y}_{j\cdot\cdot} - \bar{Y}_{\cdots}$, are not differences in average response between two separate classifications of data, but rather are the difference between the average response at one level of a factor and the overall average. The overall average includes the data used in calculating the first average. We could make t-ratios comparing averages at all possible pairs of levels of a factor, (i.e., $\bar{Y}_{1\cdot\cdot} - \bar{Y}_{2\cdot\cdot}$, and $\bar{Y}_{1\cdot\cdot} - \bar{Y}_{3\cdot\cdot}$, etc.) but there is another problem associated with this procedure. The more comparisons we make, the more likely we are to find one large t-ratio solely due to random chance. Therefore we may think we have 95% confidence in our conclusion, but actually it is less than 95%. Furthermore, it would be difficult to extend this comparison of averages to two-factor and higher way interactions.

Normally the first step in determining if there are significant factor effects is to make a simultaneous significance test of all the effects in our mathematical model using the Analysis of Variance procedure often abbreviated as ANOVA. The Analysis of Variance helps us judge the significance of all estimated effects simultaneously by comparing the variability among estimated effects to the variability among replicate observations.

11.5.1 ANOVA for One-Factor Design

To illustrate the Analysis of Variance procedure we will start with the simplest situation where we have only one factor, A, at I levels. The data will be labeled Y_{ij} to represent the jth replicate response value in the ith level of Factor A. The mathematical model would be written as:

$$Y_{ij} = \mu + A_i + \varepsilon_{ij} \tag{11.5}$$

Where i=1 to I, and j=1 to r, the number of replicate experiments in each level of Factor A. The estimated effects are $\hat{A}_i = \bar{Y}_{i\cdot} - \bar{Y}_{\cdot\cdot}$, and the numerator of the sample variance, or sum of squares, calculated from these estimated effects is:

$$SSA = \sum_{i=1}^{I} r \times \hat{A}_i^2 = \sum_{i=1}^{I} r \times (\bar{Y}_{i\cdot} - \bar{Y}_{\cdot\cdot})^2$$

This is called the sum of squares for Factor A. The sum of squares of the replicate responses in the ith level of Factor A is: $\sum_{j=1}^{r} (Y_{ij} - \bar{Y}_{i\cdot})^2$. Pooling the sum of squares from each level of Factor A,

we get the error sum of squares $SSE = \sum_{i=1}^{I} \sum_{j=1}^{r} (Y_{ij} - \bar{Y}_{i\cdot})^2$. This is the numerator of the pooled

variance s_p^2 of the random terms, ε_{ij}, in the model shown earlier. Algebraically it can be shown that total sum of squares (SST) of differences from the grand average $\bar{Y}_{\cdot\cdot}$, is equal to the sum of SSA and SSE, i.e.,

$$SST = \sum_{i=1}^{I} \sum_{j=1}^{r} (Y_{ij} - \bar{Y}_{\cdot\cdot})^2 = SSA + SSE = \sum_{i=1}^{I} r \times (\bar{Y}_{i\cdot} - \bar{Y}_{\cdot\cdot})^2 + \sum_{i=1}^{I} \sum_{j=1}^{r} (Y_{ij} - \bar{Y}_{i\cdot})^2$$

This partitioning of the total sum of squares is similar to the partitioning of the total sum of squares that was shown in Chapter 10 while quantifying the closeness to a linear regression model. If the factor effects are zero, SSA=0 and the total sum of squares will be completely due to the random errors. On the other hand, if the factor effects are large relative to the random errors, a large percentage of SST will be due to SSA.

In the Analysis of Variance procedure, the F-ratio is used to compare the variance of the estimate effects to the variance of error terms ε_{ij}, similar to the way independent variance estimates were compared in Chapter 6. A large ratio indicates the estimated effects (or differences as depicted in Figure 11.4) are larger than can be expected to occur by chance when the actual effects are truly zero, and all differences are due to random errors.

Normally the ANOVA results are displayed in a table. Table 11.5 shows an ANOVA table for a simple one-factor model.

Table 11.5 Analysis of Variance for One-Factor

Source	degrees of freedom	Sum of Squares	Mean Square	F-ratio
Factor A	I-1	$SSA = \sum_{i=1}^{I} r \times (\bar{Y}_{i\cdot} - \bar{Y}_{\cdot\cdot})^2$	MSA=SSA/(I-1)	MSA/MSE
Error	I×(r-1)	$SSE = \sum_{i=1}^{I} \sum_{j=1}^{r} (Y_{ij} - \bar{Y}_{i\cdot})^2$	MSE=SSE/I×(r-1)	
Total	I×r - 1	SST = SSA + SSE		

We will illustrate this ANOVA procedure with the data shown in Table 11.6 taken from a research study conducted at Purdue University. The purpose of the study was to quantify the effects of using different electrode shapes on metal removal rate and hardness of the cut edges. Metal removal was accomplished by an electric discharge between the electrode and the material. The factor in the study was the electrode shape. One of the responses was the Rockwell Hardness index of the metal where holes were cut. Five pieces of metal were cut using each of five different electrodes and the Rockwell Hardness results are shown in Table 11.6.

Table 11.6 Rockwell Hardness Data

Levels of Factor A = Electrode Shape

	1	2	3	4	5
	64	61	62	62	62
	63	63	62	63	62
	63	62	61	62	63
	63	63	63	63	64
	61	63	63	62	62
$\bar{Y}_{i\cdot}$:	62.8	62.4	62.2	62.4	62.6
$(\bar{Y}_{i\cdot} - \bar{Y}_{\cdot\cdot})$:	0.32	-0.08	-0.28	-0.08	-0.12
$\sum_{j=1}^{r} (Y_{ij} - \bar{Y}_{i\cdot})^2$:	4.8	3.2	2.8	1.2	3.2

The overall average $\bar{Y}_{\cdot\cdot} = (62.8 + 62.4 + 62.2 + 62.4 + 62.6)/5 = 62.48$, and an example of computing the sum of squares for replicate values in the first level of electrode shape is:

$$\sum_{j=1}^{5} (Y_{1j} - \bar{Y}_{1\cdot})^2 = (64\text{-}62.8)^2 + (63\text{-}62.8)^2 + (63\text{-}62.8)^2 + (63\text{-}62.8)^2 + (61\text{-}62.8)^2 = 4.8$$

SSE the sum of squares for error is calculated as:

$$SSE = \sum_{i=1}^{I} \sum_{j=1}^{r} (Y_{ij} - \bar{Y}_{i\cdot})^2 = 4.8 + 3.2 + 2.8 + 1.2 + 3.2 = 15.2$$

and the sum of squares due to Factor A, SSA, is computed as:

$$SSA = \sum_{i=1}^{I} r \times (\bar{Y}_{i\cdot} - \bar{Y}_{\cdot\cdot})^2 = 5(.32)^2 + 5(-.08)^2 + 5(-.28)^2 + 5(-.08)^2 + 5(-.12)^2 = 1.04$$

For this experiment it can be seen that the majority of variation among the response values is random, and electrode shape appears to have little effect on Rockwell Hardness. The ANOVA Table presented in Table 11.7 shows the formal F-ratio for testing the significance of the electrode shape effects.

We conclude that electrode shapes have no significant effect on Rockwell Hardness, since the computed F-ratio of 0.34 is not larger than 2.87, the upper 5% point of the F-Distribution with 4 and 20 degrees of freedom from Table A.6.

Table 11.7 Analysis of Variance for Rockwell Hardness Data

Source	Degrees of Freedom	Sum of Squares	Mean Square	F-ratio
Factor A	4	1.04	0.26	0.34
Error	20	15.2	0.76	
Total	24	16.24		

11.5.2 ANOVA for General Factorial Designs

The ANOVA procedure can be easily extended to factorial designs with more than one factor. The sums of squares for the main effects are calculated exactly as they were for Factor A in Table 11.5. Sums of squares for interaction effects are computed by squaring and then summing the estimated interaction effects like those shown in Table 11.4. The total sums of squares (SST) of differences from the grand average again turns out to be a sum of the sums of squares for each term in the model. For example, for a two-factor factorial with Factors A, and B, and r replicate observations in each cell, the model is:

$$Y_{ijk} = \mu + A_i + B_j + AB_{ij} + \varepsilon_{ijk} \tag{11.6}$$

the estimated effects are:

$$\hat{A}_i = \bar{Y}_{i..} - \bar{Y}_{...}$$

$$\hat{B}_j = \bar{Y}_{.j.} - \bar{Y}_{...}$$

$$\hat{AB}_{ij} = \bar{Y}_{ij.} - \hat{A}_i - \hat{B}_j$$

and the sums of squares partition as:

$$SST = SSA + SSB + SSAB + SSE,$$

where:

$$SST = \sum_{i=1}^{I} \sum_{j=1}^{J} \sum_{k=1}^{r} (Y_{ijk} - Y_{...})^2 \qquad (11.7)$$

$$SSA = \sum_{i=1}^{I} r \times J \times \hat{A}_i^2 \qquad (11.8)$$

$$SSB = \sum_{j=1}^{J} r \times I \times \hat{B}_j^2 \qquad (11.9)$$

$$SSAB = \sum_{i=1}^{I} \sum_{j=1}^{J} r \times \hat{AB}_{ij}^2 \qquad (11.10)$$

and   SSE=SST-SSA-SSB-SSAB

For a numerical example consider the example data in Table 11.8 from a chemical process experiment with two factors. Factor A =temperature, and has two levels, and Factor B = catalyst type and has three levels.

Table 11.8 Chemical Experiment

Run	Random Order	Factor A Temperature	Factor B Catalyst	Yield
1	10	50	A	45.1
2	2	50	A	44.8
3	4	60	A	33.0
4	9	60	A	32.6
5	5	50	B	45.7
6	8	50	B	45.8
7	3	60	B	45.7
8	7	60	B	45.5
9	11	50	C	44.9
10	1	50	C	44.7
11	12	60	C	53.8
12	6	60	C	54.2

Normally the computations for ANOVA are done by a computer program rather than by hand. Most statistical programs such as MINITAB and spreadsheets such as EXCEL have procedures to do all the ANOVA calculations automatically, given a data set in the format of Table 11.8.

In Table 11.9 is the output of the MINITAB balanced ANOVA routine. In addition to computing the sums of squares, mean squares, and F-ratios the program also computes a P-value, which is area under the F-Distribution density to the right of the calculated F-ratios. When P-values are printed by the program, it saves us the effort of comparing the computed F-ratios in the ANOVA table to the tabled F-ratios (in Table A.6) to determine which factors and interactions have significant effects. When the P-value for a particular factor is less than .05, we are at least 95% confident that one or more of the effects for that factor are nonzero. If the P-value is less than .01 our confidence level increases to at least 99%. For the ANOVA table below, the P-values give us at least 99.9% confidence in the effects of both factors and interaction.

Table 11.9 MINITAB ANOVA Table for Data in Table 11.8

Analysis of Variance for Yield

Source	DF	SS	MS	F	P
A	1	3.203	3.203	76.88	0.000
B	2	227.855	113.928	2734.25	0.000
A*B	2	229.082	114.541	2748.97	0.000
Error	6	0.250	0.042		
Total	11	460.390			

MEANS

A	N	Yield
1	6	45.167
2	6	44.133

B	N	Yield
1	4	38.875
2	4	45.675
3	4	49.400

A	B	N	Yield
1	1	2	44.950
1	2	2	45.750
1	3	2	44.800
2	1	2	32.800
2	2	2	45.600
2	3	2	54.000

In order to determine which factor levels are different, computer programs for ANOVA allow us to print or plot the means in each level or combination of levels of the factors, but most programs do not print out the effects as shown earlier in Table 11.4. The means option in the MINITAB balanced ANOVA procedure printed the means listed in Table 11.9. In this table we can see that on the average the yield was higher in the first level of Factor A (i.e., 50 Deg.), and that the average yield was highest in the third level of Factor B (i.e., catalyst type C). Figure 11.9 is a main effects plot that shows graphically these differences in means.

Since there is a significant interaction in this experiment, the means for the individual factor levels and Figure 11.9 are misleading and should be ignored. The means in the six combinations of factor levels listed at the bottom of Table 11.9 show that the differences between catalysts are not consistent at the two temperatures. This can be seen more clearly by examining graphs of the means. Figure 11.10 presents a much clearer picture. It is the interaction plot. It shows us that changing the catalyst type (Factor B) really has no effect on the yield if we are running the process at 50 Deg. On the other hand, using catalyst type C will result in higher yields and using catalyst type A will result in lower yields if we are running at 60 Deg.

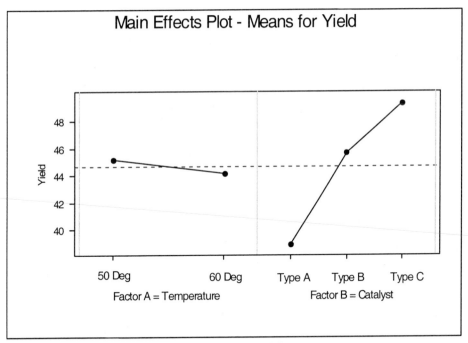

Figure 11.9 Main Effect Plots for Factors A and B

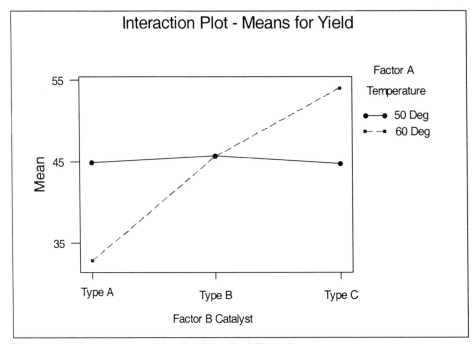

Figure 11.10 Interaction Plot for Chemical Experiment

The Analysis of Variance is easily extended to higher level factorials. The sums of squares for each factor or interaction are computed by squaring and summing the effects for that factor or interaction. For example, Table 11.10 is the ANOVA produced by MINITAB using the data from the three-factor experiment for studying vibration that was presented in Section 11.4. The significant factors and interactions (i.e., A, B, AB, and BC) are those with P-values less than .05. These are the same interactions we graphed in Section 11.4. The ANOVA table tells us there is no need to examine any other interactions since AC and ABC are nonsignificant.

Table 11.10 ANOVA for Vibration Experiment

Analysis of Variance (Balanced Designs)

Analysis of Variance for Y

Source	DF	SS	MS	F	P
A	1	145.042	145.042	19.23	0.001
B	1	135.375	135.375	17.95	0.001
C	2	1.083	0.542	0.07	0.931
A*B	1	45.375	45.375	6.02	0.030
A*C	2	11.083	5.542	0.73	0.500
B*C	2	82.750	41.375	5.49	0.020
A*B*C	2	31.750	15.875	2.10	0.165
Error	12	90.500	7.542		
Total	23	542.958			

11.6 Comparing Means after ANOVA

The F-statistic in the Analysis of Variance tells us whether any of the effects, for a factor or interaction in question, are significantly different than each other. It doesn't tell us specifically which effects are nonzero, nor whether means from different treatment combinations are significantly different. In this section we will discuss methods of comparing specific means after the ANOVA.

11.6.1 The Least Significant Difference (LSD) Method

One way to determine which means are significantly different is to do a direct comparison. The F-test in the chemical experiment (Table 11.8) analyzed at the end of the last section, showed there was a significant interaction between temperature (Factor A) and catalyst (Factor B). From the interaction plot, Figure 11.10, it appears like the means for the three catalysts are not significantly different at 50 Deg., while the mean yield for catalyst type A at 60 Deg. is lower, and the mean yield for catalyst C at 60 Deg. is higher. One way to verify this would be to use simple t-statistics. The difference in any two means in the 2×3 factor level combinations can be compared by the statistic

$$t_{I \times J \times (r-1)} = \frac{\bar{y}_{ij} - \bar{y}_{i'j'}}{\sqrt{MSE \times (\frac{1}{n} + \frac{1}{n})}}$$

where n is the number of observations averaged in $\bar{y}_{ij}$ and $\bar{y}_{i'j'}$. For example a t-statistic for comparing the means of catalyst A and B at 60 Deg. (i.e., combination i=2, j=1 to i'=2, j'=2) would be:

$$t_6 = \frac{32.8 - 45.6}{\sqrt{0.042 \times (\frac{1}{2} + \frac{1}{2})}} = -62.46$$

While the t-statistic for comparing means of catalyst A and B at 50 Deg. (i.e., combination i=1, j=1 and i'=1, j'=2) would be:

$$t_6 = \frac{44.95 - 45.75}{\sqrt{0.042 \times (\frac{1}{2} + \frac{1}{2})}} = -3.904$$

Both of these are significant at the 99% confidence level, which appears to say that even the slight differences in means represented in Figure 11.10 may be significant when compared to the variation among replicate experiments at the same conditions. However as will be explained below, using the t-statistic may result in a high number of false positives.

The t-statistic, shown above, is exactly the same formula used in Chapter 6 to compare means in a simple comparative test, and the same as the formula used in Chapter 8 to compare the means at the '-' and '+' levels of a factor or interaction. In general, any two means (at different levels of a factor, or different combinations of levels of an interaction) can be declared significantly different at confidence level $1-\alpha$ if they differ by more than the

$$LSD = t_{\nu,\ 1-\alpha} \times \sqrt{MSE \times (\frac{1}{n} + \frac{1}{n})},$$ where ν is the degrees of freedom associated with the error mean

square in the ANOVA table and n is the number of values averaged in each mean. For example, the LSD (or least significant difference) for comparing means at different levels of main effect A (temperature) would be:

$$LSD = 2.447 \times \sqrt{0.042 \times (\frac{1}{6} + \frac{1}{6})} = 0.28953$$

while the LSD for comparing means at different combinations of levels of Factors A and B in the interaction would be:

$$LSD = 2.447 \times \sqrt{0.042 \times (\frac{1}{2} + \frac{1}{2})} = 0.5014$$

The LSD method can be used if one preselected comparison of means is to be made. However, the confidence level is decreased when using the LSD method if a number of comparisons are made. This will result in a higher rate of false positives. There are [I×(I-1)/2] possible comparisons of means for an I-level factor, and [I×J×(I×J-1)/2] possible comparisons of means in combinations of levels of an I by J level interaction. If we make one comparison of means at confidence level $1-\alpha$, the probability of a Type I error for the test will be α, but if we make k comparisons at confidence level $1-\alpha$, the probability of a Type I error will be reduced to

$1-(1-\alpha)^k$. For example making six comparisons at the 95% confidence level means we really only have $(.95)^6 = .735$ or 73.5% confidence in any difference of means declared significant.

11.6.2 Tukey's Method

In order to hold the Type I error rate at α when making many comparisons of means following the ANOVA, we must increase the value of the LSD. Tukey[1] developed a method that does this by referring to the distribution of the studentized range rather than the t-distribution. Using Tukey's method we declare a difference of two means to be significant if it is greater than:

$$HSD = \frac{q(\alpha, v, k)}{\sqrt{2}} \times \sqrt{MSE \times (\frac{1}{n} + \frac{1}{n})}$$

where HSD stands for Tukey's honestly significant difference, $q(\alpha, v, k)$ is the $1-\alpha$ confidence value of the studentized range statistic with v degrees of freedom for error, and k means being compared, and finally n is the number of values averaged in any two means being compared. Table A.7 in Appendix A tabulates the 95% confidence values for the studentized range. These are used like the critical values for the t-statistic in Table A.5.

For example, returning to the means in the 2×3 combinations of levels of factors A (temperature) and B (catalyst) in the Chemical Experiment with 95% confidence,

$$HSD = \frac{5.63}{\sqrt{2}} \times \sqrt{0.042 \times (\frac{1}{2} + \frac{1}{2})} = 0.8158$$

where 5.63 is the value from Table A.7 with $v = 6$ and $k=6$. This can be seen to be larger than the LSD value shown in the last section. Table 11.11 summarizes the differences in means (from Table 11.9) for the two-way interaction, and is a convenient way to display the results of all possible comparisons of means using Tukey's Method.

---

[1]Tukey, J. W. "Comparing Individual Means in the Analysis of Variance," *Biometrics*, Vol. 5, (1949), p. 99.

Table 11.11 Differences in Means (* differences are greater than HSD and significant)

Factor Levels	A:	2	1	1	2	1	2
	B:	1	3	1	2	2	3
Means	Means:	32.8	44.8	44.95	45.6	45.75	54.00
32.8		-	12.0*	12.15*	12.8*	12.95*	21.2*
44.8			-	.15	.80	.95*	9.20*
44.95				-	.65	.80	9.05*
45.6					-	.15	8.4*
45.75						-	8.25*

This table shows us that the low mean yield of 32.8 in the reaction at 60 Deg. with catalyst A (i.e., Factor A=2 and B=1) is significantly lower than the yields in all other treatment combinations. The high mean yield of 54.0 in the reaction at 60 Deg. using catalyst C (i.e., Factor A=2 and B=3) is significantly higher than the yields in all other treatment combinations. The only other significant difference among all the means is the .95 difference between the mean yields for catalysts B and C at 50 Deg.

11.6.3 Orthogonal Contrasts

An alternative to comparing specific pairs of means, or all possible pairs of means, is to examine meaningful comparisons of several means. To illustrate the idea, consider the following data collected in an experiment designed to select a lubricant for a wire drawing process.

Two factors were studied in this experiment, one being the wire drawing speed and the other being the type of lubricant used. The wire drawing speed which could be varied over a continuous range, was studied at two levels, as were the examples in Chapters 8 - 9. The lubricant type factor was studied at three levels. These levels were comprised of two new lubricants (A and B) along with a standard (C). The new lubricants were chosen from many that are available, some containing proprietary additives and unknown chemistries. Selection of a lubricant often becomes a trial-and-error process. In a metal-forming process often 50% of the energy supplied by equipment is used to overcome friction and product quality characteristics such as surface finish, dimensions, and material strength which are directly related to friction. The presence of a lubricant and the type and amount of that lubricant can significantly alter the friction behavior. The data from the experiment and the Analysis of Variance is shown in Table 11.12 and 11.13.

Table 11.12 Data from Wire Drawing Experiment

	Factor L	Factor D	Response
Random Run Order	Lubricant Type	Wire Drawing Speed	Drawing Force
3	A	10^2	22.3
6	B	10^2	24.2
4	C	10^2	27.6
7	A	10^2	22.4
9	B	10^2	25.8
12	C	10^2	27.3
11	A	10^4	24.5
8	B	10^4	27.1
10	C	10^4	30.6
5	A	10^4	23.0
1	B	10^4	27.9
2	C	10^4	31.0

The table below is the ANOVA from MINITAB.

Table 11.13 MINITAB ANOVA for Wire Drawing Experiment

```
Analysis of Variance for Y

Source      DF        SS         MS       F       P
L            2    73.882     36.941   77.63   0.000
D            1    17.521     17.521   36.82   0.001
L*D          2     1.912      0.956    2.01   0.215
Error        6     2.855      0.476
Total       11    96.169


Means

L       N          Y
1       4     23.050
2       4     26.250
3       4     29.125

D       N          Y
1       6     24.933
2       6     27.350
```

In Table 11.13 we see that there are significant differences among the lubricants (Factor L). A reasonable comparison of the lubricants would be to compare the standard lubricant (C or the third level of Factor L i.e., L_3) to the average of the other two (i.e., L_2 and L_1). This comparison could be written as:

$$C_1 = -2 \times \hat{L}_3 + \hat{L}_2 + \hat{L}_1$$

or

$$C_1 = -2 \times (\bar{y}_{3.} - \bar{y}_{..}) + (\bar{y}_{2.} - \bar{y}_{..}) + (\bar{y}_{1.} - \bar{y}_{..}) = -2\bar{y}_{3.} + \bar{y}_{2.} + \bar{y}_{1.}$$

or

$$C_1 = \frac{1}{4} (-2 \times y_{3.} + y_{2.} + y_{1.})$$

where 4 is the number of values averaged in each mean. This comparison is called a *contrast* because the coefficients (-2, 1 and 1) sum to zero. It can be calculated from the means as:

$$C_1 = -2(29.125) + 26.250 + 23.050 = -8.95$$

or directly from the raw data as a sum of products as shown in the worksheet shown in Table 11.14:

Table 11.14 Calculation of Contrast

Factor L Lubricant Type	Coefficients	Response Drawing Force
A	1	22.3
B	1	24.2
C	-2	27.6
A	1	22.4
B	1	25.8
C	-2	27.3
A	1	24.5
B	1	27.1
C	-2	30.6
A	1	23.0
B	1	27.9
C	-2	31.0
Sum of Products	-35.8	
Standardized (i.e., $\div n_F/I$)	-35.8/4 = -8.95	

This calculation is the same as that shown for the two-level factorial in Chapter 8. In this case, the divisor for the sum of products is n_F/I where $n_F = 12$ the total runs in the factorial and I equals the number of levels for the factor. (Note: This is a generalization of the $n_F/2$ used in Chapter 8.)

Since the contrasts of means is of the same general form as that shown in Section 3.7, the

standard error of a contrast $C = \sum_{i=1}^{k} c_i \bar{Y}_{i.}$ is

$$s_C = \sqrt{\sum_{i=1}^{k} c_i^2 (\sigma^2/n)} \qquad (11.11)$$

where n is the number of data values averaged in each mean $\bar{Y}_{i.}$. To test the significance of a contrast, we can use the t-statistic formed by the ratio of the contrast of sample means to its estimated standard error, i.e.,

$$t_v = \frac{\sum_{i=1}^{k} c_i \bar{y}_{i.}}{\sqrt{(MSE/n) \sum_{i=1}^{k} c_i^2}}$$

where MSE is the mean square error from the ANOVA and v is the degrees of freedom for error. For example, the t-statistic for the contrast shown above is:

$$t_6 = \frac{-8.95}{\sqrt{(0.476/4)(-2^2+1^2+1^2)}} = \frac{-8.95}{\sqrt{.714}} = -10.59$$

which is larger than the t*=2.447 given in Table A.5 with 6 degrees of freedom and 95% confidence level. Therefore, we are 95% confident that the average drawing force required using the standard lubricant is greater than the average for the two new lubricants.

Two comparisons or contrasts of means $C_1 = \sum_{i=1}^{k} c_{1,i} \bar{Y}_{i.}$ and $C_2 = \sum_{i=1}^{k} c_{2,i} \bar{Y}_{i.}$ are called

orthogonal contrasts if $\sum_{i=1}^{k} c_{1,i} c_{2,i} = 0$, $\sum_{i=1}^{k} c_{1,i} = 0$ and $\sum_{i=1}^{k} c_{2,i} = 0$.

If a factor has I levels, then up to I-1 orthogonal contrasts can be defined comparing the means or effects of this factor. If all the comparisons of means, that are of interest, can be expressed as a set of orthogonal contrasts, then the confidence level will not be reduced when testing all of them simultaneously. For example, in the wire drawing experiment above, a second comparison of lubricant means of interest might be to compare the two new lubricants. A contrast for this comparison would be:

$$C_2 = 0 \times L_3 - 1 \times L_2 + 1 \times L_1$$

This contrast is orthogonal to C_1, given above, since the product of their coefficients sums to zero (i.e., $0 \times -2 + -1 \times 1 + 1 \times 1 = 0$). The t-statistic for this contrast is:

$$t_6 = \frac{-1 \times 26.25 + 1 \times 23.050}{\sqrt{(0.476/4)(0^2 + (-1)^2 + 1^2)}} = \frac{-3.2}{\sqrt{.238}} = -6.56$$

which is again greater than t*=2.447 given in Table A.5 with 6 degrees of freedom and 95% confidence level. Therefore we are 95% confident that new lubricant type B requires significantly more drawing force than does new lubricant type A. For this experiment, testing the two orthogonal contrasts gives all the information necessary to pick the lubricant (type A) that results in the lowest friction during wire drawing.

Let us examine what has been done in previous chapters to understand how the use of orthogonal contrasts, to compare means in an experiment, does not reduce the confidence level. For example, consider the pesticide degradation experiment (ignoring the blocks) presented in Section 9.4. If we look at the data in the format of a one-factor design, as shown in Table 11.15 it appears very similar to Table 11.6 and the four means could be compared simultaneously using the Analysis of Variance.

Table 11.15 Pesticide Degradation Data in Form of a One-Factor Design

	Factor Level			
	(1) Temp=50 Moisture=15%	(2) Temp=80 Moisture=15%	(3) Temp=50 Moisture=25%	(4) Temp=80 Moisture=25%
	6.787	5.236	5.434	4.866
	5.121	4.178	5.052	3.451
$\bar{Y}_{i.}$	5.95	4.71	5.24	4.16

In this context, the effects for temperature, moisture, and their interaction presented in Section 9.4 are a set of orthogonal contrasts, as shown in the following equations:

Temperature Effect:

$$C_1 = -1 \times \bar{Y}_{1.} + 1 \times \bar{Y}_{2.} - 1 \times \bar{Y}_{3.} + 1 \times \bar{Y}_{4.}$$
$$= -5.95 + 4.71 - 5.24 + 4.16 = -1.17$$

Moisture Effect:

$$C_2 = -1 \times \bar{Y}_{1.} - 1 \times \bar{Y}_{2.} + 1 \times \bar{Y}_{3.} + 1 \times \bar{Y}_{4.}$$
$$= -5.95 + 4.71 - 5.24 + 4.16 = -.63$$

Temperature × Moisture Interaction Effect:

$$C_3 = +1 \times \bar{Y}_{1.} - 1 \times \bar{Y}_{2.} - 1 \times \bar{Y}_{3.} + 1 \times \bar{Y}_{4.}$$
$$= -5.95 + 4.71 - 5.24 + 4.16 = -.081$$

These orthogonal contrasts could be used to partition the treatment sums of squares the Analysis of Variance as shown in Table 11.16

Table 11.16 Analysis of Variance for ln(Half-Life)

Source	DF	SS	MS	F	P
Single Factor	3	3.524	1.175	1.56	0.331
Temp	1	2.718	2.718	3.60	0.131
Moist	1	0.793	0.793	1.05	0.364
Interaction	1	0.013	0.013	0.017	0.903
Error	4	3.022	0.755		
Total	7	6.546			

In this table, the sums of squares for the single factor is just the sum of the sums of squares for each orthogonal contrast. The sums of squares for the orthogonal contrasts (i.e., temperature and moisture main effects and their interaction) were computed exactly as shown in Section 11.5. Therefore, the orthogonal contrasts can be seen to simply partition the ANOVA sums of squares for a single factor. The F-values for the three contrasts in this table are just the square of the t-statistics that would be obtained by dividing the three effects (orthogonal contrasts) by the standard error of an effect. As we have seen in the analysis of two-level factorials, the confidence level is not reduced when testing all three contrasts simultaneously.

11.6.4 Orthogonal Polynomial Contrasts

An experimenter would use more than two levels for quantitative factors if he were interested in quadratic or other higher order nonlinear effects. This would be the case if he was seeking to find the factor value that produced a maximum or minimum response. For experiments where the factor levels are equally spaced values of a quantitative factor, the orthogonal polynomial contrast coefficients can provide a quick set of comparisons that will allow the experimenter to determine if the response is related to the factor levels linearly, quadratically, or according to some higher level polynomial. Table A.8 in Appendix A shows the coefficients for these contrasts for three, four, or five-level factors. More extensive tables can be found in the Biometrika Tables for Statisticians.

As an example of the use of these tables, consider the data from the following chemical reaction experiment. The raw data is shown in Table 11.17. The experimenter is trying to determine the relationship between temperature and yield. Table 11.18 shows the summary means, variances, and calculation of the orthogonal polynomial contrasts. The sum of squares row is the sum of squares of the contrast coefficients in each column. The sum of products is the sum of products of the contrast coefficients in each column and the means in the fourth column. The pooled variance is the average of the variances in the fifth column and is the same as MSE (mean square error) that would be calculated in an ANOVA. The standard error of the contrasts is calculated as shown in Equation 11.11. In this example, the degrees of freedom for error would be 5, since there two replicates in each of 5 temperature levels. The linear and quadratic contrasts are largest, and are significant at the $\alpha = 0.01$ level.

Use of the orthogonal polynomial contrasts does not give us the equation of the quadratic equation, nor allows us to determine the temperature that is predicted to result in the highest yield. Rather it just provides us with a quick way to determine the appropriate degree (i.e., linear, quadratic, etc.) of the equation that should be fit. To determine the prediction equation, and the temperature predicted to give the highest yield, the least squares method for fitting a quadratic equation shown in Chapter 14 should be used.

Table 11.17 Chemical Reaction Experiment

Temperature	Yield
1	1.0
1	1.6
2	2.9
2	3.2
3	3.6
3	3.5
4	3.0
4	3.9
5	3.1
5	2.8

Table 11.18 Computation Worksheet for Chemical Reaction Data

Temperature	Rep 1	Rep 2	Mean	Variance	Linear	Quadratic	Cubic	Quartic
1	1	1.6	1.3	0.18	-2	2	-1	1
2	2.9	3.2	3.05	0.05	-1	-1	2	-4
3	3.6	3.5	3.55	0.01	0	-2	0	6
4	3	3.9	3.45	0.41	1	-1	-2	-4
5	3.1	2.8	2.95	0.05	2	2	1	1
Sum of Squares					10	14	10	70
Sum of Products					3.7	-5.1	0.85	-0.45
Standard Error of Contrasts					0.83	0.98	0.83	2.18
t-Statistic					4.49	-5.23	1.03	-0.21
			Pooled Variance	0.14				

11.7 Other Applications of Orthogonal Contrasts

Orthogonal contrasts are useful for things other than just comparing means after the ANOVA. In this section we will explore two of these uses.

11.7.1 Analysis of Multilevel Factorials with No Replicates

When there were no replicates (repeat experiments) in the two-level factorial presented in Chapters 8 - 9, or the screening designs presented in Chapter 12, the significance of the effects was judged by making a graph such as a Normal, Half-normal, or Pareto Diagram of the effects. Since we have seen in Section 11.6.3 that the effects are nothing more than orthogonal contrasts of the data, we can generalize the graphical analysis to factorials that have more than two levels. The idea is to standardize the contrasts so that they all have the same variance, then make the same plots of the standardized contrasts as shown in Chapters 9 and 12.

The general form of a standardized contrast is $C_i = \sum_{i=1}^{I} c_i \bar{Y}_i \Big/ \sqrt{\sum_{i=1}^{I} c_i^2}$

These standardized contrasts will all have variance σ^2 and will be comparable when plotted on the same graph. Table 11.19 is a worksheet for calculating these standardized contrasts for the first replicates of data from the wire drawing experiment in Table 11.12. In this table, the columns L1 and L2 are the orthogonal contrasts for the lubricant factor discussed in Section 11.6.3, and D1 is the contrast comparing the two drawing speeds. The coefficients for the two additional contrasts L1D1 and L1D2, are calculated by multiplying the elements of L1 and D1 in the same way as the interaction columns were created in the worksheets in Chapter 8. From these standardized contrasts, it can be seen that the largest (absolute) values are L1 and L2, the two comparisons of lubricants previously described, and D1, the comparison of drawing speeds. L1D1 and L2D1 are relatively small, indicating there is little interaction between lubricant and drawing speed.

Table 11.19 Worksheet for Calculating Standardized Orthogonal Contrasts

Lubricant	Draw Speed	Draw Force	L1	L2	D1	L1D1	L2D1
A	10^2	22.3	1	0	-1	-1	0
B	10^2	24.2	1	-1	-1	-1	1
C	10^2	27.6	-2	1	-1	2	-1
A	10^4	24.5	1	0	1	1	0
B	10^4	27.1	1	-1	1	1	-1
C	10^4	30.6	-2	1	1	-2	1
	Sum of Squares		12	4	6	12	4
	Sum of Products		-18.3	6.9	8.1	-0.90	0.10
	Standardized Contrast		-5.28276	3.45	3.306811	-0.25981	0.05

11.7.2 General Linear Model ANOVA

The formulas for computing sums of squares in the Analysis of Variance presented in Section 11.5.2 will only be valid if there are an equal number of replicates in each combination of levels of the factors. For cases where data is missing from one or more replicates of an experiment, due to causes unrelated to the factors themselves, we have what is called an unbalanced design. For example, suppose there was time to complete only the first replicate experiment for two of the treatment combinations in the wire drawing experiment, resulting in a data set like that shown in Table 11.20.

To correctly complete an Analysis of Variance for this data, we can use the orthogonal contrasts and the method of least squares. This will result in what is called the General Linear Model Analysis of Variance or GLM ANOVA. To illustrate this, in the matrix notation of Chapter 10, we will define an X matrix and Y vector as follows:

$$Y = \begin{bmatrix} 22.3 \\ 24.2 \\ 27.6 \\ 22.4 \\ 25.8 \\ 27.1 \\ 30.6 \\ 23.0 \\ 27.9 \\ 31.0 \end{bmatrix} \qquad \begin{array}{cccccc} \text{I} & \text{L1} & \text{L2} & \text{D1} & \text{L1D1} & \text{L2D2} \end{array}$$

$$X = \begin{bmatrix} 1 & 1 & 1 & -1 & -1 & -1 \\ 1 & 1 & -1 & -1 & -1 & 1 \\ 1 & -2 & 0 & -1 & 2 & 0 \\ 1 & 1 & 1 & -1 & -1 & -1 \\ 1 & 1 & -1 & -1 & -1 & 1 \\ 1 & 1 & -1 & 1 & 1 & -1 \\ 1 & -2 & 0 & 1 & -2 & 0 \\ 1 & 1 & 1 & 1 & 1 & 1 \\ 1 & 1 & -1 & 1 & 1 & -1 \\ 1 & -2 & 0 & 1 & -2 & 0 \end{bmatrix}$$

Table 11.20 - Data from Wire Drawing Experiment with Missing Replicates

	Factor L	Factor D	Response
Random Run Order	Lubricant Type	Wire Drawing Speed	Drawing Force
3	A	10^2	22.3
6	B	10^2	24.2
4	C	10^2	27.6
7	A	10^2	22.4
9	B	10^2	25.8
12	C	10^2	Not Completed
11	A	10^4	Not Completed
8	B	10^4	27.1
10	C	10^4	30.6
5	A	10^4	23.0
1	B	10^4	27.9
2	C	10^4	31.0

Using the matrix formulas presented in Chapter 10, the error sums of squares (SSE) was given by Equation 10.34, and the sums of squares for regression (SSR) could be obtained by subtraction $SSR = Y^T Y - SSE$. To get the sums of squares for the main effects due to lubricants, drawing speed, and their interaction a similar subtraction is made.

For example, to get the sums of squares for lubricants, first the regression model is fit using the X and Y matrices shown on the previous page. The regression sums of squares for this model is called SSR_F which stands for the regression sums of squares from the full model. Next, the columns in the X matrix that correspond to the orthogonal contrasts for lubricants are removed and the sums of squares for regression from this reduced model is called SSR_R for reduced sums of squares for regression. The sums of squares for lubricants is then calculated as the drop or difference in these sums of squares, i.e., $SSL = SSR_F - SSR_R$. Calculation of the sums of squares for the other two terms is similar. These computations can be completed easily using a regression program, if a GLM ANOVA program is not available. Table 11.21 shows the results of four regression runs, and the computations below the table show how the sums of squares are calculated for each term. The degrees of freedom for each sums of squares are the number of columns in the X-matrix that were removed to get the reduced regression sums of squares. Combining the information from the sums of squares and degrees of freedom, a GLM ANOVA table is constructed in Table 11.22.

Table 11.21 Regression Sums of Squares from Four Models with Data from Table 11.20

Columns in the X-matrix	SSR		SSE	Degrees of Freedom for Error
I, L1, L2, D1, L1D1, L2D1	90.424		1.685	4
I, D1, L1D1, L2D1	33.433		58.676	6
I, L1, L2, L1D1, L2D1	80.343		11.766	5
I, L1, L2, D1	88.068		4.041	6

Calculation of Sums of Squares:

$SSL = 90.424 - 33.433 = 56.991$ with 2 df

$SSD = 90.424 - 80.343 = 10.081$ with 1 df

$SSI = 90.424 - 88.068 = 2.356$ with 2 df

Table 11.22 General Linear Model ANOVA for Data in Table 11.18

Source	df	Sum of Squares	Mean Square	F-ratio	P-value
Lubricant	2	56.991	28.4955	67.65	0.0003
Drawing Speed	1	10.081	10.081	23.93	0.008
Interaction	2	2.356	1.178	2.796	0.174
Error	4	1.685	0.42125		

The GLM ANOVA will give the same results as the formulas presented in Section 11.5.2 when there are an equal number of replicates, and therefore, a more general method of analysis. Most statistical packages have a built-in GLM ANOVA routine, although general spreadsheet programs like LOTUS 1-2-3 do not at the present time. To use a GLM ANOVA program the user must specify the terms to be included in the model. For the example shown in Table 11.22 that would be:

```
Lubricant DrawSpeed Lubricant*DrawSpeed
```

We will see other applications of the GLM ANOVA in the next section.

Since we have seen that the Analysis of Variance can actually be completed using regression analysis, the assumptions of equal variance and Normal Distribution of errors can be checked using the residual plots described in Section 10.6. This should always be done in the analysis of experimental data.

11.7.3 An Example

During the 1960s and 1970s Japanese automobiles were well-known for their rust problems. Toyota Corporation's competitive position and future profitability in European markets was dependent on a solution to this problem. The following experiment[2] represents one of a series of experiments that were run by Toyota employees during the course of solving this problem. In this experiment, the response (scab test) was the diameter of a scab corrosion spot after a specified number of cycles of exposure to salt spray. The factors studied in the experiment are shown in Table 11.23. Notice that none of the factors are continuous in nature. The particular choice of levels resulted in a $3^2 \times 2^2$ factorial.

---

[2]The data presented herein is not the actual experimental data but was fabricated to match the results of an experiment described in *Implementation Manual for Three Day QFD Workshop*, American Supplier Corporation, Dearborn, Michigan.

Table 11.23 Rust Test Factors and Levels

Factors	Levels
A-Electro Coating Method	1=Anion, 2=Super Anion, or 3=Cation
B-Treatment Method	1=Spray or 2=Dip
C-Number of Coats	1=2 coats or 2=3 coats
D- Treatment Type	1=A, 2=B, or 3=C

The data are shown in Table 11.24. Here it can be seen that there are no replicates of any of the experimental conditions. To get a preliminary idea of what factors are significant, orthogonal contrasts were constructed for each of the factors. The set of contrasts for each factor are listed below:

Electro Coating Contrasts:

$$A1 = -2 \times \hat{A}_1 + \hat{A}_2 + \hat{A}_3$$

$$A2 = 0 \times \hat{A}_1 - \hat{A}_2 + \hat{A}_3$$

Treatment Method Contrast:

$$B1 = -1 \times \hat{B}_1 + \hat{B}_2$$

Number of Coats Contrast:

$$C1 = -1 \times \hat{C}_1 + \hat{C}_2$$

Treatment Type Contrasts:

$$D1 = -2 \times \hat{D}_1 + \hat{D}_2 + \hat{D}_3$$

$$D2 = 0 \times \hat{D}_1 - \hat{D}_2 + \hat{D}_3$$

384

A worksheet was set up and the standardized contrasts were computed. Also, contrasts for all two-way (such as A1B1 and A2B1), three-way (such as A1B1C1 and A2B1C1), and four-way contrasts were computed in the worksheet format shown in Table 11.19. A Normal Probability plot of these contrasts is shown in Figure 11.11.

In the plot it can be seen that the majority of contrasts fall along a straight line. The contrasts that stick out to the lower left and upper right of the plot appear to be significant. For example, the large negative value of A1 in the lower left indicates that the scab diameter is larger for Anion than for the average of Super Anion and Cation (i.e., $-2 \times \hat{A}_1 = \hat{A}_2 + \hat{A}_3 < 0$). All the contrasts that appear to be significant are labeled on the graph.

For a more formal test of significance, a GLM ANOVA[3] was completed using all contrasts in the main effects, A, B, C, and D, and interaction AB and ABD. The results are displayed in Table 11.25. Based on the ANOVA, the main effect for C (Number of Coats) was significant. Due to the fact that the contrast, C1, is negative we know that smaller scab corrosion diameters resulted when using the second level of Factor C (i.e., three coats) than the first level (two coats). The three-way interaction is significant indicating that the diameter of scab corrosion was dependent on the combination of the three factors (A-Electro Coating, B-Treatment Method, and D- Treatment Type.) Since the three-factor interaction was significant, it would be meaningless to interpret the main effects for A, B, and D nor the two-way interaction for AB alone. To understand the relation of scab corrosion diameter to the combination of factor levels, a table of means was produced in Table 11.26. Here it can be seen that use of the Dip method resulted in consistently smaller diameters, but that the Treatment method that produced the smallest diameters was dependent upon which Electro Coating method used.

---

[3]Notice for this particular example, the sums of squares in the ANOVA could be computed using the formulas in Section 11.5.2 since there were an equal number of replicates (1). However, all effects for the two-way interactions between Factors A, B, and D would have to be computed in order to calculate the sums of squares for ABD, and the error sums of squares would be calculated by subtraction i.e., $Y^T Y$ - summation of all sums of squares in the model.

Table 11.24 Data from Toyota Rust Study

A	B	C		Y=Scab Diameter		Standardized Contrasts
1	1	1	1	6.90	Mean	5.21
2	1	1	1	5.09	A1	-4.70
3	1	1	1	5.66	A2	-1.90
1	2	1	1	6.22	B1	-4.50
2	2	1	1	5.21	C1	2.05
3	2	1	1	3.17	D1	-2.50
1	1	2	1	7.66	D2	-2.50
2	1	2	1	6.62	A1B1	-1.00
3	1	2	1	6.77	A2B1	-3.20
1	2	2	1	6.60	A1C1	-0.01
2	2	2	1	5.52	A2C1	0.56
3	2	2	1	4.28	A1D1	-0.25
1	1	1	2	7.60	A2D1	-0.25
2	1	1	2	4.92	A1D2	0.18
3	1	1	2	5.74	A2D2	-0.31
1	2	1	2	3.97	B1C1	-0.36
2	2	1	2	5.03	B1D1	-0.48
3	2	1	2	2.72	B1D2	0.74
1	1	2	2	7.99	C1D1	-0.39
2	1	2	2	5.25	C1D2	-0.65
3	1	2	2	6.83	A1B1C1	-0.68
1	2	2	2	6.99	A2B1C1	-0.31
2	2	2	2	5.48	A1B1D1	-0.10
3	2	2	2	2.57	A2B1D1	-0.52
1	1	1	3	5.30	A1B1D2	-1.56
2	1	1	3	4.90	A2B1D2	-0.92
3	1	1	3	3.98	A1C1D1	-0.46
1	2	1	3	5.92	A2C1D1	0.23
2	2	1	3	3.76	A1C1D2	1.20
3	2	1	3	1.56	A2C1D2	0.51
1	1	2	3	5.62	B1C1D1	0.31
2	1	2	3	4.85	B1C1D2	-0.72
3	1	2	3	5.72	A1B1C1D1	-0.36
1	2	2	3	5.19	A2B1C1D1	-0.75
2	2	2	3	3.91	A1B1C1D2	1.08
3	2	2	3	2.1	A2B1C1D2	-0.01

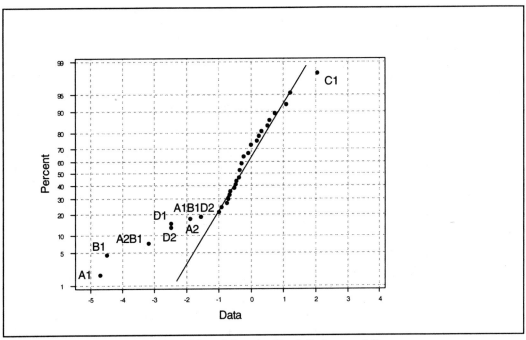

Figure 11.11 Normal Probability Plot of Standardized Orthogonal Contrasts

Table 11.25 GLM ANOVA for Rust Experiment

Source	df	Sum of Squares	Mean Square	F-Ratio	P-Value
A	2	26.24	13.12	42.88	0.000
B	1	20.55	20.55	67.16	0.000
C	1	4.20	4.20	13.72	0.001
D	2	12.70	6.35	20.75	0.000
AB	2	11.16	5.58	18.24	0.000
ABD	4	3.55	0.88	2.87	0.045
Error	23	7.04	0.31		

Table 11.26 Means for Combinations of Factors A, B and D

B=Method	1=Spray			2=Dip		
	D=Treatment			D=Treatment		
A=Electro Coat	A	B	C	A	B	C
Anion	7.28	7.79	5.46	6.41	5.48	5.55
Super Anion	5.85	5.08	4.87	5.36	5.26	3.84
Cation	6.21	6.28	4.85	3.72	2.64	1.83

After the experiments, the Super Anion method of coating was selected, mainly due to economic reasons, along with Treatment method C, the Dip method, and three coats. Having selected the basic processing technology, Toyota continued to perform further experiments to investigate process variables such as time, temperature, surface treatment, etc. before eventually finding a combination which achieved the desired level of corrosion resistance, and made Toyota automobiles competitive in European markets.

11.8 Analysis of Blocked and Split Plot Experiments with Multileveled Factors

In Chapter 9, the rational for blocked and split plot factorials was explained, but the method of analysis shown was restricted to the case where all factors have only two-levels (- and +) and the number of blocks or reps in the design is restricted to be a power of 2 (i.e., 2, 2^2, etc.). Now that the technique of ANOVA for factorial designs has been introduced, we will illustrate how it can be used to analyze blocked or split plot factorials with an arbitrary number of factor levels for each factor, and an arbitrary number of blocks or reps.

In the example analysis of a blocked factorial in Section 9.4, it was shown that all interactions with the block term were pooled to form an error. In the Analysis of Variance context, the pooling can be accomplished by adding the sums of squares and degrees of freedom separately for all block interaction terms, or by simply leaving those interactions out of the model (and full X-matrix) when using the GLM ANOVA. For example, consider the data from the four-block pesticide degradation experiment presented in Table 9.5. An ANOVA table can be constructed using the formulas presented in 11.5.2, by assigning the two-level temperature factor as A, the two-level moisture factor as B, and the four-level block factor as C. Next, the interaction terms involving the block factor C (i.e., AC, BC, and ABC) are pooled together, or summed together, to form the error sums of squares. The resulting Analysis of Variance table is shown in Table 11.27.

Table 11.27 Analysis of Variance for Blocked Factorial in Table 9.5

Source	df	Sum of Squares	Mean Square	F-ratio	P-value
A (Temperature)	1	166,601	166,601	7.83	0.021
B (Moisture)	1	55,428	55,428	2.60	0.141
AB (interaction)	1	28,499	28,499	1.34	0.277
C (Blocks)	3	164,735	54,911		
⌈ AC	3	51,720	17,240		
∣ BC	3	77,207	25,736		
⌊ ABC	3	62,588	20,863		
Error (AC +BC+ABC)	9	191,515	29,279		

Data from split plot factorial experiments can be analyzed in a similar fashion. Recall that in Section 9.6, it was shown that the whole plot error term was formed by pooling all interactions involving whole plot factors and the rep factor, and that the within plot error term was formed by pooling all other interactions involving the rep factor. Again, this pooling can be easily accomplished in the ANOVA context.

Consider the data from the split plot example shown in Section 9.7. An ANOVA table can be constructed using the method of Section 11.5.2 or the GLM ANOVA method. Assign split plot factor development time as factor A, split plot factor exposure time as factor B, whole plot factor bake time as factor C, and the rep factor as D. The whole plot error sums of squares is the sums of squares for CD, and the within plot sums of squares will be obtained by pooling the sums of squares for the interactions AD, BD, ABD, ACD, BCD, and ABCD. The resulting Analysis of Variance table is shown in Table 11.27. The F-ratio for the whole plot factor is formed by dividing the mean square for Factor C by the mean square for whole plot error (CD), while the F-ratios for the other factors and interactions are calculated by dividing their respective mean squares by the within plot error mean square.

Table 11.28 Analysis of Variance for Split Plot Design

Source	df	Sum of Squares	Mean Squares	F-ratio	P-value
C (Bake Time)	1	3.406	3.406	1.169	0.475
D (Replicate)	1	4.572	4.572		
CD (Whole Plot Error)	1	2.103	2.103		
A (Develop Time)	1	1.633	1.633	32.623	0.001
B (Expose Time)	1	0.154	0.154	3.077	0.129
AB	1	0.192	0.192	3.836	0.098
AC	1	0.028	0.028	0.060	0.814
BC	1	0.013	0.013	2.597	0.158
ABC	1	0.112	0.112	2.240	0.185
AD	1	0.137			
BD	1	0.015			
ABD	1	0.000			
ACD	1	0.006			
BCD	1	0.036			
ABCD	1	0.106			
Within Plot Error	6	0.300	0.050		

11.9 Summary

In this chapter we have discussed factorial designs with factors having more than two levels. We have presented a new form of the mathematical model for representing data from these designs, and have shown a general method called the Analysis of Variance or ANOVA for determining significant main effects and interactions. We presented three methods for comparison of specific means following the ANOVA. The LSD method should be used when one specific comparison of two means has been predetermined. Tukey's HSD method should be used when an experimenter wants to compare all pairs of means from a factor or interaction. Finally, a set of orthogonal contrasts should be used to compare means, when they can represent all comparisons of interest.

This chapter also showed that orthogonal contrasts can be used to develop Normal plots for the analysis of unreplicated designs. The least squares regression method of Chapter 10 can be used in conjunction with orthogonal contrasts to compute the sums of squares in a general linear model or GLM ANOVA. It was shown that this method gives the same results as the formulas for ANOVA sums of squares presented earlier, when the number of replicates of each experimental condition is equal. It was also explained that the GLM ANOVA method is correct when the number of replicates is different for each experimental condition, therefore this method of computing ANOVA sums of squares is more general. Finally, this chapter showed how the ANOVA method could be used to analyze data from blocked or split plot experiments when the number of factor levels is not equal to two, and the number of blocks or replicates is not a power of two.

The following is a list of the important concepts and terms from this chapter:

Tabular and Graphical Representation of Data 342, 343
Multi-level factorial 341
Definition of Effects and Mathematical Model for Multi-Level Factorials 345
Additive Model 347
ANOVA procedure 357, 359
P-Value 364
LSD Method 367
Type I error rate 369
Tukey's HSD Method 369
Orthogonal Contrasts 370
Orthogonal Polynomials 377
Standardized Contrasts 379
Normal Plot of Standardized Contrasts 379, 385, 387
GLM ANOVA procedure 380, 383
Calculation of GLM ANOVA Sums of Squares using regression 382
Analysis of Blocked Designs by ANOVA 388
Analysis of Split Plot Designs by ANOVA 389, 390

11.10 Exercises for Chapter 11

11.1   a) List the runs for a 3×5 design and show symbolic values of the response similar to Table 11.1.

b) Make a geometric representation of the design similar to Figure 11.1.

c) Sketch a hypothetical main effect plot separately for the three-level and five-level factors similar to Figure 11.2.

d) Sketch a hypothetical interaction plot similar to Figure 11.3.

11.2 Snedecor and Cochran[1] present data from a 3×4 experiment used to learn about loss of ascorbic acid in snap beans stored for different lengths and under different temperatures. The data listed below shows the averages of three experiments at each treatment combination:

Subscript (i,j)	Actual Factor Levels		Ascorbic Acid (mg/100g) $\overline{Y}_{ij}$
	Factor A Temperature °F	Factor B Weeks of Storage	
1, 1	0	2	45
2, 1	10	2	45
3, 1	20	2	34
1, 2	0	4	47
2, 2	10	4	43
3, 2	20	4	28
1, 3	0	6	46
2, 3	10	6	41
3, 3	20	6	21
1, 4	0	8	46
2, 4	10	8	37
3, 4	20	8	16

---

[1]Snedecor, G. W. and Cochran, W. G. *Statistical Methods*, 7th edition, Ames, IA: The Iowa State University Press, (1980), p. 311.

a) Calculate the main effects for Factor A.

b) Calculate the main effects for Factor B.

c) Calculate the interaction effects $\hat{AB}_{ij}$

11.3 Scheffé[2] presented data provided by Mr. John Hromi with the kind permission of U.S. Steel Corporation. The data are the results of a 4×2×2 experiment to investigate the effects of three factors on the insulation resulting from core-plate coatings on electrical steels. The factors were: A (four different coatings), B (two different curing temperatures), C (two different stress-relief annealing atmospheres). Four replicate experiments were performed at each combination of factor levels with different rolls of steel.

Run	A	B	C	Results of Franklin test (ASTM A-344-52)
		Factors		
1	1	1	1	0.25, 0.36, 0.36, 0.25
2	2	1	1	0.41, 0.28, 0.33, 0.21
3	3	1	1	0.44, 0.65, 0.42, 0.47
4	4	1	1	0.43, 0.63, 0.47, 0.52
5	1	2	1	0.30, 0.18, 0.44, 0.34
6	2	2	1	0.13, 0.06, 0.19, 0.20
7	3	2	1	0.22, 0.14, 0.17, 0.36
8	4	2	1	0.26, 0.51, 0.21, 0.32
9	1	1	2	0.16, 0.02, 0.06, 0.10
10	2	1	2	0.10, 0.04, 0.03, 0.01,
11	3	1	2	0.24, 0.08, 0.49, 0.14
12	4	1	2	0.27, 0.03, 0.28, 0.07
13	1	2	2	0.27, 0.03, 0.13, 0.04
14	2	2	2	0.06, 0.03, 0.04, 0.01
15	3	2	2	0.18, 0.36, 0.25, 0.19
16	4	2	2	0.21, 0.03, 0.25, 0.38

---

[2]Scheffé, H., *The Analysis of Variance*, New York: John Wiley and Sons, (1959), p. 141.

a) Calculate the variance of the replicates at each run, and calculate the pooled variance of the random replicates.

b) Calculate the effects of the three factors.

c) Calculate the interaction effects $\hat{AB}$, $\hat{AC}$, $\hat{BC}$, and make interaction plots.

d) Calculate the three-factor interaction effects $\hat{ABC}$

11.4 The data below[3] represent the results of an experiment in which the concentration of iron in a standard solution was determined four times by each of six analysts.

Levels of Factor - Analyst

1	2	3	4	5	6
2.963	2.958	2.956	2.948	2.953	2.941
2.996	2.964	2.945	2.960	2.961	2.940
2.979	2.955	2.963	2.953	2.961	2.931
2.970	2.932	2.950	2.944	2.953	2.942

a) Using this data compute the six Analyst effects.

b) Complete an ANOVA table including degrees of freedom, sums of squares, and mean squares for analyst and error.

c) Compute the F-statistic to determine if the analyst effect is significant.

d) Interpret the results.

11.5 a) Complete the ANOVA table for the data in Problem 3, including degrees of freedom, sums of squares, and mean squares for the main effects, two-factor interactions, three-factor interaction, and error.

b) Calculate F-statistics for all the factors and interactions, determine which are significant at the $\alpha=0.05$ level, and interpret the results.

---

[3]Bennett, C. A. and Franklin, N. L. *Statistical Analysis in Chemistry and the Chemical Industry*, New York: John Wiley and Sons, (1954), p. 331.

11.6 Given that the error sums of squares for the data in Problem 2 was SSE = 16.944 with 24 degrees of freedom:

a) Complete the ANOVA table, including degrees of freedom, sums of squares, and mean squares for the main effects and two-factor interaction.

b) Calculate F-statistics for the two factors and interaction, determine which are significant at the $\alpha=0.05$ level, and interpret the results.

11.7 a) Use the LSD method to test all pairs of analyst means for the data in Problem 4.

b) Calculate Tukey's HSD for differences of pairs of means in Problem 4, and determine if there are any significant differences in pairs of analysts using Tukey's method.

11.8 a) Use the LSD method to test all pairs of the four coating means for the data in Problem 3.

b) Calculate Tukey's HSD for differences of pairs of coating means in Problem 3, and determine if there are any significant differences in pairs of coatings using Tukey's method.

11.9 Suppose that coating 1 in the experimental data for Problem 3 represents the industry standard, and coatings 2 through 4 represent new experimental coatings.

a) Calculate the value of a contrast that would compare the industry standard coating (1) to the average of the three new coatings.

b) Calculate the standard error of the contrast you calculated in a) and use the t-statistic to test for significance.

c) Show the coefficients of another contrast among the three new coatings which is orthogonal to the contrast you calculated in a).

11.10 a) With the help of Table A.8, compute the linear and quadratic contrasts for the temperature effect for the data in Problem 2.

b) Using the sums of squares for error given in Problem 6, compute the standard error of the contrasts you computed in a) and perform t-tests to determine if they are significant at the $\alpha=0.05$ level.

c) With the help of Table A.8, compute the linear, quadratic, and cubic contrasts for the weeks of storage effect for the data in Problem 2.

d) Using the sums of squares for error given in Problem 6, compute the standard error of the contrasts you computed in c) and perform t-tests to determine if they are significant at the α=0.05 level.

11.11 Brownlee[4] showed the results of a chemical purification experiment, and the data is shown below except for two values that were left out for instructional purposes. Two replicates were planned for each combination of factor levels.

Run	Factor A (Boiling Time)	Factor B (Solvent Type)	Response (Purity)
1	1	1	3.1, (missing)
2	2	1	2.0, 1.6
3	3	1	1.3, 1.9
4	1	2	4.3, 6.1
5	2	2	(missing), 3.9
6	3	2	4.1, 2.6

a) Define the linear and quadratic orthogonal polynomial contrasts for Factor A (A_L and A_Q), and the linear orthogonal polynomial contrast for Factor B (B_L).

b) Set up an X-matrix as shown in Section 11.7.2 including the A_L, A_Q, B_L, $A_L B_L$, $A_Q B_L$ terms, and use a regression program to compute the GLM ANOVA sums of squares and mean squares for Factors A, B, the interaction AB, and error.

c) Use a statistics program that has a GLM ANOVA routine to re-do the GLM ANOVA you did in b).

11.12 Using the Data on page 350 and Table A.8

a) Set up a worksheet similar to Table 11.19 and compute the standardized linear orthogonal polynomial contrasts for Factors A and B (A_L and B_L) and the standardized linear and quadratic orthogonal polynomial contrasts for Factor C (C_L and C_Q)

---

[4]Brownlee, M. A., *Industrial Experiments*,(New York 1953): Chemical Publishing Co. Inc., , p. 90.

b) Multiplying elements, expand your worksheet to form contrast coefficients for the two-way interactions $A_L B_L$, $A_L C_L$, $A_L C_Q$, $B_L C_L$, $B_L C_Q$, and the three-way interaction $A_L B_L C_L$ and $A_L B_L C_Q$. Compute the standardized contrast for these contrasts.

c) Make a Normal Probability plot of the 11 standardized contrasts.

d) Using the results of c), eliminate the smallest contrasts and complete a GLM ANOVA like the example in Section 11.7.3.

11.13 Using only the first three blocks of data in Table 9.5, re-do the Analysis of Variance data shown in Table 11.27. How many degrees of freedom for blocks now? How many degrees of freedom for A×Blocks? How many degrees of freedom for Error (the pooled block interaction terms?)

11.14 Box[5] presents the results of a split plot experiment to improve the corrosion resistance of steel bars by applying a surface coating then baking it in a furnace for a fixed length of time. Three baking temperatures and four different coatings were tested. The experiment was conducted in the following way: The furnace was heated to the desired temperature. Four bars, each treated with a different one of the four coatings, were randomly positioned in the furnace baked for the specified length of time. Next, they were removed, cooled down, and tested for corrosion resistance. The whole plot experimental units were furnace "heats," and the subplot units were the steel bars. The whole plot factor was temperature, and its levels (360°, 370°, and 380°) were not run in random order because of the difficulty in adjusting the furnace temperature up or down. There was no problem randomizing the positions of the split plot factor levels within the furnace.

a) Complete a split plot ANOVA table using the data below and the model:

$$Y_{ijk} = \mu + A_i + R_j + AR_{ij} + B_k + ABi_k + \varepsilon$$

b) Test the coatings effect and the coating by temperature interaction using the subplot error term (ε).

c) Test the Temperature effect using the whole plot error AR.

d) What do you conclude about the coatings based on your analysis?

e) Does any qualifications have to be made about the test on temperature effect?

----

[5]Box, G. "Quality Quandaries - Split Plot Experiments," *Quality Engineering*, Vol. 8, No. 3,(1996), pp. 515-520.

Heat	Replicate (R)	Temperature (Whole Plot Factor-A)	Coating (Subplot Factor-B)	Corrosion Resistance
1	1	360°	2	73
1	1	360°	3	83
1	1	360°	1	67
1	1	360°	4	89
2	1	370°	1	65
2	1	370°	3	87
2	1	370°	4	86
2	1	370°	2	91
3	1	380°	3	147
3	1	380°	1	155
3	1	380°	2	127
3	1	380°	4	212
4	2	380°	4	153
4	2	380°	3	90
4	2	380°	2	100
4	2	380°	1	108
5	2	370°	4	150
5	2	370°	1	140
5	2	370°	3	121
5	2	370°	2	142
6	2	360°	1	33
6	2	360°	4	54
6	2	360°	2	8
6	2	360°	3	46

CHAPTER 12

Screening Designs

12.1 Introduction

Chapters 8 and 9 presented two-level factorial designs. These designs are extremely efficient and not only allow one to determine the effects of several factors, but also to determine if interactions exist between those factors. However if the number of factors, k, being examined is large, the number of experimental runs, $n_F = 2^k$, may be excessive, as shown in Table 12.1.

Table 12.1   Number of experiments needed for a Full Factorial Design

Number of Factors, k	Number of Experiments, $n_F = 2^k$
2	4
3	8
4	16
5	32
6	64
7	128
8	256
9	512
10	1024
11	2048
12	4096

The preliminary stage of experimentation, where the objective of the experiments is to determine which factors are important from a list of candidates, is called the screening stage. Often cause-and-effect diagrams, described in Chapter 1, are used in brainstorming sessions where candidate factors are chosen; the typical result is a *long* list of potentially important factors. Therefore, at this stage it is normal to experiment with a large number of factors. The strategic position of this type of experiment is illustrated in the left column of Figure 12.1. But even though screening experiments will usually involve a large number of factors, in all likelihood only a few of them will actually have large effects. Further, it is extremely unlikely that the higher order interactions (that can be estimated from 2^k designs) will have large effects. Therefore, it would be a waste of time to make as many experimental runs as required by a 2^k design at the screening stage. In order to reduce the amount of experimentation in this situation, special experimental designs constructed specifically for screening should be used.

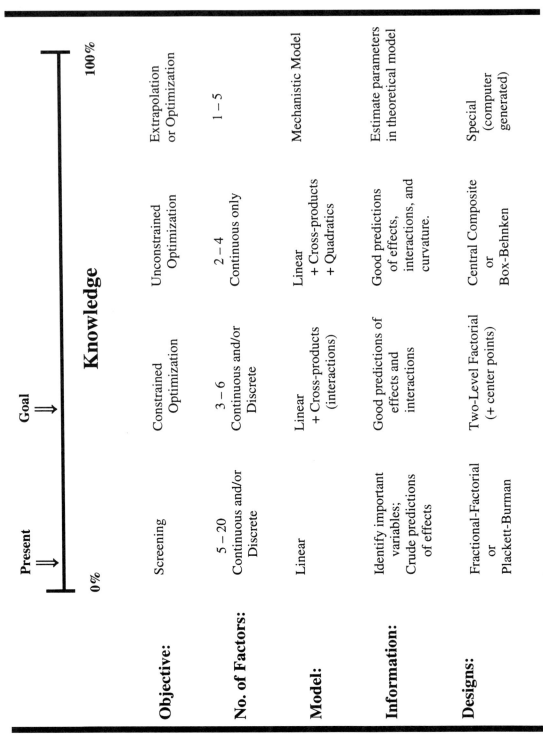

Figure 12.1 Objective of Screening Designs

400

Typically when a large number of factors are under study, even efficient researchers with some familiarity with the use of statistical experimental strategies, may abandon their thoughts of factorial designs and revert to a seat-of-the-pants approach or one-at-a-time experimentation. The one-at-a-time approach was discussed in some detail in Section 7.2. Remember that the strategy involves performing one control experiment followed by an additional experiment for each factor, varying its setting to a new value while holding all other factors constant at their control value for that experiment. As mentioned in Chapter 7, this is a poor strategy for a number of reasons. In our screening situation, the main shortcoming is that the effect of each factor is determined by the difference in response between only two runs. The result is that there is no averaging of data to reduce the noise, and so the experimental variability may overshadow the factor effects.

Another strategy often used when a large number of factors are under study is to pick a subset of the factors and do a complete factorial design. The choosing of a subset of the factors is done by intuition, hunch, opinion, experience, etc., and it may seem like there is some scientific basis for the process. But in reality, this is even a worse strategy than one-at-a-time. The whole purpose of working with the large number of factors is to *determine* which subset of the factors are most important. If you already knew which factors were important, there would be no need for the experiments at all. By guessing which factors are of practical significance, it is very likely that one or more of the most important factors will be left out of the design. Then the benefits of knowing the effects of those factors would be lost to you. The penalty for such an oversight is amplified in a competitive world, because the competition may find them.

12.2 Fractionating Factorial Designs

The best strategy for screening experiments is to choose a subset or fraction of the experimental runs required for a full 2^k design in all the factors under study. We will discuss two categories of screening designs in this chapter, both obtained by taking a subset of the full 2^k factorial design: (a) *fractional factorial* designs, and (b) *Plackett-Burman* designs. The fractional factorial designs are obtained by taking a regular fraction of a 2^k design like ½, ¼, ⅛ (i.e, where the denominator is a power of 2). For example, to study the effect of six factors, a ½ fractional factorial would consist of ½ $\times 2^6$ =32 runs. A 1/4 fractional factorial would consist of $1/4 \times 2^6 = 16$ runs. The *Plackett-Burman* designs (named after the two gentlemen who developed them) are irregular fractions of 2^k designs which are constructed in increments of 4 runs. So, for example, you could study six factors using a Plackett-Burman design with 12 runs.

Choosing the runs to be eliminated from a full factorial, in order to obtain a fractional screening design, cannot be done in a haphazard fashion. For example, if the circled runs in the 2^4 design shown in Table 12.2 were arbitrarily chosen to be eliminated, the resulting design shown in Table 12.3 has Factors 2 and 3 completely *confounded*. In other words, whenever Factor 2 is at the low level so is Factor 3, and whenever Factor 2 is at the high level so is Factor 3. Therefore, the effect of Factor 2 and the effect of Factor 3 would both be calculated as:

$$\frac{Y_3+Y_4+Y_7+Y_8}{4} - \frac{Y_1+Y_2+Y_5+Y_6}{4}$$

In this case it would be impossible to say whether the difference in these two averages was caused by the change in Factor 2 or by the change in Factor 3. In fact, what we are actually estimating by the difference in the averages above is the *sum* of the X_2 effect and the X_3 effect. Since we have no way of sorting out how much X_2 contributed to the difference in averages versus how much X_3 contributed, we say that Factors 2 and 3 are confounded.

Table 12.2   Full 2^4 Factorial Design with Arbitrarily Eliminated Runs

Run	X_1	X_2	X_3	X_4
1	-	-	-	-
2	+	-	-	-
3	-	+	-	-
4	+	+	-	-
5	-	-	+	-
6	+	-	+	-
7	-	+	+	-
8	+	+	+	-
9	-	-	-	+
10	+	-	-	+
11	-	+	-	+
12	+	+	-	+
13	-	-	+	+
14	+	-	+	+
15	-	+	+	+
16	+	+	+	+

Table 12.3 -- Runs Remaining after Arbitrary Elimination

Run	X_1	X_2	X_3	X_4	Response
1	-	-	-	-	Y_1
2	+	-	-	-	Y_2
3	-	+	+	-	Y_3
4	+	+	+	-	Y_4
5	-	-	-	+	Y_5
6	+	-	-	+	Y_6
7	-	+	+	+	Y_7
8	+	+	+	+	Y_8

An alternative, and strategic method for eliminating the runs from a 2^4 is shown in Table 12.4. By eliminating the circled runs in this table, the remaining runs shown in Table 12.5 form a design with no main effects confounded. In fact, we can observe that this design is perfectly orthogonal, by the fact that the factor level combinations for every pair of factors forms a 2^2 full factorial with two replicates at each run. See Table 12.6 for examples.

Table 12.4 Full 2^4 Factorial Design with Strategically Eliminated Runs

Run	X_1	X_2	X_3	X_4
1	-	-	-	-
2	+	-	-	-
3	-	+	-	-
4	+	+	-	-
5	-	-	+	-
6	+	-	+	-
7	-	+	+	-
8	+	+	+	-
9	-	-	-	+
10	+	-	-	+
11	-	+	-	+
12	+	+	-	+
13	-	-	+	+
14	+	-	+	+
15	-	+	+	+
16	+	+	+	+

Table 12.5 Runs Remaining after Strategic Elimination

Run	X_1	X_2	X_3	X_4	Response
1	-	-	-	-	Y_1
4	+	+	-	-	Y_4
6	+	-	+	-	Y_6
7	-	+	+	-	Y_7
10	+	-	-	+	Y_{10}
11	-	+	-	+	Y_{11}
13	-	-	+	+	Y_{13}
16	+	+	+	+	Y_{16}

Table 12.6 Factor Combinations for X_3X_4 and X_1X_2 and Associated Runs from Table 12.5

X_3	X_4	Runs	X_1	X_2	Runs
-	-	1, 4	-	-	1, 13
+	-	6, 7	+	-	6, 10
-	+	10, 11	-	+	7, 11
+	+	13, 16	+	+	4, 16

12.3 Fractional Factorial Designs

A special class of designs called *fractional factorials* provide us with a method for strategically picking a subset of runs from a full factorial without confounding main effects. To use this method, we actually start with a full factorial design with the correct number of runs and then add more factors, rather than starting with a full factorial with the correct number of factors and eliminating runs, as shown in Table 12.4. The end result will be the same. When we use this method we say that we are constructing the fractional factorial design. We will illustrate this method by showing how to construct a half fraction.

12.3.1 Constructing Half Fractions

Half the number of runs in a 2^k factorial is $\frac{1}{2}(2^k) = 2^{k-1}$. This is the number of runs in a full factorial with k-1 factors. The method of constructing a half fraction is then as follows:

1) Write down the *base design* — a full factorial in k-1 factors.

2) Add the kth factor to the design by confounding it with (i.e., making it identical to) one of the columns in the calculation matrix — the column for the highest order interaction.

3) Use the k columns to define the design.

For example, in a half fraction of a 2^4 experiment we would start with a $2^{4-1} = 2^3$ factorial and deliberately confound an added fourth factor with the three-way interaction $X_1X_2X_3$, as shown in the top half of Table 12.7. Using only the columns for Factors X_1, X_2, X_3, and X_4, in the top half of Table 12.7, we are left with the design shown in the bottom half of the table. By examination, it can be seen that this design is identical to the one shown in Table 12.5 with the runs in a different order.

Table 12.7 Construction of a Half Fraction of 2^4 Design

Mean	X_1	X_2	X_3	X_1X_2	X_1X_2	X_2X_3	X_4 $X_1X_2X_3$
+	-	-	-	+	+	+	-
+	+	-	-	-	-	+	+
+	-	+	-	-	+	-	+
+	+	+	-	+	-	-	-
+	-	-	+	+	-	-	+
+	+	-	+	-	+	-	-
+	-	+	+	-	-	+	-
+	+	+	+	+	+	+	+

$\Downarrow$

X_1	X_2	X_3	X_4	
-	-	-	-	
+	-	-	+	Runs for
-	+	-	+	half fraction
+	+	-	-	of 2^4
-	-	+	+	
+	-	+	-	
-	+	+	-	
+	+	+	+	

We can understand why main effects are not confounded with each other in fractional factorial designs, because they each represent one column in the calculation matrix for a full factorial design. However, if we were to expand the number of columns in Table 12.7 to list the full calculation matrix in $X_1 - X_4$ (i.e., a table with columns for X_1, X_2, X_3, X_4, X_1X_2, . . ., $X_1X_2X_3X_4$) we would find that the main effect columns are confounded with some interaction columns, and other interaction columns are confounded with additional interactions. When we use only a subset of the runs in a full 2^k factorial we cannot expect to be able to estimate all effects and interactions independently. There has to be some penalty for the reduction in experiments and that penalty is confounding. Let's examine the confounding in half fractions and illustrate how we deal with them in practice.

12.3.2 Confounding in Half Fractions

The confounding of main effects and interactions in fractional factorial designs is direct and easy to figure out. To facilitate our discussion on this subject, let us define the *alias structure* or *confounding pattern* for the design as the list of factor effect names, along with each effect and interaction that is confounded with it.

Obtaining an alias structure is easiest for a half fractional factorial design, so let's start with that. In Table 12.7 the X_4 column was made identical to the $X_1X_2X_3$ column when we constructed the design. We indicate the columns are the same by the notation, $X_4 = X_1X_2X_3$, or for simplicity we will usually write only the subscripts, $4 = 123$. This equation that describes how we added the extra factor to our design is called a *generator* for the design. A half fractional factorial design has one generator. A ¼ fractional factorial design would need two generators, a ⅛ fractional factorial would need three generators, and so on. Please remember that this generator (equation) does NOT mean that the effect of X_4 equals the interaction of X_1, X_2, and X_3. What it does mean is that the experimental conditions in this particular design were chosen such that the calculation column for X_4 is the same as that for $X_1X_2X_3$. Thus, the two quantities are confounded, not equal. To say it one more time: the "=" in the generator $4 = 123$ means that the *columns* in the calculation matrix are the same. The result is that when you calculate the effect of X_4, you are actually getting the sum of the effect of X_4 and the interaction of $X_1X_2X_3$.

For the eight run fractional factorial design shown in the bottom half of Table 12.4, the columns labeled X_1, X_2, X_3, and X_4 would be used to define the conditions for each of the eight experiments. When calculating the effects, however, all seven columns, from the top half of Table 12.4 will be used. The effects we calculate are often called *contrasts* as an explicit reminder that the effects are confounded with interactions. As already stated, the effect for X_4 is confounded with the $X_1X_2X_3$ interaction. But that is not the whole story; there is more confounding taking place. To elucidate the whole confounding pattern, let us start by finding the factor which is confounded with the overall mean. To do this, multiply both sides of the generator by the left hand side. We get

$$4 \times 4 = 123 \times 4, \quad \text{or } 4^2 = 1234, \quad \text{or } I = 1234,$$

where the multiplication represents element-wise multiplication of the + or - signs in the column labeled 4 and 123, and I represents a column of + signs (i.e., the sign on each data value in the sum when computing the average or mean of the data). The symbolic equation $I = 1234$, which indicates the $X_1X_2X_3X_4$ interaction is confounded with the mean, is called the *defining relation*. It is given this distinctive name because it can be used to find the rest of the confounding pattern. To find the confounding for any factor or interaction just multiply each side of the defining relation by the symbol for that factor. Multiplying I by any factor just gives that factor, and multiplying any two factors together is equivalent to replacing the two sets of subscripts by all subscripts which appear in either but not both groups. For example, to find the confounding for the X_2X_4 interaction, we multiply the defining relation by 24 and we get: $I \times 24 = 1234 \times 24$, or $24 = 12^234^2$, or $24 = 13$ (since $2^2 = I$ and $4^2 = I$). In this way, the whole confounding pattern (or alias structure), is computed from $I = 1234$ as:

$$
\begin{aligned}
I &= 1234 \\
1 &= 234 \\
2 &= 134 \\
3 &= 124 \\
4 &= 123 \\
12 &= 34 \\
13 &= 24 \\
14 &= 23
\end{aligned}
$$

By using the generator $4 = 123$ to create a half fraction of the 2^4 design, we have selected half the runs in the full factorial design in a way so that no main effect is confounded with another main effect. The other half fraction (i.e., the eight runs in Table 12.4 that we eliminated) will have the same property. The generator, $4 = 123$, generates the *principal fraction*, and the generator, $4 = -123$, (which means let the column of signs that defines X_4 be the negative of the column of signs that defines the $X_1X_2X_3$ interaction) generates the other half fraction. In practice either fraction can be used.

12.3.3 A Simple Example of a Half Fraction

In Section 1.4.2, in Chapter 1, the process for copperplating ceramic substrate circuit boards at Motorola was described. In an attempt to improve the process by decreasing variability in copper-plating thickness, a half fraction of 2^3 experiment was conducted. Table 12.8 shows the factors (identified by brainstorming and the resulting cause-and-effect diagram shown in Figure 1.11) and levels studied in the experiment. Since data collection was expensive, it was desirable to run only half of the 2^3 experiments required for a full factorial. Note: This is generally not an acceptable design, since it leaves no degrees of freedom to estimate experimental variability. This means that no significance testing is possible.

Table 12.8 Factors for Half Fraction Factorial Design

Factor	(-) Level	(+) Level
X_1 = Anode height	Up	Down
X_2 = Circuit board orientation	In	Out
X_3 = Anode placement	Spread	Tight

The plan for the half fraction factorial design was created by starting with base 2^2 design in Factors 1 and 2, and then assigning the added Factor 3 to the negative of the interaction column. Therefore, the generator is: $3 = -12$. This results in the defining relation: $I = -123$, and the confounding pattern: $1 = -23$
$2 = -13$
$3 = -12$
This means that each main effect is confounded with a two-factor interaction, and that the three-factor interaction is confounded with the mean. Even though there is confounding, we still get useful information from this experiment if we can assume the interactions are negligible.

Shown in Table 12.9 is the list of experiments and the response, which was the variance in plating thickness for replicate circuit boards coated under the same conditions. Since there are only four runs, this table is also the calculation matrix, and calculation of the effects is shown. The effects in this case are just half the sum-of-products row.

Table 12.9 2^{3-1} Fractional Factorial Design and Response

Run	Mean	X_1	X_2	X_3	s^2
1	+	-	-	-	11.63
2	+	+	-	+	5.57
3	+	-	+	+	3.57
4	+	+	+	-	7.36
$\Sigma(+)-\Sigma(-)$	28.12	-2.28	-6.28	-9.86	
Effects	7.03	-1.14	-3.14	-4.93	

Remember, with this data alone (i.e., four runs) it is impossible to say which if any of the effects are significant. Also, due to the confounding, it is impossible to know if the largest effect, -4.93, is the effect of Factor 3 (Anode Placement) or is the negative of the interaction effect between Factors 1 and 2. Similar statements could be made for the other two effects. Additional experiments could have been conducted to resolve this ambiguity. However, in this case, the two-factor interactions were assumed to be negligible, and the factors were all assumed to be significant.

Interpreting the main effects alone, it could be seen that all effects were negative. Thus, the results indicated that using the high level of each factor (i.e., anode height = down, circuit board orientation = out, and anode placement = tight) would result in a smaller response (variance of copperplating thickness). But, no experiment had been conducted at this set of factor levels. Therefore, a confirmation experiment was run at the +, +, + levels of the factors resulting in a variance of 3.13, a 70% reduction in variability. Since the result of this confirmation experiment resulted in a lower response value, as had been predicted, it justified the earlier assumption that the interactions were negligible, and supported the interpretation that all main effects had negative effects. Therefore permanent changes were implemented in the process to use this (+, +, +) configuration of the plating cells. Recall that Figure 4.5 in Chapter 4 compared histograms of the copperplating thickness before and after this process change, and shows visually the improved process consistency resulting from this simple experiment.

12.3.4 One-Quarter and Higher Fractional Factorials

In constructing half replicates of 2^k designs, only half the total experimental runs were made, therefore, each factor effect and interaction that could be estimated had one other interaction effect confounded with it (they are confounded in groups of two). If one-quarter of the total experimental runs were made, each factor effect and interaction would have three other interactions confounded with it (they would be confounded in groups of four). If one-eigth of the total experimental runs were made, the factors and interactions would be confounded in groups of eight, and so on — the greater the fractionation, the greater the confounding. This is logical. The interactions between the factors do not go away just because we do fewer experiments; instead, their impacts get added to the contrasts that can be estimated.

This will be illustrated by examining in detail a one-quarter replicate. To construct a one-quarter replicate, we begin with a full factorial in k-2 factors, called the *base design*. As with the half fraction, the base design is the full factorial design which will give us the number of runs we want to make. In the case of a one-quarter replicate, we have two factors left over whose levels still must be set. These *added factors* are taken care of by associating (confounding) them with two of the columns of the calculation matrix. For example, in constructing one-quarter replicate of a 2^5 factorial, we start by writing a 2^3 base design, then we associate the added factors, 4 and 5, with two interactions in the calculation matrix. Let us arbitrarily use the 12 and 13 interactions for X_4 and X_5 respectively. That is, our generators are: $4 = 12$ and $5 = 13$. This is shown in Table 12.10.

Table 12.10   Construction of One-quarter Replicate of 2^5 Design

X_1	X_2	X_3	X_4 X_1X_2	X_5 X_1X_3	X_2X_3	$X_1X_2X_3$
-	-	-	+	+	+	-
+	-	-	-	-	+	+
-	+	-	-	+	-	+
+	+	-	+	-	-	-
-	-	+	+	-	-	+
+	-	+	-	+	-	-
-	+	+	-	-	+	-
+	+	+	+	+	+	+

$$\Downarrow$$

X_1	X_2	X_3	X_4	X_5
-	-	-	+	+
+	-	-	-	-
-	+	-	-	+
+	+	-	+	-
-	-	+	+	-
+	-	+	-	+
-	+	+	-	-
+	+	+	+	+

From the generators we find two of the interactions confounded with the mean, which are $I = 124 = 135$. The third interaction can be found by multiplication, recognizing that if 124 and 135 are both columns of "+" signs ($= I$), then when we multiply them together we still get a column of "+" signs. Therefore, the last interaction is: $I = (124)(135) = 2345$, and the complete defining relation is: $I = 124 = 135 = 2345$.

Next, the entire confounding pattern can be obtained by multiplication as before, so that

$$
\begin{array}{llll}
1 & = 24 & = 35 & = 12345 \\
2 & = 14 & = 1235 & = 345 \\
3 & = 1234 & = 15 & = 245 \\
4 & = 12 & = 1345 & = 235 \\
5 & = 1245 & = 13 & = 234 \\
23 & = 134 & = 125 & = 45 \\
123 & = 34 & = 145 & = 25 \\
\end{array}
$$
$$\text{etc.}$$

Note that the first column corresponds to the contrasts we can estimate, which we obtain from the headings of the columns in the calculation matrix.

When we construct higher fractional replicates like 1/8 or 1/16, we will have the mean confounded with 7 or 15 other interactions, respectively. For the 1/8 replicate, we start by writing down the full factorial in k-3 factors and then associating the last three factors with any three interactions of the first k-3 factors (i.e., we need three generators). This, in effect, specifies three interactions that will be confounded with the mean, I. Then, the other four interactions confounded with the mean will be all of the possible pair-wise products, and the three-way product of the original three interactions in the defining relation. For the 1/16 replicate, it is necessary to specify four generators which will give us four interactions in the defining relation. Then, the remaining eleven interactions will be all of the possible pair-wise products, three-way products, and the four-way product of the original four.

12.3.5 Fractional Factorial Design Tables

There are usually a number of ways of associating the added p factors with interactions in a $(\frac{1}{2})^p$ fractional factorial. For example, in the one-quarter fraction of a 2^5 shown above we could use the generators 4=23 and 5=123 rather than 4=12 and 5=13 that we did use. Each of the ways of choosing the generators will result in a different design and hence a different confounding pattern. Some designs will be more desirable than others since they will have fewer low order interactions confounded with main effects. Rather than leave the choice of the generators to chance, there are two approaches that can be used. One is to use statistical software that has capabilities for automatic creation and analysis of fractional factorial designs. The illustration and use of such software will be left as a laboratory exercise. In the absence of statistical software, Appendix B.1 provides simple tables of fractional factorial designs with optimal confounding patterns.

In Appendix B.1, notice that there are two eight run fractional factorial designs, three sixteen run designs, and three thirty-two run designs. Each of these designs represents a different fraction and confounding pattern. When using these tables you should be sure to pick the appropriate one by matching the number of factors you have with the number of factors listed at the top of the table. In this way you will get the best confounding pattern possible for the number of factors you have. The confounding patterns for the 8, 16, and 32 run fractional factorials given in Appendix B.1 are also listed with the tables, omitting interaction terms higher than second order.

To illustrate interpretation of the confounding patterns listed in Appendix B.1 we will refer to a design that is the same as the one-quarter replicate of a 2^5 we constructed in the last section. This design can be found in Table B.1-5, and it is reproduced in Table 12.11 for convenience. The computation table for the experiment is shown in the body of Table B.1-5, and below the table under CONFOUNDINGS, the confounding pattern is listed. In the confounding pattern only two-way interactions are shown. The experiment constructed in the last section had only five factors where Table B.1-5 can be used for up to 7 factors. If we only had five factors, the columns labeled X_6 and X_7 would not be used when making a list of the experiments to be run, and interactions containing factor numbers greater than 5, in the confounding pattern, would be deleted (such as 67 in line 1, 36 and 57 in line 2, etc.). When calculating effects the entire body of Table B.1.5 would be used for a computation table.

Table 12.11   Copy of Appendix Table B.1-5

Run No.	X_1	X_2	X_3	X_4	X_5	X_6	X_7	CONFOUNDINGS
1	−	−	−	−	+	+	−	$X_1 = 1 + 24 + 35 + 67$
2	+	−	−	+	−	+	+	$X_2 = 2 + 14 + 36 + 57$
3	−	+	−	+	+	−	+	$X_3 = 3 + 15 + 26 + 47$
4	+	+	−	−	−	−	−	$X_4 = 4 + 12 + 37 + 56$
5	−	−	+	+	−	−	+	$X_5 = 5 + 13 + 27 + 46$
6	+	−	+	−	+	−	−	$X_6 = 6 + 17 + 23 + 45$
7	−	+	+	−	−	+	−	$X_7 = 7 + 16 + 25 + 34$
8	+	+	+	+	+	+	+	

12.3.6 An Example of Fractional Factorial in Process Improvement

A chemical process for manufacturing a new herbicide product was devised and tested in the laboratory. When the process was scaled up to run in the pilot plant, the levels of a certain impurity (by product of the reactions) became a problem. In the laboratory there was less than 0.8% of this impurity in the final product. But, typical runs in the pilot plant had more than 2%. To make the process cost effective, and to eliminate the need for further process steps to purify the product, it was desired to reduce the level of the impurity to less than 0.75%. To find a combination of factor levels that would do this, some experiments in the pilot plant were needed.

At first, trial-and-error tests were planned. However, due to limited time available in the pilot plant, it was decided to use a statistical plan to get the maximum information in a limited time. Figure 12.2 is a diagram of the pilot plant, and Table 12.12 is a list of the potential factors that was made in a brainstorming session. Before the brainstorming session, the lead chemist had planned to conduct a simple 2^3 design with two replicates using the factors PH, Pumper on Loop, and RPM. But, after the brainstorming his eyes were opened by other's opinions about many other potential causes for the impurities. Since all these factors could be studied in the same number of experiments (16) that he had originally planned on running, he decided to conduct a fractional factorial experiment.

411

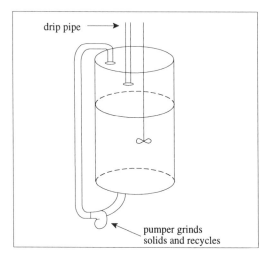

Figure 12.2   Pilot Plant

Figure 12.2   Pilot Plant Diagram

Table 12.12   Result of Brainstorming

List of Potential Factors in Pilot Plant
1. PH
2. Temperature
3. H_2O Level (Full Charge vs Partial +Makeup)
4. Addition Time
5. Agitation Rate
6. Drip Pipe Height
7. PH probe accuracy
8. Source of Raw Material
9. Pumper on Loop
10. 2/3 Batch or Whole
11. Close Couple vs Solid Cake
12. Stir at Completion

Two of the factors listed in Table 12.12 (Drip Pipe Height and PH probe accuracy) could not be varied or controlled so they were eliminated from the study, and two other factors were dropped

412

because they were felt to be less important. Therefore with seven remaining factors, a 2^{7-3} 16-run fractional factorial experiment was planned. The first seven columns of appendix Table B.1-8, were used to define the runs. Table 12.13 shows these experimental conditions and the response data resulting from the experiments. The experiments were run in the random order shown in Column 2 (taken from Table 8.3) to prevent biases from changes in levels of factors that couldn't be controlled and other background factors.

Table 12.13 Pilot Plant Experimental Results

Run	Random Order	PH X_1	Temp X_2	Add Time X_3	Stir X_4	H_2O X_5	RPM X_6	Pump X_7	% Imp. Y
1	5	-	-	-	-	-	-	-	2.60
2	4	+	-	-	-	+	-	+	0.89
3	14	-	+	-	-	+	+	-	1.82
4	7	+	+	-	-	-	+	+	0.49
5	15	-	-	+	-	+	+	+	1.83
6	2	+	-	+	-	-	+	-	1.26
7	1	-	+	+	-	-	-	+	2.23
8	3	+	+	+	-	+	-	-	1.98
9	8	-	-	-	+	-	+	+	0.67
10	11	+	-	-	+	+	+	-	0.07
11	12	-	+	-	+	+	-	+	0.51
12	16	+	+	-	+	-	-	-	0.72
13	10	-	-	+	+	+	-	-	1.61
14	6	+	-	+	+	-	-	+	0.08
15	9	-	+	+	+	-	+	-	0.53
16	13	+	+	+	+	+	+	+	0.16

After completing the experiments and making some initial exploratory plots, as described in Chapter 8, the next step was to analyze the data completely to determine which factors appear to have significant effects. There are two ways the effects can be calculated. We will illustrate both of these methods on the following pages. The first way is to use the worksheet method described in Chapter 8 (or equivalently the least squares method described in Chapter 10). To use the worksheet or least squares method, copy the entirety of Table B.1-8 (not just the first seven columns) to use as the computation table. Equivalently, the columns of Table B.1-8 can be used as the columns of the X-matrix used in the least squares method. Table 12.14 shows the results of these computations.

Table 12.14 Effects from Pilot Plant Study Calculated by Worksheet

Run	Mean	X1	X2	X3	X4	X5	X6	X7	X8	E1	E2	E3	E4	E5	E6	E7	Y
1	1	-1	-1	-1	-1	-1	-1	-1	-1	1	1	1	1	1	1	1	2.6
2	1	1	-1	-1	-1	1	-1	1	1	-1	-1	-1	1	-1	1	1	0.89
3	1	-1	1	-1	-1	1	1	-1	1	-1	1	1	-1	-1	1	-1	1.82
4	1	1	1	-1	-1	-1	1	1	-1	1	-1	-1	-1	1	1	-1	0.49
5	1	-1	-1	1	-1	1	1	1	-1	1	-1	1	-1	-1	-1	1	1.83
6	1	1	-1	1	-1	-1	1	-1	1	-1	1	-1	-1	1	-1	1	1.26
7	1	-1	1	1	-1	-1	-1	1	1	-1	-1	1	1	1	-1	-1	2.23
8	1	1	1	1	-1	1	-1	-1	-1	1	1	-1	1	-1	-1	-1	1.98
9	1	-1	-1	-1	1	-1	1	1	1	1	1	-1	1	-1	-1	-1	0.67
10	1	1	-1	-1	1	1	1	-1	-1	-1	-1	1	1	1	-1	-1	0.07
11	1	-1	1	-1	1	1	-1	1	-1	-1	1	-1	-1	1	-1	1	0.51
12	1	1	1	-1	1	-1	-1	-1	1	1	-1	1	-1	-1	-1	1	0.72
13	1	-1	-1	1	1	1	-1	-1	1	1	-1	-1	-1	1	1	-1	1.61
14	1	1	-1	1	1	-1	-1	1	-1	-1	1	1	-1	-1	1	-1	0.08
15	1	-1	1	1	1	-1	1	-1	-1	-1	-1	-1	1	-1	1	1	0.53
16	1	1	1	1	1	1	1	1	1	1	1	1	1	1	1	1	0.16
SumProd	17.45	-6.15	-0.57	1.91	-8.75	0.29	-3.79	-3.73	1.27	2.67	0.71	1.57	0.81	0.41	-1.09	-0.450	
Effects		-0.77	-0.07	0.24	-1.09	0.04	-0.47	-0.47	0.16	0.33	0.09	0.196	0.101	0.05	-0.136	-0.06	

The second way of computing the effects is based on the fact that the first four columns of Table B.1-8 represent a full 2^4 factorial. Therefore we can use Yates algorithm, described in Appendix 8.A. This reduces the work if the computations are being done on a spreadsheet like EXCEL, because you don't have to copy the entire computation table. But, the confounding pattern given with Table B.1-8 must be used to determine the names of the computed effects. Table 12.15 Illustrates the computations. Following four cycles of the algorithm presented in Appendix 8.A, results in the contrast sums. Dividing the contrasts sums by half the number of experiments ($n_F/2 =$ 8) results in the effects shown in the sixth column of the table. The effect names given in the seventh column (labeled (1)) are the standard Yates order of effects for a full 2^4 factorial. Note $1=X_1$, etc. in this notation. The effect labels given in the last column (labeled (2)) were obtained from the confounding pattern at the bottom of Table B.1-8. As examples, consider the following: From the CONFOUNDINGS, $E_1 = 12 + 35 + 48 + 67$. Therefore the third effect that is labeled 12 can be recognized to be E_1. From the Defining Relation $I = 1235 = 2346 = 1347 = 1248$ which tells us that $5 = 123$ is a generator, so the seventh effect labeled 123 can be recognized to be X_5.

Table 12.15 Computation of Pilot Plant Effects Using Yates Algorithm

Y	1	2	3	4	Effects	(1)	(2)
2.6	3.49	5.8	13.1	17.45	1.090625	Mean	
0.89	2.31	7.3	4.35	-6.15	-0.76875	1	X1
1.82	3.09	1.97	-3.86	-0.57	-0.07125	2	X2
0.49	4.21	2.38	-2.29	2.67	0.33375	12	E1
1.83	0.74	-3.04	-0.06	1.91	0.23875	3	X3
1.26	1.23	-0.82	-0.51	0.71	0.08875	13	E2
2.23	1.69	-0.39	0.7	0.81	0.10125	23	E4
1.98	0.69	-1.9	1.97	0.29	0.03625	123	X5
0.67	-1.71	-1.18	1.5	-8.75	-1.09375	4	X4
0.07	-1.33	1.12	0.41	1.57	0.19625	14	E3
0.51	-0.57	0.49	2.22	-0.45	-0.05625	24	E7
0.72	-0.25	-1	-1.51	1.27	0.15875	124	X8
1.61	-0.6	0.38	2.3	-1.09	-0.13625	34	E6
0.08	0.21	0.32	-1.49	-3.73	-0.46625	134	X7
0.53	-1.53	0.81	-0.059	-3.79	-0.47375	234	X6
0.16	-0.37	1.16	0.35	0.41	0.05125	1234	E5

Once the effects were calculated from the data, the next step was to determine which effects are significant and interpret the results. Since there were no replicate experiments, no standard error of an effect (s_E) could be calculated, and the significant effects had to be judged using graphical methods like a Normal Probability plot of effects described in Section 8.10. Figure 12.3 shows a Normal plot of the effects. This graph shows clearly that the significant effects appear to be $X_1 =$ PH, $X_4 =$ Stir at Completion, $X_6 = H_2O$ Level, and $X_7 =$ Agitation Rate (RPM). All of the effects

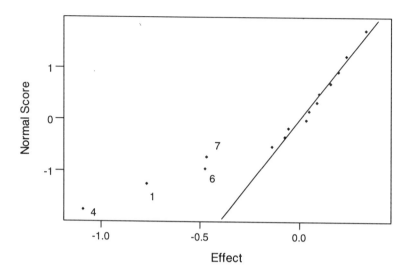

Figure 12.3 Normal Probability Plot of Effects from Pilot Plant Study

were negative indicating that the high level of each would produce the lowest impurity level.

Since none of the unassigned effects (X_8 or E_1–E_8) appear to be significant in this experiment, a simple linear prediction equation can be written to represent the results. This equation is:

$$Y(\%impurities) = 1.091 - 0.38X_1 - 0.54X_4 - 0.24X_6 - 0.23X_7$$

It can be seen from this equation that the lowest impurities should result when all four of the significant factors are at their high level. However the predicted value at those conditions is negative! This negative prediction, along with the fact that the response data spans over 1½ orders of magnitude (i.e., 0.07 to 2.7) may indicate that the underlying relationship is nonlinear (see Section 10.7). A better prediction equation (the calculation of the coefficients is left as an exercise) is:

$$\ln(Y) = -0.346 - 0.56X_1 - 0.73X_4$$

which results in reasonable predicted values with the high levels of X_1 = PH and X_4 = Stir. These results were utilized, with no further experimentation, and the process continued to produce product with acceptably low impurities. Statistical experimentation, as shown in this example, has been a valuable tool in chemical process development that tremendously reduces the trial and error experimentation needed to make processes work efficiently.

An interesting note about this experiment is, that the 2^3 experiment initially planned by the lead chemist did not include Factor 4 (Stir). But, this factor turned out to have the largest effect and was the key to finding a combination of factor levels that resulted in low impurity levels and a cost-effective process. In screening experiments, it is very wise to study as many factors listed during brainstorming as possible. The number of experiments required to study 11 factors is no more than the number required to study four factors when using a fractional factorial design. But, on the other hand if an important factor is left out of a screening experiment the results may be inconclusive.

In the example above, four or less factors appear to be significant in a 16-run screening experiment. Since continued plant operation at the conditions predicted to be best from the simple linear prediction equation (i.e., no interactions included) were successful, the results of this experiment were conclusive. However, many times after screening experiments the results will not be conclusive, due to interactions among the important factors. In these cases, additional follow-up experiments with the important factors may be required after the screening experiments.

12.3.7 Advantages of Fractional Factorial Designs

Fractional factorial designs have two advantages over the other screening designs that will be presented later in this chapter. The first advantage is called the *projection property*. After the data collection and analysis are complete, and only a few of the many factors in the design are found to be important, we may be left with a full factorial design in these key variables (which would allow us to estimate all interactions between those factors, as well as their main effects). The second advantage is that confoundings between the variables (main effects and interactions) are much simpler to determine than they are for other types of screening designs.

To illustrate the first advantage, consider the eight run design for four factors in Table B.1-5. Any three columns selected from the first four will form a full factorial design. This is illustrated in Table 12.16 which shows that by reordering the run numbers the same full factorial pattern can be seen for any subset of the four factors. Therefore, if we use the design in Table B.1-5 to screen four factors and at least one of the four factors turns out to be unimportant, we will have a full factorial in the key variables if we ignore the column(s) for the unimportant factor(s). If we want to study the interaction effects involving the important variables there will be no need to do follow-up experiments — we already have a full factorial design.

The second advantage of fractional factorial designs — being able to determine the confoundings quite easily — can help in deciding which follow-up experiments to perform when they are necessary. To be specific let us consider an example. Suppose that in the experiment presented in the last section, the unassigned factor E_1 had been large (in the range 0.7 to 1.1 like Factors 1 and 4) and was considered to be significant. This unassigned effect represents a string of possible two-factor interactions. From the CONFOUNDINGS listed with Table B.1-8 we see that $E_1 = 12 + 35 + 67$. (Note: We drop 48 from the interactions in the table because there were only seven factors in the experiment.) With only the 16 experiments run in the fractional factorial design it is impossible to determine which of the three potential interactions this effect represents. This can be seen clearly by examining the calculation columns for these three interactions shown in Table 12.17. The effect that would be computed for each of these interactions would be the exactly the same using the 16 experiments in the table. In order to de-confound these three interactions, we need to run additional experiments.

Table 12.16 Full Factorials formed from subsets of a Fractional Factorial

Run	X_1	X_2	X_3	X_4
1	-	-	-	-
2	+	-	-	+
3	-	+	-	+
4	+	+	-	-
5	-	-	+	+
6	+	-	+	-
7	-	+	+	-
8	+	+	+	+

Run	X_1	X_2	X_4
1	-	-	-
6	+	-	-
7	-	+	-
4	+	+	-
5	-	-	+
2	+	-	+
3	-	+	+
8	+	+	+

Run	X_1	X_3	X_4
1	-	-	-
4	+	-	-
7	-	+	-
6	+	+	-
3	-	-	+
2	+	-	+
5	-	+	+
8	+	+	+

Run	X_2	X_3	X_4
1	-	-	-
4	+	-	-
6	-	+	-
7	+	+	-
2	-	-	+
3	+	-	+
5	-	+	+
8	+	+	+

The additional experiments shown in Table 12.18 are one set that will work. This table was obtained by writing the levels for the three confounded interactions first. This was done by writing a full factorial in X_1X_2, X_3X_5, and X_6X_7, which can be seen by examining the last three columns of Table 12.18. This pattern is used because we know that it will result in the three interactions being independently estimable. Next, any combination of levels for Factors 1–7 can be chosen for each run whose products will result in the appropriate levels for the three interactions. For example, in Run 17, any combination of levels for X_1–X_7 that will result in $X_1X_2 = -$, $X_3X_5 = -$, and $X_6X_7 = -$, will work. The combination $X_1 = -$, $X_2 = +$, $X_3 = -$, $X_4 = +$, $X_5 = +$, $X_6 = -$, $X_7 = +$ was used, but the combination $X_1 = +$, $X_2 = -$, $X_3 = +$, $X_4 = -$, $X_5 = -$, $X_6 = +$, $X_7 = -$ will work just as well. Using this logic, the levels for Factors 1–7 were completed for runs 17–24. If the experiments listed in Table 12.18 were completed in a random order, there would be sufficient data to estimate the main effects and the interactions confounded with the E_1 effect in the fractional factorial, but the analysis would have to be done by regression.

When we combine the two sets of experiments, the columns will no longer be orthogonal (i.e., will not form a full factorial in every pair of columns). Therefore, we cannot analyze the

Table 12.17 Calculation Columns for Confounded Interactions in Pilot Plant Example

Run	PH X_1	Temp X_2	Add Time X_3	Stir X_4	H_2O X_5	RPM X_6	Pump X_7	X_1X_2	X_3X_5	X_6X_7
1	−	−	−	−	−	−	−	+	+	+
2	+	−	−	−	+	−	+	−	−	−
3	−	+	−	−	+	+	−	−	−	−
4	+	+	−	−	−	+	+	+	+	+
5	−	−	+	−	+	+	+	+	+	+
6	+	−	+	−	−	+	−	−	−	−
7	−	+	+	−	−	−	+	−	−	−
8	+	+	+	−	+	−	−	+	+	+
9	−	−	−	+	−	+	+	+	+	+
10	+	−	−	+	+	+	−	−	−	−
11	−	+	−	+	+	−	+	−	−	−
12	+	+	−	+	−	−	−	+	+	+
13	−	−	+	+	+	−	−	+	+	+
14	+	−	+	+	−	−	+	−	−	−
15	−	+	+	+	−	+	−	−	−	−
16	+	+	+	+	+	+	+	+	+	+

Table 12.18 Some Additional Runs that Allow Estimation of Confounded Interactions

Run	PH X_1	Temp X_2	Add Time X_3	Stir X_4	H_2O X_5	RPM X_6	Pump X_7	X_1X_2	X_3X_5	X_6X_7
17	−	+	−	+	+	−	+	−	−	−
18	+	+	−	+	+	−	+	+	−	−
19	−	+	+	−	+	−	+	−	+	−
20	+	+	+	−	+	−	+	+	+	−
21	+	−	+	+	−	−	−	−	−	+
22	−	−	−	−	+	+	+	+	−	+
23	−	+	+	+	+	+	+	−	+	+
24	+	+	−	−	−	+	+	+	+	+

419

combined data from the two sets of experiments by the worksheet method or Yates algorithm. Instead the least squares, regression analysis must be used. An X-matrix would be defined containing columns for X_1-X_7, X_1X_2, X_3X_5, and X_6X_7 and one additional column, X_8 , representing a block of experiments (X_8 = - for Runs 1–16 and X_8 = + for Runs 17–24). This X matrix would define a linear model with 12 coefficients (the 11 terms plus the intercept). Since there are 24 data points in the combined experiments, there would be 12 degrees of freedom left for estimating the standard error of the regression coefficients and calculating t-statistics.

12.3.8 Resolution of Fractional Factorial Designs

Since there are many ways of generating a $(½)^p$ fraction of a 2^k factorial design, it is useful to have a measure of how "good" a particular design (or fraction) is. One measure that is used is called *resolution*. Resolution is defined as the length of the shortest word in the defining relation. For a half fraction there is only one word in the defining relation (other than the letter I). For example, consider the half fraction of a 2^5 factorial design with the generator, 5 = 1234. The defining relation, is I = 12345. Since the shortest (only) word (12345) has five factors in it, the resolution is five. Resolution is indicated by a Roman numeral, in this case V. Had the generator for the design been 5 = 234, then the defining relation would have been I = 2345, which is a resolution IV. Or, had the generator for the design been 5 = 23, the defining relation would have been I = 235 which is resolution III. The resolution V design is most desirable, because the higher the resolution the better.

Why is higher resolution better? Because a resolution V design has main effects and two factor interactions confounded only with three factor interactions and higher. For the example with I=12345, we see that 5=1234, 1=2345, etc. and 12=345, 13=245, etc. For all practical purposes, all main effects and two-factor interactions are not confounded with anything, because it is *very rare* that a three-factor interaction exists. Also, a resolution V design has the projection property that every subset of four factors forms a full 2^4 factorial, in the sense described in Section 12.3.7. So if four or less factors are important, a full factorial in the important factors is imbedded in the resolution V design. Tables B.1-9 and B.1-13 are resolution V designs in five and six factors respectively.

A resolution IV design has main effects clear of any confounding except three-factor (or higher) interactions. But some or all of the two-factor interactions are confounded with other two-factor interactions. For our resolution IV example of 5 = 234 (or I=2345) we have 1=12345, 2=345, etc., which are great, but 23=45, 24=35, etc. Resolution IV designs have the projection property that every subset of three factors forms a full 2^3 design. Tables B.1-6, B.1-8 and B.1-11 are resolution IV.

A resolution III design has (some) main effects confounded with two-factor interactions. For our resolution III example of 5 = 23 (or I = 235) we have 1 = 1235, 2 = 35, 3 = 25, etc. Resolution III designs have the projection property that every pair of factors in a forms a full 2^2 design, and they are the lowest resolution that is generally acceptable. Their use requires assuming that two-factor interactions are essentially negligible. This assumption is acceptable in screening designs because it allows us to experiment with a large number of factors with a minimal number of runs. Note that a resolution II design would have main effects confounded with other main effects which is unacceptable. So, useful resolutions are III, IV, and V. Resolution III is used mainly for screening. Tables B.1-5, B.1-7, and B.1-10 are resolution III.

12.4 Plackett-Burman Screening Designs

As an alternate way to avoid the problem of confounding the (main) effects of factors under study, Plackett and Burman[1] in 1946 developed a set of tables to be used for screening designs. Their tables are resolution III designs. Table 12.19 is an example of an eight run Plackett-Burman design. This design can be used to screen up to seven factors. This table is a subset of the 2^7 runs necessary for a full factorial in seven factors and is similar to the fractional factorial designs in that no factors are confounded in this table. Specifically, at the high (+) level of each of the factors there are an equal number of experiments conducted at the high (+) and low (-) levels of each of the other factors. The same is true, of course, at the low level of each of the factors.

Table 12.19 A Plackett-Burman Design for Seven Factors

Run	X_1	X_2	X_3	X_4	X_5	X_6	X_7
1	+	+	+	-	+	-	-
2	-	+	+	+	-	+	-
3	-	-	+	+	+	-	+
4	+	-	-	+	+	+	-
5	-	+	-	-	+	+	+
6	+	-	+	-	-	+	+
7	+	+	-	+	-	-	+
8	-	-	-	-	-	-	-

To verify that this is true, the data below show a tally of runs at all combinations of Factors 1 and 2, similar to Table 12.6.

X_1	X_2	Run #'s at these levels
-	-	3, 8
+	-	4, 6
-	+	2, 5
+	+	1, 7

A full factorial in seven factors would take $2^7 = 128$ experiments or runs. The Plackett-Burman design allows seven factors to be examined in only eight runs. This is a *huge* reduction in the amount of experimentation, but it does not come without a penalty. There is no free lunch! The price we pay is that we cannot estimate any of the interactions. This is a resolution III design and they are confounded with the main effects of the factors. To see that this is true, let us create the column for computing the X_1X_2 interaction by multiplying the column of signs for X_1 times the column of signs for X_2 in Table 12.19 (which is done in Table 12.20).

Table 12.20 Confounding of Interactions in a Plackett-Burman Design

X_1	X_2	...	X_6	$X_1 \cdot X_2$	Response
+	+		-	+	Y_1
-	+		+	-	Y_2
-	-		-	+	Y_3
+	-		+	-	Y_4
-	+		+	-	Y_5
+	-		+	-	Y_6
+	+		-	+	Y_7
-	-		-	+	Y_8

It can be seen that the resulting interaction column is exactly the opposite of the column of signs that would define the X_6 factor. In other words the difference in averages,

$$\frac{Y_2+Y_4+Y_5+Y_6}{4} - \frac{Y_1+Y_3+Y_7+Y_8}{4}$$

would be used to estimate either the effect of Factor 6, or the negative of the X_1X_2 interaction. Therefore, the X_6 effect is confounded with the X_1X_2 interaction. In fact, every main factor effect calculated from data derived from the design in Table 12.4 will be confounded with three two-factor interactions.[2]

Thus, even though the Plackett-Burman tables were carefully constructed to avoid confounding of the effect of one factor with another, it is impossible to avoid confounding of interactions with the main effects of factors. In other words, some confounding is the price we pay for the great reduction in experimental effort. In order to use Plackett-Burman screening designs, all interactions must be assumed to be negligible. The saving grace is that this assumption is not too critical if follow up experiments with the important factors are planned. Interactions normally only exist between important factors. Therefore, screening designs should be used to determine the important factors, and follow-up experiments can be conducted to study interactions of the important variables.

12.4.1 Tables of Plackett-Burman Designs

Although Plackett-Burman designs are available in every multiple of four runs from 4 to 100, tables are presented in Appendix B.1 for those designs which are most useful, that is, designs with 12, 20, 24, or 28 runs. Tables B.1-1, B.1-2, B.1-3, and B.1-4 show these designs. The eight run Plackett-Burman design shown in Table 12.4 is not included because there are better eight run fractional factorial designs which were discussed in Section 12.3. Notice that the 16 run Plackett-Burman design is also omitted (for the same reason). The 12, 20, 24, and 28 run Plackett-Burman designs can handle up to 11, 19, 23, or 27 factors respectively, although that many are not generally recommended. The factors are identified across the top of the page and the run number is identified along the left margin.

12.4.2 Using the Tables of Plackett-Burman Designs

To use a Plackett-Burman design, all you need to do is determine how many runs are necessary. Then you simply select the design with that many runs. The number of runs needed depends on two considerations: (a) how many factors are there to be studied, and (b) how much precision is wanted.

It seems reasonable that the more factors that you want to investigate, the more runs you need. Specifically, every run gives you one independent piece of information on a particular response (or, as we say in statistical jargon, it gives us one degree of freedom). If you are studying k factors, you want to know a minimum of k+1 pieces of information which are the k effects of the factors and the overall average response. This is most easily seen if you think in terms of the model you are fitting to your data, either explicitly or implicitly. It is the simple linear model:

$$Y = b_0 + b_1 X_1 + b_2 X_2 + b_3 X_3 + \ldots + b_k X_k$$

which has k+1 coefficients, and it requires one degree of freedom to estimate each coefficient. This model can be fit from the experimental data by regression, as described for fractional factorials in Section 12.3.7.

In general, you will not have an estimate of experimental error, σ, that you are comfortable using to determine the significance of each of the effects. Therefore, in addition to the coefficients in your model, you will also need to have some degrees of freedom to estimate σ from your data. From the t-table in Appendix A (Table A-5), you can see that any significance testing will suffer greatly if you have less than four degrees of freedom for your estimate of σ. This means that you need four more runs in your design, over and above the k+1. That is why it is recommended that no more than seven factors be studied using the 12 run Plackett-Burman design, no more than 15 be studied using the 20 run design, and no more than 23 be studied using the 28 run design.

In summary, the information we want from our data requires that:

$$n_F \geq k + 5 \quad \text{(for information)} \tag{12.1}$$

Before proceeding, we should also check to see that the design we select meets the desired precision, δ. Remember that δ is the smallest effect we do not want to overlook (as discussed in Section 9.2). The relationship derived in Chapter 9 between n_F and δ is:

$$n_F \geq (8\sigma/\delta)^2 \quad \text{(for precision)} \tag{12.2}$$

It should be kept in mind that n_F required for precision is negotiable, because δ is a compromise between cost (n_F) and benefits received from the information. That is, the sponsor may decide to settle for a bigger δ and pay less (smaller n_F). *But* the n_F required for information is not negotiable, and it cannot be reduced (except by reducing the number of factors to be studied).

The full procedure to determine the appropriate number of runs is to:
(a) Calculate the number of runs, n_F, required for information purposes.
(b) Calculate the number of runs, n_F, required for adequate precision.
(c) Pick the larger of the two n_F values.
(d) Round *up* to the nearest multiple of four runs.

Once the number of factors and runs has been decided upon, the appropriate table is used by assigning one factor to each column, then reading off the experiments to be run. Before actually running the experiments, it is, of course, important to randomize the run order. This can be done using any randomization technique including the use of the random number tables discussed previously.

Although the use of Equation 12.2 for determining the number of runs for precision was not mentioned in Section 12.3, it should be used in the same way when planning a fractional factorial design. However, there are only fractional factorial designs with 8, 16, or 32 runs, so Step (d) above would have to be modified, or include the possibility of running a Plackett-Burman design. For example, if $n_F = 20$ was calculated from Equation 12.2, you could either run a 20 run Plackett-Burman design or a 32 run fractional factorial to get the necessary precision (a 16 run fractional factorial would not give the desired precision). The 32 run fractional factorial design would have a clearer confounding pattern, but at a large cost in terms of the number of runs.

If there are fewer factors than columns in the design (which is usually the case), the extra columns are not needed to define the list of experiments. The extra columns will be used, however, when computing the factor effects. At that stage, the entire table from Appendix B.1 should be used as a worksheet for calculating effects. The effects calculated for unassigned columns will represent interaction effects, although it is difficult to say exactly what specific interactions they represent. (For a discussion on how to determine what combination of interactions they correspond to, see the appendix of Hunter's paper[3].)

Once the number of factors and runs are determined, the Plackett-Burman tables are used for a screening design by copying the complete computation table for the design used from the appropriate table in Appendix B.1. Use only the columns for which you have factors to determine the experimental runs. For example, if you are studying 12 factors in a 20 run design, use columns X_1–X_{12} to define the experiments. Randomize the run order then run the experiments to collect the data.

When analyzing the data, use the entire table from the appendix as the computation table or X-matrix for a regression. Yates algorithm cannot be used for calculating effects from Plackett-Burman designs. The relative importance of the factor effects can be obtained by examination of their magnitudes. The usual assumption in the analysis of screening designs is that only a few of the factor effects will be important. Therefore, these vital few can be identified by plotting the effects. Pareto Diagrams of the absolute values of effects, Normal Probability plots, or Half-Normal plots (as described in 8.10) are useful tools for identifying important effects.

Another method for identifying important effects from a screening design (either Plackett-Burman or fractional factorial) is to calculate a standard error of effects by using all the unassigned (or interaction) effects from the calculation worksheet, and then calculating t-statistics to compare to Table A-5. An equivalent way of doing this is to perform a regression analysis including only the factors used to define the experiments. By doing this, the unassigned factors will automatically be lumped together as an estimate of error. Caution must be taken when using this approach. If one of the interactions used to calculate the error term is large it may actually be important. Combining an important effect into the error will inflate s_E, or s_β, and reduce the significance of the other main effects. A rule of thumb is not to include any interaction term in the error calculation if it is one of the largest effects in absolute value. We will illustrate this method with the next example.

12.4.3 An Example

At one step in the manufacture of multilayered printed circuit boards, holes (drilled in the board to connect various layers) are copperplated by electrolysis. The manufacturer was concerned with getting a uniform copperplating of the correct thickness. Figure 12.4 shows a cross section of a drilled hole. A problem was being experienced in the plating cells — the copper plate was not uniform. It was thicker at top and bottom than in the middle. Also, the target thickness was not being met.

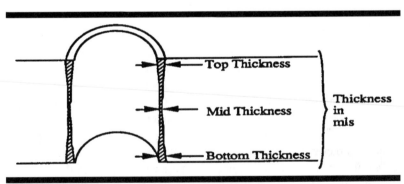

Figure 12.4 Cross Section of Copperplated Hole

The process engineers wanted to fix the problem, but were not sure which factors were most important in determining the thickness or uniformity of the copper plate. Figure 12.5 shows a cause-and-effect diagram of factors they felt might be important. The factors listed at the top of Table 12.21 are those they decided to include in the experiment.

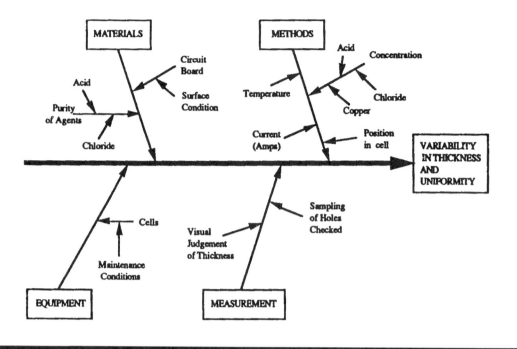

Figure 12.5 Cause and Effect Diagram of Factors Influencing Copperplating

A full 2^7 design would require 128 experiments which was impractical. The use of a Plackett-Burman design was a much more viable approach. For information, 12 runs were required: $n_F \geq k + 5 = 7 + 5 = 12$. This number of runs gave acceptable precision as well (the calculations are not shown), so a 12 run Plackett-Burman Design was selected.

The first seven columns from Table B.1-1 were copied into Table 12.21 to define the twelve test conditions. The experiments were run in the random order shown in Table 12.21 and the resulting data were collected as shown on the right of the table.

Measurements of the copperplating thickness were made at the top, middle, and bottom of five sample drill holes on a circuit board plated under each of the conditions listed in each row of the design. The average of the five measurements at each location in the hole are presented in the table as Top, Mid, and Bot. The average of these values, $\bar{y}$, is a summary statistic which indicates the average thickness of the copperplating. The variance, s^2, of the three averages is a measure of the uniformity of the plating thickness. Since the variance s^2 ranges over nearly an order of magnitude, the linear model implicit in factorial designs was thought to be inappropriate, and the $\log_e(s^2)$ was felt to be a better response to represent uniformity.

Table 12.21 -- Copper Plating Experiment

Factors	Levels low (-)	high (+)
X_1 - Copper concentration	16 gm/L	19 gm/L
X_2 - Chloride concentration	65 PPM	85 PPM
X_3 - Acid (H+) concentration	198 gm/L	225 gm/L
X_4 - Temperature	72°	78°
X_5 - Total current	180 amp-hours	192.5 amp-hours
X_6 - Position in cell	right	left
X_7 - Surface condition	smooth	slightly rough

Run order	X_1	X_2	X_3	X_4	X_5	X_6	X_7	Top	Mid	Bot	$\bar{y}$	s^2
9	+	+	-	+	+	+	-	2.74	2.13	2.68	2.52	.113
2	+	-	+	+	+	-	-	2.71	2.15	2.72	2.53	.106
4	-	+	+	+	-	-	-	2.02	1.67	2.06	1.92	.046
5	+	+	+	-	-	-	+	1.84	1.53	1.77	1.71	.026
12	+	+	-	-	-	+	-	1.79	1.49	1.71	1.66	.024
1	+	-	-	-	+	-	+	2.42	1.78	2.39	2.20	.1327
7	-	-	-	+	-	+	+	2.05	1.80	2.10	1.98	.026
8	-	-	+	-	+	+	-	2.44	1.93	2.38	2.25	.077
10	-	+	-	+	+	-	+	2.65	2.19	2.70	2.51	.079
3	+	-	+	+	-	+	+	1.98	1.61	2.06	1.88	.058
11	-	+	+	-	+	+	+	2.40	1.70	2.30	2.13	.143
6	-	-	-	-	-	-	-	1.80	1.43	1.75	1.66	.040

Table 12.22 and Table 12.23 are summaries of the calculations performed to get the factor effects. Notice that in these two tables all the columns from Table B.1-1 were copied, and the effects were calculated for each column. The effects calculated for the unassigned factor columns X_8, X_9, X_{10}, and X_{11} represent interactions, although we cannot associate any one column to a specific interactions as we could the unassigned columns in the fractional factorial tables. Similar to the way the standard error of an effect was calculated using block interactions in Section 9.5.1, a standard error of the effects, s_E, was calculated using the unassigned effects X_8, X_9, X_{10}, and X_{11} in Table 12.22. In Table 12.23, only X_8, X_{10}, and X_{11} were used to calculate the error, since the X_9 effect was relatively large. In order to assess statistical significance, t-values were calculated by dividing each effect by the calculated s_E. Also, Figures 12.6 and 12.7 are Normal plots of the effects. These diagrams graphically display the relative importance of the effects, and are used to help us screen out the vital few effects from the trivial or negligible ones. We can easily see that for the average thickness response, the Factors X_4 (temperature) and X_5 (total current) have the most dramatic effects, with thicker plating resulting at the higher temperature and higher total current.

Table 12.22 Calculation of Effects for $\bar{y}$

	Mean	X1	X2	X3	X4	X5	X6	X7	X8	X9	X10	X11	$\bar{y}$
	+	+	+	−	+	+	+	−	−	−	+	−	2.52
	+	+	−	+	+	+	−	−	−	+	−	+	2.53
	+	−	+	+	+	−	−	−	+	−	+	+	1.92
	+	+	+	+	−	−	−	+	−	+	+	−	1.71
	+	+	+	−	−	−	+	−	+	+	−	+	1.66
	+	+	−	−	−	+	−	+	+	−	+	+	2.20
	+	−	−	−	+	−	+	+	−	+	+	+	1.98
	+	−	−	+	−	+	+	−	+	+	+	−	2.25
	+	−	+	−	+	+	−	+	+	+	−	−	2.51
	+	+	−	+	+	−	+	+	+	−	−	−	1.88
	+	−	+	+	−	+	+	+	−	−	−	+	2.13
	+	−	−	−	−	−	−	−	−	−	−	−	1.66

	Mean	X1	X2	X3	X4	X5	X6	X7	X8	X9	X10	X11
Σ(+)−Σ(−)	24.95	0.05	−0.05	−0.11	1.73	3.33	−0.11	−0.13	−0.11	0.33	0.21	−0.11
Effects	2.08	0.0083	−0.0083	−0.018	0.288	0.555	−0.018	−0.021	−0.018	0.055	0.035	−0.018
t-values		0.24	−0.24	−0.51	8.32	15.86	−0.51	−0.60				

$$s_E = \sqrt{\frac{(-0.018)^2 + (0.055)^2 + (0.035)^2 + (-0.018)^2}{4}} = 0.035$$

Table 12.23 Calculation of Effects for $\log_e(s^2)$

	Mean	X1	X2	X3	X4	X5	X6	X7	X8	X9	X10	X11	$\log_e(s^2)$
	+	+	+	−	+	+	+	−	−	−	+	−	−2.180
	+	+	−	+	+	+	−	−	−	+	−	+	−2.244
	+	−	+	+	+	−	−	−	+	−	+	+	−3.079
	+	+	+	+	−	−	−	+	−	+	+	−	−3.650
	+	+	+	−	−	−	+	−	+	+	−	+	−3.729
	+	+	−	−	−	+	−	+	+	−	+	+	−2.100
	+	−	−	−	+	−	+	+	−	+	+	+	−3.650
	+	−	−	+	−	+	+	−	+	+	+	−	−2.564
	+	−	+	−	+	+	−	+	+	+	−	−	−2.538
	+	+	−	+	+	−	+	+	+	−	−	−	−2.847
	+	−	+	+	−	+	+	+	−	−	−	+	−1.945
	+	−	−	−	−	−	−	−	−	−	−	−	−3.219

	Mean	X1	X2	X3	X4	X5	X6	X7	X8	X9	X10	X11
Σ(+)−Σ(−)	−33.745	0.245	−0.497	1.087	0.669	6.603	−0.084	0.285	0.031	−3.005	−0.701	0.251
Effects	−2.812	0.041	−0.083	0.181	0.111	1.100	−0.041	0.048	0.005	−0.501	−0.117	0.042
t-values		0.57	−1.15	2.51	1.54	15.31	−0.19	0.67		−6.97		

$$s_E = \sqrt{\frac{(0.005)^2 + (-0.117)^2 + (0.042)^2}{3}} = 0.0718$$

Normal Probability Plot of the Effects

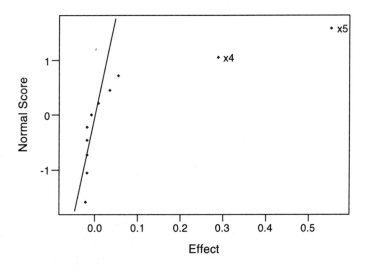

Figure 12.6--Normal Plot of Effects on Average Thickness

Normal Probability Plot of the Effects

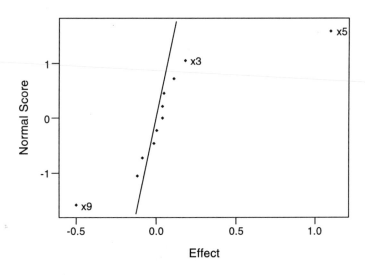

Figure 12.7--Normal Plot of Effects on $\log_e(s^2)$

429

For the uniformity response, $\log_e$ (variance), it can be seen that the most important factor is X_5 (total current), and the lower total current tends to produce more uniform (i.e,. smaller variance) copperplating. However, the unassigned effect, X_9, also appears to be relatively large as affirmed by the significantly large t-value in Table 12.23, and the point on the left in the Normal plot of effects — Figure 12.7. The X_9 effect represents an interaction, but the difference in averages represented by X_9 is actually a confounding of all 21 interactions. More specifically:

$$X_9 = (1/3)X_1X_2 + (1/3)X_1X_3 + (1/3)X_1X_4 + (1/3)X_1X_5 + (1/3)X_1X_6 - (1/3)X_1X_7$$
$$- (1/3)X_2X_3 - (1/3)X_2X_4 + (1/3)X_2X_5 + (1/3)X_2X_6 + (1/3)X_2X_7 + (1/3)X_3X_4$$
$$+ (1/3)X_3X_5 + (1/3)X_3X_6 - (1/3)X_3X_7 + (1/3)X_4X_5 - (1/3)X_4X_6 + (1/3)X_4X_7$$
$$- (1/3)X_5X_6 + (1/3)X_5X_7 - (1/3)X_6X_7$$

This was determined by using the Alias matrix (see the appendix of Hunter's[3] paper).

With X_9 representing so many possible interaction effects, interpretation would be very difficult without further experiments to reduce the confusion. Use of the fractional factorial designs discussed in the last section will significantly reduce the confounding of unassigned effects and also provide the experimenter with an easy way of determining what is confounded with what.

An alternate way of obtaining the t-statistics in Tables 12.22 and 12.23 would be to use regression analysis. For example, Table 12.24 shows the results of a regression analysis performed by MINITAB Version 11, using the seven columns of Table 12.21 as factors and the mean, $\bar{y}$, as the response. The t-statistics in this table can be seen to be the same as those shown in Table 12.22.

Table 12.24 Regression Analysis of Plackett-Burman Design

```
The regression equation is
y bar = 2.08 + 0.0042 X1 - 0.0042 X2 - 0.0092 X3 + 0.144 X4 + 0.278 X5
             - 0.0092 X6 - 0.0108 X7

Predictor        Coef        StDev            T          P
Constant      2.07917      0.01754       118.54      0.000
X1            0.00417      0.01754         0.24      0.824
X2           -0.00417      0.01754        -0.24      0.824
X3           -0.00917      0.01754        -0.52      0.629
X4            0.14417      0.01754         8.22      0.001
X5            0.27750      0.01754        15.82      0.000
X6           -0.00917      0.01754        -0.52      0.629
X7           -0.01083      0.01754        -0.62      0.570

S = 0.06076      R-Sq = 98.8%      R-Sq(adj) = 96.6%

Analysis of Variance

Source        DF          SS           MS          F          P
Regression     7     1.17732      0.16819      45.56      0.001
Error          4     0.01477      0.00369
Total         11     1.19209
```

From the results of this experiment, it could probably be assumed that no interactions are important in affecting the average thickness, and the simple linear model:

$$\bar{y} = 2.08 + .144X_4 + .278X_5$$

derived from the significant effects could be used to identify intermediate conditions for temperature and total current that would result in an average thickness close to the target value. However, its more difficult to determine the conditions that would result in the most consistent copperplating thickness (i.e., smallest variance), since there are potentially significant interactions, represented by X_9, that affect the response. There are a couple of options that could be followed as the next logical step.

Number one, if time for further experimentation were available, a full factorial in the most important factors, X_3, X_4, and X_5, could be run to try and identify the important interactions affecting $\log_e$(variance). As a second alternative, the interactions could be assumed to be less important and the low level of Factor 5, (total current = 180 amp-hours) could be chosen to minimize the log variance, and the level of X_4 = temperature could be adjusted to yield the desired average thickness. Then, the copperplating cell could be run at these conditions during production to see if the resulting average thickness and uniformity were acceptable.

12.5 Other Applications of Fractional Factorials

Although Plackett-Burman designs are used strictly for screening, in some situations fractional factorial designs may be used for much more than just screening. If we are willing to assume all higher order interactions (perhaps three-way and above) are negligible, it is possible to estimate quite a few interactions from certain fractional replicate designs. This includes blocked factorials discussed in Chapter 9. Also, fractional factorial designs can be used to create screening designs for factors with more than two levels, as discussed in Chapter 11. In this section we will illustrate the use of fractional factorials in situations where it is desirable to estimate some interactions. In the Section 12.6 we will show how fractional factorials can be used to create screening designs with multiple level factors.

12.5.1 Fractional Factorial Designs for Estimating Some Interactions

With regard to interactions, it is possible to construct a half replicate of a 2^6 design so that all two-way interactions are aliased with only three-way and higher interactions (see Table B.1-13 in Appendix B). This is a resolution V design as described in Section 12.3.8. Therefore, if we assume all ≥three-way interactions were negligible (a very reasonable assumption), it is possible to use the half replicate to estimate all main effects and two-way interactions — something usually reserved for full factorial designs. The half fraction of a 2^6 design is particularly efficient (see Table B.1-9).

In general, resolution V designs have the shortest word in the defining contrast equal to five and allow estimation of all main effects and two-factor interactions. If it is desired to create a design to estimate only a few of the two-factor interactions, a resolution V design may be larger than necessary. More economical designs can be created for estimating main effects and a few specific interactions using the algorithm of Franklin and Bailey[4], or the interaction graphs of Kackar and Tsui[5]. Turiel[6] has published a computer program which implements Franklin and Bailey's algorithm. In simple situations, designs can be derived for estimating main effects and a few specific interactions using the CONFOUNDINGS in the Appendix B.1 tables. The next example illustrates this.

Consider a situation where four factors are under study. To estimate the main effects and all interactions would require $2^4 = 16$ experiments. However, if some of the variables in the experiment are believed to be less important and less inclined to have interactions, a very economical fractional factorial design can be created using the tables in Appendix B.1. If Factors X_1 and X_2 were considered potentially important, and likely to interact with each other, while X_3 and X_4 are thought to be less important and less likely to interact, then the 8 run half replicate of a 2^4 presented in Table B.1-6 could be used with no penalty (except a loss in precision). From this design, all main effects, X_1–X_4, and $E_1 = 12$ (the interaction between X_1 and X_2) could be estimated if the interaction between X_3 and X_4 were assumed negligible. One practical situation where it is useful to estimate specific interactions with less experiments than required by a full factorial is in robust product design studies that will be described in Chapter 18.

12.5.2 Designs for Blocked Factorials in Fractional Arrangements

Another situation where the ideas of fractional factorials and confounding are very useful is in blocked designs. Recall, the blocked factorial designs presented in Section 9.5 required a complete replication of a full factorial within each block. This could result in an extensive number of experiments. If we can assume some higher order interactions are negligible, we can create much more economical blocked factorials by running only a fraction of the factorial in each block.

Table 12.25 shows examples of how to set up various blocked designs by running fractions of a 2^k factorial as blocks. This table refers to the tables in Appendix B.1 and B.2. In these designs some interaction terms can be estimated, and others are confounded with block effects or with the block interactions used to estimate error. When an interaction is confounded with blocks, we cannot separate its effect from the differences in blocks. If we are willing to assume the confounded interaction is negligible, there is a reduction in the number of experiments, but no loss of information. When an interaction is confounded with a block interaction used to estimate error, we should examine its effect before automatically pooling it to estimate an error term (similar to the way unassigned effects in Plackett-Burman designs were examined before creating the error term in Section 12.4.4).

The designs in Table 12.25 should be used when the desired number of factors is larger, or the desired number of runs, or blocks size, is less than that shown in the table of blocked factorials in Chapter 9.

Table 12.25 Blocked Factorial Designs

Factors	Appendix Table	Number of Runs	Number of Blocks	Block Size	Estimable Interactions	Interactions Defining Blocks	Interactions that Estimate Block Effects	Block Interactions Used for Error Terms
1, 2, 3	B.2-3	8	2	4	none	123	123	12, 13, 23
1, 2, 3	B.2-3	8	4	2	none	12, 13	12, 13, 23	123
1, 2, 3, 4	B.2-4	16	2	8	12, 13, 14, 23, 24, 34	1234	1234	123, 134, 124, 234
1, 2, 3, 4	B.2-4	16	4	4	none	124, 134	124, 134, 23	12, 13, 14, 24, 34, 123, 234, 1234
1, 2, 3, 4	B.2-4	16	8	2	none	12, 23, 34	12, 13, 14, 23, 24, 34, 1234	123, 124, 134, 234
1, 2, 3, 4, 5	B.2-5	32	2	16	12, 13, 14, 15, 23, 24, 25, 34, 35, 45	12345	12345	1234, 1235, 1245, 1345, 2345, 123, 124, 125, 134, 135, 145, 234, 235, 245, 345
1, 2, 3, 4, 5	B.2-5	32	4	8	12, 13, 14, 15, 23, 24, 25, 34, 35, 45	123, 345	123, 345, 1245	124, 125, 145, 245, 235, 135, 134, 234, 1234, 1235, 1345, 2345, 12345
1, 2, 3, 4, 5	B.2-5	32	8	4	none	125, 235, 345	125, 235, 345, 13, 24, 145, 1234	12, 14, 15, 23, 25, 34, 35, 45, 123, 124, 134, 135, 234, 245, 1235, 1245, 1345, 2345, 12345
1, 2, 3, 4, 5	B.2-5	32	16	2	none	12, 13, 34, 45	12, 13, 34, 45, 14, 15, 23, 24, 25, 35, 1234, 1235, 1245, 1345, 2345	123, 124, 125, 134, 135, 145, 235, 234, 245, 345, 12345
1, 2, 3, 4, 5, 6	B.1-13	32	2	16	12, 13, 14, 15, 16, 23, 24, 25, 26, 34, 35, 36, 45, 46, 56	E_1	E_1	$E_2, E_3, E_4, E_5, E_6, E_7, E_8, E_9, E_{10}$

433

Consider an example of a 2^4 design in two blocks of eight taken from Johnson and Leone[7]. Four factors listed below were being investigated for their effect on the velocity of a projectile in a ballistics test:

X_1: propellent charge in lbs.
X_2: projectile weight in lbs
X_3: propellent web
X_4: weapon

Only eight tests could be performed in a day. Therefore it required two days to perform all 16 tests. Since extraneous factors may change between days, it was desirable to treat days as blocks to prevent possible bias. The blocked designs of Chapter 9 will not work in this situation, because they require a full 2^4 design repeated in each block. To create a blocked design with 8 runs (a half fraction of the 2^4) in each block, refer to the third row of Table 12.25. This defines a 16 run design with four factors in blocks of size eight. The table shows us that this design is created using Table B.2-4, treating the 1234 interaction column as blocks (days). Table 12.26 shows the resulting design. All combinations of factor levels for Day 1 were run first followed by those for Day 2. Normally this would be done in a random order within each day.

In the right columns of Table 12.26, we see the effects calculated by Yates algorithm. The standard error of the effects is $s_E = \sqrt{[(-7.125)^2+(-0.125)^2+0.875^2+(-3.875)^2]/4} = 4.08$, which was calculated by pooling the block interactions that estimate error terms listed in the last column of Table 12.25. Before calculating this standard error, one should examine the effects listed as block interactions used for error, to make sure none are unusually large. One way to do this is to make a Normal or Half-Normal plot of the all the effects except those that estimate block effects listed in the second to last column of Table 12.25. This was done in Figure 12.8.

Table 12.26 2^4 Design in Two Blocks from Johnson and Leone

Run	X_1	X_2	X_3	X_4	$X_1X_2X_3X_4$ Block	Velocity (coded)	Effects	t-statistic
1	−	−	−	−	Day 2	97		
2	+	−	−	−	Day 1	151	1= 68.375	16.76*
3	−	+	−	−	Day 1	68	2=−24.875	−6.10*
4	+	+	−	−	Day 2	150	12= 4.375	1.07
5	−	−	+	−	Day 1	39	3=−62.375	−15.29*
6	+	−	+	−	Day 2	100	13= −5.125	1.26
7	−	+	+	−	Day 2	15	23=−10.875	2.66
8	+	+	+	−	Day 1	66	123= −7.125	error
9	−	−	−	+	Day 1	75	4=−13.875	−3.40*
10	+	−	−	+	Day 2	145	14= 6.375	1.53
11	−	+	−	+	Day 2	53	24= −2.875	0.70
12	+	+	−	+	Day 1	141	124= −0.125	error
13	−	−	+	+	Day 2	26	34= −0.875	0.21
14	+	−	+	+	Day 1	97	134= 0.875	error
15	−	+	+	+	Day 1	−16	234= −3.875	error
16	+	+	+	+	Day 2	54	1234= 2.375	Blocks

434

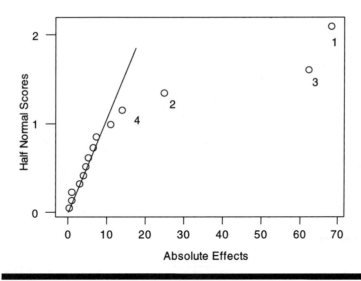

Figure 12.8 Half-Normal Plot of Effects from Table 12.18

In this figure none of the block interactions used to estimate error appear large. Only the main effects stick out from a straight line drawn through the line of points that rise from the origin. Therefore, the standard error of effects calculated above was used to calculate the t-statistics in the last column of Table 12.26 with the formula $t = Effect/s_E$. The degrees of freedom for these t-statistics is four, since four block interaction terms were pooled to calculate the standard error. The critical $t^*_{4,(95\%)} = 2.776$ from Table A-5. The significant effects (whose t-statistics exceed the critical value) are asterisked in Table 12.26. It can be seen that all of the main effects appear to be significant. These are the same factors that appear to be large in Figure 12.8.

All six two-factor interactions were estimable in this experiment, but since none appear to be significant the interpretation is simple. To get the maximum projectile speed we should set the factor that has a positive effect (X_1) at its high level, and the factors that have negative effects (X_2, X_3, and X_4) at their low levels.

12.6 Screening Designs with Multiple Level Factors

In Chapter 11 we discussed factorial experiments with multiple (>2) levels for one or more factors in the design. These designs are appropriate when using qualitative factors with more than two levels like the factor, cover type, in the motor noise experiment described in Section 11.4. It is not difficult to create screening designs for multiple level factorials using the fractional factorial tables in Appendix B and simple rules for combining and collapsing columns.

12.6.1 Combination of Factors Method (Pseudo Factors)

As an example, consider an experiment with one four-level factor and four two-level factors. The number of experiments required for a full factorial would be $4 \times 2^4 = 64$. However, an 8 run screening design could easily be created using the two-level fractional factorial in Table B.1-5. To do this, first let the levels of factors X_1 and X_2 define the four-level factor as shown below:

Level of X_1	Level of X_2		Levels of 4-level Factor
-	-	➡	1
+	-	➡	2
-	+	➡	3
+	+	➡	4

Next, assign the four two-level factors to columns X_3, X_5, X_6, and X_7. Column X_4 is not used because it is confounded with the interaction between X_1 and X_2, as can be seen in the list of confoundings below Table B.1-5. The interaction between X_1 and X_2 is part of the four-level factor and should not be confounded with other factors. The design would appear as:

Run	A (X_1, X_2)	B (X_3)	C (X_5)	D (X_6)	E (X_7)
1	1	1	2	2	1
2	2	1	1	2	2
3	3	1	2	1	2
4	4	1	1	1	1
5	1	2	1	1	2
6	2	2	2	1	1
7	3	2	1	2	1
8	4	2	2	2	2

The analysis of a design like this would be identical to the analysis of the half fraction in Section 12.3.3. Calculate all the effects from Table B.1-5 and identify the significant factor effects using a Pareto diagram or Half-Normal plot. The effects for the two-level factors are interpreted as usual. The X_1 effect is a comparison between levels 1 and 3 with 2 and 4 of the four-level factor.

Likewise the X_2 effect is a comparison of Levels 1 and 2 with 3 and 4, while the $X_1 \times X_2$ interaction (or X_4 effect) is a comparison of Levels 1 and 4 with 2 and 3.

The previous example illustrates the method of combining two-level factors to create a factor with more levels. A four-level factor can always be created by combining any two, two-level columns in a two-level fractional factorial design, as long as the interaction between the two columns is not assigned to any other two-level factor. The fractional factorial tables listed in Appendix B.1 can be used when combining two-level factors, because the CONFOUNDINGS show us which column is confounded with the interaction of the columns we are combining, and therefore should not be assigned to another factor. Plackett-Burman designs cannot be used to create combined factors, because it is difficult to tell which columns are confounded with two-factor interactions.

The two-level factors that are combined are referred to as the *pseudo factors* and actually represent comparisons between different combinations of levels of the four-level factor as described previously. An eight-level factor can be created by assigning levels of the eight-level factor to the combinations of levels of three two-level pseudo factors. In this case none of the three two-way interactions between pseudo factors nor the three-way interaction among pseudo factors can be assigned to any other two-level factor or used to define any other four or eight-level factors.

In general a 2^p level factor can be created by assigning its levels to all combinations of p two-level pseudo factors. Interactions of all levels among the pseudo factors must not be used in defining any other factors.

12.6.2 Collapsing Levels (Dummy Levels)

When the number of levels of a multiple level factor is not a power of 2, the pseudo factor method cannot be used to generate the levels. Another simple method that will work is the method of collapsing levels or dummy levels. To use this method, simply assign the unneeded levels of a 2^p-level factor to previous levels. For example, suppose we want to create a screening design for a 3×2^4 factorial experiment. Start by creating the 4×2^4 experiment described above using columns X_1 and X_2 of Table B.1-5 to define the four-level factor and columns X_3, X_5, X_6 and X_7 to represent the two-level factors. Next collapse the levels of the four-level factor to three as shown below:

four-level factor		three-level factor
1	➡	1
2	➡	2
3	➡	3
4	➡	3

This would result in the following eight run 3×2^4 design:

Run	A (X_1, X_2)	B (X_3)	C (X_5)	D (X_6)	E (X_7)
1	1	1	2	2	1
2	2	1	1	2	2
3	3	1	2	1	2
4	3	1	1	1	1
5	1	2	1	1	2
6	2	2	2	1	1
7	3	2	1	2	1
8	3	2	2	2	2

Since the number of repeats of each level is no longer equal (i.e., two 1's, two 2's, and four 3's) after collapsing levels, the standard method of analysis will not work for mixed-level screening designs created in this way. The way these designs are typically analyzed is to pool interactions to form an error term (similar to the example with a Plackett-Burman design in Section 12.4.4) and test the main effects using the GLM ANOVA that was discussed in Section 11.8.

12.6.3 The L_{18} Orthogonal Array

Another tabled design which is very handy for creating screening designs with some two-level factors and other three-level factors is the L_{18} orthogonal array shown in Table B.1-15. This design table lists one two-level factor and seven three-level factors. Any of the three-level factors can be collapsed by the dummy level technique to create two-level factors. For example, to create an 18 run screening design for a $2^3 \times 3^5$, simply collapse the levels of Columns 2 and 3 to two-levels resulting in the design in Table 12.27.

In Table B.1-15, the interaction between Columns 1 and 2 is not confounded with any of the other columns in the design. The interactions between all other columns are partially confounded with main effects similar to the Plackett-Burman designs.

Table 12.27 Screening Design for $2^3 \times 3^5$ Created by Collapsing Columns 2 and 3 of Table B.1-15

				Column				
Run	1	2	3	4	5	6	7	8
1	1	1	1	1	1	1	1	1
2	1	1	2	2	2	2	2	2
3	1	1	2	3	3	3	3	3
4	1	2	1	1	2	2	3	3
5	1	2	2	2	3	3	1	1
6	1	2	2	3	1	1	2	2
7	1	2	1	2	1	3	2	3
8	1	2	2	3	2	1	3	1
9	1	2	2	1	3	2	1	2
10	2	1	1	3	3	2	2	1
11	2	1	2	1	1	3	3	2
12	2	1	2	2	2	1	1	3
13	2	2	1	2	3	1	3	2
14	2	2	2	3	1	2	1	3
15	2	2	2	1	2	3	2	1
16	2	2	1	3	2	3	1	2
17	2	2	2	1	3	1	2	3
18	2	2	2	2	1	2	3	1

Since the interaction between Columns 1 and 2 is not confounded with any other column, these two columns can be combined using the pseudo factor method to create one six-level factor. This six-level factor can in turn be collapsed using dummy levels to create a five-level or four-level factor. Thus 18 run fractions of $6 \times 2^p \times 3^q$ or $5 \times 2^p \times 3^q$ or $4 \times 2^p \times 3^q$ type designs can be easily created using the L_{18} table.

Mixed-level designs created with the L_{18} table will not be balanced, in general, and like the designs created by the collapsing levels method, they must be analyzed using the GLM ANOVA method described in Section 11.8 by pooling interactions as an error term. One exception is when all columns are used exactly as they appear in Table B.1-15 for an 18 run fraction of a 2×3^7 design. In this case, orthogonal contrasts can be formed for each main effect and the interaction between Columns 1 and 2, as shown in Section 11.7, and a Half-Normal plot or Pareto diagram of the standardized contrasts can be used to identify important effects.

12.7 Sequential Experimentation

When planning experiments with many factors, it is likely that only a few factors will have significant effects. This principle is called *effect sparsity* and it can help us in planning experiments at the screening stage. It is generally wasteful to plan large, full factorial designs or resolution V fractions with many factors. It is better to start small. Begin with an economical screening design and add to it later if needed. With this in mind, we can plan a conditional sequence of experimental designs that will allow us to estimate the important main effects and two-factor interactions with the minimal number of experiments.

Usually no more than 25% of the budget for a research program should be expended on the initial screening experiment. In an initial screening experiment, things may not work out as planned due to poor choice of factor levels or other procedural problems. Even after successful completion of a screening experiment, follow up experiments may be needed to identify significant interaction effects and allow for fitting a prediction equation. For this reason, plenty of time should be left to repeat botched experiments, and conduct follow-up trials with important variables. By planning in advance, we can do this.

Since the confounding patterns for fractional factorials can be used for identifying the interactions confounded with any large, unassigned effects, these designs are ideal for defining a conditional sequence of experiments. Figure 12.9 is a flow diagram that illustrates a simple sequence that starts with the resolution IV, half fraction (2^{4-1}) design shown in Table B.1-6 . The initial experiment in this sequence only involves four factors. If , after running this design, the data analysis shows that no unassigned effects (that represent strings of confounded two-factor interactions) are significant, experimentation stops and only the main effects are considered when interpreting the effects and defining a prediction model. If , on the other hand, one or more unassigned or interaction effects appear to be significant, a second step consisting of running the remaining experiments needed for a full factorial should be completed.

The sequence shown in Figure 12.9 can be illustrated by returning to the example used in Section 12.5.2. This was a blocked design, but the runs made on Day 2 are exactly the resolution IV half fraction design for four factors listed in Table B.1-6. So if these runs had been completed on Day 1 instead of Day 2, Step 1 from Figure 12.9 would have been followed. The next step would be to check to see if any of the unassigned effects were large. Table 12.28, shows the analysis of the Day 2 data.

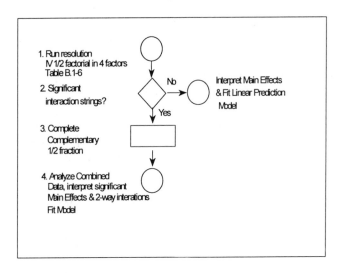

Table 12.28 Analysis of Half-Fraction Consisting of Day 2 Data from Table 12.26

Term	Effect	t-value	P-value
Constant	80.00		
X_1	64.50	11.37	0.001
X_2	-24.00	-4.23	0.024
X_3	-62.50 -11.02		0.002
X_4	-21.00	-3.70	0.034
E_1=12+34	3.50	1.75	
E_2=13+24	-8.00	-4.00	
E_3=14+23	-4.50	-2.25	
s_E = 5.673			

The standard error of the effects, s_E , was calculated by pooling the three unassigned effects (none of which appear large) as was done in Section 12.4.4. All four main effects appear to be significant, while none of the interactions strings are significant. Therefore, following the flow diagram in Figure 12.9, no further experimentation would be necessary, and the main effects could be interpreted alone. Notice that the main effects calculated from the half fraction in Table 12.28, are very similar to the effects calculated from the full factorial in Table 12.26. The conclusions and interpretation after eight experiments would be the same as that given on page 435, and we see that by assuming effect sparsity and using a sequential approach the experimental effort could have been cut in half.

Admittedly, the plan for a sequence of experiments that follows the flow diagram in Figure 12.9 is a simple case, because it starts with only four factors. However, Figure 12.10 is a similar diagram that is much more general. It can be followed to plan a sequence of experiments starting with any number of factors. This flow diagram has two entry points. The first entry point is Step 1, where the initial design is a resolution III fractional factorial like those shown in Tables B.1-5, B.1-7, and B.1-10. The second entry point is Step 3, where the initial experiment is a resolution IV fractional factorial like those shown in Tables B.1-8 or B.1-11. The decision on what entry point to use depends upon the number of factors understudy and the number of runs, n_F, which is calculated on the basis of information and precision described in Section 12.4.2. We will first discuss the use of the flow diagram as if beginning at Step 3. Later we will show an example of following the flow diagram from Step 1.

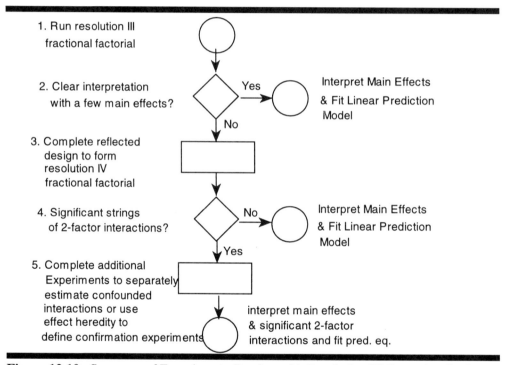

Figure 12.10 Sequence of Experiments Starting with Resolution III Screening Design

The chemical process experiment presented in Section 12.3.6 was a ⅛ fractional factorial of resolution IV (from Table B.1-8). Planning a sequence of experiments by starting with a resolution IV experiment like this would follow the flow diagram in Figure 12.10 by entering at step 3. The next step would be to analyze the data and determine if any unassigned effects (that represent strings of two-factor interactions) are significant. In the example chemical process experiment, there were

442

no apparent interactions, and therefore the next step according to Figure 12.10 would be to stop experimentation and interpret the main effects and develop a prediction equation. This is exactly what was done on page 416. Because there were no significant interactions, the sequential plan allowed essentially the same information to be obtained with a 16 run $\frac{1}{8}$ fraction as would have been obtained with a 128 run 2^7 full factorial.

If some of the interaction strings appear to be significant in a resolution IV design, the next step on the flow diagram would be number 5. At Step 5, additional experiments are completed to allow separate estimation of each of the interactions confounded with any significant unassigned factor, or if the experimenter feels that he can guess which interaction (of the confounded string) is the important one, a set of confirmation experiments should be run. An example of how to choose additional experiments to allow separate estimation of confounded interactions was shown in Table 12.18 and discussed on page 418. The second alternative at Step 5, guessing, may seem to be contradictory to the scientific principles we are trying to establish with experimentation, but experience has shown that certain patterns usually emerge in experimental data.

This experience can be summarized by a paradigm called *effect heredity*. By effect heredity, we mean first that if several interactions are confounded with a significant unassigned effect, one of the lowest order interactions will normally be the important one. The order of interactions is like the number of planets lining up; two in a row is not unusual but three or four is rare. Secondly by effect heredity we mean that if one or more interactions of the same order are confounded with a significant unassigned effect, then the interaction that involves the largest and most significant main effects is usually the important one, because interactions usually don't exist between insignificant factors.

To illustrate the use of the effect heredity paradigm to choose a set of confirmatory experiments, let's return to the example discussed on page 418. In that example, a resolution IV 2^{7-3} experiment was completed finding that the significant main effects (in order of magnitude) were 4, 1, 6, and 7. Then it was supposed that the unassigned effect E_1 was significant, and the procedure for choosing additional experiments to de-confound the interactions confounded with E_1 was explained. Instead of separately estimating the interactions confounded with E_1, we could use effect heredity paradigm to guess which of the confounded interactions was the important one.

When using only the first four columns of Table B.1-8 to define the factors, the generators for the design in the example were $5 = 123$, $6 = 234$, and $7 = 134$. This results in the defining relation $I = 1235 = 2346 = 1347 = 1456 = 2457 = 1267 = 2457$. Multiplying on both sides of the defining relation by 12, we can find the interactions confounded with E_1. That is $E_1 = 12 + 35 + 67 + 1346 + 2347 + 2456 + 1457$. (Note: Only the two-factor interactions were listed in the CONFOUNDINGS below Table B.1-8.) Looking at this string of interactions that are confounded with E_1, through the eyes of effect heredity, we would first dismiss the four-factor interactions because they are unlikely to be important. Second we would dismiss the 35-interaction, since neither main effect 3 nor 5 were significant. We are left with the conclusion that either the 12 or 67 interactions are responsible. Since main effect 1 was much larger than either 6 or 7, we might conclude that 12 was the culprit.

Afer guessing which interaction is important in this manner, we would next write the prediction equation:

$$y = \hat{b}_0 + \hat{b}_1 X_1 + \hat{b}_4 X_4 + \hat{b}_6 X_6 + \hat{b}_7 X_7 + \hat{b}_{12} X_1 X_2$$

where $\hat{b}_{12} = \hat{E}_1/2$ is half the estimate of the unassigned effect. With this prediction equation, levels of the process variables (X_1, X_2, X_4, X_6, and X_7) that would yield the lowest predicted impurities could be defined. Next, a set of confirmatory experiments would be conducted at these levels to see if satisfactory results are obtained. If the answer is yes, no further experiments would be needed and the problem would be solved. Otherwise, the experimenter would have to return to Step 5 in the flow diagram.

Because the effect heredity paradigm has been true in the majority of cases, its use often reduces the number follow-up experiments needed. Recall the simple example of a half fraction presented in Section 12.3.3. There again the effect heredity paradigm was followed in assuming the main effects were responsible rather than the two-factor interactions. The confirmation experiments showed the assumption was correct, and eliminated the need for further experiments.

Now that we have discussed the sequence of experiments that can be planned by starting at step 3, in Figure 12.10, let's continue by discussing how a sequence of experiments can be planned starting at step 1. At step 1, we start with a resolution III fractional factorial (i.e., Table B.1-5, B.1-7, or B.1-10). Resolution III designs are the most economical possible. After running a resolution III design, if there appears to be only one or two main effects significant, we move to the right at step 2. There, it is usually safe to assume that no interactions are important. So experimentation stops, an interpretation is made of the main effects, and a simple linear prediction equation is written. When more than two main effects appear significant in a resolution III design, the assumption that interactions are negligible may not be a safe bet. To see why, let's consider a specific case.

In resolution III designs, each main effect that we can estimate is confounded with a string of two-factor interactions. For example, looking at the confoundings from Table B.1-5, we see:

$$X_1 = 1 + 24 + 35 + 67$$
$$X_2 = 2 + 14 + 36 + 57$$
$$X_3 = 3 + 15 + 26 + 47$$
$$X_4 = 4 + 12 + 37 + 56$$
$$X_5 = 5 + 13 + 27 + 46$$
$$X_6 = 6 + 17 + 23 + 45$$
$$X_7 = 7 + 16 + 25 + 34$$

If Factors 1 and 2 had the largest effects but Factor 4 also had what seemed to be a significant effect, then by the effect heredity paradigm, it would be hard to say whether the three main effects were important or whether Factors 1 and 2 were important and the fourth factor looked significant because it was confounded with the interaction of Factors 1 and 2. The only way we can answer a question like this is to complete additional experiments.

If a resolution III fractional factorial is augmented by its *mirror image* or *reflected design*, Box, Hunter, and Hunter[8] have shown that the main effects can be de-confounded from the strings of two-factor interactions if they are confounded within the original design. A mirror image or reflected design is a design with all the signs reversed from the original design. In other words, replace each - with a + in the original design and each + with a -. When we combine an 8 run resolution III fractional factorial with its mirror image, we will have a 16 run design. Likewise when we combine a 16 run or 32 run resolution III design with their mirror images we get a 32 or 64 run design respectively. A combined design with double the number of runs in the original resolution III design will be a resolution IV design, that allows estimation of: 1) the main effects (clear of two-factor interactions), and 2) the strings of two-factor interactions that were previously confounded with main effects. There will also be one extra degree of freedom that can be used to estimate 3) the block effect that represents the average difference between the two sets of experiments (original resolution III design and mirror image design).

After the reflected design is completed and combined with the data from the original resolution III design, we are again at step 3 in Figure 12.10. To illustrate the whole sequence in Figure 12.10 starting at step 1, we will present an excellent example of sequential experiments published by Box and Hunter[9]

12.8 An Application of Sequential Experimentation in a Process Start-up

In the start up of a new manufacturing unit, considerable difficulty was experienced at the filtration stage. Similar units operated satisfactorily at other sites, but this particular unit, although apparently similar in most major respects to the other units, gave a crude product which required much longer filtration times. Meetings were held to discuss possible explanations and to consider ways of curing the trouble. The following variables were proposed as possibly being responsible:

1. The Water Supply. The new plant used piped water from the local municipal reservoir. An alternate, but somewhat limited supply of water was available from a local well. It was proposed that the effect of changing to the well water should be tried since it was argued that the well water corresponded more closely to the water used at other sites.

2. Raw Material. The raw material used was manufactured on the site and it was suggested that this might be in some way deficient. It was proposed that raw material which had been satisfactorily used in manufacturing the product at another site should be shipped in and tested locally.

3. Temperature of Filtration. This was not thought to be a critical factor over the range involved, and no special attempt to control this temperature had been made. However, the physical arrangement of the new process was such that filtration was accomplished at a somewhat lower temperature than had been experienced at other plants. By temporarily covering pipes and equipment, provision could be made to raise the temperature to the level experienced elsewhere.

4. Recycle. The only major difference between production facilities at the other plants and the present one lay in the introduction of a recycle stage which slightly increased conversion of the reagents prior to precipitation and filtration. Arguments were advanced which accounted for the longer filtration time in terms of this recycle stage. Arrangements could be made to temporarily eliminate the recycle stage.

5. Rate of Addition of Caustic Soda. Immediately prior to filtration, a quantity of caustic soda liquor was added resulting in precipitation of the product. The addition rate was somewhat faster with the new plant, but it was possible to reduce the rate of addition.

6. Type of Filter Cloth. The filter cloths employed in this plant were very similar to those used at the other sites. However, they did come from a more recently supplied batch, and it was suggested that their performance should be compared with cloths from previously supplied batches which were still available.

7. Hold-up Time. Prior to filtration, the product was held in a stirred tank. The average period of hold-up in the new plant was somewhat less than that used in the other plants, but it could easily be increased.

In the following list of factors, the minus version corresponds to the usual operation for the new plant and the plus version to the proposed level. Thus, we have:

		Level	
	Factor	-	+
1)	Water	Town	Well
2)	Raw material	On site	Other
3)	Temperature of filtration	Low	High
4)	Recycle	Included	Omitted
5)	Rate of addition NaOH	Fast	Slow
6)	Filter cloth	New	Old
7)	Hold-up time	Low	High

A series of tests were run wherein each of these variables was changed to the proposed level while all other variables were held constant. However, the results were disappointing, the improvement sought was not found in any of the tests. There was increased pressure from management to solve the problem and the process engineers, being out of ideas, called in a consultant on the project. The consultant suggested they vary all of the seven variables or factors simultaneously in a 2^{7-3} 16 run fractional factorial design. The engineers were not willing to commit to 16 experiments since they were seeking a quick solution; however, with no other ideas they compromised and agreed to run a 2^{7-4} 8 run fractional factorial design.

The 1/16 replicate of the 2^7 is shown in Table 12.29. It was the same design as that given in Table B.1-5 in Appendix B, obtained by using the generators: $4 = 12$, $5 = 13$, $6 = 23$, and $7 = 123$. The defining relations from the generators are: $I = 124 = 135 = 236 = 1237$. The pair-wise products are 2345, 1346, 347, 1256, 257 and 167. The three-way products are 456, 1457, 2467, and 3567. The four-way product is 1234567. So, the complete defining relation is:

$$I = 124 = 135 = 236 = 1237 = 2345 = 1346 = 347 = 1256 = 257 = 167 = 456 = 1457 = 2467 = 3567$$
$$= 1234567$$

The eight experimental runs to be made were determined, run in random order, and gave the filtration times listed in Table 12.29.

446

Table 12.29 2^{7-4} (8 run) Design and Results

Run No.	Factor							Filtration Time
	1	2	3	4	5	6	7	
1	-	-	-	+	+	+	-	68.4
2	+	-	-	-	-	+	+	77.7
3	-	+	-	-	+	-	+	66.4
4	+	+	-	+	-	-	-	81.0
5	-	-	+	+	-	-	+	78.6
6	+	-	+	-	+	-	-	41.2
7	-	+	+	-	-	+	-	68.7
8	+	+	+	+	+	+	+	38.7

The filtration times for Runs 6 and 8 were comparable to those experienced at other similar units and the process engineers were ready to conclude the project and recommend running at the conditions listed in Run 8. However, the consultant argued that maybe not all factors were important, and that the plus level of each factor was generally more expensive than the minus level (usual operation). In the long run, it would pay them to find out which factors were really important and which were not, since they could then set the unimportant variables to their (minus) or cheaper level.

The analysis of the data is shown below:

	Factors						
	X_1	X_2	X_3	X_4	X_5	X_6	X_7
$\Sigma(+) - \Sigma(-)$	-43.5	-11.1	-66.3	12.7	-91.3	-13.7	2.1
Effect	-10.9	-2.8	-16.6	3.2	-22.8	-3.4	0.5

and the confounding patterns shown in Table B.1-5 reveal:

Water	$X_1 = 1 + 24 + 35 + 67 = -10.9$
Raw material	$X_2 = 2 + 14 + 36 + 57 = -2.8$
Temperature	$X_3 = 3 + 15 + 26 + 47 = -16.6$
Recycle	$X_4 = 4 + 12 + 37 + 56 = 3.2$
Rate of addition NaOH	$X_5 = 5 + 13 + 27 + 46 = -22.8$
Filter cloth	$X_6 = 6 + 17 + 23 + 45 = -3.4$
Hold up time	$X_7 = 7 + 16 + 25 + 34 = 0.5$

From the Pareto Diagram of absolute effects in Figure 12.11 it can be seen that the estimates -10.9, -16.6, and -22.8 are suspiciously large when compared to the others. The simplest interpretation of the results would be that the main effects of the Factors 1, 3, and 5 were important. However, due to the confounding and the effect heredity paradigm, other interpretations are possible. For example, possibly the main effects of Factors 3 and 5 along with their interaction (which is confounded with 1) were significant and responsible for the observed results. Maybe Factor 1 was not important and could be run at its cheaper level. An equivalent interpretation is that the main effects of Factors 1 and 5 could be significant along with their interaction, or Factors 1 and 3 with their interaction.

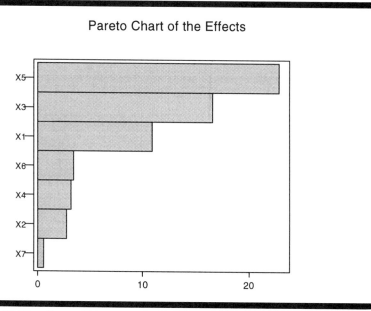

Pareto Chart of the Effects

Figure 12.11   Pareto Diagram of Effects in Table 12.29

There was no way to determine which of these interpretations was correct without additional experiments. Therefore, the answer to the question at Step 2, in the flow diagram in Figure 12.10 is no, and the next step was Step 3, where a reflected design would be run to de-confound the strings of two-factor interactions.

In this experiment, unlike the copperplating example presented in Section 12.3.3, there was a higher cost to run at the high level of each factor. Therefore, it would not have been economical in the long run to run the filtration unit with Factors 1, 3, and 5 set at their high levels, if unnecessary. Therefore, it was worth further investigation to find out which of the three factors (1, 3, or 5) were significant and must be run at their high levels, and which (if any) was insignificant and could be run at its low level.

Table 12.30 shows the factor settings and results of the reflected design that was run to de-confound the main effects.

Table 12.30 Reflected Design for Filtration Unit

Run No.	1	2	3	4	5	6	7	Filtration
1	+	+	+	-	-	-	+	66.7
2	-	+	+	+	+	-	-	65.0
3	+	-	+	+	-	+	-	86.4
4	-	-	+	-	+	+	+	61.9
5	+	+	-	-	+	+	-	47.8
6	-	+	-	+	-	+	+	59.0
7	+	-	-	+	+	-	+	42.6
8	-	-	-	-	-	-	-	67.6

If the two tables of data are combined, the main effects can be estimated free of two-factor interactions. The following is a list of the effects that can be estimated with the combined data:

Mean	b_o
Water	X_1
Raw material	X_2
Temperature	X_3
Recycle	X_4
Rate of addition NaOH	X_5
Filter cloth	X_6
Hold up time	X_7

$X_1 X_2 = 12 + 37 + 56$
$X_1 X_3 = 13 + 27 + 46$
$X_1 X_4 = 14 + 36 + 57$
$X_1 X_5 = 15 + 26 + 47$
$X_1 X_6 = 16 + 25 + 34$
$X_1 X_7 = 17 + 23 + 45$
$X_2 X_4 = 24 + 35 + 67$
Block= difference between average response in Table 12.29 and 12.30

The first eight effects are the same as those that could be estimated after the original 8 run resolution III design. The next seven effects are the strings of two-factor interactions that were confounded with main effects after 8 runs, but can now be estimated separately. The final effect is the block effect. Table 12.31 is a worksheet for calculating the effects. The first seven columns are copied from

Table 12.31   Worksheet for Calculating Effects from Combined Resolution III and Reflected Designs

Run	X1	X2	X3	X4	X5	X6	X7	12	13	14	15	16	17	24	Block	F.Time
1	-1	-1	-1	1	1	1	-1	1	1	-1	-1	-1	1	-1	-1	68.4
2	1	-1	-1	-1	-1	1	1	1	-1	-1	-1	1	1	1	-1	77.7
3	-1	1	-1	-1	1	-1	1	-1	1	1	1	1	-1	-1	-1	66.4
4	1	1	-1	1	-1	-1	-1	1	-1	1	1	-1	-1	1	-1	81
5	-1	-1	1	1	-1	-1	-1	1	1	-1	-1	-1	-1	-1	-1	78.6
6	1	-1	1	-1	1	-1	-1	-1	1	-1	1	-1	-1	1	-1	41.2
7	-1	1	1	-1	-1	1	-1	-1	-1	1	1	-1	1	-1	-1	68.7
8	1	1	1	1	1	1	1	1	1	1	1	1	1	1	-1	38.7
9	1	1	1	1	-1	-1	-1	-1	-1	-1	-1	-1	1	-1	1	66.7
10	-1	1	1	1	-1	-1	-1	-1	-1	-1	-1	1	1	1	1	65
11	1	-1	1	1	1	1	1	1	1	1	-1	1	-1	-1	1	86.4
12	-1	-1	1	-1	1	1	-1	1	-1	-1	-1	-1	-1	1	1	61.9
13	1	1	-1	-1	1	1	-1	1	-1	-1	1	1	-1	-1	1	47.8
14	-1	1	-1	1	-1	1	1	-1	1	1	1	-1	-1	1	1	59
15	1	-1	-1	-1	1	-1	1	-1	-1	1	1	-1	1	-1	1	42.6
16	-1	-1	-1	-1	-1	-1	-1	1	1	1	1	1	1	1	1	67.6
Sumprod	-53.5	-31.1	-3.30	21.7	-153.7	-0.50	-34.5	3.70	-28.9	8.90	-129.3	38.7	-26.90	-33.5	-23.70	
Effects	-6.69	-3.89	-0.41	2.71	-19.21	-0.06	-4.31	0.46	-3.61	1.11	-16.16	4.84	-3.36	-4.19	-2.96	

Tables 12.29 and 12.30. The next seven columns, representing interaction strings, were obtained by multiplying (element-wise) the first two pairs of columns represented in each string of interactions. For example, the column labeled 12 was obtained by multiplying the X_1 column by the X_2 column, and it represents the sum of the 12, 37, and 56 interactions. The last contrast column labeled Block is -1 for each of the 8 runs in the original resolution III design, and +1 for each run in the reflected design. The effects are calculated by the worksheet method explained in Chapter 8. The row labeled SUMPROD is the sum of products of the column of contrasts and the column of response values, which are labeled F.Time, for filtration time. The effects were calculated by dividing the sum of products by 8.

Similar results could be obtained using a regression program. The dependent variable for regression is filtration time, and the 15 columns of contrasts in Table 12.31 are the independent variables. The calculated regression coefficients would be exactly ½ the effects shown in Table 12.31.

Figure 12.12 is a Normal plot of the effects from Table 12.31, where it can be seen that Factors 1, 5, and the interaction string that represents the sum of 15 + 26 + 47 appear to be the only significant effects. In the flow diagram in Figure 12.10, the answer to the question at Step 4 is yes. Step 5 is next, where either additional experiments will be run to de-confound the interactions, or the effect heredity paridigm will be employed to define acceptable operating conditions for confirmation experiments.

It is possible the interaction effect was due to either 26 or 47, but in this case the plant engineers thought it was easier to believe the effect heredity idea that main effects 1 and 5 and their two-factor interaction were important. Assuming the interaction was due to 15, no additional experiments were needed to de-confound the three interactions, and the two-way table of average values shown in Figure 12.13 helped the engineers discover an acceptable operating condition.

According to the representation of the results in Figure 12.13, if the high level of Factor 1 (well water) and the high level of Factor 5 (slow addition of caustic) were used, the filtration time should be equivalent to what it was at other similar plants. All other factors could be left at their lower (less expensive level). This representation also explains why the desirable operating condition was not discovered with the earlier one-at-a-time trials that were performed. The two-way diagram shows the interaction to be very important, and both Factor 1 and Factor 5 need to be at their high levels in order to achieve the desired result. Interaction effects cannot be seen in one-at-a-time trials.

Of the possible explanations for the interaction, the 15 interaction effect seemed by far the most likely. Nevertheless, the fact that none of the Factors 2, 4, 6, and 7 have main effects does not, of course, preclude the possibility that their interactions exist. The crucial test was whether the trouble would be cured by using well water and the slow addition rate of caustic soda while leaving the other variables at their usual levels.

A number of confirmation trials were run in the plant where the only modifications made were the use of well water with a slow rate of addition of caustic. These runs did give satisfactorily short filtration times in the neighborhood of forty minutes. It confirmed the assumption that the interaction was between Factors 1 and 5 and validated the efficacy of the proposed change in operating conditions. The modification was adopted for all future production.

451

Normal Probability Plot of the Effects

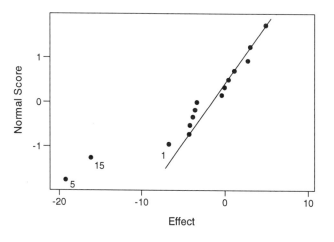

Figure 12.12 Normal Plot of Effects from Combined Filtration Experiments

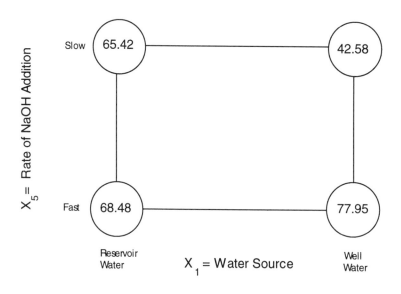

Figure 12.13 Two-Way Table of X_1 and X_5 Means

This example illustrates the sequential use of fractional factorial designs starting at Step 1 in Figure 12.10, and continuing at Step 3 by running a reflected second design to de-confound main effects from two-factor interactions. It also illustrates the use of the effect heredity paradigm for the interpretation of a confounded string of two-way interactions by picking the most likely single important interaction based on which main effects are significant. Confirmation runs validated the choice.

Another important concept illustrated by this example is the fact that finding out which variables or factors are not important may be just as valuable as finding out which variables are important. In that way unimportant variables can be set to their least expensive levels, since they don't significantly effect the results. This may identify great areas of cost savings when experimenting with prototype products and production process designs, or selection of materials and components.

12.9 Summary

The following is a list of the important terms and concepts covered in this chapter.

screening experiments 399
one-at-a-time experiments 400
fractional factorial design 401, 404
Plackett-Burman design 401, 421
confounding 401, 411, 417
half fraction 404
base design 404, 409
added factors 404, 409
alias structure 405
confounding pattern 405
generator 406
contrast 406
defining relation 406
principle fraction 407
¼ fraction 408
fractional factorial design tables 410
Yates algorithm 415, 434
projection property of fractional factorial designs 417
regression 420, 425, 430
resolution of fractional factorial 420
Plackett-Burman design tables 423
n_F for information 423
n_F for precision 424
blocked factorials 432
screening designs for multiple level factors 436
psuedo factors 436
dummy levels 437
L_{18} orthogonal array 438
effect sparsity 440
effect heredity 443
mirror image design 444
reflected design 444

References

1. Plackett, R. L. and Burman, J. P., "The Design of Optimum Multifactor Experiments," *Biometrika*, Vol. 33, (1946) p. 305.

2. Nelson, L.S., "Extreme Screening Designs" *Journal of Quality Technology*, Vol. 14, (1982), p. 99.

3. Hunter, J. S., "Statistical Design Applied to Product Design", *Journal of Quality Technology*, Vol. 17, (1985), p. 210.

4. Franklin, M. F. and Bailey, R. A., "Selection of Defining Contracts and Confounded Effects in Two-Level Experiments," *Applied Statistics*, Vol. 26, No. 3, (1977).

5. Kackar, R. N. and Tsui, K., "Interaction Graphs: Graphical Aids for Planning Experiments," *Journal of Quality Technology*, Vol. 22, No. 1, (1990).

6. Turiel, T. P., "A FORTRAN Program to Generate Fractional Factorial Experiments," *Journal of Quality Technology*, Vol. 20, No. 1, (1988).

7. Johnson, N. L., and Leone, F. C. *Statistics and Experimental Design*, Volume II, (New York, John Wiley and Sons, 1964), p. 196.

8. Box, G. E. P., Hunter, W. G., and Hunter, J. S. *Statistics for Experimenters*, (New York, John Wiley and Sons, 1978), chapter 12.

9. Box, G. E. P., and Hunter, J. S., "The 2^{k-p} Fractional Factorial Designs Part I", *Technometrics*, Vol. 3, (1961) p. 331.

12.10 Exercises for Chapter 12

12.1 The following data[1] was recorded for an 8-run Plackett-Burman screening design.

Run	X_1	X_2	X_3	X_4	X_5	X_6	X_7	Yield
1	+	+	+	-	+	-	-	1.1
2	-	+	+	+	-	+	-	6.3
3	-	-	+	+	+	-	+	1.2
4	+	-	-	+	+	+	-	0.8
5	-	+	-	-	+	+	+	6.0
6	+	-	+	-	-	+	+	0.9
7	+	+	-	+	-	-	+	1.1
8	-	-	-	-	-	-	-	1.4

a) Calculate the effects for factors $X_1 - X_7$.

b) Make a Normal Probability plot of the effects and a Pareto diagram of the absolute effects

c) Which main effect is the $X_1 X_2$ interaction confounded with? $X_1 X_6$? $X_2 X_6$?

d) What would you recommend after the analysis of this data?

12.2 In the fall of 1986, durability tests of GM trucks at the desert proving grounds[2] revealed that the shoe friction material was loosening from the shoe assembly on the rear brakes. An immediate but costly solution was implemented by adding a preconditioning process to the assembly. Next a screening design was undertaken to discover the cause of the problem. Eight factors (four process factors and four product design factors) were selected for the screening experiment. The experiments were conducted and the response, force to initiate lateral slip, was recorded on 16 experimental brake shoes. The experimental design was given by Table B.1-8 and the resulting data is listed on the next page.

---

[1]Nelson, L. S., "Extreme Screening Designs," *Journal of Quality Technology*, (1982) pp. 99-100.

[2]Gibbons, N. M., "Applications of Experimental Designs to a Brake Riveting Process," Presented at the ASA Annual Meeting, (1988).

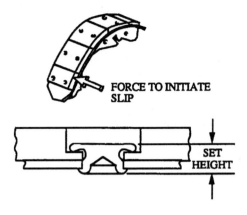

FORCE TO INITIATE
SLIP

SET
HEIGHT

	Levels	
	-	+
Process Factors:		
X_1 - Spring Force	Min	Max
X_2 - Anvil and Hammer Design	Delco	GMC
X_3 - Anvil and Hammer Condition	New	Worn
X_4 - Method	End Start	Center Start
Design Factors:		
X_5 - Shoe Surface Finish	Varnish	Sandblast
X_6 - Rivet Venders	Acme	Townsend
X_7 - Rivet Length	Short	Long
X_8 - Counter Bore Depth	0.05"	0.06"

Data in Standard order of Table B.1-8

Run	Y = Force to initiate lateral slip
1	0.0
2	162.5
3	90.0
4	90.0
5	40.0
6	92.5
7	60.0
8	70.0
9	50.0
10	90.0
11	90.0
12	112.5
13	80.0
14	100.0
15	40.0
16	137.5

a) Calculate the 15 effects as shown in Table B.1-8 and make a
Normal plot to judge which effects are significant.

b) Are there any relatively lare unassigned effects? If so interpret.
(i.e., what interactions might they represent)

12.3   In an experimental program to investigate the properties of a thin film plastic coating on ceramic based resistors, six factors were identified.

X_1 - Supplier of a basic component of the plastic (there were 2)
X_2 - Viscosity of the coating at time of application
X_3 - Thickness of coating
X_4 - Temperature of the first bake (dry) cycle
X_5 - Temperature of the second bake (cure) cycle
X_6 - Speed of the conveyer

a) List the experiments and a random list of run orders for a half faction design with these six factors using symbolic - and + signs for the levels. Show the complete confounding table.

b) List the experiments and random run orders if the half fraction is to be run in 2 blocks. What is the block effect confounded with?

c) Describe how you would analyze the data in each of the situations defined in Parts (a) and (b).

12.4   An experiment is to be performed in order to determine which factors affect the wear of a slider pump[3] . It was desired to study five factors at two levels each in eight runs, and two interactions were thought to be important.

Factors:
A. Material
B. Weight
C. Surface roughness
D. Clearance
E. Slide material

Determine which columns of Table B.1-5 Factors A-E should be assigned to in order to estimate all effects in the requirement set (A, B, C, D, E, AB, AC). Determine which columns of Table B.1-5 will be used to estimate the interaction effects AB and AC.

---

[3]Taguchi, Genichi, *Introduction to Quality Engineering*, Asian Productivity Association, Tokyo, chapter 7.

12.5 You are trying to help an ice cream vendor to be able to predict how many ice cream cones he will sell in a given afternoon (so that he knows how much ice cream to buy). He decides, with your help, to run a half fraction of a 2^5 to study the influence of several factors on sales. The data are given below:

X_1	X_2	X_3	X_4	X_5	No. of Cones Sold
-	-	-	-	+	105
+	-	-	-	-	122
-	+	-	-	-	92
+	+	-	-	+	149
-	-	+	-	-	106
+	-	+	-	+	153
-	+	+	-	+	111
+	+	+	-	-	113
-	-	-	+	-	100
+	-	-	+	+	146
-	+	-	+	+	105
+	+	-	+	-	125
-	-	+	+	+	105
+	-	+	+	-	126
-	+	+	+	-	94
+	+	+	+	+	156

The factors studied and their levels are:

Factor	Low Value (-)	High Value (+)
X_1 = temperature (F)	75	85
X_2 = weather	Cloudy	Sunny
X_3 = # of flavors	5	10
X_4 = cone type	Regular	Waffle
X_5 = personality	Surly	Happy

From the data given above, is a prediction equation justified? If so, what is it? Justify your answers.

12.6 An important property of electrical tape is the Percent Elongation at break. You decide to study a number of factors that may influence this property using a quarter fraction of a 2^6 design. The data are given below:

X_1	X_2	X_3	X_4	X_5	X_6	Percent Elongation
-	-	-	-	-	-	96
+	-	-	-	+	-	89
-	+	-	-	+	+	97
+	+	-	-	-	+	91
-	-	+	-	+	+	100
+	-	+	-	-	+	95
-	+	+	-	-	-	105
+	+	+	-	+	-	95
-	-	-	+	-	+	112
+	-	-	+	+	+	106
-	+	-	+	+	-	114
+	+	-	+	-	-	108
-	-	+	+	+	-	120
+	-	+	+	-	-	116
-	+	+	+	-	+	120
+	+	+	+	+	+	118

The factors studied and their levels are:

Factor		Low Value (-)	High Value (+)
X_1 =	% of Plasticizer	20%	40%
X_2 =	Time of Compounding	5 min	15 min
X_3 =	Temp of Compounding	200 F	350 F
X_4 =	Extruder Speed	200 rpm	400 rpm
X_5 =	Thickness of Tape	5 mil	8 mil
X_6 =	Source of Plasticizer	A	B

(a) Analyze the data above and determine if any of the factors have a significant effect on the response.

(b) From your analysis, what operating conditions would give a maximum Percent Elongation? What operating conditions would give a minimum Percent Elongation?

12.7   Consider the 2^{5-2} fractional factorial design with generators: $4 = 12$ and $5 = 123$. Specify the factor level combinations for this design and determine the complete confounding pattern. Note the resolution of the design.

12.8   The 2^{7-4}_{III} design in Table B.1-5 is one particular fraction of the full 2^7 factorial design with the generators: $4 = 12,\ \ 5 = 13,\ \ 6 = 23,\ \ $ and $\ 7 = 123$. Consider another fraction of the design which is obtained by changing all the signs in Table B.1-5.

(a)   What are the generators for the new design? What are the defining relations?

(b)   Ignoring all interactions of third order or higher, what is the confounding pattern for the new design?

12.9 In creating an 8 run fractional 4×2^4 design using the method of pseudo factors and Table B.1-5:

   a) Show why a two-level factor cannot be assigned to column X_4 if columns X_1 and X_2 are used to define the four-level factor.

   b) Show why a two-level factor cannot be assigned to column X_1 if columns X_6 and X_7 are used to define the four level factor.

12.10 Using Table B.1-7 and the method of pseudo factors:

   a) Create a 16 run fraction of a $4 \times 4 \times 2^2$ design.

   b) Create a 16 run fraction of a 8×2^3 design.

   c) Using the dummy levels method with the design you created in Part (a), create a 16 run $4 \times 3 \times 2^2$ design.

   d) Using the dummy levels method with the design you created in Part (b), create a 16 run 6×2^3 design.

12.11 Using the pseudo factor method along with the L18 (Table B.1-15):

   a) Create an 18 run 6×3^3 design.

   b) Using dummy levels, change the design you created in Part (a) to a 18 run $6 \times 3 \times 2^2$ design.

Part IV
Optimization Experiments

This part is a continuation of the technical manual for determining optimal conditions for maximizing or minimizing a response. New experimental designs are presented that will allow an experimenter to fit a polynomial approximation equation relating the factor variables to the response. Using this equation an experimenter can interpolate within his experimental region, and find the combination of factor levels that is predicted to produce the maximum or minimum response. This combination may be different than any tested.

CHAPTER 13
Response Surface Methodology

13.1 Response Surface Concepts and Methods

Response surface designs are normally used at the last stage of experimentation — if they are required at all. The important factors have already been determined by screening experiments, experience, or sound theory. The goal at this stage of experimentation is to describe *in detail* the relationship between the factors and a response. It is now known (or strongly anticipated) that a simple linear model — even with interactions — is not good enough to adequately represent that relationship. If we look at Figure 13.1, we see that we want to move far to the right on our Knowledge axis.

As a practical matter, since we want to know a lot about our system, it will require us to take quite a bit of data. (There is NO FREE LUNCH!) This is almost certainly prohibitive with more than four independent continuous factors. Although five, six, or even more factors can be studied, they would require more experiments (as we will soon see) than can usually be justified. In all cases, efficiency is very important so as to minimize the number of experiments that we must run. The experimental designs discussed in this chapter were derived with this practical objective in mind. They also have other important characteristics which we will discuss.

The term "Response Surface Methodology"[1](RSM) refers to the complete package of statistical design and analysis tools which generally are used for the following three steps:

(1) Design and collection of experimental data which allow fitting a general quadratic equation for smoothing and prediction

(2) Regression analysis to select the best equation for description of the data

(3) Examination of the fitted surface via contour plots and other graphical and numerical tools

In this chapter and the next, all three of the above steps will be described.

13.2 Empirical Quadratic Model

Up to this point, we have been discussing designs that allow only the estimation of linear effects of the X_i's on the response, Y. In graphical form these relationships would plot as straight lines, as shown for two X's in Figure 13.2(a). Even if interactions are estimated, they only allow the slope of the plot (of Y versus one of the independent variables, X_1) to change depending on the value of the other independent variables, as shown in Figure 13.2(b). But it is important to remember that, even with an interaction, the plot of Y versus X_1 is always a straight line. Clearly, that simple representation of the true situation is not always adequate. When we have already run some experiments and found curvature to be present (presumably using a factorial design plus center points), or when we *a priori* expect a plot of Y versus one of the X_i's to be curved, then we must run enough experiments to fit a more complicated model that allows for this curvature.

---

[1] Box, G. E. P. and Draper, N. R. *Empirical Model Building and Response Surfaces*, (New York: John Wiley and Sons), 1987.

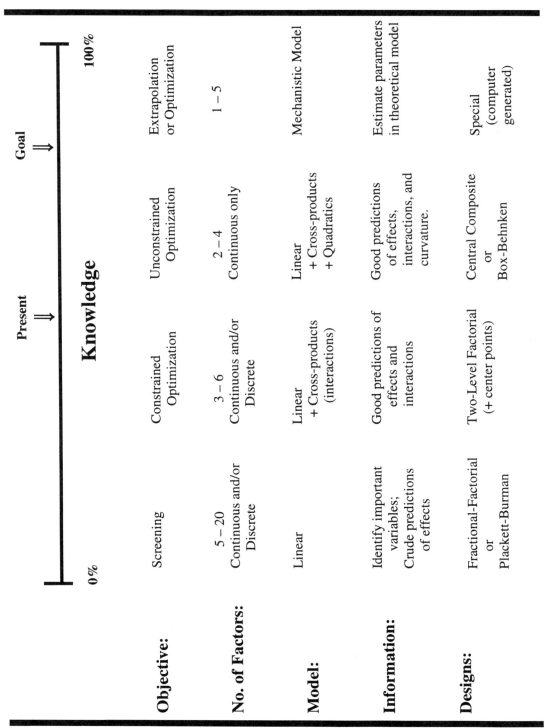

Figure 13.1 Objective of Response Surface Methods

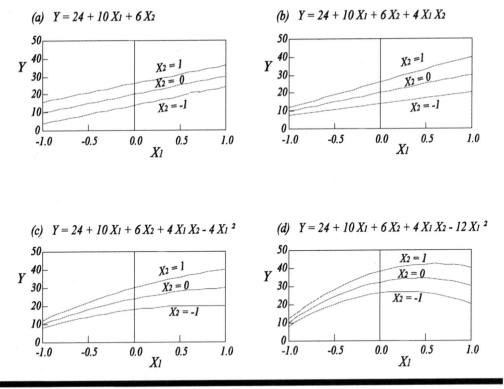

Figure 13.2 Impact of Terms Added to Polynomial Equation

An example with a small amount of curvature is shown in Figure 13.2(c). The optimum is still at or near the boundary, but a curved line is needed to predict Y reasonably well over the whole region of X_1. Therefore, a linear model (with interactions) would still have been adequate if the goal was simply to maximize Y, but not if we were interested in knowing Y for all values of the X's. A more severely curved response is shown in Figure 13.2(d). In this case, a straight line model (even with interactions) is not adequate for any purpose. Not only would the response be poorly predicted over the whole X_1 region, but the optimum would have been badly missed. In this case, a full quadratic model is needed to do anything.

The general quadratic model for k independent variables is:

$$\hat{Y} = b_0 + \sum_{i=1}^{k} b_i X_i + \sum_{i=1}^{k} b_{ii} X_i^2 + \sum_{i=1}^{k-1} \sum_{j=i+1}^{k} b_{ij} X_i X_j \qquad (13.1)$$

For one independent variable, this equation becomes

$$\hat{Y} = b_0 + b_1 X_1 + b_{11} X_1^2 \tag{13.2}$$

while for two independent variables, the equation is

$$\hat{Y} = b_0 + b_1 X_1 + b_2 X_2 + b_{11} X_1^2 + b_{22} X_2^2 + b_{12} X_1 X_2 \tag{13.3}$$

This is the simplest model which will still allow curvature in a graph of $\hat{Y}$ versus one of the independent variables, X_i. In fact, it may appear that it is much too simple to be able to describe any realistic situation with some degree of accuracy, but that is not true. First of all, the equation is quite flexible, and with the appropriate coefficients, it can describe a wide variety of different surfaces (see Figure 13.3).[2] These surfaces include hilltops, valleys, ridges, rising ridges, and falling ridges. These are the most commonly seen surfaces in practice. But the equation will even describe more unusual surfaces like saddle points if necessary. So it is very useful indeed.

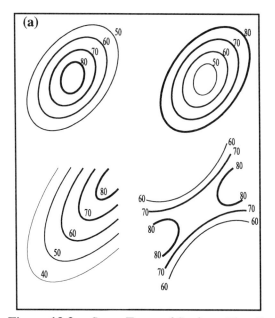 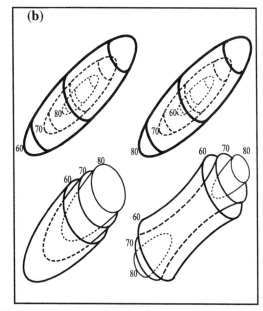

Figure 13.3   Some Types of Surfaces That Can Be Described by a Quadratic Equation:
   (a) Two-Dimensional Surfaces,   **(b)** Three-Dimensional Surfaces

---

[2]   Box, G. E. P., "Some Considerations in Process Optimization," *Journal of Basic Engineering*, (1960), p. 113 – 119.

As further justification of the utility of the simple quadratic equation, consider the following argument. Any situation, if we know enough about it, could be described by a mathematical model. The quadratic equation is somewhat equivalent to expanding the true model in a Taylor series about the center of the region and then dropping all terms of order greater than two. The difference is that the Taylor series fits exactly at the center of the region and gets progressively worse as we move away from the center, but the quadratic equation (which we obtained via a least squares fit to data over the whole region) spreads the error in fitting to all points. That is, it gives up some accuracy in the center if necessary to do a better job over the rest of the region.

If the region of interest is of modest size, then the polynomial (quadratic) fit is quite good. For example, let us say that we are interested in the chemical reaction:

$$\text{Reactant} \overset{k_1}{\Rightarrow} \text{Product} \overset{k_2}{\Rightarrow} \text{Decomposition Products}$$

and the reactions follow simple, first-order kinetics. Then the concentration of product, [P], can be modelled as a function of time by the highly nonlinear model:

$$[P] = [R]_0 \{\exp(-k_1 t) - \exp(-k_2 t)\} k_1 / (k_1 - k_2)$$

If k_1 and k_2 can be given as functions of temperature by the Arhenius expressions:

$$k_1 = 0.5 \exp[-10,000 (1/T - 1/400)] \text{ and}$$
$$k_2 = 0.2 \exp[-12,500 (1/T - 1/400)]$$

then contours of yield of product as a function of time and temperature can be predicted as shown in Figure 13.4. Over the region of time from 5 to 15 hours and temperature from 380 to 400 °K, the quadratic fit, as shown in Figure 13.5, is seen to be quite good. It is clearly not perfect, but the quadratic equation certainly describes the main features of the response contour plot.

Before we end our discussion about how good and useful quadratic equations are, two words of caution are in order. First of all, a quadratic equation is usually not able to describe adequately what is going on if the region of interest is too large such as that shown in Figure 13.6. If it were necessary to predict [P] over such a large span of temperature and time, then a more complicated model, perhaps a theoretical one with mechanistic interpretation, would have to be used. But mechanistic models shall not be discussed at any length in this text.

Secondly, it must also be pointed out that the polynomial model, while being a good mathematical French curve for smoothing data in the experimental region, is atrocious for extrapolation of any sizeable extent away from where the data were taken. As can be seen from Figure 13.7, the predictions are quite likely to be utter nonsense. Thus, it must be remembered that the quadratic equation should not be used outside of the experimental region, even though it is mathematically easy to do so.

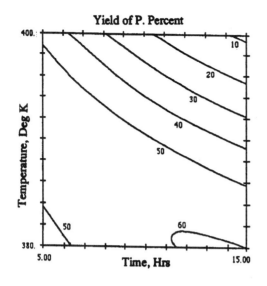

Figure 13.4   Actual Yield Contours

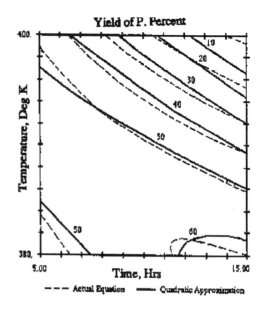

Figure 13.5   Quadratic Model Contours

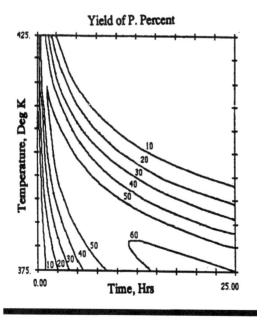

Figure 13.6   Actual Yield Contours

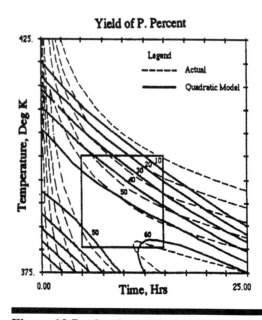

Figure 13.7   Quadratic Model Contours

13.3 Design Considerations

So now that we have decided that we will use a quadratic equation to describe the relationship between our response, Y, and the independent variables, X_i's, we must next decide what experiments we will run in order to accomplish that goal. As in previous chapters, we call that collection of experiments an *experimental design*. Some of the important properties that we would like our second order experimental design to have are (in order of priority):

- Allow the coefficients in the quadratic model, Equation 13.1, to be estimated
- Have a small number of runs
- Allow sequential buildup (i.e., first order design + some more points = second order design)
- Allow detection of lack-of-fit
- Permit blocking

The first of these criteria — being able to estimate all the terms in a full quadratic model — is of course critical. If the experiments don't permit estimating the full quadratic equation, we do not have a second order design. In order to satisfy this criterion, we must have at least three levels of each X_i variable. The simplest design that meets this requirement is the 3^k factorial design. In this design, all possible combinations of the three levels (-, 0, +) for each independent variable are run as our experiments.

However, for larger values of k, this design does not meet our second objective — to have a small number of runs — as can be seen from Table 13.1. The number of runs that is a theoretical minimum would be one run for every coefficient to be estimated. However, if this were all the experiments that were run, the equation would go right through all of the data points. There would be no smoothing taking place, there would be no estimate of error, and there would be no opportunity to test for lack of fit. Thus, even with efficient designs, the number of experiments performed is usually between 1.5 to 2.0 times the number required for estimation of the coefficients alone (unless more are necessary for the desired precision of estimation). But even by this more realistic standard, 3^k designs for more than three factors are seen to be very wasteful.

Table 13.1 Number of Runs for a 3^k Design

Number of Factors, k	No. of Runs in 3^k Design, N	No. of Coefficients in Quadratic Equation
2	9	6
3	27	10
4	81	15
5	243	21
6	729	28
7	2,187	36

13.4 Central Composite Designs

A class of designs which is more frugal with experiments is called *central composite* designs. Central composite designs build upon the two-level factorial design that was discussed in Chapter 8. Remember that the model used to fit the data from a 2^k-factorial design was of the form:

$$\hat{Y} = b_0 + \sum_{i=1}^{k} b_i X_i + \sum_{i=1}^{k-1} \sum_{j=i+1}^{k} b_{ij} X_i X_j \qquad (13.4)$$

with interaction terms of higher order usually neglected. Equation 13.4 reminds us that the 2^k-factorial design allows the estimation of all main effects and two-factor interactions. The only terms missing to give us a full quadratic equation are the squared terms in each X_i. In order to permit the estimation of these terms, the central composite design adds a set of axial points (called *star points*) and some (more) center points. The axial points combined with the center points are essentially a set of one-at-a-time experiments, with three levels of each of the independent variables, denoted by $-\alpha$, 0, and α (where α is the distance from the origin to the axial points in coded units). With the three levels of each X_i, the quadratic coefficients can be obtained. A central composite design for two factors is shown in Figure 13.8 and one for three factors is shown in Figure 13.9.

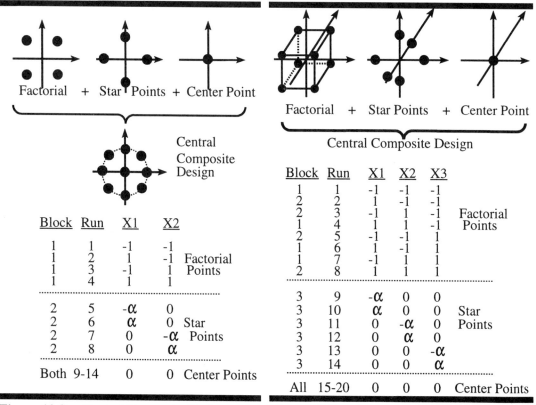

Figure 13.8 Central Composite Design for 2 Factors

Figure 13.9 Central Composite Design for 3 Factors

It should be noticed from the figures that α is not 1.0, so that the star points extend out beyond the cube of the factorial points. The only things to decide are how large α should be and how many replicated center points should be included in the design. To answer these questions, statisticians have brought in two additional criteria: *rotatability* and *uniform precision*. Rotatability implies that the accuracy of predictions from the quadratic equation only depends on how far away from the origin the point is, not the direction. This criterion fixes α. The other criterion, uniform precision, means that the variance of predictions should be as small in the middle of the design as it is around the periphery. This fixes the number of center points. The detailed calculations were done by others[3] and will not be discussed here. However, the recommended values are given in Table 13.2 and Appendix B.3.

It can be seen from Figure 13.8 that for two factors, rotatability dictates that $\alpha = \sqrt{2}$, which puts all the points (except for the center points) on a circle of radius, α. The value of α for all the designs was actually calculated using the formula: $\alpha = \sqrt[4]{\text{number of factorial points}}$. For three factors, rotatability gives $\alpha = 1.68 \approx \sqrt{3}$, which puts all the points (besides the center points) close to the surface of a sphere of radius α. This can be generalized to four factors as well (with $\alpha = \sqrt{4}$). But beyond four factors, that way of visualizing the design breaks down.

Table 13.2 Number of Runs for a Central Composite Design

Number of Factors, k	No. of Runs in Central Composite Design (Factorial + Star + Center)	No. of Coefficients in Quadratic Equation
2	$2^2 + 4 + 6 = 14$ ($\alpha = 1.41$; Two Blocks)	6
3	$2^3 + 6 + 6 = 20$ ($\alpha = 1.68$; Three Blocks)	10
4	$2^4 + 8 + 6 = 30$ ($\alpha = 2.00$; Three Blocks)	15
5	$2^4 + 10 + 7 = 33$ ($\alpha = 2.00$; Two Blocks)	21
6	$2^5 + 12 + 10 = 54$ ($\alpha = 2.38$; Three Blocks)	28
7	$2^6 + 14 + 12 = 90$ ($\alpha = 2.83$; Nine Blocks)	36

---

[3] Box, G. E. P. and J. S. Hunter, "Multifactor Experimental Designs for Exploring Response Surfaces," *Annals of Mathematical Statistics,* **28**, (1957), pp. 195–241.

The number of factorial points, star points, and center points necessary to run a central composite design are given in the second column of Table 13.2, along with the total of all three types of points. The total number of runs can be seen, via comparison to Table 13.1, to be much less than that for a 3^k design for three or more factors. The number of runs per estimated coefficient is seen to be in the desired range of 1.5 to 2.0, except for the design for seven factors ($k = 7$) which is not of great practical importance, since running a response surface design with seven factors is extremely unusual. The five-factor design is particularly efficient, since it requires only three more runs than the four-factor design.

Thus, central composite designs are looking good so far. They have been seen to meet the first two of our design considerations — they allow fitting the full quadratic equation, and they manage to do it in a small number of runs. The third consideration (to allow sequential build-up of the design) is also met — the designs were constructed by starting with a first order design, the 2^k factorial design, and adding to that some star and center points to obtain the full design. The first order design can be run first, and the second order piece can be run next if needed. This is also related to the consideration of blocking, discussed next.

The fourth consideration (to allow detection of lack-of-fit) is also met and will be discussed in more depth in Chapter 14. The last consideration (to allow blocking) is also met. All the designs can be run in blocks. The factorial portion with center points constitutes one or more of the blocks, and the star points with some more center points is another block. This gives two big advantages to the design. First, the factorial portion can be run *and analyzed* first. If curvature is found to be negligible, then the star points need not be run at all (a big savings). Secondly, blocking is an important tool whose main purpose is to increase the precision of the results. This has been discussed in depth elsewhere in this book.

So, central composite designs are seen to be very good indeed. They have no major shortcomings, which explains why they are the most commonly used response surface designs in the industry. The actual designs for two to seven factors, with the runs and blocking totally spelled out, are given in Appendix B.3.

13.5 An Example of a Central Composite Design: Polypropylene Pyrolysis

This example is based loosely on work that was done at the Energy and Environmental Research Center at the University of North Dakota. The study is real, but the data and results were fabricated to protect confidentiality. Broadly, this example is concerned with the economic recycling of plastics. Specifically, polypropylene pyrolysis (i.e., heating) to crude fuel oil had been studied extensively in a fluid bed process, and the thought was to see if a cheaper, kiln process could be used. The main difference between the two reactors is the fluidization velocity. In order to see if the fluidization velocity was important, and thereby determine the feasibility of using a kiln, the liquid yield from pyrolysis needed to be studied at different fluidization velocities. Since reactor temperature was known to be very important, it was also included in the tests, primarily to see if the effect of fluidization velocity depended on temperature.

Since the effects of two factors needed to be determined, the central composite design for two factors was used (given in Table B.3-1). The design was run in blocks; the factorial portion was

Table 13.3 Analysis of the Factorial Experiments for Example 1 (polypropylene pyrolysis)

Run No.	Run Order	Temp.	Veloc.	X_1	X_2	X_1X_2	(Curvature) X^2	Yield	
1	5	425	0.25	-1	-1	1	1	73.2	
2	3	625	0.25	1	-1	-1	1	51.1	Factorial
3	2	425	0.75	-1	1	-1	1	76.8	Avg = 66.3
4	6	625	0.75	1	1	1	1	64.1	
5	1	525	0.50	0	0	0	0	74.7	Center Point
6	7	525	0.50	0	0	0	0	76.8	Avg = 74.9
7	4	525	0.50	0	0	0	0	73.2	

Effects: -17.40 8.30 4.70 -8.60

t-Statistics: -9.62 4.59 2.60 -6.23 $t^*(2, .95) = 4.303$

$s =$ 1.81 (from center points)

$s(E) =$ 1.81 ($= s \ \sqrt{1/2 + 1/2}$))

$s(C) =$ 1.38 ($= s \ \sqrt{1/4 + 1/3}$))

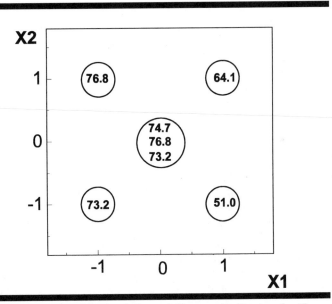

Figure 13.10 First Block of Central Composite Design for Two Factors (polypropylene pyrolysis)

run first and the star points were to be added only if needed. The first block of the design is also shown in Table 13.3. The designs in the back of this book are always given in terms of the coded (and scaled) factors, denoted as X_1 and X_2 here. In order to run the experiments, the nominal ranges for the actual variables had to be decided. These were chosen by the engineers based on prior work to be 425 to 625 °C for reactor temperature and 0.25 to 0.75 ft/sec for fluidization velocity. Once the actual factor levels were determined, the experiments were run. It is worth repeating that the experiments are *not* to be run in the order listed in the table, but should be randomized to minimize the chance of the conclusions being biased. That is why Run 5 was the first actual experiment, Run 3 was the second, etc. The resulting yields are given in Table 13.3 and also shown graphically in Figure 13.10.

The results were analyzed using the methods described in detail in Chapter 8. A summary of the analysis is given at the bottom of Table 13.3. Both reactor temperature (X_1) and fluidization velocity (X_2) were found to be statistically significant. This was determined by comparing their t-statistics to the critical two-sided t-statistic for two degrees of freedom and 95% confidence of 4.303 (denoted by t*). As expected, curvature was found to be present, and the star point block was needed in order to adequately describe the response.

The star point block was therefore run (in random order, of course). In order to run the experiments, the settings for temperature and velocity had to be determined. This was done using the relationships between the uncoded and coded factors discussed in Section 9.2:

X_i ≡ Coded and scaled value for Factor$_i$
    = [Factor value - Center]/["High" value - Center]

Therefore,

X_1 = (Temp - 525) / (625 - 525) = (Temp - 525) / 100
and
X_2 = (Veloc - 0.50) / (0.75 - 0.50) = (Veloc - 0.50) / 0.25

Since we know that the X's are to be $+\alpha$ and $-\alpha$ (or 0), and we want to know the appropriate values for temperature and velocity, we must rearrange the equations. When we do so, we get

Temperature = 525 + 100 X_1 and
Velocity = 0.50 + 0.25 X_2

So, for example, when we want X_1 to be $-\alpha$ (or -1.41), we must set the temperature to 525 + 100(-1.41) = 525 - 141 = 384 °C. The whole design (including the first block) is given in Table 13.4. The second block of experiments was run, and the crude oil yields were recorded. The yields are given in Table 13.3 and also shown graphically in Figure 13.11.

Once the data have been collected, a quadratic equation must be fit to the response. This task, called *regression analysis*, can be done by hand, but it is a tedious task much better suited to computers. The details of what these computer programs do is discussed in Chapter 14.

Table 13.4 Analysis of Complete Central Composite Design for Example 1
(polypropylene pyrolysis)

Run No.	Run Order	Temp. °C	Veloc. ft/sec	X1	X2	X1²	X2²	X1X2	Block	Yield
1	5	425	0.25	-1	-1	1	1	1	-1	73.2
2	3	625	0.25	1	-1	1	1	-1	-1	51.1
3	2	425	0.75	-1	1	1	1	-1	-1	76.8
4	6	625	0.75	1	1	1	1	1	-1	64.1
5	1	525	0.50	0	0	0	0	0	-1	74.7
6	7	525	0.50	0	0	0	0	0	-1	76.8
7	4	525	0.50	0	0	0	0	0	-1	73.2
8	13	384	0.50	-1.41	0	2	0	0	1	80.6
9	9	666	0.50	1.41	0	2	0	0	1	61.0
10	12	525	0.15	0	-1.41	0	2	0	1	57.2
11	10	525	0.85	0	1.41	0	2	0	1	66.5
12	11	525	0.50	0	0	0	0	0	1	72.2
13	14	525	0.50	0	0	0	0	0	1	74.5
14	8	525	0.50	0	0	0	0	0	1	76.9

Regression Output:

Constant	74.72					
Std Err of Y Est	1.908					
R Squared	0.974					
No. of Observations	14					
Degrees of Freedom	7	$t^*(7, 0.95) = 2.365$				

X Coefficients	-7.81	3.72	-1.96	-6.44	2.35	-0.07
Std Err of Coef.	0.674	0.674	0.702	0.702	0.954	0.510
t-Statistics	-11.59	5.51	-2.80	-9.17	2.46	-0.14
	X1	X2	X1²	X2²	X1X2	Block

Regression Output (with Block Effect Deleted):

Constant	74.72				
Std Err of Y Est	1.787				
R Squared	0.974				
No. of Observations	14				
Degrees of Freedom	8	$t^*(8, 0.95) = 2.306$			

X Coefficients	-7.81	3.72	-1.96	-6.44	2.35
Std Err of Coef.	0.632	0.632	0.658	0.658	0.893
t-Statistics	-12.37	5.89	-2.99	-9.79	2.63
	X1	X2	X1²	X2²	X1X2

F (Lack of Fit) = 0.452
F* (4, 4, 0.95) = 6.39

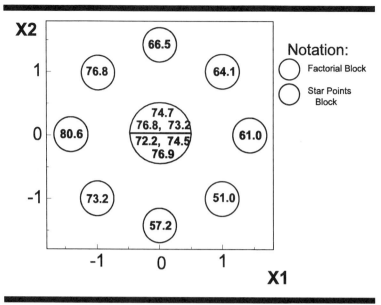

Figure 13.11   Complete Central Composite Design for
Two Factors in Two Blocks

The most rudimentary, and also probably the most available programs that do regression analysis are spreadsheets. Of course, numerous statistical packages are available that do a more thorough job and do it virtually automatically. But the regression equation (or more precisely, the set of coefficients in the quadratic equation) is the same regardless of which program is used. The output from LOTUS 1-2-3 is shown at the bottom of Table 13.4. The output was supplemented by a row of t-statistics, which were calculated by dividing the coefficients by their respective standard errors (standard deviations). It can be seen that all the coefficients in the equation were important (i.e., statistically significant) except for a block effect. Therefore, blocks were ignored in the final equation. In other words, a second regression analysis was performed with the blocking column omitted, and it is shown at the bottom of Table 13.4. As an aside, the student may want to notice what happened to the other coefficients when blocking was deleted — absolutely nothing! This was not a fluke. The blocking was deliberately chosen so that even if there was a major shift in response from one block to the other, it would only impact the estimate of the mean (and the variability) not the other coefficients in the model. This is called *orthogonal blocking*. For other designs (more factors), the blocks cannot always be made to be orthogonal, but the blocking is still chosen to make it as nearly orthogonal as possible.

A natural question at this point is what would have been done if other coefficients had turned out not to be statistically significant — would they have been dropped also? Some statisticians prefer to leave all the terms in the model, regarding the entire equation as a mathematical French curve. Others (the authors included) prefer to simplify the model by deleting all terms that have insignificant t-ratios (then refitting the model). This question will be covered fully in Chapter 14.

Before accepting this equation as gospel, the data should be checked for bad points (called *outliers*) via residual plots, and the equation should be checked for lack of fit. Both these diagnostics are very important, and are discussed fully in Chapter 14. For now, suffice it to say that the residual plots (not shown) looked fine, so all the data were assumed to be representative. Likewise, the lack-of-fit ratio (simply given at the bottom of Table 13.4 without explanation) showed absolutely no problem with our quadratic equation.

13.6   Graphical Interpretation of Response Surfaces

Once we are satisfied that the equation is a good representation of our system (over the region studied), then the only thing that remains is to use it to come to some conclusions. The final equation for the polypropylene pyrolysis example discussed in Section 13.5 is:

$$\hat{Y} = 74.72 - 7.81X_1 + 3.72X_2 - 1.96X_1^2 - 6.44X_2^2 + 2.35X_1X_2 \qquad (13.5)$$

This equation "says it all"; all we have to do is understand what it is "saying." It is a rare individual that can visualize the relationships between the response and the factors directly from the equation. Therefore, the equation is typically reported graphically (at least for a small number of factors).

The best way to summarize the equation graphically is largely a matter of taste. A commonly used graph to show a response as a function of two factors is a contour plot. Figure 13.12 is a contour plot of yield versus the two factors in Example 1. It very clearly shows the nature of the relationships and the location of the highest yield in the region of experimentation.

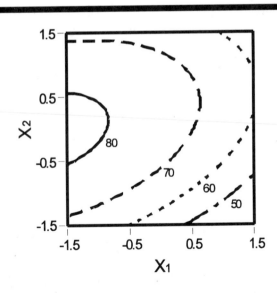

Figure 13.12   Contours of Yield vs. Coded Temperature (X_1)
and Coded Fluidization Velocity (X_2)

479

In this example, however, the relationship between yield and fluidization velocity was the main focus. Temperature was brought into the study because it was thought that an interaction may exist. Therefore, a better summary graph for this situation is a simple yield versus velocity graph. Since temperature was important, the graph is actually a collection of curves, each one at a different temperature. These are shown in Figure 13.13. The curves were obtained by taking Equation 13.5, putting in a specific value for X_1 (the coded temperature) and then simplifying. For example when temperature is 525, $X_1 = 0$, and the equation simplifies to:

$$\hat{Y} = 74.72 + 3.72 X_2 - 6.44 X_2^2 \tag{13.6}$$

Likewise, when temperature is 625, $X_1 = +1$, and the equation becomes:

$$\hat{Y} = 74.72 - 7.81(1) + 3.72 X_2 - 1.96(1)^2 - 6.44 X_2^2 + 2.35(1)$$
$$\hat{Y} = 64.95 - 6.07 X_2 - 6.44 X_2^2 \tag{13.7}$$

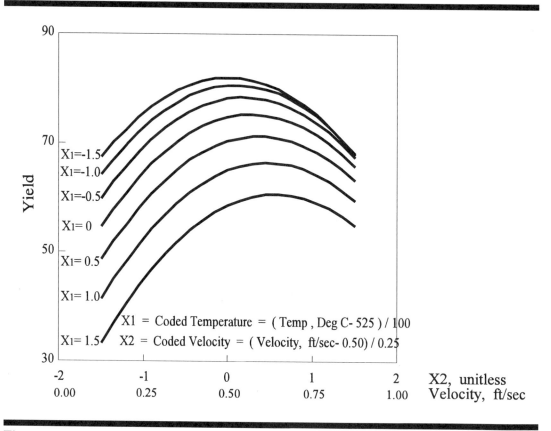

Figure 13.13   X-Y Plots of Yield vs. Fluidization Velocity (at several temperatures)

When there are more than two factors, it becomes increasingly difficult to visualize the response surface equation by using response curves like Figure 13.13; contour plots are usually better. Since there are more than two independent variables in the equation, several contour plots can be made at fixed levels of the independent variables not represented on the axes. For example, let us say that we studied the effect of catalyst concentration, reaction time and temperature on the yield of a chemical reaction, and we obtained the following response surface equation:

$$\hat{Y} = 85.83 - 0.48X_1 - 0.75X_2 - 0.85X_3 - 1.29X_1^2 + 0.06X_2^2 + 0.66X_3^2 \tag{13.8}$$
$$- 0.30\,X_1\,X_2 - 0.10\,X_1\,X_3 + 1.10\,X_2\,X_3$$

where:
X_1 = (catalyst, mole % - 0.03) / 0.01,
X_2 = (time, min - 90) / 30,
X_3 = (temperature, C - 65) / 15,
and Y = yield of the reaction.

Figures 13.14 –3.16 show contour plots with X_2 fixed at -1, 0, and 1 respectively (i.e., time = 60, 90, and 120). These can be thought of as three slices through the three-dimensional response contours shown in Figure 13.17. Contour plots can be used for the same purposes as simple graphs of the response versus each independent variable. When more than one response variable results from each experiment (e.g., Y_1 and Y_2), contour plots of the equations for each response are often overlaid in order to graphically determine a constrained maximum or minimum. For example, if the cost of the chemical reaction experiments could be expressed as a function of the three independent factors (catalyst, time, and temperature), a contour plot for cost could be drawn and overlaid on Figures 13.14 – 13.16. Then the maximum yield for a fixed cost level could be determined. If only the yield were important, then its maximum is readily apparent from the contour plots as being at short times, low temperatures, and middle catalyst amounts.

For a larger number of factors, we can still take one or two dimensional slices through the surface and show them as curves or contour plots. But we end up with an exponentially increasing number of plots. If we are looking for a maximum (or minimum) or we just want an overall feeling for what the response looks like, analytical or numerical methods may come to our rescue. These methods are discussed in Chapter 14.

13.7   Other Response Surface Designs

13.7.1 Box-Behnken Designs

Another class of commonly used designs for full response surface estimation are called *Box-Behnken*[4] designs, and they have two advantages over the central composite designs.

---

[4]   Box, G. E. P. and Behnken, D. W., "Some New Three Level Designs for the Study of Quantitative Variables," *Technometrics*, Vol. 2, (1960), pp. 455–460.

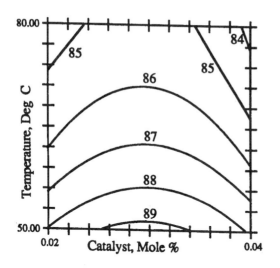

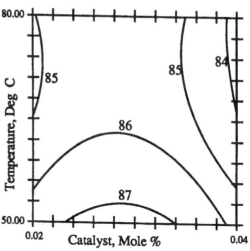

Figure 13.14   Yield Contours for
Time = 60 min ($X_2 = -1$)

Figure 13.15   Yield Contours for
Time = 90 min ($X_2 = 0$)

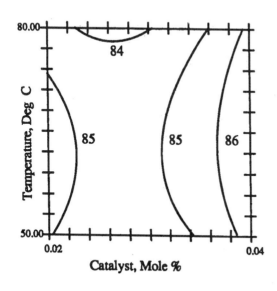

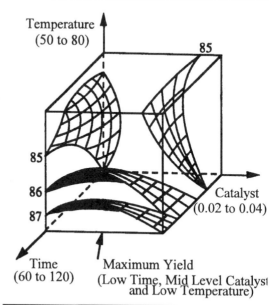

Figure 13.16   Yield Contours for
Time = 120 min ($X_2 = +1$)

Figure 13.17   Yield Contours in Three
Dimensions

482

First of all, they are more sparing in the use of runs, particularly for the very common three- and four-factor designs. The number of runs and the breakdown between center points and other points for the Box-Behnken designs are shown in the third column of Table 13.5. The exceptions to the rule can be seen from the table to be the five-factor design (the Box-Behnken design actually requires 13 more runs) and the six-factor design (which does not save any runs).

The second advantage is that the Box-Behnken designs are only three-level designs (i.e., each factor is controlled at only -1, 0, or +1), whereas the central composite designs are five-level designs (i.e., each factor will have been set at $-\alpha$, -1, 0, +1, or $+\alpha$ over the course of the set of experiments). On the surface this may seem to be totally inconsequential. But in an industrial setting, keeping the number of levels down to the bare minimum can often be a great help in the practical administration of an experimental program.

The Box-Behnken design for three factors is shown in Figure 13.18. The design consists of running all possible pairs of 2^2 designs, with the factor not considered being held at zero or its mid-level. Replicated center points are also added to complete the design. This pattern is the same for three to five factors; for six to nine factors the designs consist of sets of 2^3 designs plus center points. The specific designs for three to seven factors are given in Appendix B.4.

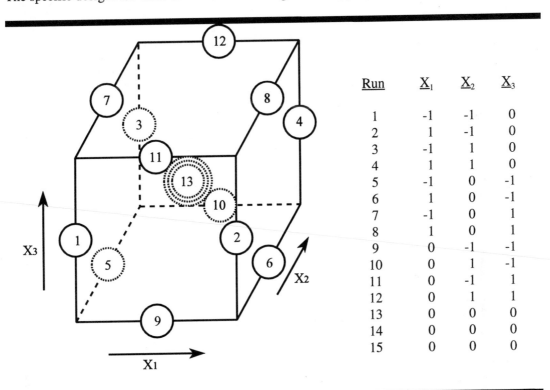

Run	X_1	X_2	X_3
1	-1	-1	0
2	1	-1	0
3	-1	1	0
4	1	1	0
5	-1	0	-1
6	1	0	-1
7	-1	0	1
8	1	0	1
9	0	-1	-1
10	0	1	-1
11	0	-1	1
12	0	1	1
13	0	0	0
14	0	0	0
15	0	0	0

Figure 13.18   Box-Behnken Design for Three Factors

Table 13.5 Number of Experiments Required for Response Surface Designs

Number of Factors	Central Composite Designs (Factorial + Star + Center = Total) Section 13.4	Box-Behnken Designs (Pattern + Center = Total) Section 13.7.1	Small Composite Designs (Unblocked) Section 13.7.2	Number of Coefficients in Quadratic Models
2	4 + 4 + 6 = 14 (Two Blocks)	No Design	6	6
3	8 + 6 + 6 = 20 (Three Blocks)	12 + 3 = 15 (One Block)	10	10
4	16 + 8 + 6 = 30 (Three Blocks)	24 + 3 = 27 (Three Blocks)	16	15
5	16 + 10 + 7 = 33 (Two Blocks)	40 + 6 = 46 (Three Blocks)	22	21
6	32 + 12 + 10 = 54 (Three Blocks)	48 + 6 = 54 (Three Blocks)	28	28
7	64 + 14 + 12 = 90 (Nine Blocks)	56 + 6 = 62 (Three Blocks)	40	36

It should also be stated that Box-Behnken designs also meet the criteria of rotatability (or nearly do), uniform precision of predictions, and can be broken down into two or more blocks when k, the number of factors, is between four and seven.

In fact, Box-Behnken designs meet all the considerations for a good design discussed in Section 13.3 except for one — they do not build upon the 2^k factorial design. This means that the full response surface design must always be run, whether or not it is needed. Remember that a big advantage of central composite designs is that the limited response surface portion (factorial + center points), can be run first and analyzed to determine if further work to separate the quadratics for each of the factors is justified. The star point block may never need to be run. That, of course, is a hefty savings! No such potential exists for the Box-Behnken designs. Conversely, if a limited response surface design was run first, there are no runs that could be added to it to make it into a Box-Behnken. You would have to start all over again. Naturally, you would never do that. Rather, you would just add star points (and center points) and be happy with a central composite design.

In summary, unless you are sure that a full response surface is necessary before beginning, or unless there is a distinct advantage to having only three levels of each factor, the central ccmposite design is the design of choice.

13.7.2 Small Composite Designs

Sometimes the number of runs required to complete a central composite or Box-Behnken design may seem prohibitive. If the researcher is willing to assume his experimental region is restricted enough to ensure that the quadratic model will provide a good approximation to the underlying function, then *small composite designs* are available.

One situation where small composite designs would be desirable is when the experiments or runs are very expensive or time consuming. Another case is where a difficult to compute deterministic function is being approximated by a quadratic in order to reduce computations necessary to locate the approximate maximum or minimum.

Example 2: Coal Gasification Modeling

For example, at the combustion research lab at BYU a computer program had been written to simulate the local fluid flow, chemical properties, and species concentrations in a coal gasification or combustion chamber. This simulation model required specification of 75 or more inputs or factors (X's) and took 10 to 15 CPU hours (on a VAX 11/750) computer to make one prediction. Even though the simulations were quite repeatable at the same input conditions (i.e., there was essentially no experimental error) it would have been very difficult to locate the optimum conditions through direct simulation, because of the long computations. Therefore, a response surface[5] approach was used to maximize the cold gas efficiency (CGE), which is defined as gross caloric value of the product gas divided by the gross caloric value of the coal fed, in a gasification simulation. Table 13.6 shows the conditions and results of 13 simulation

---

[5] Hurst, T. N., W. A. Sowa, J. C. Free, and P. J. Smith. "Statistical Optimization of Feed Streams for an Entrained-Flow Coal Gasifier using the PCGC-2 Simulator," *ASME Computer Aided Engineering of Energy Systems-Optimization* , Dec. 1986.

experiments conducted at conditions defined by a Box-Behnken experimental design. After a quadratic model was fit to the data in Table 13.6, the maximum CGE was predicted to be 0.8428 at $X_1 = 10$ atm, $X_2 = 0.787$, and $X_3 = 0.2322$. This was confirmed by running an actual simulation at these conditions which resulted in CGE = 0.8486.

In this particular case (three-factors) there were 10 coefficients in the quadratic model and 13 simulation experiments were required. If the researchers had been totally confident in the quadratic model approximation, only 10 simulation experiments would really have been needed and 30-45 hours of CPU time could have been saved.

Table 13.6 Example 2: Coal Gasifier System Modeling

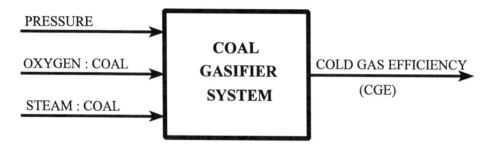

Factor Ranges

X_1 : $1 \leq$ Pressure $\leq$ 10 atm
X_2 : $0.7 \leq$ O2 : Coal $\leq$ 1.15
X_3 : $0.0 \leq$ Steam : Coal $\leq$ 0.3

Simulation Results

Run	X_1	X_2	X_3	CGE
1	1	0.7	0.15	0.6375
2	10	0.7	0.15	0.8274
3	1	1.15	0.15	0.6167
4	10	1.15	0.15	0.6555
5	1	0.925	0.0	0.6734
6	10	0.925	0.0	0.7826
7	1	0.925	0.3	0.6396
8	10	0.925	0.3	0.7887
9	5.5	0.7	0.0	0.7275
10	5.5	1.15	0.0	0.6570
11	5.5	0.7	0.3	0.7843
12	5.5	1.15	0.3	0.6560
13	5.5	0.925	0.15	0.7882

The approach of fitting a quadratic model to a few data points obtained by evaluating a difficult-to-compute physical model, works well whenever the underlying physical relationship is smooth and the experimental region is not too large (like that shown in Figure 13.7).

As shown in Table 13.5, the number of experiments (or function evaluations in deterministic situations) required for a central composite or Box-Behnken design is approximately 1.5 to 2 times as large as the number of coefficients fit in the empirical quadratic model, while the number of experiments for the small composite designs [6,7] listed in the fourth column of Table 13.5 is seen to be nearly equal to the number coefficients in the model to be fit. These designs were derived by taking a fraction of a factorial design and adding star points on the boundary of the experimental region for each factor as shown in Figure 13.19 for the three-factor case. Compare this to Figure 13.9 in Section 13.4, which shows the central composite design for three factors, and you will see that there are only two differences: a fractional factorial was used, and there are no center points. A listing of the small composite designs is shown in Appendix B.5.

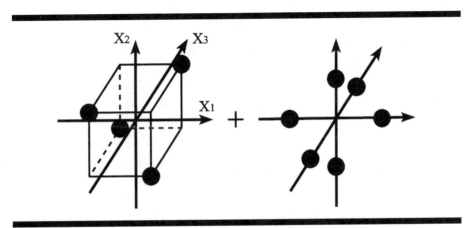

Figure 13.19   A Small Composite Design in Three Factors

The only disadvantage to using the small composite designs is the fact that the goodness-of-fit of the model, residual plots, etc., cannot be checked. Therefore, the researcher must be sure that his experimental region is small enough to ensure that the quadratic model will provide a good approximation. And, of course, the number of experiments used must be sufficient to provide the precision desired when experimental error is present.

---

[6] Hartley, H. O. "Smallest Composite Designs for Quadratic Response Surfaces," *Biometrics*, (1959) pp. 611 – 624.

[7] Westlake, W. J. "Composite Designs Based on Irregular Fractions of Factorials", *Biometrics*, (1965), pp. 324 – 336.

13.8 Summary

13.8.1 Procedure for Design of Experiments for RSM

The following is a summary in outline form of the procedure for running a response surface experimental design.

● List the important independent variables and their usual ranges (it should be remembered that if a central composite design will be used, the star points will extend beyond these ranges).

● List the dependent variables and their units.

● Select a class of RSM designs, keeping in mind that central composite designs allow building on a 2^k design which may be run first, that Box-Behnken designs require only three levels which may be simpler to administer, and that small composite designs should be used in deterministic situations when no estimate of experimental error is needed.

● Obtain the design in coded X's from Appendix B.3 for central composite designs, B.4 for Box-Behnken designs or B.5 for small composite designs.

● Decide whether or not the design should be run in blocks. This should normally be done in the case of central composite designs if the factorial part is to be analyzed first to determine if it is necessary to complete the second block of experiments. Blocking should be done for either central composite or Box-Behnken if it is likely that experimental uniformity is better within a block. The blocking is indicated in the appendices along with the design.

● Write out the experimental design in uncoded X's as a data collection worksheet. To do this, the coded X's are first copied from the appropriate table in Appendix B. For Example 1, Columns 5 and 6 of Table 13.3 were copied from Table B.3-1. Next, the high, low, and mid-level of the uncoded factor ranges are equated to +1, -1, and 0 respectively. For Factor 1 (temperature) this is shown as:

	Uncoded Factor (T)	Coded Factor (X_1)
High	625	+1
Low	425	-1
Center	525	0

Lastly, the uncoded factor levels of the star points are determined, if needed, as follows:
The uncoded and coded factor levels are related by the general formula,

$$X = (\text{Factor - Mid}) / (\text{Half of Difference between High and Low})$$

which becomes, for this example,

$$X_1 = (\text{Temperature - 525}) / \tfrac{1}{2}(625 - 425) = (T - 525) / 100$$

Solving for the uncoded factor level results in,

$$T = 100\,X_1 + 525$$

This equation can be used to determine the uncoded factor levels of the star points, $\pm \alpha$. For Factor 1, the $+\alpha$ level is

$$T = 100(+\alpha) + 525 = 100 (1.414) + 525 = 666$$

rounded to nearest integer.

● Review each of the experimental conditions for feasibility. If one or more runs does not appear operable, move them toward the center of the design by cutting down on all factors that are not zero (in coded units) in the same proportions. If the run is moved appreciably, it should be replicated to maintain the balance and precision of the design. If an experiment that was originally thought to be operable is found not to be, the same procedure applies.
(Note: If a point, j, was moved by reducing all the coded factor levels for that point by an amount c $(0 < c < 1)$, then the number of replicates that should be run at the new conditions, r_j , is given approximately by: $r_j = 1/c$. Of course, this number must be rounded to the nearest integer.)

● Assign random run order numbers to each run within each block. In cases where complete randomization is impractical (very expensive), it may be compromised a bit, but each level of the factors should be experienced at least twice. The exception to randomization is the center points. It may be preferable to spread them uniformly throughout the block, rather than to randomly run them, in order to get as wide a variation as possible and to facilitate checking for time trends.

● Run the experiments and collect the data. Often, as a practical matter, it is wise to use a data collection worksheet in which the experiments to be run are listed in the order to be run with the factor levels given in real (uncoded) units. This worksheet minimized the chance of mistakes, particularly if the runs are to be made by someone else.

13.8.2 Important Terms and Concepts

The following is a list of the important terms and concepts covered in this chapter.

13.9 Exercises for Chapter 13

13.1   You studied a process to manufacture electrical tape and you found (via a screening design) that only three factors significantly impacted the response — the percent elongation of the tape (which is how much it stretches before it breaks). The three factors and limits are given below:

Factor	Description	Low Value	Mid Value	High Value
X_1	% Plasticizer in Formulation	-1 = 10%	0 = 25%	1 = 40%
X_2	Temperature of Compounding	-1 = 200 F	0 = 275 F	1 = 350 F
X_3	Extruder Speed	-1 = 300 rpm	0 = 450 rpm	1 = 600 rpm

(a)   Write out the next set of experiments, in both coded and uncoded form, if a central composite design is to be used. Be sure to include the run order.

(b)   Write out the next set of experiments, in both coded and uncoded form, if a Box-Behnken design is to be used. Be sure to include the run order.

490

13.2   You studied the effects of three factors of interest on the yield of a chemical reaction using a full (two-level) factorial design plus three center points. You found that curvature was important (as well as all three factors), and want to augment the design with star points and three more center points to make it a full central composite design. List the additional runs that are needed in the table below, in both coded and uncoded form. Note: $\alpha = 1.68$ for this design.

Variable Coding	Low (-1)	Mid (0)	High (+1)
X_1 (Temperature)	130	150	170
X_2 (Pressure)	800	1000	1200
X_3 (Catalyst)	30	40	50

13.3   You have the same system described in Problem 13.2, but you have severe budget constraints. Therefore, you are thinking about running a small composite design with the three factors.

(a) Write out the whole central composite design in three factors, using coded factors. Then put a check in front of each run that would still be required for the small composite design" and put an X in front of each run that would not be required. How many runs would be saved?

(b) Under what conditions is the small composite design justified?

(c) If you use the small composite design, how many degrees of freedom are available to estimate error once you fit the full quadratic equation?

13.4 Please answer the following questions about RSM designs in outline form.

(a) What is the main drawback of using 3^k designs for Response Surface designs when $k > 2$?

(b) What are the advantages of a Box-Behnken design over a central-composite design?

(c) Are there any disadvantages of a Box-Behnken design versus a central-composite design?

13.5   Make a "Family of Curves" plot, similar to Figure 13.13, for the Polypropylene Pyrolysis example, but change the x-axis for the plot to temperature (X_1), and make a curve for $X_2 = -1.5$, -1.0, -0.5, 0.0, 0.5, 1.0, and 1.5. Put all seven curves on one graph. The final equation for the example is given in Table 13.4 and repeated as Equation 13.5. Are the conclusions from the graph (the prediction of the optimum conditions) consistent with Figure 13.13?

CHAPTER 14

Response Surface Model Fitting

14.1   Introduction

As was explained in the last chapter, the goal of response surface methodology is to describe in detail the relationship between a response variable (Y) and the independent variables (X_1, X_2, ..., X_k). A general quadratic equation is used as the mathematical French curve to accomplish this goal once data have been collected using one of the RSM designs discussed in Chapter 13. The primary objectives of this chapter are to find the best equation to describe the system under study, check its adequacy, and use it to see what the surface looks like.

Remember that the general quadratic model for k independent variables is:

$$\hat{Y} = b_0 + \sum_{i=1}^{k} b_i X_i + \sum_{i=1}^{k} b_{ii} X_i^2 + \sum_{i=1}^{k-1} \sum_{j=i+1}^{k} b_{ij} X_i X_j \tag{14.1}$$

Finding the best equation means that we must determine the best values for the coefficients in the quadratic equation (b_0, b_1, . . ., b_k, b_{11}, . . ., b_{kk}, b_{12}, . . ., $b_{k-1,k}$) for a particular set of experimental data. We should also simplify our model to the extent that is warranted by dropping out terms that are not justified (i.e., significant).   We will use the method of least squares, which was described extensively in Section 10.2 to find the coefficients. Once you have your equation, you should not use it for anything until you check to see if it actually describes your data adequately. This was discussed in Sections 10.5 and 10.6, but we will add one more check (Section 14.4). Trimming the model down to the smallest adequate equation is discussed in Section 14.5; our philosophy will always be "the simpler, the better" unless greater complexity is actually shown to be needed. Some techniques for exploring the fitted equation (to determine optimum conditions, etc.) are discussed in Section 14.6. Finally, quantifying the precision of the model predictions is revisited in Section 14.7.

14.2   Estimation of Coefficients in Quadratic Model

As stated in the introduction, our goal now is to find the best equation that describes a set of data. The measure of goodness we use (as discussed in detail in Chapter 10) is the sum of the squared vertical distances from the data points to the line, or the SSE (Sum of the Squared Errors) for short. In equation form:

$$SSE = \sum_i (Y_i - \hat{Y}_i)^2$$

Remember that since we are making the sum of the squared distances as small as possible, the whole procedure is often called the *method of least squares*.

We found in Section 10.4 that the values of the coefficients that minimize SSE are given by Equation 10.37 which is repeated here for ease of reference as Equation 14.2.

$$\hat{\mathbf{B}} = (\mathbf{X}^T\mathbf{X})^{-1}\mathbf{X}^T\mathbf{Y} \tag{14.2}$$

Equation 14.2 is a simple but very powerful matrix equation which defines the least squares estimates for *all* the regression coefficients simultaneously. To review how it is used, we will calculate the coefficients in the quadratic model for the polypropylene pyrolysis example in Chapter 13. (A much more extensive discussion of regression analysis is available in a number of texts. See, for example, Draper and Smith[1].) The data are given in Table 13.4. Since we want to fit a full quadratic equation in two X variables to some data, the equation we will use is:

$$\hat{Y} = b_o + b_1 X_1 + b_2 X_2 + b_{11} X_1^2 + b_{22} X_2^2 + b_{12} X_1 X_2 \tag{14.3}$$

where X_1 & X_2 are the two independent variables. Using the coded values of the X's, we get the following **X** and **Y** matrices:

I	X_1	X_2	X_1^2	X_2^2	$X_1 X_2$		Yields
1	-1	-1	1	1	1		73.2
1	1	-1	1	1	-1		51.1
1	-1	1	1	1	-1		76.8
1	1	1	1	1	1		64.1
1	0	0	0	0	0		74.7
1	0	0	0	0	0		76.8
1	0	0	0	0	0		73.2
1	-1.414	0	2	0	0		80.6
1	1.414	0	2	0	0		61.0
1	0	-1.414	0	2	0		57.2
1	0	1.414	0	2	0		66.5
1	0	0	0	0	0		72.2
1	0	0	0	0	0		74.5
1	0	0	0	0	0		76.9

X = (left matrix), **Y** = (right matrix)

The least squares normal equation (Equation 10.36) for this example is:

$$
\overset{\mathbf{X^TX}}{\begin{bmatrix} 14 & 0 & 0 & 8 & 8 & 0 \\ 0 & 8 & 0 & 0 & 0 & 0 \\ 0 & 0 & 8 & 0 & 0 & 0 \\ 8 & 0 & 0 & 12 & 4 & 0 \\ 8 & 0 & 0 & 4 & 12 & 0 \\ 0 & 0 & 0 & 0 & 0 & 4 \end{bmatrix}} \overset{\mathbf{\hat{B}}}{\begin{bmatrix} \hat{b}_0 \\ \hat{b}_1 \\ \hat{b}_2 \\ \hat{b}_3 \\ \hat{b}_4 \\ \hat{b}_5 \end{bmatrix}} = \overset{\mathbf{X^TY}}{\begin{bmatrix} 978.800 \\ -62.519 \\ 29.752 \\ 548.400 \\ 512.600 \\ 9.400 \end{bmatrix}}
$$

and the solution (Equation 14.2) is

$$
\overset{\mathbf{\hat{B}}}{\begin{bmatrix} \hat{b}_0 \\ \hat{b}_1 \\ \hat{b}_2 \\ \hat{b}_3 \\ \hat{b}_4 \\ \hat{b}_5 \end{bmatrix}} = \overset{\mathbf{(X^TX)^{-1}}}{\begin{bmatrix} 0.167 & 0 & 0 & -0.080 & -0.080 & 0 \\ 0 & 0.125 & 0 & 0 & 0 & 0 \\ 0 & 0 & 0.125 & 0 & 0 & 0 \\ -0.083 & 0 & 0 & 0.135 & 0.010 & 0 \\ -0.083 & 0 & 0 & 0.010 & 0.135 & 0 \\ 0 & 0 & 0 & 0 & 0 & 0.250 \end{bmatrix}} \overset{\mathbf{X^TY}}{\begin{bmatrix} 978.800 \\ -62.519 \\ 29.752 \\ 548.400 \\ 512.600 \\ 9.400 \end{bmatrix}} = \begin{bmatrix} 74.717 \\ -7.815 \\ 3.719 \\ -1.965 \\ -6.440 \\ 2.350 \end{bmatrix}
$$

These values can be compared to those given in Table 13.4 and reproduced in Table 14.1 (which were calculated using a spreadsheet computer program). Naturally, they are the same.

Table 14.1   Spreadsheet Regression Output for Polypropylene Pyrolysis
Example (reproduced from the bottom of Table 13.4)

The following is standard spreadsheet program output:

Regression Output:

Constant	74.717				
Std Err of Y Est	1.7869				
R Squared	0.9736				
No. of Observations	14				
Degrees of Freedom	8				

X Coefficient(s)	-7.81	3.72	-1.96	-6.44	2.35
Std Err of Coef.	0.632	0.632	0.658	0.658	0.893

14.3 Checking Model Assumptions (Residual Plots)

We typically invest quite a bit of time and money in our modeling efforts in order to come up with a model we can use to make our system better. But before we use it, we need to be careful that it doesn't have any major shortcomings, as was discussed at length in Chapter 10 (Section 10.6). This amounts to checking our assumptions:

(1) The errors are Normally distributed.
(2) The errors have the same variance.
(3) The errors are independent (or the experiment was randomized).
(4) The errors have a mean of zero (i.e., the model is adequate).

If one or more of our assumptions turns out to be incorrect, we must reanalyze our data taking this into account.

Since all four of our assumptions deal directly with the errors in our data, we examine our estimates of the errors, the residuals, $\varepsilon_i = Y_i - \hat{Y}_i$. The residuals for our polypropylene pyrolysis example are given again in Table 14.2 (along with the calculations for Normal plots).

Table 14.2 Residuals for the Polypropylene Pyrolysis Example and Detailed Calculations Required for the Full-Normal and Half-Normal Plots

Run	Y_i	$\hat{Y}_i$	ε_i	$\varepsilon_{(i)}$	$\dfrac{[(i) - \frac{1}{2}]}{n}$	z	$\lvert\varepsilon_i\rvert$	$\lvert\varepsilon_{(i)}\rvert$	$\dfrac{0.5 + [(i) - \frac{1}{2}]}{2n}$	z
1	73.2	72.76	0.44	-2.52	0.036	-1.80	0.44	0.02	0.518	0.04
2	51.1	52.43	-1.33	-1.52	0.107	-1.24	1.33	0.22	0.554	0.13
3	76.8	75.50	1.30	-1.33	0.179	-0.92	1.30	0.44	0.589	0.23
4	64.1	64.57	-0.47	-1.24	0.250	-0.67	0.47	0.47	0.625	0.32
5	74.7	74.72	-0.02	-0.60	0.321	-0.46	0.02	0.60	0.661	0.41
6	76.8	74.72	2.08	-0.47	0.393	-0.27	2.08	0.62	0.696	0.51
7	73.2	74.72	-1.52	-0.22	0.464	-0.09	1.52	1.24	0.732	0.62
8	80.6	81.84	-1.24	-0.02	0.536	0.09	1.24	1.26	0.768	0.73
9	61.0	59.74	1.26	0.44	0.607	0.27	1.26	1.30	0.804	0.85
10	57.2	56.58	0.62	0.62	0.679	0.46	0.62	1.33	0.839	0.99
11	66.5	67.10	-0.60	1.26	0.750	0.67	0.60	1.52	0.875	1.15
12	72.2	74.72	-2.52	1.30	0.821	0.92	2.52	2.08	0.911	1.35
13	74.5	74.72	-0.22	2.08	0.893	1.24	0.22	2.18	0.946	1.61
14	76.9	74.72	2.18	2.18	0.964	1.80	2.18	2.52	0.982	2.10

The plots to check our assumptions are shown in Figure 14.1 (dot plot & Normal plot to check for outliers), Figure 14.2 (residuals vs. $\hat{Y}_i$ to check on constant variance), and Figure 14.3 (residual plotted against independent variables, run order, and potential lurking variables to check that the model is adequate).

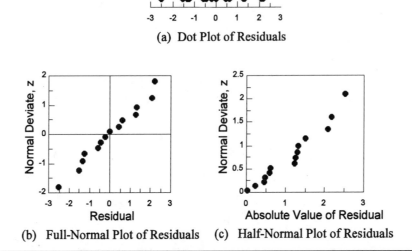

(a) Dot Plot of Residuals

(b) Full-Normal Plot of Residuals   (c) Half-Normal Plot of Residuals

Figure 14.1   Plots of Residuals to Check Normality of Errors for the Polypropylene Pyrolysis Example

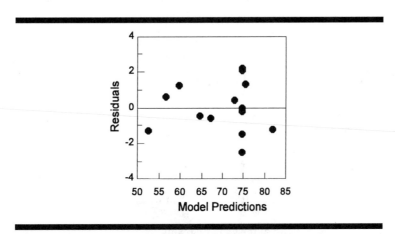

Figure 14.2   Plot of Residuals vs. Model Predictions for Polypropylene Pyrolysis Example

The plots all look fine, so our assumptions do not seem to be violated. But, in addition to the plots to check model adequacy, there is a quantitative test we can and should use which is discussed in the next section.

497

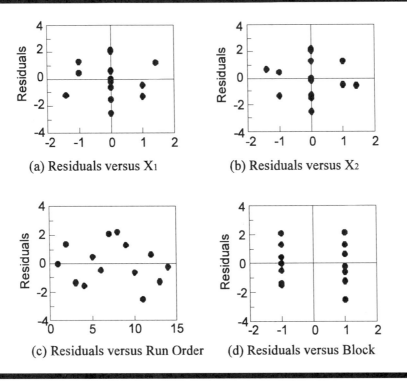

Figure 14.3   Plots of Residuals vs. Independent Variables, Run Order, and
Blocks to Check Model Adequacy

14.4 Statistical Check of Model Adequacy (LoF)

An objective statistical test is available (in addition to the residual plots) to check whether or not a quadratic model is giving an adequate representation of the data, but it requires some runs at the same conditions (replicates). The goal is exactly the same as the curvature test used for factorial designs. But the impact is even more severe. If the model is not adequate, the (quadratic) equation cannot be used reliably *anywhere* in the experimental region. Remember that for factorial designs, if the model was poor, it was still valid at the corners, but could not be used for interpolation inside the region.

The test is based on the sum of the squared residuals, SSE. The philosophy of the test is very simple and very basic. If the model is adequate, then SSE contains only experimental variability, and so it can be used to estimate the experimental error variance, σ^2. That is,

$$s^2_R = SSE / \nu_R = SSE / [n - (p+1)] \approx \sigma^2 \qquad (14.4)$$

(The subscript, R, on s^2_R is used to denote that it is the estimate of σ^2 based on the regression model.) If the model is not adequate, at least some of the residuals will be too big. That will make SSE bigger than it ought to be, and s^2_R will be an overestimate of σ^2. Thus s^2_R can be compared to

σ^2, and if it is significantly bigger (tested using a χ^2 distribution), then the model is deemed to be inadequate. If s_R^2 is not significantly bigger than σ^2, then the model is presumed to be fine.

This test looks great on paper, but there is one big catch — we usually do not know the value of σ^2. But if we have some replication, we can estimate σ^2 without using any model (other than the average at each set of conditions). We denote this estimate of σ^2 as s_{PE}^2. The subscript, PE, stands for pure error, because the only reason that replicates deviate from each other is the pure random variation in the data. Since all response surface designs have replication (at the center point at a minimum), the test described in this section should be used routinely with these designs.

The first step in using the test, is to calculate s_{PE}^2. This is done by first identifying all replicates and calculating the average value at each replicated set of conditions. The sum of squared deviations from the average is then calculated at any point, j, at which there is replication (m_j is the number of replicates at point, j).

$$SS_{PE,j} = \sum_{i=1}^{m_j} (Y_{ij} - \bar{Y}_j)^2 \qquad (14.5)$$

If only the center point was replicated, then the sum of squares of the center points around their average is the total SS_{PE}. If there is replication at several points, the various sums of squares are simply added together to get SS_{PE}.

$$SS_{PE} = \sum_j SS_{PE,j} \qquad (14.6)$$

In order to calculate s_{PE}^2, we also need to determine the degrees of freedom for pure error, DF_{PE} (also denoted by ν_{PE}). The degrees of freedom for any set of replicates is $\nu_{PE,j} = (m_j - 1)$. The total degrees of freedom for pure error is just the sum of the degrees of freedom for each set of replicates.

$$\nu_{PE} = \sum_j \nu_{PE,j} \qquad (14.7)$$

Once both SS_{PE} and ν_{PE} are calculated, we can determine s_{PE}^2 from the ratio,

$$s_{PE}^2 = SS_{PE} / \nu_{PE} \qquad (14.8)$$

And, since s_{PE}^2 is a perfectly valid estimate of σ^2, we could compare s_R^2 to it. But that is not what is usually done. The reason is that s_R^2 is based on a total sum of squared errors which includes a contribution from SS_{PE}. This contribution will mitigate any model inadequacies.

In order to make the test more sensitive, what is typically done is to break the total sum of squares into two pieces — SS_{PE} and whatever is left over which we denote by SS_{LoF}. The subscript, LoF, stands for lack-of-fit (of the model). We, likewise, break the total degrees of freedom into two pieces — ν_{PE} and whatever is left over, which we denote by ν_{LoF}. Since we already have the totals, and the pure error pieces, the remaining calculation is to get the lack-of-fit pieces by difference.

$$SS_{LoF} = SSE - SS_{PE} \qquad (14.9)$$

$$\nu_{LoF} = \nu_R - \nu_{PE} \qquad (14.10)$$

It is important for the student to realize that *both* parts of the original sum of squares, SSE, can be used to estimate the experimental error variance, σ^2. It has already been stated that SS_{PE}/ν_{PE} $(= s^2_{PE})$ is a pure estimate of σ^2. But if the model is adequate, so is the quantity, SS_{LoF}/ν_{LoF} $(= s^2_{LoF})$. As justification of this statement, remember that SSE is used to estimate σ^2 by dividing by ν_R. Therefore SSE is approximately equal to $\nu_R \sigma^2$. If we remove the amount, $\nu_{PE}\sigma^2$ $(= SS_{PE})$ from SSE, then what remains is $(\nu_R-\nu_{PE})\sigma^2$. Therefore, $SS_{LoF} \approx \nu_{LoF}\sigma^2$, or $\sigma^2 \approx SS_{LoF}/\nu_{LoF}$.

Thus, our test becomes a comparison of s^2_{LoF} to s^2_{PE} (which is our best estimate of σ^2). If our model is adequate, s^2_{LoF} should be the same as s^2_{PE} within experimental variability. If our model is not adequate, s^2_{LoF} will be too big. Therefore our test asks the question: is s^2_{LoF} significantly bigger than s^2_{PE}? Since two variances are being compared, the ratio of the two estimates of the (same) variance is distributed according to an F-Distribution,

Table 14.3 Detailed Calculations Required for the Lack-of-Fit Test for the Polypropylene Pyrolysis Example

Run No.	X1	X2	Block	Yield	j	Avg	d	d²	ν
1	-1	-1	-1	73.2					
2	1	-1	-1	51.1					
3	-1	1	-1	76.8					
4	1	1	-1	64.1					
5	0	0	-1	74.7	1	74.90	-0.20	0.04	
6	0	0	-1	76.8	1	74.90	1.90	3.61	2
7	0	0	-1	73.2	1	74.90	-1.70	2.89	
8	-1.41	0	1	80.6					
9	1.41	0	1	61.0					
10	0	-1.41	1	57.2					
11	0	1.41	1	66.5					
12	0	0	1	72.2	2	74.53	-2.33	5.44	
13	0	0	1	74.5	2	74.53	-0.03	0.00	2
14	0	0	1	76.9	2	74.53	2.37	5.60	
							Totals =	17.59	4

$SSE = 25.54 \qquad \nu_R = 8$

$SS_{PE} = 17.59 \qquad \nu_{PE} = 4 \qquad\qquad s^2_{PE} = 4.40$

$SS_{LoF} = 7.957 \qquad \nu_{LoF} = 4 \qquad\qquad s^2_{LoF} = 1.99$

F (Lack of Fit) = s^2_{LoF}/s^2_{PE} = 1.99 / 4.40 = 0.452
F* (4, 4, 0.95) = 6.39

$$F = \frac{s^2_{LoF}}{s^2_{PE}} = \frac{SS_{LoF}/v_{LoF}}{SS_{PE}/v_{PE}} \qquad (14.11)$$

and the yardstick is the tabled F-Distribution (with v_{LoF} and v_{PE} degrees of freedom).

The lack-of-fit calculations for our polypropylene pyrolysis example are shown in Table 14.3. The pure error estimate of variance, s^2_{PE}, was found to be 4.40, while the lack of fit estimate, s^2_{LoF}, was found to be 1.99. Since s^2_{LoF} came out to be smaller than s^2_{PE} (strictly due to chance), the question about s^2_{LoF} being significantly bigger than s^2_{PE} becomes mute, and the conclusion is that our model is fine. If s^2_{LoF} had been bigger, we would have used the F-statistic (with 4 and 4 degrees of freedom) to tell us how much bigger than 1.0 the ratio, s^2_{LoF}/s^2_{PE}, can be just due to chance. In this case, s^2_{LoF} would need to be 6.39 times bigger than s^2_{PE} before we have enough evidence to say that the model is not adequate.

In our example, it should be noted that the center points were treated as two groups of replicates, one group in each block. This was done to be on the safe side in case there was actually a shift in the response from one block to the next (even though no shift was found to be significant in our regression analysis).

14.5 Trimming Insignificant Terms from a Model

14.5.1 Justification

The overall goal of this section is to achieve model parsimony. The underlying principle is that the best model is the simplest model that "does the job." Although this philosophy is open to some debate, it is the firm belief of the authors.

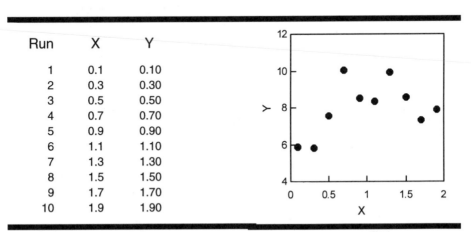

Run	X	Y
1	0.1	0.10
2	0.3	0.30
3	0.5	0.50
4	0.7	0.70
5	0.9	0.90
6	1.1	1.10
7	1.3	1.30
8	1.5	1.50
9	1.7	1.70
10	1.9	1.90

Figure 14.4   A Sample Set of X - Y Data

To justify this position a bit, let us look at the simple situation of a response as a function of a single independent variable, and let us suppose that we have the data shown in Figure 14.4. We can fit numerous models of varying complexities to this data. If we limit ourselves to polynomials, we can still fit anything from a simple average to a complex ninth order polynomial. Since there seems to be a trend in the data, the simplest model actually used was straight line. The least squares fits for that line and several higher order polynomials are shown in Figure 14.5. Which is best?

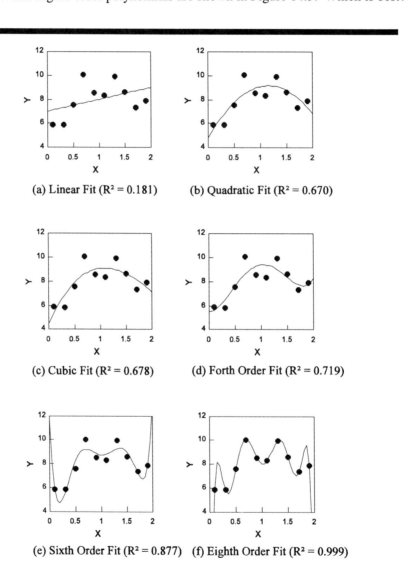

(a) Linear Fit ($R^2 = 0.181$) (b) Quadratic Fit ($R^2 = 0.670$)

(c) Cubic Fit ($R^2 = 0.678$) (d) Forth Order Fit ($R^2 = 0.719$)

(e) Sixth Order Fit ($R^2 = 0.877$) (f) Eighth Order Fit ($R^2 = 0.999$)

Figure 14.5   Several Polynomial Fits to a Set of X - Y Data

We can see from the figure, that as we increase the complexity of the model, it fits the data more and more closely. In fact, the eighth order polynomial fits the data almost exactly (and a ninth order polynomial *would fit exactly*), as the R^2 value indicates. Does that mean that we can conclude that the bigger a model is, the better it is? No! The problem is that the big models have gone past the point of describing the underlying relationship between X and Y, and they are describing the errors in the data. Since the errors are random, they will be different the next time some data is collected. So the big models, are actually worse. Another way of stating the conclusion is that a model that is too big is better at "predicting" the past (the data that we are analyzing), but what we want is a model that is good at predicting the *future* (new data that we may collect). The simplest adequate model is the best for that purpose; that is the quadratic fit in our example.

The next section describes how to trim away excess terms in a model in order to arrive at the simplest adequate equation.

14.5.2   Checking Terms in the Model for Significance

The general procedure for trimming a response surface model is to start by fitting the full quadratic equation (via least squares, of course). The residuals from the full model are checked via residual plots to ensure that the assumptions are valid before investing any more time in the model. If the residual plots look about right, then the model is trimmed down. Each coefficient in the model is examined for significance by comparing its t-statistic (the ratio of the coefficient to its standard error) to the critical t-value from a table.

Example of Calculation of s_b :

These computations can be done manually using Equation 10.42 which is repeated below as Equation 14.12, remembering to replace σ^2 with s^2, its estimate.

$$\text{Var}(\hat{\mathbf{B}}) = (\mathbf{X}^T\mathbf{X})^{-1}\sigma^2 \qquad (14.12)$$

As a demonstration of the use of Equation 14.12, let us continue with the polypropylene pyrolysis example in Chapter 13. The estimate of s^2 is $SSE/v_R = 25.544/(8)$ = 3.193, and our estimate of s is $\sqrt{3.193}$ = 1.787. This value of s can be compared to the value obtained by using the least squares regression function of a spreadsheet program to do the calculations (the output is shown at the bottom of Table 13.4 and reproduced for convenience in Table 14.1). The estimate of s is labeled "Std Err of Y Est," and it is seen to be identical to the value we calculated.

Since we already know $(\mathbf{X}^T\mathbf{X})^{-1}$, we can now use Equation 14.12 to estimate Var ($\hat{\mathbf{B}}$):

$$\text{Var}(\hat{\mathbf{B}}) = \begin{bmatrix} 0.1667 & 0 & 0 & -0.083 & -0.083 & 0 \\ 0 & 0.125 & 0 & 0 & 0 & 0 \\ 0 & 0 & 0.125 & 0 & 0 & 0 \\ -0.083 & 0 & 0 & 0.135 & 0.010 & 0 \\ -0.083 & 0 & 0 & 0.010 & 0.135 & 0 \\ 0 & 0 & 0 & 0 & 0 & 0.250 \end{bmatrix} \times 3.193,$$

or

$$\text{Var}(\hat{\mathbf{B}}) = \begin{bmatrix} 0.532 & 0 & 0 & -0.265 & -0.265 & 0 \\ 0 & 0.399 & 0 & 0 & 0 & 0 \\ 0 & 0 & 0.399 & 0 & 0 & 0 \\ -0.265 & 0 & 0 & 0.431 & 0.032 & 0 \\ -0.265 & 0 & 0 & 0.032 & 0.431 & 0 \\ 0 & 0 & 0 & 0 & 0 & 0.798 \end{bmatrix}$$

As mentioned earlier, the diagonal terms in this matrix give us the variances of each of the coefficients in our model. We actually want the standard deviations, so we must take the square root of each diagonal element. If we do this, we get the standard deviations listed in Table 14.4 which are labeled "Std Err of Coef." For example, the standard deviation of $\hat{b}_1$ is $\sqrt{0.399} = 0.632$. It should be pointed out that the standard deviation of $\hat{b}_0$ (which is $\sqrt{0.532} = 0.729$) is not calculated by the spreadsheet program. The reason (presumably) is that it is unusual to want to test the significance of the intercept.

Table 14.4   Spreadsheet Regression Output for Polypropylene Pyrolysis
Example (reproduced from the bottom of Table 13.4)

The following is standard spreadsheet program output:

X Coefficient(s)	-7.81	3.72	-1.96	-6.44	2.35
Std Err of Coef.	0.632	0.632	0.658	0.658	0.893

The following are extra (calculated) statistics:

t-Statistic(s)	-12.37	5.89	-2.99	-9.79	2.63
	X1	X2	X1²	X2²	X1X2

504

Once we have all of the standard deviations of the coefficients, all that remains is to calculate the t-statistics, $t_i = \hat{b}_i/s_{b_i}$ and compare them to the critical (tabled) value with the appropriate degrees of freedom. The t-statistics for our example are listed at the bottom of Table 14.4. For example $t_1 = \hat{b}_1/s_{b_1} = -7.81/0.632 = -12.37$. The critical value of t for eight degrees of freedom and a 95% confidence level (two-sided) is 2.306. Since all of the t-statistics are greater than this value, all of the coefficients are significant and should be kept in the model.

If any coefficients are not significantly different from zero, then those coefficients should be dropped, *one-at-a-time* starting with the coefficient that has the smallest t-value. The regression is repeated until only statistically significant coefficients remain. If there are many coefficients that are not significant, this procedure will require many iterations, and it will be tedious. But doing the analysis in steps is necessary. The reason that only one term is dropped at a time (and the regression analysis repeated) is that the remaining coefficients may change and their significance will likely change (as well as the estimate of s). This may affect your conclusions about the significance of the remaining terms. It should be noted that most statistical software packages will do this type of stepwise regression automatically. Therefore, a statistics package is probably a worthwhile investment if this type of analysis is done reasonably often.

One further bit of advice on dropping terms from the model that are not statistically significant: It may happen that a linear term in a model is not significant even though a quadratic term or an interaction term with the same variable is significant. In this event, which is not very common, it is recommended that you *do not* drop the linear term. Again, this recommendation is open to debate, but it is the opinion of the authors that it is a good one. The best analogy is to trimming a tree — you trim the branches not the trunk. The same reasoning applies to a mathematical model. You trim a particular variable in a model starting with its higher order terms (the quadratic and interactions) not with the main term (the linear term).

This procedure was carried out for our polypropylene pyrolysis example, and the output from the regression analysis was shown in Chapter 13, in Table 13.4. After fitting the full model, including a term for the block, the only coefficient that was not significant was for the block term. This term was dropped from the model and the regression repeated. The results from the reduced model (with the blocking term dropped) are shown in Table 14.4. All the coefficients in the reduced model were significant, so the regression analysis was complete.

Sometimes, in the analysis of experimental data, there is a variable that looks questionable. In other words, most of the terms in the model for that variable are not significant, but there is a term (or two) that is just barely statistically significant, and you wonder if the variable is really needed. After all, if a variable appears in four terms in the model, it is roughly four times as likely that one of the terms will be statistically significant, so α is not really 0.05 for that variable. In a case like that, the whole variable should be tested for statistical significance. This is done by fitting the model with and without the variable under scrutiny. If m terms are dropped when the variable is deleted (i.e., all interactions involving the variable in addition to the linear and quadratic terms), the model will no longer fit as closely to the data, and SSE will get bigger. This happens whether the variable is significant or not. The statistical test is based on the fact that SSE increases by σ^2 (on the average) for every unnecessary term in the model that is dropped. When m terms are dropped, SSE should increase by $m\sigma^2$. If the terms are needed in the model, then SSE will increase by more than that. Thus, the increase in SSE is our gauge of how important the terms are to our model.

The statistical test is to estimate σ^2 by dividing the increase in SSE by m. This estimate is compared to our estimate of σ^2 from the regression of our *bigger* model. The ratio of the two estimates is distributed as an F with m and v_R degrees of freedom. An example of this procedure is shown in Table 14.5 for a response surface study with three independent variables. It can be seen in the example that X_3 was found to be unimportant, even though two of the terms involving X_3 were marginally significant. A final regression was run deleting the not significant X_1^2 term (in addition to all X_3 terms).

Table 14.5 Checking the Significance of X_3 in a Three-Factor Response Surface Study

REGRESSION ANALYSIS (Full Model)

15.14	Constant								
2.934	Std Err of Y Est				Variance of Y =		8.611		
0.957	R Squared				Sum of Squared Resid's =		129.16		
25	No. of Observations								
15	Degrees of Freedom								

-7.70	-11.7	-1.87	0.72	5.17	-0.28	5.21	2.41	0.81	X Coefficients
0.85	0.85	0.85	1.30	1.30	1.30	1.04	1.04	1.04	Std Err of Coef.
-9.09	-13.8	-2.20	0.55	3.96	-0.22	5.02	2.33	0.78	t-Statistics
1	2	3	1^2	2^2	3^2	12	13	23	$t^* = 2.131$

REGRESSION ANALYSIS (X3 Deleted)

15.08	Constant					
3.428	Std Err of Y Est			Variance of Y =		11.748
0.925	R Squared			SSE =		223.22
25	No. of Observations		Extra Sum of Squares (Delete X3)			94.06
19	Degrees of Freedom		Extra Degrees of Freedom, m =			4

-7.70	-11.7	0.64	5.09	5.21	X Coefficients	F(X3)=(94.06/4)/8.611
0.99	0.99	1.47	1.47	1.21	Std Err of Coef.	= 2.73
-7.78	-11.8	0.44	3.46	4.30	t-Statistics	
1	2	1^2	2^2	12	$t^* = 2.093$	F*(4,15,0.95)=3.06

REGRESSION ANALYSIS (Final)

15.28	Constant
3.358	Std Err of Y Est
0.924	R Squared
25	No. of Observations
20	Degrees of Freedom

-7.70	-11.7	5.32	5.21	X Coefficients
0.97	0.97	1.34	1.19	Std Err of Coef.
-7.94	-12.1	3.96	4.39	t-Statistics
1	2	2^2	12	$t^* = 2.083$

14.6 Exploring the Response Surface

After all of the regression analysis and model checking has been done for a given set of data, you are left with a mathematical model that (hopefully) describes your system adequately. What remains to be done is to use that model to understand the system under study and perhaps optimize it. The best way to get an intuitive feeling for what the model is saying about how your system behaves, is to make a plot or plots of the response. People are much more capable of grasping the content of a graph than an equation. The graphical interpretation of responses was already discussed in Section 13.6, and so it will not be delved into again.

On the other hand, when there are four or more factors or independent variables in a response surface equation, it becomes increasingly difficult to visualize the relationships with contour plots or other graphs, since there would be so many possible "slices" of the full surface to show. In this situation, a mathematical interpretation of the response surface may yield more satisfactory results.

14.6.1 — Analytical Interpretation of Response Surfaces

It is known from elementary calculus that the maxima or minima of a simple quadratic equation in one variable, x, can be obtained by differentiating the equation with respect to x, setting the result equal to zero and solving for x. This procedure can be extended to higher dimensional quadratic equations also. If the quadratic equation is written in the form

$$Y = b_o + \mathbf{x}\,\mathbf{B}_L + \mathbf{x}\,\mathbf{B}_Q\,\mathbf{x}^T \qquad\qquad (14.13)$$

where $\mathbf{x} = [x_1\ x_2\ ...\ x_k]$ is the vector of the coded level of each of the independent variables,

$$\mathbf{B}_L = \begin{bmatrix} b_1 \\ b_2 \\ \vdots \\ b_k \end{bmatrix} \quad \text{is the vector of coefficients of the linear terms, and}$$

$$\mathbf{B}_Q = \begin{bmatrix} b_{11} & \tfrac{1}{2}b_{12} & \cdots & \tfrac{1}{2}b_{1k} \\ & b_{22} & \cdots & \tfrac{1}{2}b_{2k} \\ & & \ddots & \\ \text{sym.} & & & b_{kk} \end{bmatrix}$$

is a symmetric matrix with the coefficients of the squared terms on the diagonal and half the coefficients of the cross product terms on the off-diagonals. Then the solution for $x_1, x_2, ..., x_k$, after differentiating Equation 14.13 with respect to each x ($x_1, x_2, ..., x_k$), and setting each result equal to zero, is obtained by solving the k simultaneous linear equations for the stationary point.

The result, in matrix form is:

$$\mathbf{x}_S = -\tfrac{1}{2}\,\mathbf{B}_Q^{-1}\,\mathbf{B}_L \tag{14.14}$$

As an example, consider the two variable quadratic equation

$$Y = 74.33 + 1.99\,X_1 + 0.45\,X_2 - 3.15\,X_1^2 - 2.12\,X_1X_2 - 1.88\,X_2^2 \tag{14.15}$$

where: Y = Yield of a chemical reaction
X_1 = (Aqueous to organic wt. ratio - 0.5) / 0.5
and X_2 = (Temperature - 113 °C) / 18

In this case $\mathbf{B}_L = \begin{bmatrix} 1.99 \\ 0.45 \end{bmatrix}$ $\mathbf{B}_Q = \begin{bmatrix} -3.15 & -1.06 \\ -1.06 & -1.88 \end{bmatrix}$ and

$$\mathbf{x}_S = -\tfrac{1}{2}\begin{bmatrix} -3.15 & -1.06 \\ -1.06 & -1.88 \end{bmatrix}^{-1}\begin{bmatrix} 1.99 \\ 0.45 \end{bmatrix} = \tfrac{1}{2}\begin{bmatrix} -0.3918 & 0.2209 \\ 0.2209 & -0.6565 \end{bmatrix}\begin{bmatrix} 1.99 \\ 0.45 \end{bmatrix} = \begin{bmatrix} 0.3401 \\ -0.0721 \end{bmatrix}$$

These are the coordinates of the maximum as can be seen in Figure 14.6.

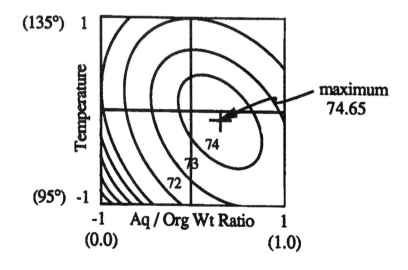

Figure 14.6 Contours of Equation 14.15

For a three variable example, consider Equation 13.8 that was discussed in Section 13.6.

In this case $\mathbf{B_L} = \begin{bmatrix} -0.475 \\ -0.750 \\ -0.850 \end{bmatrix}$ and $\mathbf{B_Q} = \begin{bmatrix} -1.29 & -0.15 & -0.05 \\ -0.15 & -0.058 & 0.55 \\ -0.05 & 0.55 & 0.66 \end{bmatrix}$

so that $\mathbf{x_s}$ is calculated to be:

$$\mathbf{x_s} = -\tfrac{1}{2} \begin{bmatrix} -1.29 & -0.15 & -0.05 \\ -0.15 & -0.058 & 0.55 \\ -0.05 & 0.55 & 0.66 \end{bmatrix}^{-1} \begin{bmatrix} -0.474 \\ -0.750 \\ -0.850 \end{bmatrix} = -\tfrac{1}{2} \begin{bmatrix} -0.7908 & 0.2140 & -0.2383 \\ 0.2140 & -2.5558 & 2.1461 \\ -0.2383 & 2.1416 & -0.2913 \end{bmatrix} \begin{bmatrix} -0.474 \\ -0.750 \\ -0.850 \end{bmatrix}$$

$$\mathbf{x_s} = \begin{bmatrix} -0.2090 \\ 0.0045 \\ 0.6244 \end{bmatrix}$$

Notice (by comparison with Figure 13.17) that this is not the coordinates of the maximum yield, which occurs at approximately $x_1 = 0$, $x_2 = -1$, and $x_3 = -1$. In general, $\mathbf{x_s}$ is called the stationary point of the response surface, and it must be determined if it is a maximum, minimum, or saddle point by taking the second derivatives of the quadratic equation. Examination of these second derivatives is equivalent to examination of the roots of the characteristic equation for $\mathbf{B_Q}$. The characteristic equation for $\mathbf{B_Q}$ is

$$|\mathbf{B_Q} - \lambda\mathbf{I}| = 0 \tag{14.16}$$

where the | | refers to the determinant of the matrix ($\mathbf{B_Q} - \lambda\mathbf{I}$). The roots of this equation are the eigenvalues, λ_1, λ_2, ..., λ_p. If all of the roots eigenvalues) are positive, the stationary point, $\mathbf{x_s}$, represents a minimum of the response surface. If all of the roots are negative, it indicates that the stationary point, $\mathbf{x_s}$, is the maximum of the response surface. Finally, if some of the roots are positive while others are negative, it indicates that the stationary point, $\mathbf{x_s}$, is a saddle point in the response surface (which is neither a minimum nor a maximum).

For Equation 14.15, the characteristic equation is

$$\begin{vmatrix} -3.15 - \lambda & -1.06 \\ -1.06 & -1.88 - \lambda \end{vmatrix} = (-3.15 - \lambda)(-1.88 - \lambda) - (-1.06)^2 = 0$$

The roots of this quadratic equation are $\lambda_1 = -3.75$ and $\lambda_2 = -1.28$, indicating that $\mathbf{x}_S$ is a maximum. The eigenvalues of the matrix, $\mathbf{B}_Q$, for Equation 13.8 (Section 13.6) are $\lambda_1 = -1.31$, $\lambda_2 = -0.26$ and $\lambda_3 = +0.99$, which indicates that the stationary point, $\mathbf{x}_S$, is a saddle point. Computer programs for fitting quadratic equations of the form of Equation 14.13 automatically compute $\mathbf{x}_S$ and the eigenvalues of $\mathbf{B}_Q$.

If the purpose of constructing a response surface is to locate the maximum or minimum response, the stationary point may be very helpful. If $\mathbf{x}_S$ is within the experimental region (i.e., each component is within -1 to +1 for a Box-Behnken design or within $-\alpha$ to $+\alpha$ for a central composite design), and it is a maximum or minimum (whatever is sought), the purpose has been accomplished. This was the case when computing $\mathbf{x}_S$ for Equation 14.15. If the stationary point is out of the region, or not the maximum or minimum being sought, like it was for Equation 13.8, additional work may be necessary. If the maximum is on a boundary of the experimental region as in Figure 13.17, the stationary point of the reduced equation (after fixing time, $x_2 = 1$ and temperature, $x_3 = -1$) may be useful; otherwise numerical techniques described in the next section can be used.

14.6.2 Numerical Methods for Interpreting Response Surfaces

With the availability of modern computers, function minimization is often accomplished by numerical methods rather than analytically (as described in the last section). If the response is described as a linear function of the independent variables or factors, minimization of maximization within the experimental region is a linear programming problem which can be solved using the well-known simplex algorithm. When the response is described by a quadratic equation, then maximization, minimization or constrained maximization or minimization can be accomplished using nonlinear programming[2]. Most of these algorithms will work well on quadratic functions like Equation 14.57. Some commercially available computer programs[3,4,5] for response surface analysis also have nonlinear programming algorithms.

510

14.7 Variability of Predictions

Determining the precision of a prediction calculated using our model was discussed in Chapter 10, Section 10.4.3. In summary, our model prediction at some set of conditions, $\mathbf{X_o}$, is calculated by $\hat{\mathbf{Y}} = \mathbf{X_o}\hat{\mathbf{B}}$. Although an equation has the appearance of being very precise, it clearly has some potential error because $\hat{\mathbf{B}}$ is not known exactly. The variability in $\hat{\mathbf{Y}}$ is given by Equation 10.48, which is repeated below as Equation 14.17:

or
$$\mathrm{Var}(\hat{\mathbf{Y}}) = \mathbf{X_o}(\mathbf{X^T X})^{-1}\mathbf{X_o^T}\sigma^2 \tag{14.17}$$

Remember that $\mathbf{X}$ is the n sets of conditions at which experiments were run, and thus it does not change. On the other hand, $\mathbf{X_o}$ is any point at which we would like to calculate the variance of $\hat{\mathbf{Y}}$. It is totally up to us; we can choose any set of conditions of interest to us. The only thing to keep in mind, is that the danger is great when extrapolating outside of the region of experimentation. The polynomial model is merely a flexible curve that should only be expected to be useful in the vicinity of our data. If we are interested in the precision at more than one point, we can just use the equation again for each set of conditions that we care about.

Continuing with our polypropylene pyrolysis example, if we are interested in the curve of $\hat{\mathbf{Y}}$ versus fluidization velocity, when the temperature is 525, we can calculate predictions (the curve) using the regression equation and the variability in those predictions using Equation 14.17. The results of those calculations are shown numerically in Table 14.6 and graphically in Figure 14.7.

Table 14.6   Calculation of Error Limits for Polypropylene Pyrolysis

Fluidization Velocity	X_2	$s_{\hat{Y}}^2$	$s_{\hat{Y}}$	$\hat{Y}$	$\hat{Y} - 2s_{\hat{Y}}$	$\hat{Y} + 2s_{\hat{Y}}$
0.15	-1.4	1.93	1.39	56.89	54.11	59.67
0.25	-1.0	0.83	0.91	64.56	62.73	66.38
0.35	-0.6	0.54	0.74	70.17	68.70	71.64
0.45	-0.2	0.53	0.73	73.72	72.26	75.17
0.55	0.2	0.53	0.73	75.20	73.75	76.66
0.65	0.6	0.54	0.74	74.63	73.16	76.10
0.75	1.0	0.83	0.91	72.00	70.17	73.82
0.85	1.4	1.93	1.39	67.30	64.52	70.08

To make sure that the use of Equation 14.17 is clear, let us calculate the variability of $\hat{\mathbf{Y}}$ at a fluidization velocity of 0.15 and a temperature of 525. Remembering our coding:

$$X_1 = [\text{temperature}-525]\,/\,100 = (525-525)\,/\,100 = 0$$
$$X_2 = [\text{velocity}-0.50]\,/\,0.25 = (0.15-0.50)\,/\,0.25 = -1.4$$
$$X_3 = X_1^2 = (0)^2 = 0$$
$$X_4 = X_2^2 = (-1.4)^2 = 1.96$$

and $\quad X_5 = X_1X_2 = (0)(-1.4) = 0$

The $\mathbf{X_o}$ matrix (being careful to include a "1" for the constant term) then becomes:

$$\mathbf{X_o} = \begin{bmatrix} 1 & 0 & -1.40 & 0 & 1.96 & 0 \end{bmatrix}$$

When we use this $\mathbf{X_o}$ matrix in Equation 14.17, and substitute s^2 for σ^2, we get:

$$\text{Var}(\hat{\mathbf{Y}}) = \begin{bmatrix} 1 & 0 & -1.40 & 0 & 1.96 & 0 \end{bmatrix} \begin{bmatrix} 0.532 & 0 & 0 & -0.265 & -0.265 & 0 \\ 0 & 0.399 & 0 & 0 & 0 & 0 \\ 0 & 0 & 0.399 & 0 & 0 & 0 \\ -0.265 & 0 & 0 & 0.431 & 0.032 & 0 \\ -0.265 & 0 & 0 & 0.032 & 0.431 & 0 \\ 0 & 0 & 0 & 0 & 0 & 0.798 \end{bmatrix} \begin{bmatrix} 1 \\ 0 \\ -1.40 \\ 0 \\ 1.96 \\ 0 \end{bmatrix}$$

or $\quad \text{Var}(\hat{\mathbf{Y}}) = \begin{bmatrix} 1.932 \end{bmatrix}$

The error limits on the equation values are typically reported as plus or minus twice the standard deviation of $\hat{\mathbf{Y}}$, which are approximate 95% confidence limits on the curve. For these values of X_1 and X_2, the error limits are $\pm 2\sqrt{1.932} = \pm 2.78$. The other error limits in Table 14.6 were calculated in a similar fashion, changing $\mathbf{X_o}$ appropriately.

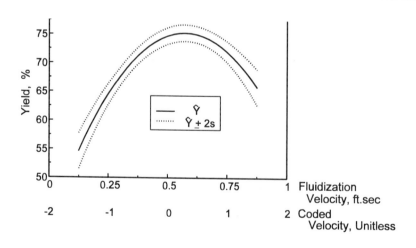

Figure 14.7   Error Limits on Polypropylene Pyrolysis Model (at temperature = 525 °C)

14.8 Summary

14.8.1 General Procedure for RSM Analysis

- **FIT FULL QUADRATIC MODEL**
  Use a computer program that is based on minimizing the residual sum of squares.

- **CHECK DATA FOR OUTLIERS**
  a. Calculate residuals
  b. Construct a Half-Normal Plot of the residuals
  c. If the residuals look OK (fall on a straight line), proceed with the next step (bullet).
  d. If any residuals are too big, throw out the worst data point and go back to the first step (fit the full quadratic equation again to the reduced set of data).

- **TRIM MODEL DOWN TO THE SMALLEST ADEQUATE FUNCTION**
  a. Check each independent variable to see if it can be deleted.
    This is done by fitting the model with and without the variable under scrutiny. If m terms are dropped when the variable is deleted, you must see if SSE increased by significantly more than $m \times s^2$. Note: s^2 is estimated from the fit of the bigger model (with ν degrees of freedom) and significance is judged by comparing the ratio of $(SS_{Increase}/m)/s^2$ to an F-Distribution with m & ν degrees of freedom.
  b. Delete any *high order terms* (for the important variables) that are not significant. This should be done one term at a time, starting with the term that has the smallest t-value. Note: It may happen that a linear term is not significant even though an interaction or a quadratic term with the same variable *is* significant. In this case (not very common), it is recommended that you do *not* drop the linear term.

- **CHECK MODEL ADEQUACY**
  a. Check lack-of-fit to ensure that a quadratic model is adequate. This is done by breaking the residual sum of squares, SSE, and its associated degrees of freedom, ν_R, into two pieces:
    i. sum of squares from pure error (from replicates), SS_{PE}, and
    ii. sum of squares due to lack-of-fit, SS_{LoF}, by difference
      $(SS_{LoF} = SSE - SS_{PE})$. Note: $\nu_{LoF} = \nu_R - \nu_{PE}$.
      The lack-of-fit variance, s^2_{LoF}, can then be calculated $(s^2_{LoF} = SS_{LoF}/\nu_{LoF})$ and checked to see if it is significantly bigger than $s^2 = s^2_{PE} = SS_{PE}/\nu_{PE}$
      The significance is determined by comparison to an F-distribution with ν_{LoF} and ν_{PE} degrees of freedom.
  b. Check residuals for trends.
    This is done be plotting the residuals against any variable that makes sense (e.g. run order, X_1, etc.). If there are any trends, it is an indication that something should be added to the model.

- **DISPLAY FINAL MODEL IN GRAPHICAL FORM**
  This is usually best done via contour plots if there are only two (or perhaps three) important variables.

14.8.2   Important Terms and Concepts

The following is a list of the important terms and concepts covered in this chapter:

References

[1] Draper, N. R. and H. Smith, "Applied Regression Analysis," New York: John Wiley & Sons, (1966).

[2] Hillier, F. S., and Lieberman, G. J. *Introduction to Operations Research*, New York: McGraw-Hill, (1990), chapter 14.

[3] COED available through COMPU-SERVE.

[4] XSTAT distributed by John Wiley & Sons, NY.

[5] RSI/Discover BBN Software.

14.9 Exercises for Chapter 14

14.1   The following data were taken to study the effects of three factors on the yield of a chemical reaction. Determine the best model for describing the system based on the data. Use that model to determine optimum operating conditions via graphical display of the equation. Verify that any assumptions in your analysis are reasonable (via plots of residuals), and that there is no lack-of-fit for your model. What is your expected yield at the optimum conditions, and what are its error limits? (Note: The goal is to maximize the yield = 100% - % impurity.)

Run	Run Order	Block	X1	X2	X3	% Impurity
1	6	-1	-1	-1	-1	8.55
2	2	-1	1	-1	-1	31.33
3	4	-1	-1	1	-1	8.24
4	9	-1	1	1	-1	30.89
5	7	-1	-1	-1	1	27.79
6	5	-1	1	-1	1	29.13
7	8	-1	-1	1	1	27.37
8	1	-1	1	1	1	30.77
9	11	-1	0	0	0	19.27
10	3	-1	0	0	0	17.37
11	10	-1	0	0	0	17.76
12	14	1	-1.68	0	0	10.93
13	17	1	1.68	0	0	31.25
14	20	1	0	-1.68	0	20.74
15	18	1	0	1.68	0	19.61
16	12	1	0	0	-1.68	25.52
17	13	1	0	0	1.68	37.06
18	15	1	0	0	0	20.10
19	19	1	0	0	0	18.53
20	16	1	0	0	0	20.83

14.2   List the assumptions that are made in regression analysis, and what plot(s) you would make (if any) to check each assumption.

14.3   The Y data below were fit using a linear model in X using a spreadsheet. Is the model adequate?   Use a lack-of-fit test to determine your answer.

X	Y
-2	31
-2	33
-1	53
-1	53
0	71
0	69
1	83
1	83
2	93
2	91

SUMMARY OUTPUT

Regression Statistics	
Multiple R	0.987
R Square	0.974
Adjusted R Square	0.971
Standard Error	3.841
Observations	10

ANOVA

	df	SS	MS	F	Sig F
Regression	1	4500	4500	305.08	1.2E-07
Residual	8	118	14.75		
Total	9	4618			

	Coeff	Std Error	t Stat	P-value
Intercept	66.00	1.214	54.344	1.46E-11
X	15.00	0.859	17.467	1.18E-07

14.4   You fit the data from a central composite design in three factors with a full quadratic equation.   The sum of the squared errors from the regression was:

$$SSE = 175 \qquad \nu_R = 10 \qquad (s^2_R = 17.50)$$

When factor X_2 was dropped from the model (4 terms), the sum of squares increased to

$$SSE = 323 \qquad \nu_R = 14 \qquad (s^2_R = 23.07)$$

Is X_2 needed in the model?

14.5   You took the following data to study the effects of three factors on the percent elongation of electrical tape (which is how much it stretches before it breaks). The three factors and limits are given below:

Factor	Description	Low Value	Mid Value	High Value
X_1	% Plasticizer in Formulation	-1 = 10%	0 = 25%	1 = 40%
X_2	Temperature of Compounding	-1 = 200 F	0 = 275 F	1 = 350 F
X_3	Extruder Speed	-1 = 300 rpm	0 = 450 rpm	1 = 600 rpm

Run	Run Order	X1	X2	X3	Percent Elongation
1	7	-1	-1	-1	102
2	5	1	-1	-1	106
3	10	-1	1	-1	100
4	9	1	1	-1	106
5	12	-1	-1	1	115
6	3	1	-1	1	119
7	18	-1	1	1	115
8	15	1	1	1	120
9	6	-1.68	0	0	108
10	11	1.68	0	0	115
11	2	0	-1.68	0	119
12	4	0	1.68	0	117
13	13	0	0	-1.68	118
14	17	0	0	1.68	100
15	14	0	0	0	119
16	8	0	0	0	116
17	1	0	0	0	118
18	16	0	0	0	120

(a) Analyze the experimental results to find the best equation.

(b) Use that model to determine optimum operating conditions (maximum elongation) via graphical display of the equation.

(c) Verify that any assumptions in your analysis are reasonable (via plots of residuals), and that there is no lack-of-fit for your model.

(d) What is your expected response at the optimum conditions, and what are its error limits?

CHAPTER 15
Mixture Experiments

15.1 Introduction

Many products, such as textile fiber blends, explosives, paints, polymers, ceramics, etc., are made by mixing two or more ingredients together. The qualities of the product are not dependent upon the total amount of the ingredients in the mixture, but only on the relative proportions in which they are present.

If the proportion of the ith component is X_i and there are k components in the mixture, then the proportions must satisfy the constraints

$$0 \le X_i \le 1.0 \quad \text{for each component,} \quad \text{and} \quad \sum_{i=1}^{k} X_i = 1.0$$

For example, in a two-component mixture problem, $0.0 \le X_1 \le 1.0$, $0.0 \le X_2 \le 1.0$ and $X_1 + X_2 = 1.0$. These constraints prevent us from using the experimental designs we have already described for studing the relationship of product quality to mixture components. The 2^2 factorial design would include all the corners of a square, while only two of these points would be permissible in studying mixture problems as shown in Figure 15.1.

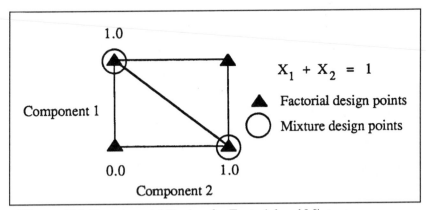

Figure 15.1 Experimental Region for Factorial and Mixture Experiments

The entire experimental region for a nonmixture experiment would consist of all the area inside a square or rectangle while for the mixture design it consists of only the points on the line $X_1 = 1 - X_2$. Therefore, the experimental region for the mixture problem (a line) is one dimension less than the experimental region for the normal problem which is a two-dimensional plane.

For the three-component mixture problem, the experimental region consists of a two-dimensional surface of the triangular plane shown below in Figure 15.2, rather than the normal three-dimensional cube. In the four-component mixture problem, the region consists of the three-dimensional volume of a regular tetrahedron.

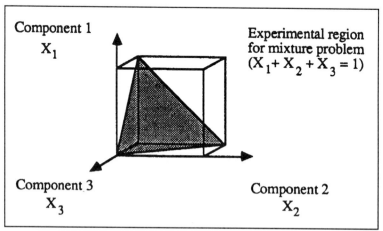

Figure 15.2 Experimental Region for Three-Component Mixture Experiment

The recommendend experimental designs for mixture experiment differ depending on the purpose, just as they did for the screening and response surface designs discussed earlier in the book. Different designs for mixture experiments will be presented after a discussion of the linear, quadratic, and cubic models used for these problems.

15.2 Models for Mixture Problems

As in all response surface analysis, the objective of the statistical analysis of mixture problems is to determine a model to predict the response as a function of the mixture components. This can be done with a linear model like:

$$Y = b_o + \sum_{i=1}^{k} b_i X_i$$

where Y is the response and X_i is the mixture components. Or, it can be done with a quadratic model:

$$Y = b_o + \sum_{i=1}^{k} b_i X_i + \sum_{i<j=2}^{k} b_{ij} X_i X_j + \sum_{i=1}^{k} b_{ii} X_i^2$$

The linear model would generally be used for cases where the mixture components are additive and the product quality is defined as a simple linear combination of the component proportions. The quadratic model would be used if there were some interaction (synergism or antagonism) between components, and the product quality or response from a mixture of components was better or worse than would be predicted by the straight linear combination of the mixing proportions. For the three-component case, the linear model is written as

$$Y = b_o + b_1 X_1 + b_2 X_2 + b_3 X_3$$

and the quadratic model is written as

$$Y = b_o + b_1 X_1 + b_2 X_2 + b_3 X_3 + b_{11} X_1^2 + b_{22} X_2^2$$

$$+ b_{33} X_3^2 + b_{12} X_1 X_2 + b_{13} X_1 X_3 + b_{23} X_2 X_3$$

At first glance these look like the normal linear or response surface models. However, due to the mixture constraint $X_1 + X_2 + X_3 = 1$, it can be seen that once the value of X_1 and X_2 are known, X_3 is automatically fixed to be $1 - X_1 - X_2$. Therefore, the prediction models are redundant, and Y is not really a function of X_1, X_2, and X_3 but only of X_1 and X_2. Therefore, the models could really be written as

$$Y = b_o + b_1 X_1 + b_2 X_2, \text{ or}$$

$$Y = b_o + b_1 X_1 + b_2 X_2 + b_{11} X_1^2 + b_{12} X_1 X_2 + b_{22} X_2^2$$

for the three-component case. This is called the slack variable form of the model, and is used when X_3 is the major component proportion-wise, and is the most inert of the three components. The slack variable model in its general form is

$$Y = b_o + \sum_{i=1}^{k-1} b_i X_i + \sum_{i=1}^{k-1} b_{ii} X_i^2 + \sum_{i<j=2}^{k-1} b_{ij} X_i X_j \qquad \textbf{(15.1)}$$

This model looks like the normal quadratic model used in response surface analysis, with the kth component omitted. However, the coefficients in the model cannot be interpreted in the same way that they are for nonmixture models. Insignificant coefficients, in general, should not be removed from a fitted equation. Figure 15.3 illustrates the meaning of the coefficients in the linear model.

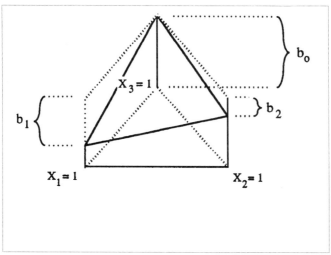

Figure 15.3 Graphical Interpretation of Coefficients in
Linear Model

As can be seen in Figure 15.3, b_o is the height of the response plane above the vertex at $X_3 = 1$, and b_1 and b_2 are the differences in heights of vertices at $X_3 = 1$ and $X_1 = 1$, and $X_3 = 1$ and $X_2 = 1$, respectively. The coefficients in the quadratic terms in the model do not have such a simple interpretation. However, there is a better way of writing the model that gives us a simple interpretation for the coefficients in the quadratic model. To rewrite the model notice that:

$$1 = (X_1 + X_2 + X_3)$$

and

$$X_1^2 = X_1X_1 = X_1(1-X_2-X_3) = X_1 - X_1X_2 - X_1X_3$$

$$X_2^2 = X_2X_2 = X_2(1-X_1-X_3) = X_2 - X_1X_2 - X_2X_3$$

$$X_3^2 = X_3X_3 = X_3(1-X_1-X_2) = X_3 - X_1X_3 - X_2X_3.$$

The redundant quadratic model

$$Y = b_o + b_1X_1 + b_2X_2 + b_3X_3 + b_{11}X_1^2 + b_{22}X_2^2 + b_{33}X_3^2 + b_{12}X_1X_2 +$$

$$b_{13}X_1X_3 + b_{23}X_2X_3$$

can be written in the form:

$$Y = b'_1X_1 + b'_2X_2 + b'_3X_3 + b'_{12}X_1X_2 + b'_{13}X_1X_3 + b'_{23}X_2X_3$$

Where:

$$b'_1 = b_0 + b_1 + b_{11}$$

$$b'_2 = b_0 + b_2 + b_{22}$$

$$b'_3 = b_0 + b_3 + b_{33}$$

$$b'_{12} = b_{12} + b_{11} + b_{22}$$

$$b'_{13} = b_{13} + b_{11} + b_{33}$$

$$b'_{23} = b_{23} + b_{22} + b_{33}$$

This is the Scheffé[1] form of the mixture model, and like the slack variable model, is not redundant. In most situations this is a more reasonable model to use since all the mixture components are represented. This linear form of the Scheffé model is:

$$Y = b'_1X_1 + b'_2X_2 + b'_3X_3$$

The interpretation of the linear coefficients b'_1, b'_2, and b'_3 are simply the heights of the vertices at $X_1 = 1$, $X_2 = 1$ and $X_3 = 1$, respectively. The coefficients, b'_{ij}, in the quadratic Scheffé model also have a simple interpretation in that they represent the amount of curvature along the face of the line connecting the point $X_i = 1$ and $X_j = 1$. For example, in the three-component case, see Figure 15.4.

---

[1]Scheffé, H. "Experiments with Mixtures," *Journal of the Royal Statistical Society*, B, Vol. 20, (1958), pp. 344–360.

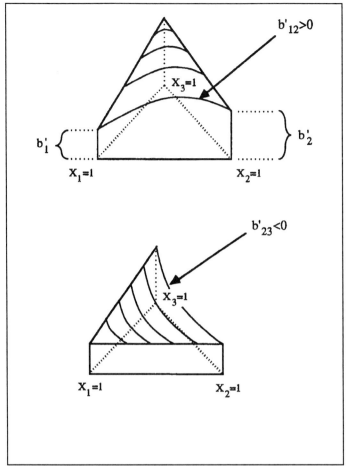

Figure 15.4 Surfaces Described by Equations with Cross Product Terms

The general form of the Scheffé quadratic model is

$$Y = \sum_{i=1}^{k} b_i X_i + \sum_{i<j=2}^{k} b_{ij} X_i X_j \qquad (15.2)$$

Frequently in mixture experiments a special cubic model is needed to represent the interaction between mixing components. The form of the special cubic model is:

$$Y = \sum_{i=1}^{k} + \sum_{i<j=2}^{k} b_{ij} X_i X_j + \sum_{i<j<k=3}^{k} b_{ijk} X_i X_j X_k \qquad (15.3)$$

524

An example of a special cubic model for k = 3 components is

$$Y = b_1X_1 + b_2X_2 + b_3X_3 + b_{12}X_1X_2 + b_{13}X_1X_2 + b_{23}X_2X_3 + b_{123}X_1X_2X_3$$

and the cubic term b_{123} represents the difference in the response at the mixture $X_1 = 1/3$, $X_2 = 1/3$, $X_3 = 1/3$ and the value predicted at that point by the quadratic model.

15.3 Experimental Designs for Mixture Problems

15.3.1 Unconstrained Problems

Unconstrained mixture problems are encountered when studying mixtures which may be 100 percent one component. For example, if when studying the properties of a lubricant made up of mixtures of various types of mineral and synthetic oils, there would be no restriction on the proportion of one type oil in the mixture (i.e., it could be 100 percent mineral oil, 100 percent synthetic, or anything in between). For this type of problem, the best experimental designs available are the simplex designs.

The simplex designs for linear models consist of one observation taken at each vertex. For the three-component mixture, this would consist of the points shown in Table 15.1. The responses to be recorded at the three vertices are symbolically labeled Y_1, Y_2, and Y_3.

Table 15.1 Simplex Design in Three Components

Run No.	X_1	X_2	X_3	Symbolic Response
1	1	0	0	Y_1
2	0	1	0	Y_2
3	0	0	1	Y_3

Since there are three parameters to be estimated in the linear model, this design has no extra runs to estimate experimental error or test the goodness of fit. This can be compensated for by replicating the design or by adding a replicated center point. Adding a replicated center point allows for making a lack-of-fit test to determine if the linear model is adequate. The lack-of-fit test is determined by comparing the variation between the observed and predicted responses at the center point to the variation among the replicate responses at the center point. An example of a three-component simplex design with a replicated center point is shown in Table 15.2.

Table 15.2 Simplex Design with Replicated Center Point

Run No.	X_1	X_2	X_3	Symbolic Response
1	1	0	0	Y_1
2	0	1	0	Y_2
3	0	0	1	Y_3
4	1/3	1/3	1/3	$Y_{123(1)}$
5	1/3	1/3	1/3	$Y_{123(2)}$
6	1/3	1/3	1/3	$Y_{123(3)}$

The simplex designs for quadratic models include points at the midpoint of the lines connecting each vertex in order to allow estimation of the nonlinear effects. For example, the three-component simplex design for a quadratic model is shown in Table 15.3.

Table 15.3 Simplex Quadratic Design

Run No.	X_1	X_2	X_3	Symbolic Response
1	1	0	0	Y_1
2	0	1	0	Y_2
3	0	0	1	Y_3
4	½	½	0	Y_{12}
5	½	0	½	Y_{13}
6	0	½	½	Y_{23}
7	1/3	1/3	1/3	$Y_{123(1)}$
8	1/3	1/3	1/3	$Y_{123(2)}$
9	1/3	1/3	1/3	$Y_{123(3)}$

The replicated center points in the quadratic simplex design allow for both testing the lack of fit from a quadratic model and fitting the special cubic model to the data.

These designs are shown graphically in Figures 15.5 and 15.6 along with formulas for computing the coefficients in the Scheffé models. These formulas are defined as the solution to the linear equations defined by Equation 15.2 and Table 15.3.

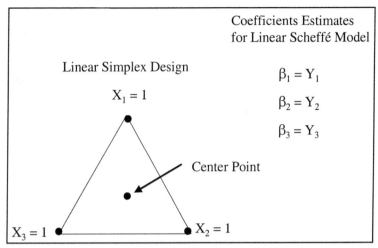

Figure 15.5 Linear Simplex Design and Scheffé Model Coefficient Estimates

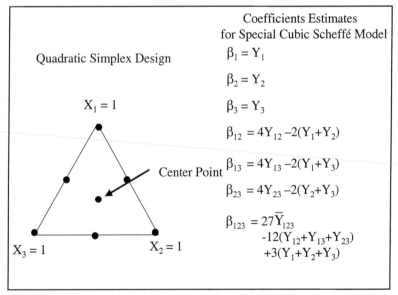

Figure 15.6 Quadratic Simplex Design and Scheffé Special Cubic Model Coefficient Estimates

The generalization of these designs to more components is straight forward. In Appendix B, Tables B.6-1 to B.6-5 show quadratic mixture designs for 3–7 components along with formulas for estimating the coefficients in the special cubic model. A simplex linear design can be created from these tables by including only the vertex points and overall center points.

15.3.2 An Example

In agricultural field tests two or more herbicides are often mixed together in so-called "tank mixes" in attempt to find a mixture which is more effective than individual components in controlling pest weeds. In a specific test, various combinations of herbicide A (formulated to control broad leaf weeds), herbicide B (formulated to control seedling grasses), and a general purpose herbicide C were mixed together in the proportions shown in Table 15.4. Each mixture was applied to a randomly selected plot of corn and the weed control index in percent was recorded as the response.

Table 15.4 Proportion of Herbicide

Mixture	A	B	C	% Weed Control
1	1.0	0.0	0.0	$y_1 = 73$
2	0.0	1.0	0.0	$y_2 = 68$
3	0.0	0.0	1.0	$y_3 = 80$
4	0.5	0.5	0.0	$y_{12} = 77$
5	0.5	0.0	0.5	$y_{13} = 86$
6	0.0	0.5	0.5	$y_{23} = 75$
7	0.33	0.33	0.33	$y_{123(1)} = 92$
8	0.33	0.33	0.33	$y_{123(2)} = 93$
9	0.33	0.33	0.33	$y_{123(3)} = 88$

The design can be seen to be a quadratic simplex design with three replicates of the center point. The coefficients in the quadratic Scheffé model can be estimated as follows:

$$\hat{b}_1 = 73, \ \hat{b}_2 = 68, \ \hat{b}_3 = 80, \ \hat{b}_{12} = 26, \ \hat{b}_{13} = 38, \ \hat{b}_{23} = 4$$

using the formulas shown in Figure 15.6. The predicted values and differences between actual and predicted values are shown in Table 15.5.

Table 15.5--Experimental Design, Data and Results

Mixture	A	B	C	Actual Response	Predicted	Residual
1	1	0	0	73	73	0
2	0	1	0	68	68	0
3	0	0	1	80	80	0
4	.5	.5	0	77	77	0
5	.5	0	.5	86	86	0
6	0	.5	.5	75	75	0
7	.33	.33	.33	92	81.22	10.78
8	.33	.33	.33	93	81.22	11.78
9	.33	.33	.33	88	81.22	6.78

The sum of squares which represents the variation between the actual and predicted values at the center point is

$$S_{LOF}^2 = n_c(\overline{Y}_{123} - \hat{Y}_{123})^2 = 3(91 - 81.22)^2 = 286.815$$

and has one degree of freedom. The sum of squares which represents the variation in replicated center points is:

$$S_{PE}^2 = \sum_{i=1}^{n_c}(Y_i - \overline{Y}_c)^2/(n_c - 1) = \frac{[(92 - 91)^2 + (93 - 91)^2 + (88 - 91)^2]}{2} = \frac{14}{2}$$

which has two degrees of freedom. These two sums of squares can be compared using the F-ratio

$$F_{1,2} = \frac{286.82}{7} = 40.974$$

In this case, the computed F-ratio is larger than the tabulated 5% point for one and two degrees of freedom, indicating there is a significant lack of fit to the quadratic Scheffé model. Therefore, the special cubic model

$$Y = b_1X_1 + b_2X_2 + b_3X_3 + b_{12}X_1X_2 + b_{13}X_1X_3 + b_{23}X_2X_3 + b_{123}X_1 X_2X_3$$

provides a better fit to the data.

When the average of replicated center points represented as $\bar{Y}_{123} = \left(\sum\limits_{i=1}^{n_c} Y_{123(i)}\right)/n_c$, the coefficient of the special cubic term is calculated as shown in Figure 15.6 as:

$$\hat{b}_{123} = 27(91) - 12(77 + 86 + 75) + 3(73 + 68 + 80) = 264$$

and the special cubic equation is

$$Y = 73X_1 + 68X_2 + 80X_3 + 26X_1X_2 + 38X_1X_3 + 4X_2X_3 + 264X_1X_2X_3$$

Figure 15.7 shows a contour plot of the fitted special cubic equation. In this case it can be seen that a mixture of the three components will provide more effective weed control than any one of the three components (vertex points) or mixture of any two (which would lie along the line connecting any two vertices). The exact coordinates of the best mixture is $X_1 = 0.374$, $X_2 = 0.234$, and $X_3 = 0.393$, where the weed control is predicted to be 91.8%. This can be verified by interpolating between the dotted grid lines on the graph.

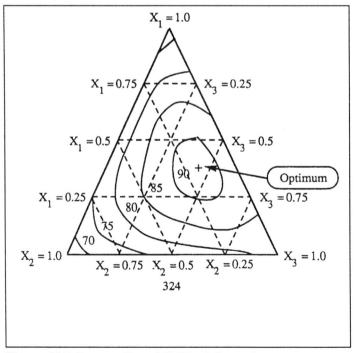

Figure 15.7 Contour Plot of % Weed Control

15.3.3   Constrained Mixture Problems

In some mixture problems it is impossible to have a product which is 100 percent pure in one component. For example, a rocket propellant which is a mixture of binder, oxidizer, and fuel cannot be 100 percent binder or any other component for that matter. The proportion of the three components can be varied, but must remain within certain constraints in order for the propellant to work at all. In this case, the experimental region is not the entire triangular plane of the unconstrained three-component mixture experiment. Instead, it is some subset of the plane. Kurotori[2] described an example where the binder in a rocket propellant must be at least 20 percent, the oxidizer must be at least 40 percent and the fuel must be at least 20 percent. In this case, the experimental region is the shaded area shown below in Figure 15.8.

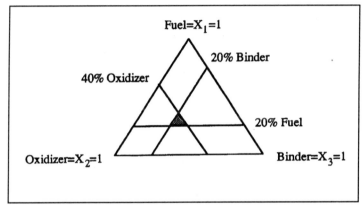

Figure 15.8 Experimental Region for Constrained Mixture Experiment

It can be clearly seen that the simplex designs will be inappropriate in this situation, since all design points except the center point are not even in the experimental region.

---

[2]Kurotori, I. S. "Experiments with Mixtures of Components Having Lower Bounds," *Industrial Quality Control*, Vol 22, (May 1966), pp. 592–596.

15.3.3.1 *Pseudo components*

In cases where each mixture component has only a lower constraint, appropriate experimental designs for constrained mixture problems can be defined by using pseudo components. Pseudo components are not pure components but a mixture of several components. If the lower constraints for the mixture component X_i is L_i (i.e., $X_i \geq L_i$) then the ith pseudo component X_i' is defined by

$$X_i' = \frac{X_i - L_i}{1 - \sum_{j=1}^{k} L_j} \tag{15.4}$$

and i^{th} pure component can be written in terms of the pseudo components

$$X_i = L_i + \left(1 - \sum_{j=1}^{k} L_j\right) X_i' \tag{15.5}$$

For example, in rocket propellant described in the last section, the upper vertex of the experimental region shown in Figure 15.8 is

Fuel	$X_1 = .40$	
Oxidizer	$X_2 = .40$	
Binder	$X_3 = .20$	

and the lower constraints for each component are

Fuel	$L_1 = .20$	
Oxidizer	$L_2 = .40$	
Binder	$L_3 = .20$	

Therefore, the pseudo components can be defined as:

$$X_1' = \frac{X_1 - L_1}{1 - \sum_{j=1}^{k} L_j} = \frac{.40 - .20}{1 - .80} = 1.0$$

$$X_2 = \frac{X_2 - L_2}{1 - \sum_{j=1}^{k} L_j} = \frac{.40 - .40}{1 - .80} = 0.0$$

$$X_3 = \frac{X_3 - L_3}{1 - \sum_{j=1}^{k} L_j} = \frac{.20 - .20}{1 - .80} = 0.0$$

All of the pseudo components will be between 0 and 1 for any mixture in the constrained experimental region, and therefore a simplex design in the pseudo components can be defined. Once this is done the actual experimental mixtures can be determined by using Equation 15.5 to convert the pseudo components back to the pure components. For example, a quadratic simplex design in three pseudo components is shown in Table 15.6. The pure components are obtained from the pseudo components; for example from Run 1,

$X_1 = .20 + (1 - .80)(1) = .40$
$X_2 = .40 + (1 - .80)(0) = .40$
$X_3 = .20 + (1 - .80)(0) = .20$

And an experimental design in the pure components is created as shown on the right half of Table 15.6. The response for this experiment was the elasticity of the mixture.

Table 15.6 Experimental Design and Data

	Pseudo component X_1' X_2' X_3'			Pure Component X_1=Fuel X_2=Oxidizer X_3=Binder			Response Elasticity
1	1	0	0	.400	.400	.200	2350
2	0	1	0	.200	.600	.200	2450
3	0	0	1	.200	.400	.400	2650
4	½	½	0	.300	.500	.200	2400
5	½	0	½	.300	.400	.300	2750
6	0	½	½	.200	.500	.300	2950
7	1/3	1/3	1/3	.266	.466	.266	3000

This design is shown graphically in Figure 15.9 (which is an enlargement of the shaded region in Figure 15.8).

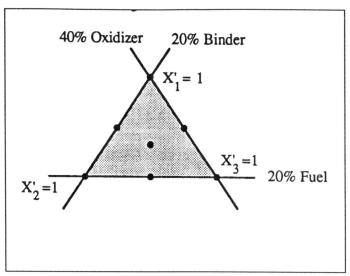

Figure 15.9 Simplex Design for Constrained Mixture
Experiment

To analyze the data in Table 15.6 the special cubic model is fit to the pseudo components. This
will be left for an exercise.

15.3.3.2 Extreme Vertices Design

In many mixture problems there are both upper and lower constraints on each component, i.e.,
$L_i \le X_i \le U_i$ for each component X_i. For example, coke briquettes, which are used as a reducing
agent in metal production in an electric smelter furnace, are made by mixing certain proportions
of coal calcinate, tar solids, and solids free of tar. These components are pressed into fixed size
briquettes, and then baked at a constant temperature. It is known that the fixed carbon of the final
briquettes is a function of the mixture components, and it is desirable to determine what mixture
proportions will produce the highest fixed carbon.

This is a constrained mixture problem because the three components must be within the
following limits in order for satisfactory coking:

80 percent $\le$ calcinate $\le$ 90 percent
 8 percent $\le$ solid-free tar $\le$ 15 percent
 0 percent $\le$ tar solids $\le$ 5 percent

In this problem, the experimental region is irregular and the pseudo components cannot be used
to define an experimental design. A useful experimental design for this situation is the
extreme vertices designs of Mclean and Anderson[3]. These designs consist of all the extreme

---

[3]Mclean, R. and Anderson, V. "Extreme Vertices Designs of Mixture Experiments,"
Technometrics, Vol 8, (1966), pp 447-454.

vertexes of the irregular experimental region. For example, the constrained experimental region and design points for the coke briquette problem are shown in Figure 15.10.

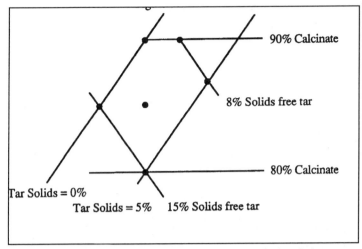

Figure 15.10 Constrained Experimental Region for Coke Briquette Experiment

A listing of these vertices is shown in Table 15.7.

Table 15.7 Extreme Vertices for Coke Briquette Experiment

Run No.	Calcinate	Tar	Tar Solids	
1	.90	.10	0.00	
2	.90	.08	0.02	
3	.87	.08	0.05	
4	.80	.15	0.05	
5	.85	.15	0.00	
6	.864	.112	0.024	(Centroid)

The centroid point is formed by averaging the components from all the vertices. Often midpoints of the sides are added to the list of experimental mixtures in order to fit the coefficients in the quadratic or special cubic model. These midpoints are found by averaging the components of the two corresponding vertices.

When there are more than three mixture components, it is difficult to visualize the experimental region or determine the extreme vertices graphically. Snee and Marquart[4] have developed an

---

[4]Snee, R. D. and Marquart, D. W. "Extreme Vertices Designs for Linear Mixture Models," *Technometrics*, Vol. 16, (1974), pp. 399–408.

algorithm for locating the extreme vertices in these more complicated situations. The algorithm is a five-step procedure which is illustrated with the coke briquette problem.

1. Rank the components in order of increasing range $U_i - L_i$.

Component	U_i	L_i	Range	Rank
X_1	.90	.80	.10	3
X_2	.15	.08	.07	2
X_3	.05	.00	.05	1

2. Set up a two-level factorial design in the k-1 components with the smallest ranges using the upper and lower constraint values as factor levels.

Run	$\underline{X_3}$	$\underline{X_2}$
1	.00	.08
2	.05	.08
3	.00	.15
4	.05	.15

3. Calculate the value of the omitted component X_k (with largest range) as

$$1.0 - \sum_{i=1}^{k-1} X_i \text{ for each of the } 2^{k-1} \text{ runs listed in step 2.}$$

Run	X_3	X_2	X_1
1	.00	.08	$\underline{.92}$
2	.05	.08	$\underline{.87}$
3	.00	.15	$\underline{.85}$
4	.05	.15	$\underline{.80}$

4. If the calculated value of X_k in step 3 falls within its upper and lower constraint (as runs 2, 3, and 4 do) these runs are extreme vertices. The other runs, such as 1, must be adjusted.

5. For each of the runs where X_k is out of its lower to upper constraint range, adjust X_k to its nearest limit, and form k-1 new runs where each of the other k-1 components are in turn adjusted by the same magnitude but in the opposite direction as X_k, so that $\sum_{i=1}^{k-1} X_i = 1.0$.

536

Finally the new values for the first k-1 components are checked to see if they fall within their constraints. If they all are within their constraints, the adjusted k-1 runs are added to the list of extreme vertices, in place of the original run where X_k was out of its constraint range.

Run No.	X_1	X_2	X_3	
1	.02	.08	.90	(Adjusting X_1 and X_3)
	.00	.10	.90	(Adjusting X_2 and X_3)
2	.05	.08	.87	
3	.00	.15	.85	
4	.05	.15	.80	

This algorithm may duplicate some vertices, but it will find all of the extreme vertices. Piepel[5] has published FORTRAN code for a similar algorithm that can be used for locating extreme vertices in up to 12 components.

When fitting a quadratic or special cubic model in a constrained region, face centroids and the over all centroid are usually added to the extreme vertices design. The face centroids can be found by locating all extreme vertices that have k-3 components equal, and averaging the levels of the other three components. For example, Anderson and McLean[6] present a constrained mixture experiment where the formula for a flare is obtained by mixing four chemicals, X_1 = Magnesium, X_2 = Sodium Nitrite, X_3 = Strontium Nitrate, and X_4 = Binder, together with the following constraints.

$$.40 \le X_1 \le .60$$
$$.10 \le X_2 \le .50$$
$$.10 \le X_3 \le .50$$
$$.03 \le X_4 \le .08$$

In this case k=4 and the constrained region is shown in Figure 15.11. Table 15.8 shows the eight extreme vertices (Runs 1 - 8) of the constrained region. It can be seen that Runs 1 - 4 all have the same values for k - 3 = 4 - 3 = 1 component (namely X_2). Therefore, the average of the

---

[5]Piepel, G. F., "Programs for Generating Extreme Vertices and Centroids of Linearly Constrained Experimental Regions," *Journal of Quality Technology*, Vol. 20, (1988), pp. 125–139.

[6]Anderson, V. L. and McLean, R. A. (1974) *Design of Experiments: a Realistic Approach*,(New York: Marcel Dekker 1974), chapter 13.

components for these 4 runs is Run 9, the face centroid of the side defined by the vertices 1, 2, 3 and 4. See Figure 15.11. Runs 5 through 8 all have a common value for X_3 (lower constraint)

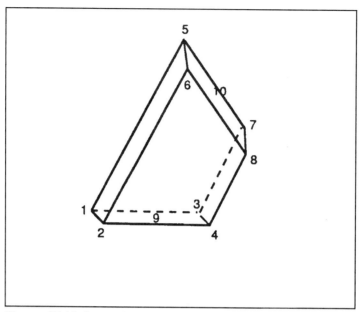

Figure 15.11 Constrained Experimental Region Flare Experiment

and therefore Run 10 (the average of the component values for these runs) is the face centroid for the side defined by the vertices 5, 6, 7, and 8. The other face centroids are defined as shown in Table 15.8, and the overall centroid (Run 15) is the average of the components for all the vertices.

Table 15.8 Extreme Vertices Design for Flare Experiment

					Illumination	
Run	X_1	X_2	X_3	X_4	Y	
1	.40	.1000	.4700	.030	75	⎫
2	.40	.1000	.4200	.080	180	⎪
3	.60	.1000	.2700	.030	195	⎬
4	.60	.1000	.2200	.080	300	Extreme Vertices
5	.40	.4700	.1000	.030	145	⎪
6	.40	.4200	.1000	.080	230	⎪
7	.60	.2700	.1000	.030	220	⎪
8	.60	.2200	.1000	.080	350	⎭
9	.50	.1000	.3450	.055	220	(1+2+3+4)/4
10	.50	.3450	.1000	.055	260	(5+6+7+8)/4
11	.40	.2725	.2725	.055	190	(1+2+5+6)/4
12	.60	.1725	.1725	.055	310	(3+4+7+8)/4
13	.50	.2350	.2350	.030	260	(1+3+5+7)/4
14	.50	.2100	.2100	.080	410	(2+4+6+8)/4
15	.50	.2225	.2225	.055	425	Centroid

15.4 Data Analysis and Model Fitting for General Constrained Mixture Problems

When an extreme vertices design is used there are no longer simple formulas to calculate estimates of the b's in Equation 15.1 or 15.2 from the observed experimental data. In this case, a linear regression routine must be used to estimate the b-parameters of the mixture model. To check how well the model fits the data, the R^2, t-statistic and F-statistic discussed in Sections 14.4 - 14.5, and the goodness-of- fit F-statistic and residual plots discussed in Section 14.6 are appropriate. Their application to mixture models will be discussed in Section 15.4.1. The method of contour plotting and numerical analysis of a fitted model discussed in Section 14.9, is also applicable to the mixture model, and will be discussed in Section 15.4.2.

15.4.1 Least Squares Model Fitting Examples

The method of least squares can be used to fit both the slack variable form of the model, Equation 15.1, or the Scheffé form of the model, Equation 15.2. For example, assume the mixture experiments were run for the coke briquetting problem and the following crushing strengths were observed:

Run No.	X_1	X_2	X_3	Crushing Strength
1	.90	.10	.00	1122
2	.90	.08	.02	1109
3	.87	.08	.05	1103
4	.80	.15	.05	1118
5	.85	.15	.00	1130
6	.85	.12	.03	1114

To fit the slack variable linear model to this data, the X and Y matrices, as defined in Section 14.4, would be

$$
Y = \begin{bmatrix} 1122 \\ 1109 \\ 1103 \\ 1118 \\ 1130 \\ 1114 \end{bmatrix} \qquad X = \begin{bmatrix} 1 & .10 & .00 \\ 1 & .08 & .02 \\ 1 & .08 & .05 \\ 1 & .15 & .05 \\ 1 & .15 & .00 \\ 1 & .12 & .03 \end{bmatrix}
$$

where the major component (calcinate = X_1) is designated as the slack variable.

The normal equations are

$$\text{X}^\text{T}\text{X} \qquad\qquad \text{B} \quad \text{X}^\text{T}\text{Y}$$

$$\begin{bmatrix} 6 & .68 & .15 \\ .68 & .0822 & .0167 \\ .15 & .0167 & .0063 \end{bmatrix} \begin{bmatrix} b_o \\ b_2 \\ b_3 \end{bmatrix} = \begin{bmatrix} 6696 \\ 760.04 \\ 166.65 \end{bmatrix}$$

and the solution is $\hat{\text{B}} = (\text{X}^\text{T}\text{X})^{-1}\,\text{X}^\text{T}\text{Y}$

$$\begin{bmatrix} \hat{b}_o \\ \hat{b}_2 \\ \hat{b}_3 \end{bmatrix} = \begin{bmatrix} 3.0637 & -22.807 & -12.4872 \\ -22.807 & 196.1538 & 23.0769 \\ -12.4872 & 23.0769 & 394.8718 \end{bmatrix} \begin{bmatrix} 6696 \\ 760.04 \\ 166.65 \end{bmatrix} = \begin{bmatrix} 1098.908 \\ 210.231 \\ -269.385 \end{bmatrix}$$

Therefore, the fitted slack variable linear model is

y = 1098.908 + 210.231X$_2$ - 269.385X$_3$

Using this equation the predicted values can be found for each observation as:

$$\begin{bmatrix} 1119.932 \\ 1110.339 \\ 1102.258 \\ 1116.974 \\ 1130.443 \\ 1116.055 \end{bmatrix} = \begin{bmatrix} 1 & .10 & .00 \\ 1 & .08 & .02 \\ 1 & .08 & .05 \\ 1 & .15 & .05 \\ 1 & .15 & .00 \\ 1 & .12 & .03 \end{bmatrix} \begin{bmatrix} 1.098.908 \\ 210.231 \\ -269.385 \end{bmatrix}$$

$$\hat{Y} \quad = \quad X \quad \hat{B}$$

The residuals are

$$
Y \quad - \quad \hat{Y} \qquad\qquad Y - \hat{Y}
$$

$$
\begin{bmatrix} 1122 \\ 1109 \\ 1103 \\ 1118 \\ 1130 \\ 1114 \end{bmatrix}
-
\begin{bmatrix} 1119.932 \\ 1110.339 \\ 1102.258 \\ 1116.974 \\ 1130.443 \\ 1116.055 \end{bmatrix}
=
\begin{bmatrix} 2.068 \\ -1.339 \\ 0.742 \\ 1.026 \\ -0.443 \\ -2.054 \end{bmatrix}
$$

$\bar{Y} = 1116$

$SSE = (2.068)^2 + (-1.339)^2 + \ldots + (-2.054)^2 = 12.094$

$SST = (1122^2 + 1109^2 + \ldots 1114^2) - 6(1116)^2 = 7,473,194 - 7,472,736 = 458$ and

$SSR = SST - SSE = 458 - 12.094 = 445.906$

The overall F-statistic, $F_{2,3} = \dfrac{445.906/2}{12.094/3} = 55.31$ is significant (i.e., $55.31 > F_{2,3(.05)} = 9.55$ from

Table A.6), and the fact that $R^2 = \dfrac{445.906}{458} = .916$ indicates that 91.6% of the variability in

crushing strength can be explained by the relationship with the three mixing proportions X_1, X_2, and X_3. The estimate of the error standard deviation $s = \sqrt{SSE/(n-p-1)} = \sqrt{12.094/3} = 2.008$,

and the standard errors of the regression coefficients $\hat{b}_0$, $\hat{b}_2$, and $\hat{b}_3$ are shown in Table 15.9.

Table 15.9 Worksheet for Calculating t-Statistics

Coefficient	Value	c_{ii} - Diagonal of $(X'X)^{-1}$	Standard Error of Coefficient $s_b = s \sqrt{c_{ii}}$	t-statistic
$\hat{b}_0$	1098.908	3.0637	3.5144	312.69
$\hat{b}_2$	210.2308	196.1538	28.1203	7.48
$\hat{b}_3$	-269.3846	394.8718	38.8979	-6.75

From the t-ratios it can be seen that both $\hat{b}_2$ and $\hat{b}_3$ are significant (since $t > t_{3\,(.95\%)} = 3.182$ in Table A.5). This indicates that the difference between the response values at the vertex $(X_1 = 0, X_2 = 1, X_3 = 0)$ and the vertex $(X_1 = 1, X_2 = 0, X_3 = 0)$ and the difference between the response values at the vertex $(X_1 = 0, X_2 = 0, X_3 = 1)$ and the vertex $(X_1 = 1, X_2 = 0, X_3 = 0)$ are significantly different from zero. In other words, the response plane above the simplex region, as shown in Figure 15.4 in Section 15.2, is not parallel to the simplex plane. The significance of the b_0 coefficient indicates that the response value above the vertex $(X_1 = 1, X_2 = 0, X_3 = 0)$ is significantly greater than zero.

To fit the Scheffé form of the model, the X-matrix would be of the form

$$X = \begin{bmatrix} .90 & .10 & .00 \\ .90 & .08 & .02 \\ .87 & .08 & .05 \\ .80 & .15 & .05 \\ .85 & .15 & .00 \\ .85 & .12 & .03 \end{bmatrix}$$

and the Normal equations would be

$$
\begin{array}{ccc}
\text{X}^\text{T}\text{X} & \text{B} & \text{X}^\text{T}\text{Y}
\end{array}
$$

$$
\begin{bmatrix} 4.4619 & .5811 & .127 \\ .5811 & .0822 & .0167 \\ .127 & .0167 & .0063 \end{bmatrix}
\begin{bmatrix} b_1 \\ b_2 \\ b_3 \end{bmatrix}
=
\begin{bmatrix} 5769.31 \\ 760.04 \\ 166.65 \end{bmatrix}
$$

and the fitted model is summarized in Table 15.10

Table 15.10 Summary of Fitted Coefficients

Coefficient	Value	Standard Error of Coefficient	t-statistic
$\hat{b}_1$	1098.908	3.5144	312.69
$\hat{b}_2$	1309.139	24.884	52.61
$\hat{b}_3$	829.524	38.775	21.39

In this case all the coefficients $\hat{b}_1$, $\hat{b}_2$, and $\hat{b}_3$ are all significant (i.e., > 3.182 from Table A.5). The interpretation of the Scheffé coefficients is that the responses above the pure components are significantly different from zero as described in Section 15.2. Better ways of interpreting the fitted model will be explained in Section 15.4.2.

Computer programs are normally used to fit equations to mixture data. Table 15.11 shows a portion of the results of fitting the Scheffé quadratic model,

$$y = b_1X_1 + b_2X_2 + b_3X_3 + b_4X_4 + b_{12}X_1X_2 + b_{13}X_1X_3 + b_{14}X_1X_4 + b_{23}X_2X_3 + b_{24}X_2X_4 + b_{34}X_3X_4$$

to the flare data from Table 15.7, using the MINITAB regression procedure. The no-intercept option was used to fit this model.

Table 15.11 Multiple Regression Report

Regression Analysis

```
* NOTE *        X1 is highly correlated with other  predictor variables
* NOTE *        X2 is highly correlated with other  predictor variables
* NOTE *        X3 is highly correlated with other  predictor variables

*         X4 is highly correlated with other X variables
*         X4 has been removed from the equation

* NOTE *        X1 is highly correlated with other  predictor variables
* NOTE *        X2 is highly correlated with other  predictor variables
* NOTE *        X3 is highly correlated with other  predictor variables
* NOTE *      X1X2 is highly correlated with other  predictor variables
* NOTE *      X1X3 is highly correlated with other  predictor variables
```

The regression equation is
Y = - 1533 X1 - 2298 X2 - 2373 X3 + 7999 X1X2 + 7776 X1X3 + 9332 X1X4
         + 3055 X2X3 - 911 X2X4 - 1040 X3X4

Predictor	Coef	Stdev	t-ratio	p
Noconstant				
X1	-1533.3	817.0	-1.88	0.110
X2	-2298.0	895.4	-2.57	0.043
X3	-2373.1	895.4	-2.65	0.038
X1X2	7999	3321	2.41	0.053
X1X3	7776	3321	2.34	0.058
X1X4	9332	4433	2.11	0.080
X2X3	3055	1723	1.77	0.127
X2X4	-911	5217	-0.17	0.867
X3X4	-1040	5217	-0.20	0.849

s = 55.09
Analysis of Variance

SOURCE	DF	SS	MS	F	p
Regression	9	1057193	117466	38.71	0.000
Error	6	18207	3034		
Total	15	1075400			

The regression output in Table 15.11 illustrates one of the problems that often occurs when fitting prediction equations to data from constrained mixture designs by the least squares regression procedure. The problem is a high degree of correlation between the independent variables x_1, x_2, x_3, etc. The first ten lines in the MINITAB results are notes to the user of the program, indicating these correlations. The correlations are due to the highly irregular, experimental region illustrated in Figure 15.11. When there is too much correlation between the independent variables in the X^TX matrix, the normal equations will be singular within machine precision, and a computer program will not be able to calculate the regression coefficients. Even in cases where the X^TX^{-1} matrix does exist, a high degree of correlation between the independent variables (often refered to as multicollinearity) can cause difficulties. The variance of the estimated regression coefficients is inflated by multicollinearity, and calculated values may have the wrong magnitude and/or sign. Inconsistent results are often reached when two different people try to fit a reduced model to the data by eliminating insignificant terms, or when automatic computer routines are used to select a subset of the independent variables (i.e., stepwise regression).

The regression procedure in MINITAB automatically eliminated one of the independent variables (x_4) in an attempt to resolve the problem of multicollinearity. Since x_4 is highly correlated with the other dependent variables in the equation, adding it to the equation will not improve the accuracy of fitting the 15 data points, but it will cause the difficulties due to multicollinearity described above. By removing x_4 from the list of independent variables, we have no coefficient to indicate the level of the predicted surface above the vertex where $x_4 = 1$, as described in Section 15.2, and the coefficients of x_1x_4, x_2x_4, etc. will have less meaning. However, if the purpose of the mixture experiment is to develop a prediction equation within the constrained experimental region shown in Figure 15.11 rather than interpret the actual coefficients, this will be an acceptable solution. The regression equation found in Table 15.11 will be accurate for predicting within the experimental region shown in Figure 15.11, but it could be very inaccurate for extrapolating outside the experimental region and should not be used for that purpose.

Some computer regression programs will calculate regression coefficients as long as the $(X^TX)^{-1}$ matrix exists, and will not warn the user of potential problems due to multicollinearity. In this case the user should examine the standard errors of the regression coefficients or the diagonal elements of $(X^TX)^{-1}$. Extremely large values indicate a problem. A better indicator of multicollinearity are the VIF (variance inflation factors) that are optionally printed out by some computer regression programs. St. John[7] describes the use of VIF's to detect multicollinearity in mixture experiments and illustrates a solution to the problem using ridge regression. Ridge regression will provide an equation that is not quite as accurate for predicting the actual experimental data within the experimental region, but the coefficients are less likely to have the wrong sign or magnitude and the prediction equation will be more stable for extrapolations slightly outside the constrained experimental region.

15.4.2 A Procedure for Choosing the Correct Model

Due to synergism or antagonism of mixing components, highly nonlinear response surfaces often result in mixture experiments. If an estimate of the experimental variance is available from

---

[7]St. John, R. C. "Experiments with Mixtures, Ill-Conditioning, and Ridge Regression," *Journal of Quality Technology*, Vol. 16, No. 2, April 1984.

replicated experimental design points, Khuri and Cornell[8] have recommended the following model fitting procedure.

1. Fit the linear model

$$Y = \sum_{i=1}^{k} b_i X_i$$

and test for lack of fit. If the lack of fit is significant try the quadratic model, Equation 15.2; otherwise assume the linear model is adequate.

2. Fit the quadratic model

$$Y = \sum_{i=1}^{k} b_i X_i + \sum_{i<j=2}^{k} b_{ij} X_i X_j$$

and test for lack of fit. If the lack of fit is significant try the special cubic model; otherwise assume the quadratic model is adequate.

3. Fit the special cubic model

$$Y = \sum_{i=1}^{k} b_i X_i + \sum_{i<j=2}^{k} b_{ij} X_i X_j + \sum_{i<j<k=3}^{k} b_{ijk} X_i X_j X_k$$

and test for lack of fit. If the lack of fit is significant use the full cubic model; otherwise assume the special cubic model is adequate.

4. Fit the full cubic model

$$Y = \sum_{i=1}^{k} b_i X_i + \sum_{i<j=2}^{k} b_{ij} X_i X_j + \sum_{i<j<k=3}^{k} b_{ijk} X_i X_j X_k + \sum_{i<j=2}^{k} c_{ij} X_i X_j (X_i - X_j)$$

---

[8]Khuri, A. I., and Cornell, J. A. , *Response Surfaces*, (New York: Marcel Dekker, 1987), chapter 9.

15.4.3 Interpretation of Fitted Models

Interpretation of the coefficients in the Scheffé mixture models fit by least squares is different than the interpretation of nonmixture quadratic models, such as those fit in Chapter 14. When a linear coefficient $\hat{b}_i$ is not significantly different from zero, as determined by the t-statistic $t = \hat{b}_i/s_{\hat{b}_i}$, it does not mean that the component X_i has no significant effect on the response. A linear coefficient b_i, that is zero implies that the predicted response is zero for the pure mixture $X_i = 1.0$.

Each linear coefficient, b_i, in a nonmixture model is a direct measure of the effect, or change in the response, Y, that will be caused by a one-unit change in the coded independent variable X_i while all other independent variables or factors are held constant. In the mixture model case it is impossible to change one component while holding all others constant because all components must sum to 100%. Therefore, Snee and Marquart[9] suggested a more appropriate measure of the effect of a mixture component X_i. Their measure is called the *total effect*, and it is calculated as

$$E_i = \hat{b}_i - (k-1)^{-1} \sum_{j \neq i}^{k} \hat{b}_j.$$ This quantity measures the difference in the predicted values at the

vertex, $X_i=1.0$, and the point $X_i=0$, $X_j= \dfrac{1}{k-1}$, for $j \neq i$. This is illustrated for a three-component

case in Figure 15.12.

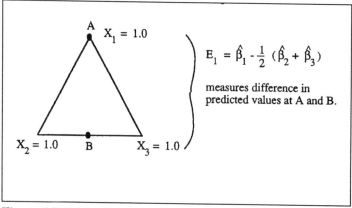

$$E_1 = \hat{\beta}_1 - \frac{1}{2}(\hat{\beta}_2 + \hat{\beta}_3)$$

measures difference in predicted values at A and B.

Figure 15.12 The Total Effect of X_1

[9]Snee, R. D. and Marquart, D. W. , "Screening Concepts and Designs for Experiments with Mixtures," *Technometrics*, Vol. 18, (1976), pp. 19–29.

The total effect, E_i, is a good way of expressing the effects of the variables in linear models fit in unconstrained mixture regions. However if there are constraints of the form $L_i \leq X_i \leq U_i$ or if higher order terms such as quadratic or special cubic are included in the model, Piepel[10] has suggested a simple Effect plot which provides a more accurate way of comparing the effects of several components. The plot is constructed by calculating the predicted values at two or more points along the line connecting the centroid of the experimental region and the vertex for the ith pseudo component. This is illustrated using the constrained experimental region for the coke briquette example of Section 15.3.3.2 in Figure 15.13.

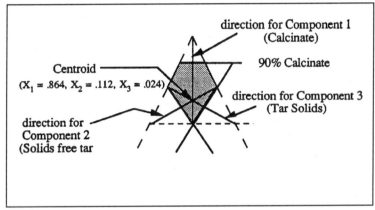

Figure 15.13 Directions for Effect Plot

Legend

Region Outlined by – – – – represents pseudo component

space defined by $\qquad X_i' = \dfrac{X_i - L_i}{1 - L}$

Shaded region represents constrained experimental region

Since the linear model $Y = \sum\limits_{i=1}^{k} \hat{b}_i X_i$ was fit to the briquette data, predicted values need to

[10]Piepel, G. "Measuring Component Effects in Constrained Mixture Experiments," *Technometrics*, Vol. 24, (1982), pp. 29–39.

be calculated only at the centroid and pseudo component verticies (represented in terms of the pure components as shown in Section 15.3.3.1).This is illustrated in the worksheet shown in Table 15.12. The predicted values are obtained using the equation coefficients shown in Table 15.10. The Effect plot is then constructed by plotting the predicted value versus the percentage change from the centroid for the ith component as shown in Table 15.12. The plot for this example is shown in Figure 15.14.

Table 15.12 Worksheet for Effect Plot

	$\underline{X}_1$	$\underline{X}_2$	$\underline{X}_3$	Predicted Value	% Change from Centroid $100\%(X_i - X_{ic})/(U_i - L_i)$		
Centroid	0.864	.112	.024	1115.988	X_1	X_2	X_3
X'_1 vertex	0.920	0.08	0.0	1115.726	+56%		
X'_2 vertex	0.080	0.20	0.0	1140.954		+126%	
X'_3 vertex	0.800	0.08	.012	1083.400			+192%

The X_i' in the worksheet represents the X_i component in the X_i pseudo component vertex, X_{ic} represents the X_i component in the centroid, and U_i and L_i are the upper and lower constraints for component I. The percent change $100\% (X_i - X_{ic}) / (U_i - L_i)$ is calculated for component 1 as $(100\%)(.92 - .864) / (.90 - .80) = 56\%$. In Figure 15.14, it can be seen that the effect of X_1 (Calcinate) is minor, while the effects of X_2 (Solids free tar) and X_3 (Tar solids) are stronger with opposite signs.

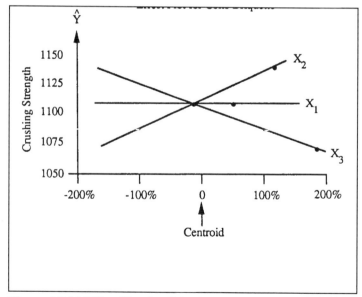

Figure 15.14 Effect Plot for Coke Briquette

550

If a quadratic or higher order model has been fit to the data, the predicted values should be calculated at least four points along the line connecting the centroid and the vertex in order to show the curvature. This is illustrated in Table 15.13 for the Herbicide example of Section 15.3.2. In the Effect plot for this example, shown in Figure 15.15, it can be seen that X_2 has a strong negative and nearly linear effect, while X_1 and X_3 have quadratic effects. This indicates that there is synergism between X_1 and X_3, and that a lower percent of X_2 is desirable.

Table 15.13 Worksheet for Effect Plot for Quadratic Model
(Herbicide Example)

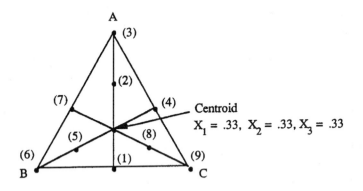

Centroid
$X_1 = .33$, $X_2 = .33$, $X_3 = .33$

Worksheet

		X_1	X_2	X_3	Predicted Value	%Change from Centroid $100\%(X_i - X_{ic})/(U_i - L_i)$		
						$\underline{X}_1$	$\underline{X}_2$	$\underline{X}_3$
	Centroid	.333	.333	.333	81.21			
Direction for A=X_1	(1)	0.00	0.5	0.5	75.00	-33%		
	(2)	.667	.167	.167	80.65	+33%		
	(3)	1.0	0.0	0.0	73.00	+67%		
Direction for B=X_2	(4)	0.5	0.0	0.5	86.00		-33%	
	(5)	.167	.667	.167	75.31		+33%	
	(6)	0.0	1.0	0.0	68.00		+67%	
Direction for C=X_3	(7)	0.5	0.5	0.0	77.00			-33%
	(8)	.167	.167	.667	82.31			+33%
	(9)	0.0	0.0	1.0	80.00			+67%

551

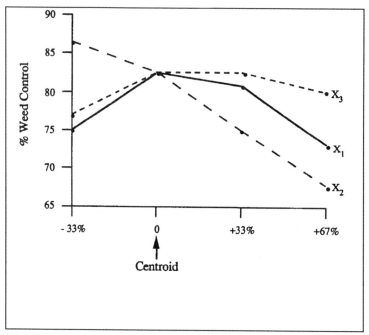

Figure 15.15 Effect Plot for Herbicide

15.4.4 Identifying Optimum Conditions

When mixture designs are used in the response surface situation, the ultimate goal is to identify the optimum mixture. This can be done in the same way that it is done in nonmixture response surfaces, through use of contour plots, or numerical methods.

Contour plots over the simplex mixture experimental region have been illustrated previously in this chapter (e.g., Figure 15.7). Computer programs[11,12] have been published for producing these plots. When more than three mixture components are involved and constraints are placed on the experimental region, contour plots at various slices may be used to identify the optimum mixture. For example, considering the constrained experimental region for the flare mixture problem, contour plots can be made at three slices $X_1 = 0.40$, $X_1 = 0.50$, and $X_1 = 0.60$ as shown in Figure 15.16.

---

[11]Hare, L. B. and Brown, P. L., "Plotting Response Surface Contours on a Three Component Mixture Space," *Journal of Quality Technology*, Vol. 9, (1977), pp. 193–197.

[12]Koons, G. F. and Heasly, R. H., "Response Surface Contour Plots for Mixture Problems," *Journal of Quality Technology*, Vol. 13, (1981), pp. 207–214.

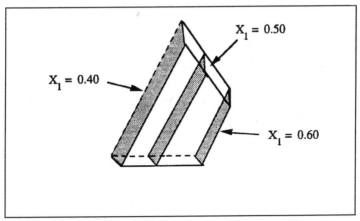

Figure 15.16 Slices Through Constrained Region Where
Contour Plots Are Made

The resulting contour plots with the experimental region shaded are shown in Figure 15.17. From the contour plots it can be seen that the greatest illumination within the shaded experimental region is somewhat over 300 in the middle or bottom graphs where $X_1 = .5$ or .6 and along the right-most boundary where $X_4 = .08$.

When more than four mixture components are involved, it becomes increasingly more difficult to identify the optimum mixture through contour plots. Determination of the stationary point as described in Section 14.9 for nonmixture response surfaces is also more complicated for mixture response surfaces, due to the cubic terms in the response equations. Therefore, one of the most common methods of locating the optimum mixture is through the use of nonlinear programming as mentioned in Section 14.9.2. Using nonlinear programming, the optimum mixture for the flare is $X_1 = .523$, $X_2 = .230$, $X_3 = .164$, and $X_4 = .08$ where illumination is predicted to be 392.

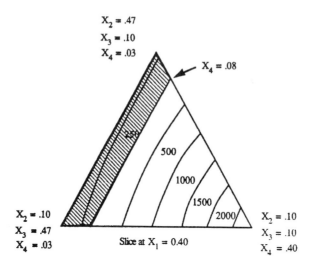

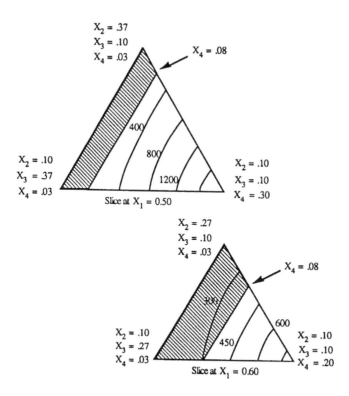

Figure 15.17 Contour Plots for Flare Mixture

554

15.5 Screening Experiments with Mixture Components

15.5.1 Designs for Screening Experiments with Mixtures

At the beginning of any program of experimentation, the experimental environment should be diagnosed as explained in Section 2.2.3. How many factors or mixture components are being studied? What is the cost of experiments, and how much prior knowledge is there about the relationship of the response to mixture components? When there are a large number of mixture components under study with little understanding as to which ones are really important, it is wasteful to conduct a response surface type experiment. A screening experiment aimed at identifying the most important components is the natural first step and may reduce the amount of experimentation required in the long run. A screening experimental design should allow for fitting a linear model with as few experiments as possible.

For nonmixture problems the Plackett-Burman design and fractional factorial design were used in screening situations to allow fitting a linear model. In mixture problems these designs cannot be used directly because of the constraints on the experimental region. Extreme vertices designs have been described in Section 15.3.3.2, and without edge midpoints or face centroids, these designs can be used to fit a linear model. However, in a screening situation where there may be 10 or more mixture components, a constrained experimental region may contain 100 or more extreme vertices. Therefore, a subset of the vertices from the McLean-Anderson Extreme Vertices design is normally used for a screening design[13].

The subset of extreme vertices which minimizes the variances of the estimated regression coefficients, b_i's, is the subset which maximizes the determinant of the X'X matrix used in the regression calculations. This is the subset that should be used for a screening design.

To illustrate the principle, consider the six extreme vertices for the coke briquetting example described in Sections 15.3.3.2 and 15.4. If a four-vertex subset were to be selected for fitting the linear model, there would be five possible designs. One such design would be the first four vertices. The X matrix for this design would be:

$$X = \begin{bmatrix} 0.90 & 0.10 & 0.00 \\ 0.90 & 0.80 & 0.02 \\ 0.87 & 0.08 & 0.05 \\ 0.80 & 0.15 & 0.05 \end{bmatrix}$$

[13]Snee, R .D. and Marquart, D.W., "Screening Concepts and Designs for Experiments with Mixtures," *Technometrics*, Vol. 18, (1976), pp. 19–29.

the X'X matrix would be

$$X^TX = \begin{bmatrix} 3.0169 & 0.3516 & 0.1015 \\ 0.3516 & 0.0453 & 0.0131 \\ 0.1015 & 0.10131 & 0.0054 \end{bmatrix}$$

and its determinant is Det (X^TX) = .00002102. The vertices and determinant of X'X for all five possible designs are shown in Table 15.14, and the best four-vertex design for fitting the linear model would be Design 4 since it has the maximum determinant of (X^TX) = 3.7×10^{-5}.

Table 15.14 Four Subset Designs for Coke Briquetting Example

Design	Vertices Included	Determinant (X'X)
1	1, 2, 3, 4	2.102×10^{-5}
2	1, 2, 3, 5	1.202×10^{-5}
3	1, 2, 4, 5	2.35×10^{-5}
4	1, 3, 4, 5	3.7×10^{-5}
5	2, 3, 4, 5	3.332×10^{-5}

In general, the computations are very time consuming to determine the subset of vertices with the maximum determinant of X'X. However, there are computer programs available which can automate the process. Nachtsheim[14] reviewed several commercially available experimental design programs and indicates that E-CHIP, ACED, COED-RSM and RS/DISCOVER, all contain determinant maximizing routines. Since Nachtsheim's article additional programs such as SAS QC Proc Opdex, NCSS 2000, and MINITAB version 13 also contain determinant maximizing routines. To use these programs, a list of all the vertices in the constrained region would be obtained using the algorithm described in Section 15.3.3.2. Next the program would be used to choose the subset which had the maximum Det (X'X).

---

[14]Nachtsheim, C. J. "Tools for Computer Aided Design of Experiments," *Journal of Quality Technology*, Vol. 19, (1988), pp. 132–160.

If the above mentioned computer programs are not available, Piepel[15] has shown that a design developed from the nonmixture screening designs presented in Chapter 11 will be almost as good. His procedure is a slight modification of the algorithm described in Section 15.3.3.2 and can be outlined for the k component case as follows:

1. Rank the k components in order of increasing range $U_i - L_i$ and call X_i the component with the ith largest range.

2. Set up a fractional factorial or Plackett-Burman design in the k - 1 components with the smallest ranges using the upper and lower constraint values as factor levels.

3. Calculate the value of the omitted component X_k (the one with the largest range)

as $1.0 - \sum_{i=1}^{k-1} X_i$ for each of the runs in the fractional factorial design obtained in step 2.

4. If the calculated value of X_k in step 3 falls within its upper and lower constraints the run is a design point. Otherwise set X_k equal to the violated limit (i.e., U_k if the computed X_k from step 3 is $> U_k$ or L_k if the computed $X_k < L_k$) and adjust, (by an amount equal to the difference between the computed value of X_k and its violated limit) the first possible component (working from X_{k-1} to X_1) such that a value within the computed limits is obtained.

5. If the adjustment required in step 4 cannot be accommodated by a single component, make as much of the adjustment as possible on X_{k-1}, then as much of what remains as possible on X_{k-2}, etc. Note that each adjusted component, except possibly the last one adjusted, will be at its upper or lower constraint.

6. Add a center point to the design.

This algorithm is illustrated with a gasoline blending example described by Cornell[16]. The six components are:

X_1 = Straightrun
X_2 = Reformate
X_3 = Thermally cracked naphtha
X_4 = Catalytically cracked naphtha
X_5 = Polymer
X_6 = Alkylate
X_7 = Natural Gasoline

---

[15]Piepel, G. F., "Screening Designs for Constrained Mixture Experiments Derived from Classical Screening Designs," *Journal of Quality Technology*, Vol. 22, (1990), pp. 23–33.

[16]Cornell, J. A., *Experiments with Mixtures*, (New York: John Wiley and Sons, 1981), p. 182.

As in step 1, the variables are ranked according to their range as shown below:

Component	Lower Limit	Upper Limit	Range	Rank
X_1	0	.21	.21	4
X_2	0	.62	.62	5
X_3	0	.12	.12	3
X_4	0	.62	.62	6
X_5	0	.12	.12	2
X_6	0	.74	.74	7
X_7	0	.08	.08	1

Next, according to Step 2, an appropriate screening design should be selected. Since seven components are to be screened, a reasonable design is a 12 run Plackett-Burman design as described in Chapter 11. An 8 run fractional factorial design would leave only 1 degree of freedom for error and would probably not provide enough information. A 16 run fractional factorial would be another reasonable choice. In this example, we illustrate with the 12 run Plackett-Burman design.

The first six columns are copied from Table B.1-1 in the appendix into Table 15.15. The six components with the smallest ranges are copied in succession to the columns and the + and - signs in the table are replaced by the upper and lower ranges of the components respectively as shown in Table 15.15. Next, according to Step 3, the value of X_6 is calculated as $1 - (X_7 + X_5 + X_3 + X_1 + X_2 + X_4)$ in the table. For Runs 1, 2, 8, 10, and 11 it can be seen that the computed value for X_6 exceeds its lower constraint of 0.0 and for Run 12 it exceeds its upper constraint of .74. According to Step 4, the computed value of X_6, in Run 1 is replaced by its lower constraint 0.0 and the difference $(-.65 - 0.0) = -.65$ must be adjusted (i.e., added to the component with the next largest range (X_4). Since adding -.65 to .62 would exceed the lower constraint for X_4, X_4 is set to its lower constraint 0.0 and the remainder $-.65 + .62 = .03$ must be adjusted from the next component to the left, X_2, according to Step 5. Thus the altered component values for Run 1 would be $X_7 = .08$, $X_5 = .12$, $X_3 = 0.0$, $X_1 = .21$, $X_2 = .59$, $X_4 = 0.0$, $X_6 = 0.0$. This procedure is repeated for Runs 2, 8, 10, and 11 where the computed value of X_6 also exceeded the lower constraint. In these runs only single adjustments to components 2, 4, 4, and 4 respectively were required. For Run 12 where the computed value of X_6 exceeded the upper constraint, it is replaced by its upper constraint, .74, and in this case the positive difference $(1.0 - 0.74) = 0.26$ is again added to the component with the next largest range, X_4. Finally according to Step 6 a center point is added to the design by averaging each of the components over the 12 runs. The resulting screening design in 7 mixture components is shown in Table 15.16.

Table 15.15 Plackett-Burman Design for Gasoline Blending

Run	X_7		X_5		X_3		X_1		X_2		X_4		Computed X_6
1	+	.08	+	.12	−	0.0	+	.21	+	.62	+	.62	−.65
2	+	.08	−	0.0	+	.12	+	.21	+	.62	−	0.0	−.03
3	−	0.0	+	.12	+	.12	+	.21	−	0.0	−	0.0	.55
4	+	.08	+	0.0	+	.12	−	0.0	−	0.0	−	0.0	.68
5	+	.08	+	.12	−	0.0	−	0.0	−	0.0	+	.62	.18
6	+	.08	−	0.0	−	0.0	−	0.0	+	.62	−	0.0	.3
7	−	0.0	−	.12	−	0.0	+	.21	−	0.0	+	.62	.17
8	−	0.0	−	0.0	+	.12	−	0.0	+	.62	+	.62	−.36
9	−	0.0	+	.12	−	0.0	+	.21	+	.62	−	0.0	.05
10	+	.08	−	0.0	+	.12	+	.21	−	0.0	+	.62	−.03
11	−	0.0	+	.12	+	.12	−	0.0	+	.62	+	.62	−.48
12	−	0.0	−	0.0	−	0.0	−	0.0	−	0.0	−	0.0	1.0

Table 15.16 Complete Screening Design for Gasoline Blending

Run	Natural Gasoline X_7	Polymer X_5	T. Cracked naphtha X_3	Straightrum X_1	C. Cracked Reformate X_2	naphtha X_4	Alkylate X_6
1	.08	.12	0.0	.21	.59	0.0	0.0
2	.08	0.0	.12	.21	.59	0.0	0.0
3	0.0	.12	.12	.21	0.0	0.0	.55
4	.08	.12	.12	0.0	0.0	0.0	.68
5	.08	.12	0.0	0.0	0.0	.62	.18
6	.08	0.0	0.0	0.0	.62	0.0	.30
7	0.0	0.0	0.0	.21	0.0	.62	.17
8	0.0	0.0	.12	0.0	.62	.26	0.0
9	0.0	.12	0.0	.21	.62	0.0	.05
10	.08	0.0	.12	.21	0.0	.59	0.0
11	0.0	.12	.12	0.0	.62	.14	0.0
12	0.0	0.0	0.0	0.0	0.0	.26	.74
Center→13 Point	.04	.06	.06	.105	.305	.2075	.2225

Once a screening design has been selected, the experiments are run, and the linear model is fit to the data using a regression program. The Effect plots described in Section 15.4.3 are used to determine which components are most important and should be used in further response surface type experiments. The effect plots take the place of the Normal Probability plots or Pareto Diagrams used to identify the important factors in nonmixture screening experiments. The mixture components having the steepest lines in the effect plots are most important and those whose lines have slopes near zero (like X_1, in Figure 15.14) can be dropped from further experimentation.

15.5.2 An Example

Snee and Marquart[17] presented an example of screening an experiment in eight mixture components. Economic requirements dictated that four of the components (X_1 X_2 X_5 and X_6) should be included in the mixture and the effects of the other four components were unknown. The factor ranges, were as follows:

$$.10 \leq X_1 \leq .45$$
$$.05 \leq X_2 \leq .50$$
$$0.0 \leq X_3 \leq .10$$
$$0.0 \leq X_4 \leq .10$$
$$.10 \leq X_5 \leq .60$$
$$.05 \leq X_6 \leq .20$$
$$0.0 \leq X_7 \leq .05$$
$$0.0 \leq X_8 \leq .05$$

and using the algorithm described in Section 15.3.3.2, 188 extreme vertices were found in the constrained experimental region.

It was decided to run a design consisting of 16 vertices plus four replicates of a centroid. A computer algorithm was used to select the subset of 16 vertices to maximize the determinant of the X^TX matrix and the resulting design plus observed responses are shown in Table 15.17.

---

[17]Snee, R .D. and Marquart, D. W., "Screening Concepts and Designs for Experiments with Mixtures," *Technometrics*, Vol. 18, (1976) pp. 19–29.

Table 15.17 Snee and Marquart's Example

Blend	Run Order	X_1	X_2	X_3	X_4	X_5	X_6	X_7	X_8	Y
1	18	.10	.50	0.0	0.0	.10	.20	.05	.05	30
2	3	.10	.05	0.0	0.0	.55	.20	.05	.05	113
3	11	.10	.50	0.0	.10	.10	.20	0.0	0.0	17
4	17	.15	.05	0.0	.10	.60	.05	.05	0.0	94
5	1	.10	.05	.10	0.0	.55	.20	0.0	0.0	89
6	13	.10	.50	.10	.10	.10	.05	0.0	.05	18
7	7	.10	.05	.10	.10	.55	.05	0.0	.05	90
8	12	.40	.05	.10	.10	.10	.20	.05	0.0	20
9	16	.35	.05	.10	.10	.10	.20	.05	.05	21
10	8	.30	.50	0.0	0.0	.10	.05	0.0	.05	15
11	6	.10	.50	.10	0.0	.20	.05	.05	0.0	28
12	14	.45	.05	0.0	0.0	.45	.05	0.0	0.0	48
13	4	.45	.20	0.0	.10	.10	.05	.05	.05	18
14	19	.45	.15	0.0	.10	.10	.20	0.0	0.0	7
15	2	.45	.25	.10	0.0	.10	.05	.05	0.0	16
16	9	.45	.10	.10	0.0	.10	.20	0.0	.05	19
17-20	5,10, 15,20	.259	.222	.05	.05	.244	.125	.025	.025	38, 30, 35, 40

The results of a MINITAB regression run to fit the linear model $Y = b_1X_1 + b_2X_2 + b_3X_3 + b_4X_4 + b_5X_5 + b_6X_6 + b_7X_7 + b_8X_8$ are shown in Table 15.18. Although the t-statistics for b_2, b_3, and b_4 show these coefficients are not significantly different from zero, they should not be dropped from the equation, as they would be in a nonmixture screening experiment. These coefficients only represent the predicted value at the pure component mixtures. In order to judge which coefficients should be left in the model it is necessary to examine the Effects plot. The Effects plot is used in screening designs for mixtures in the same way that Normal or Half Normal plots of effects are used in the analysis of unreplicated factorials in the nonmixture case. The worksheet in Table 15.19 was constructed to develop an Effect plot.

Table 15.18 MINITAB Equation Fit to Snee and Marquart's Example Data

Regression Analysis

```
The regression equation is
Y = - 33.3 x1 - 10.3 x2 - 2.7 x3 - 19.7 x4 + 150 x5 + 46.6 x6 + 165 x7 + 189
x8

Predictor        Coef        Stdev       t-ratio          p
Noconstant
x1            -33.320        7.620        -4.37        0.001
x2            -10.258        6.438        -1.59        0.137
x3             -2.70        27.36        -0.10        0.923
x4            -19.74        27.45        -0.72        0.486
x5            150.397        5.978        25.16        0.000
x6             46.55        17.04         2.73        0.018
x7            165.45        55.72         2.97        0.012
x8            188.65        56.37         3.35        0.006

s = 5.690
Analysis of Variance

SOURCE          DF           SS           MS           F          p
Regression       8        49063.5       6132.9       189.45      0.000
Error           12          388.5         32.4
Total           20        49452.0
```

In the worksheet, the centroid point, X_c, is converted to pseudocomponents, X'_c. Next a point X'_{pi} halfway between the centroid and the pure ith component vertex, in pseudo component space, is calculated by averaging the centroid with the pure ith pseudocomponent vertex. Finally the point X'_{pi} is converted back to the original component space, X_{pi}, and the predicted response and percent change from the centroid is calculated for the Effect plot.

Figure 15.18 is the Effect plot. From this plot it can be seen that X_6 has a very minor effect since its effect line has a near zero slope. The mixture system can therefore be simplified by considering X_6 to be inert and refitting the equation in the seven components $Y = \sum_{i=1}^{7} b_i X_i$,

where $X'_i = X_i/(1-X_6)$ for i = 1 to 5, and $X'_i = X_{i-1}/(1-X_6)$ for i = 7 and 8.

Table 15.19 Effect Plot Worksheet

	X_1	X_2	X_3	X_4	X_5	X_6	X_7	X_8	%change from Centroid	Predicted Value
Centroid X_c	.259	.222	.05	.05	.244	.125	.025	.025		
X'_c	.2271	.2457	.0714	.0714	.2057	.1071	.0357	.0357		
X'_{p1}	.6135	.1229	.0357	.0357	.1028	.0536	.0179	.0179		
X_{p1}	.5295	.1360	.0250	.0250	.1720	.0875	.0125	.0125	+77.3%	14.77
X'_{p2}	.1136	.6229	.0357	.0357	.1029	.0536	.0179	.0179		
X_{p2}	.1795	.4860	.0250	.0250	.1720	.0875	.0125	.0125	+58.7%	22.83
X'_{p3}	.1136	.1228	.5357	.0357	.1029	.0536	.0179	.0179		
X_{p3}	.1795	.1360	.3750	.0250	.1720	.0875	.0125	.0125	+325%	25.49
X'_{p4}	.1136	.1229	.0357	.5357	.1029	.0536	.0179	.0179		
X_{p4}	.1795	.1360	.0250	.3750	.1720	.0875	.0125	.0125	+325%	19.54
X'_{p5}	.1136	.1229	.0357	.0357	.6029	.0536	.0179	.0179		
X_{p5}	.1795	.1360	.0250	.0250	.5220	.0875	.0125	.0125	+55.6%	79.07
X'_{p6}	.1136	.1229	.0357	.0357	.1028	.5357	.0178	.0178		
X_{p6}	.1795	.1360	.0250	.0250	.1720	.4250	.0125	.0125	+200%	42.16
X'_{p7}	.1136	.1229	.0357	.0357	.1029	.0536	.5179	.0179		
X_{p7}	.1795	.1360	.0250	.0250	.1720	.0875	.3625	.0125	+67.5%	84.36
X'_{p8}	.1136	.1229	.0357	.0357	.1029	.0536	.0179	.5179		
X_{p8}	.1795	.1360	.0250	.0250	.1720	.0875	.0125	.3625	+67.5%	92.44

Snee and Marquart[18] also noticed that, within the computed standard errors of the coefficients shown in Table 15.18, that the following coefficients were nearly equal $b_1 \cong b_4$, $b_2 \cong b_3$, and $b_5 \cong b_7 \cong b_8$. Therefore, the mixture component system could be further simplified to a three-component system

$$Z_1 = (X_1 + X_4) / (1 - X_6)$$
$$Z_2 = (X_2 + X_3) / (1 - X_6)$$
$$Z_3 = (X_5 + X_7 + X_8) / (1 - X_6)$$

and with 20 data points the quadratic equation,

$$Y = -11.6Z_1 + 3.6Z_2 + 158.3Z_3 + 28.7Z_1Z_2 - 70.9Z_1Z_3 - 80.0Z_2Z_3$$

was fit to the data. Figure 15.19 shows a contour plot of this equation. The optimum mixture can be seen to be near the vertex where $Z_3 = 1$ or $(X_5 + X_7 + X_8) / (1 - X_6) = 1$. Within the constraints of the system this implies that components X_1, X_2, X_3, X_4 and X_6 should be set to their lower constraints, and highest amounts of X_5, X_7, and X_8 should be used for the remaining part of the mixture.

This example illustrates, much like the nonmixture example shown in Section 12.4, how the result of a screening experiment can quickly lead to optimization when many factors are involved. It is always better to start a research project with more components and a screening type design, than it is to optimize with respect to a few variables, and later find out that results are not conclusive since some important component was left out of the study.

In mixture screening experiments, the Effect plot helps to identify important components, and equality of some regression coefficients can also lead to simplification of the fitted model.

---

[18]Snee, R. D. and Marquart, D. W., "Screening Concepts and Designs for Experiments with Mixtures," *Technometrics*, Vol. 18, (1976), pp. 19–29.

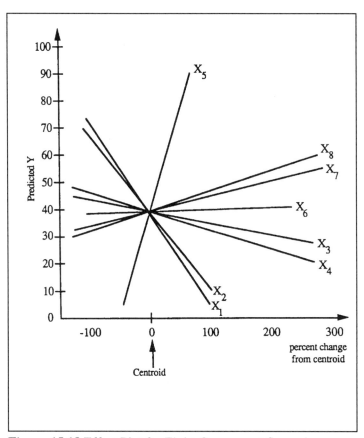

Figure 15.18 Effect Plot for Eight-Component Screening
Example

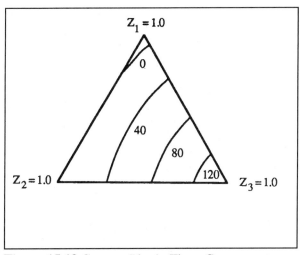

Figure 15.19 Contour Plot in Three Components

15.6 Mixture Experiments with Process Variables

In mixture experiments, the qualities or characteristics of the product are sometimes influenced by process variables in addition to proportions of the mixing components. For example, in the production of silicon carbide ceramics for use in advanced heat engines, not only do the proportions of SiC powder and Boron influence the strength of the parts, but also the process variables like the method of mixing, sintering time, and sintering temperature.

15.6.1 Designs for Mixture Experiments with Process Variables

Designs for mixture experiments with process variables are created by combining the nonmixture designs discussed in Chapters 8, 11, and 13 with the mixture designs discussed in Sections 15.3 and 15.5. For example, Figure 15.20 shows the combination of a linear simplex mixture design in three components with a 2^2 factorial design in two process variables. The experiments are listed in Table 15.20. It can be seen that the total number of runs in the experiment is equal to the product of the number of runs in the design for process variables (i.e., four) and the number of runs in the mixture design (i.e., again four). One of the most common situations in mixture experiments is when the total amount of the mixture may influence the product qualities, as well as the proportion of individual components. Of course before running a list of experiments like that shown in Table 15.20, the run order would be randomized to prevent biasing the estimates of either the process variable effects or mixture component coefficients.

In general, any mixture design can be combined with any design for process variables by just repeating all the runs in the mixture design for each combination of runs in the process variable design.

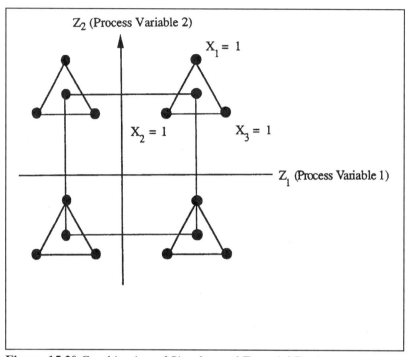

Figure 15.20 Combination of Simplex and Factorial Design

Table 15.20 Combination of Simplex and Factorial Design

Run Number	Mixture Components			Process	Variables
	X_1	X_2	X_3	Z_1	Z_2
1	1	0	0	-	-
2	0	1	0	-	-
3	0	0	1	-	-
4	1/3	1/3	1/3	-	-
5	1	0	0	+	-
6	0	1	0	+	-
7	0	0	1	+	-
8	1/3	1/3	1/3	+	-
9	1	0	0	-	+
10	0	1	0	-	+
11	0	0	1	-	+
12	1/3	1/3	1/3	-	+
13	1	0	0	+	+
14	0	1	0	+	+
15	0	0	1	+	+
16	1/3	1/3	1/3	+	+

15.6.2 Models for Mixture Experiments with Process Variables

The models $Y = f(X_1, X_2, X_3, Z_1, Z_2)$ to be fit to the data resulting from a mixture experiment with process variables can be determined from the separate models appropriate for the specific mixture design and process variable design used. For example, the mixture design used to form Table 15.20 is a linear simplex design and the appropriate model for this design is the linear model $Y = b_1X_1 + b_2X_2 + b_3X_3$. The process variable design used for Table 15.20 is the 2^2 design and an appropriate model for this design is the linear model with interaction $Y = a_0 + a_1Z_1 + a_2Z_2 + a_{12}Z_1Z_2$. The model for the combined and mixture process variable experiment is determined by multiplying the two separate models together, i.e.

$$Y = (b_1X_1 + b_2X_2 + b_3X_3)(a_0 + a_1Z_1 + a_2Z_2 + a_{12}Z_1Z_2)$$

$$= b_1a_0X_1 + b_1a_1Z_1X_1 + b_1a_2Z_2X_1 + b_1a_{12}Z_1Z_2X_1 + b_2a_0X_2 + b_2a_1Z_1X_2 + b_2a_2Z_2X_2 + b_2a_{12}Z_1Z_2X_2 +$$

$$b_3a_0X_3 + b_3a_1Z_1X_3 + b_3a_2Z_2X_3 + b_3a_{12}Z_1Z_2X_3$$

Letting $b_i a_j = d_j^i$, we can write the model as

$$Y = d_0^1 X_1 + d_1^1 Z_1 X_1 + d_2^1 Z_2 X_1 + d_{12}^1 Z_1 Z_2 X_1 + d_0^2 X_2 + d_1^2 Z_1 X_2 + d_2^2 Z_2 X_2 + d_{12}^2 Z_1 Z_2 X_2 +$$

$$d_0^3 X_3 + d_1^3 Z_1 X_3 + d_2^3 Z_2 X_3 + d_{12}^3 Z_1 Z_2 X_3 \qquad (15.6)$$

In general, this multiplication procedure can be used to determine the appropriate model for any mixture and process variable design. These models can then be fit to the data using the regression procedure outlined in Chapter 14, and optimum conditions can be found through use of contour plots or nonlinear optimization techniques.

15.6.3 An Example

Cornell[19] describes an experiment involving the mixture of three plasticizers to produce vinyl for automobile seat covers. The three plasticizers (X_1, X_2, and X_3) were subject to the following constraints

$$0.47 \leq X_1 \leq 0.85$$
$$0 \leq X_2 \leq .25$$
$$0.15 \leq X_3 \leq 0.28$$

and the extreme vertices design plus a centroid shown in Table 15.21 was used to study them. This 5-point design allows fitting the linear model $Y = b_1 X_1 + b_2 X_2 + b_3 X_3$ which was thought appropriate for the mixture components.

Table 15.21 Extreme Verticies for Plasticizer Components

Blend Number	Plasticizer Proportions		
	X_1	X_2	X_3
1	0.85	0.0	0.15
2	0.72	0.0	0.28
3	0.60	0.25	0.15
4	0.47	0.25	0.28
Centroid	0.66	0.125	0.215

[19]Cornell, J. A., "Analyzing Mixture Experiments Containing Process Variables: A Split Plot Approach," *Journal of Quality Technology*, Vol. 20, (1988), pp. 2–25.

The experiment also included two process variables: rate of extrusion (Z_1) and temperature at drying (Z_2). Both of these variables were studied at two levels in a 2^2 factorial design, which supported the model $Y = a_0 + a_1Z_1 + a_2Z_2 + a_{12}Z_1Z_2$. Two replicate experiments were performed, at each of the 5 X 4 combinations of conditions obtained by combining the mixture and process variables designs, to obtain a total of 40 runs. The **X**-matrix and the **Y**-vector of measured responses (scaled thickness value) that were used to fit the combined model in the form of Equation 15.5 are shown below. In the table the first twenty runs and the second twenty runs are repeat experiments at the same conditions. Part of the results from fitting the equation via MINITAB are shown in Table 15.22.

Table 15.22 Fitted Model for Plasticizer Components

```
The regression equation is
Y = 15.9 X1 + 14.4 X2 - 16.8 X3 + 2.95 X1Z1 - 0.55 X2Z1 - 2.82 X3Z1 - 0.57 X1Z2
        - 1.57 X2Z2 + 2.31 X3Z2 + 0.26 X1Z1Z2 + 1.26 X2Z1Z2 - 0.70 X3Z1Z2

Analysis of Variance
```

SOURCE	DF	SS	MS	F	p
Regression	12	3270.00	272.50	60.08	0.000
Error	28	127.00	4.54		
Total	40	3397.00			

$$
X = \begin{bmatrix}
0.85 & 0.000 & 0.150 & -0.85 & 0.000 & -0.150 & -0.85 & 0.000 & -0.150 & 0.85 & 0.000 & 0.150 \\
0.72 & 0.000 & 0.280 & -0.72 & 0.000 & -0.280 & -0.72 & 0.000 & -0.280 & 0.72 & 0.000 & 0.280 \\
0.60 & 0.250 & 0.150 & -0.60 & -0.250 & -0.150 & -0.60 & -0.250 & -0.150 & 0.60 & 0.250 & 0.150 \\
0.47 & 0.250 & 0.280 & -0.47 & -0.250 & -0.280 & -0.47 & -0.250 & -0.280 & 0.47 & 0.250 & 0.280 \\
0.66 & 0.125 & 0.215 & -0.66 & -0.125 & -0.215 & -0.66 & -0.125 & -0.215 & 0.66 & 0.125 & 0.215 \\
0.85 & 0.000 & 0.150 & 0.85 & 0.000 & 0.150 & -0.85 & 0.000 & -0.150 & -0.85 & 0.000 & -0.150 \\
0.72 & 0.000 & 0.280 & 0.72 & 0.000 & 0.280 & -0.72 & 0.000 & -0.280 & -0.72 & 0.000 & -0.280 \\
0.60 & 0.250 & 0.150 & 0.60 & 0.250 & 0.150 & -0.60 & -0.250 & -0.150 & -0.60 & -0.250 & -0.150 \\
0.47 & 0.250 & 0.280 & 0.47 & 0.250 & 0.280 & -0.47 & -0.250 & -0.280 & -0.47 & -0.250 & -0.280 \\
0.66 & 0.125 & 0.215 & 0.66 & 0.125 & 0.215 & -0.66 & -0.125 & -0.215 & -0.66 & -0.125 & -0.215 \\
0.85 & 0.000 & 0.150 & -0.85 & 0.000 & -0.150 & 0.85 & 0.000 & 0.150 & -0.85 & 0.000 & -0.150 \\
0.72 & 0.000 & 0.280 & -0.72 & 0.000 & -0.280 & 0.72 & 0.000 & 0.280 & -0.72 & 0.000 & -0.280 \\
0.60 & 0.250 & 0.150 & -0.60 & -0.250 & -0.150 & 0.60 & 0.250 & 0.150 & -0.60 & -0.250 & -0.150 \\
0.47 & 0.250 & 0.280 & -0.47 & -0.250 & -0.280 & 0.47 & 0.250 & 0.280 & -0.47 & -0.250 & -0.280 \\
0.66 & 0.125 & 0.215 & -0.66 & -0.125 & -0.215 & 0.66 & 0.125 & 0.215 & -0.66 & -0.125 & -0.215 \\
0.85 & 0.000 & 0.150 & 0.85 & 0.000 & 0.150 & 0.85 & 0.000 & 0.150 & 0.85 & 0.000 & 0.150 \\
0.72 & 0.000 & 0.280 & 0.72 & 0.000 & 0.280 & 0.72 & 0.000 & 0.280 & 0.72 & 0.000 & 0.280 \\
0.60 & 0.250 & 0.150 & 0.60 & 0.250 & 0.150 & 0.60 & 0.250 & 0.150 & 0.60 & 0.250 & 0.150 \\
0.47 & 0.250 & 0.280 & 0.47 & 0.250 & 0.280 & 0.47 & 0.250 & 0.280 & 0.47 & 0.250 & 0.280 \\
0.66 & 0.125 & 0.215 & 0.66 & 0.125 & 0.215 & 0.66 & 0.125 & 0.215 & 0.66 & 0.125 & 0.215 \\
0.85 & 0.000 & 0.150 & -0.85 & 0.000 & -0.150 & -0.85 & 0.000 & -0.150 & 0.85 & 0.000 & 0.150 \\
0.72 & 0.000 & 0.280 & -0.72 & 0.000 & -0.280 & -0.72 & 0.000 & -0.280 & 0.72 & 0.000 & 0.280 \\
0.60 & 0.250 & 0.150 & -0.60 & -0.250 & -0.150 & -0.60 & -0.250 & -0.150 & 0.60 & 0.250 & 0.150 \\
0.47 & 0.250 & 0.280 & -0.47 & -0.250 & -0.280 & -0.47 & -0.250 & -0.280 & 0.47 & 0.250 & 0.280 \\
0.66 & 0.125 & 0.215 & -0.66 & -0.125 & -0.215 & -0.66 & -0.125 & -0.215 & 0.66 & 0.125 & 0.215 \\
0.85 & 0.000 & 0.150 & 0.85 & 0.000 & 0.150 & -0.85 & 0.000 & -0.150 & -0.85 & 0.000 & -0.150 \\
0.72 & 0.000 & 0.280 & 0.72 & 0.000 & 0.280 & -0.72 & 0.000 & -0.280 & -0.72 & 0.000 & -0.280 \\
0.60 & 0.250 & 0.150 & 0.60 & 0.250 & 0.150 & -0.60 & -0.250 & -0.150 & -0.60 & -0.250 & -0.150 \\
0.47 & 0.250 & 0.280 & 0.47 & 0.250 & 0.280 & -0.47 & -0.250 & -0.280 & -0.47 & -0.250 & -0.280 \\
0.66 & 0.125 & 0.215 & 0.66 & 0.125 & 0.215 & -0.66 & -0.125 & -0.215 & -0.66 & -0.125 & -0.215 \\
0.85 & 0.000 & 0.150 & -0.85 & 0.000 & -0.150 & 0.85 & 0.000 & 0.150 & -0.85 & 0.000 & -0.150 \\
0.72 & 0.000 & 0.280 & -0.72 & 0.000 & -0.280 & 0.72 & 0.000 & 0.280 & -0.72 & 0.000 & -0.280 \\
0.60 & 0.250 & 0.150 & -0.60 & -0.250 & -0.150 & 0.60 & 0.250 & 0.150 & -0.60 & -0.250 & -0.150 \\
0.47 & 0.250 & 0.280 & -0.47 & -0.250 & -0.280 & 0.47 & 0.250 & 0.280 & -0.47 & -0.250 & -0.280 \\
0.66 & 0.125 & 0.215 & -0.66 & -0.125 & -0.215 & 0.66 & 0.125 & 0.215 & -0.66 & -0.125 & -0.215 \\
0.85 & 0.000 & 0.150 & 0.85 & 0.000 & 0.150 & 0.85 & 0.000 & 0.150 & 0.85 & 0.000 & 0.150 \\
0.72 & 0.000 & 0.280 & 0.72 & 0.000 & 0.280 & 0.72 & 0.000 & 0.280 & 0.72 & 0.000 & 0.280 \\
0.60 & 0.250 & 0.150 & 0.60 & 0.250 & 0.150 & 0.60 & 0.250 & 0.150 & 0.60 & 0.250 & 0.150 \\
0.47 & 0.250 & 0.280 & 0.47 & 0.250 & 0.280 & 0.47 & 0.250 & 0.280 & 0.47 & 0.250 & 0.280 \\
0.66 & 0.125 & 0.215 & 0.66 & 0.125 & 0.215 & 0.66 & 0.125 & 0.215 & 0.66 & 0.125 & 0.215
\end{bmatrix}
\qquad
Y = \begin{bmatrix}
8 \\ 6 \\ 10 \\ 4 \\ 11 \\ 12 \\ 9 \\ 13 \\ 6 \\ 15 \\ 7 \\ 7 \\ 9 \\ 5 \\ 9 \\ 12 \\ 10 \\ 14 \\ 6 \\ 13 \\ 7 \\ 5 \\ 11 \\ 5 \\ 10 \\ 10 \\ 8 \\ 12 \\ 3 \\ 11 \\ 8 \\ 6 \\ 10 \\ 4 \\ 7 \\ 11 \\ 9 \\ 12 \\ 5 \\ 9
\end{bmatrix}
$$

The residuals from the fitted model were plotted versus the mixture proportions X_1, X_2, and X_3, the results are shown below in Figures 15.21, 15.22, and 15.23. From these figures it can be seen that there is some curvature in the mixture components that has not been accounted for

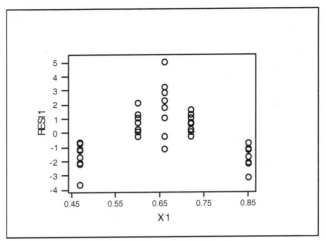

Figure 15.21 Residuals vs. X_1

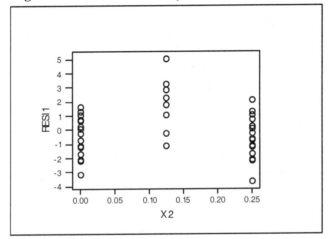

Figure 15.22 Residuals vs. X_2

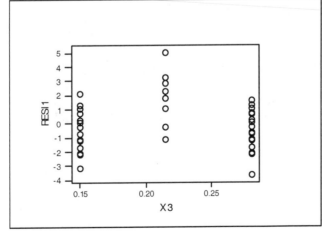

Figure 15.23 Residuals vs. X_3

in the fitted model. Since curvature in the Scheffé mixture model is accounted for by the cross product terms X_1X_2, X_1X_3, and X_2X_3 three additional regression models were fit to the data each adding a different cross product term. It was impossible to add all three cross product terms to the regression model simultaneously because there were only five extreme verticies in the design. The model with the X_1X_2 term added fit the data best, and the MINITAB regression report for this model is shown in Table 15.23[20].

Table 15.23 Expanded Model for Plasticizer Components

Regression Analysis

```
The regression equation is
Y = 11.5 X1 - 69.5 X2 - 2.63 X3 + 2.95 X1Z1 - 0.55 X2Z1 - 2.82 X3Z1
          - 0.570 X1Z2 - 1.57 X2Z2 + 2.31 X3Z2 + 0.257 X1Z1Z2 + 1.26 X2Z1Z2
          - 0.70 X3Z1Z2 + 149 X1X2
```

Predictor	Coef	Stdev	t-ratio	p
Noconstant				
X1	11.482	1.044	10.99	0.000
X2	-69.54	12.27	-5.67	0.000
X3	-2.630	3.463	-0.76	0.454
X1Z1	2.9529	0.8219	3.59	0.001
X2Z1	-0.547	1.794	-0.30	0.763
X3Z1	-2.816	2.798	-1.01	0.323
X1Z2	-0.5702	0.8219	-0.69	0.494
X2Z2	-1.570	1.794	-0.88	0.389
X3Z2	2.314	2.798	0.83	0.415
X1Z1Z2	0.2567	0.8219	0.31	0.757
X2Z1Z2	1.257	1.794	0.70	0.490
X3Z1Z2	-0.705	2.798	-0.25	0.803
X1X2	148.64	21.49	6.92	0.000

```
s = 1.303
Analysis of Variance
```

SOURCE	DF	SS	MS	F	p
Regression	13	3351.19	257.78	151.95	0.000
Error	27	45.81	1.70		
Total	40	3397.00			

The significant positive coefficient for X_1Z_2 indicates that high levels of the process variable, Z_2 (temperature at drying) increases Y (scaled thickness) at mixture combinations with high proportions of plasticizer 1 (i.e., near the pure component $X_1 = 1.0$), and the significant coefficient for X_1X_2 indicates that there is nonlinear blending with mixtures of X_1 and X_2 that produces higher values of Y when mixtures of X_1 and X_2 are used.

The optimum conditions can be located by first producing four prediction equations by fixing

---

[20]Note in Cornell's article he used a split plot model to compute the standard errors differently than that shown here.

(Z_1, Z_2) at the combinations $(-1,-1)$, $(-1,1)$, $(1,-1)$, $(1,1)$. Next make contour plots of Y versus X_1, X_2, and X_3 at each of the four Z_1, Z_2 combinations. Examination of the four contour plots will reveal the optimum.

15.7 Summary

In this chapter we have discussed the ideas of experimental design and data analysis in the context of mixture experiments. We have discussed the experimental designs, and mathematical models appropriate for both screening experiments and optimization experiments. The following is a list of important concepts we have discussed in this chapter.

mixture component 519
experimental region for mixture experiments 519, 520
slack variable model 521
Scheffé model – linear, quadratic and special cubic 523
interpretation of coefficients in Scheffé model 524
simplex design, linear and quadratic 524 to 525
formulas for calculating model coefficients from simplex designs 526
constrained mixture experimental region 531
pseudo components 532
extreme vertices design 534, 535
Snee and Marquart's algorithm for finding extreme vertices 536, 537
multicollinearity 546
effect plot 549, 550
screening designs for mixture experiments 556 to 560
mixture-process variables experiments and models 567 to 570

15.8 Exercises for Chapter 15

15.1 A mixture experiment in three components X_1, X_2, and X_3 was run.

a) If the fitted slack variable model is: $Y=12.0+3.1X_1 + 4.4X_2 + 0.7X^2_1 + 0.3X^2_2+1.4X_1X_2$, then write the model in the Scheffe= form.

b) What is the predicted response from the model given in a) athe the 100% X_1 mixture.

c) Which components are synergystic (i.e., a mixture of the two results in a higher response than the pure blends).

15.2 Beloto, Jr. et al.[1] Studied the relation between Y= Solubility of phenobarbital and mixture components X_1 = ethanol, X_2 = propylene gycol and X_3 = water.

a) List the experiments required to fit a linear model

b) List the experiments needed to fit a quadratic model

c) list the experiments needed to fit a special cubic model

15.3 a) List the experiments or mixtures that would be required to fit a special cubic model in three components X_1, X_2, and X_3.

b) How would this list change if the components were subject to the following constraints:

$$0.20 \leq X_1 \leq 1.0$$
$$0.25 \leq X_2 \leq 1.0$$
$$0.15 \leq X_3 \leq 1.0$$

c) List the experiments required to fit the linear model in five components X_1, X_2, X_3, X_4,and X_5.

---

[1]Belloto Jr., R.J., Dean, A.M, Moustafa, M.A. , Molokhia, A.M., gouda, M. W., and Sokoloski, T.D. (1985) AStatistical Techniques Applied to Solubility Predictions and Pharmaceutical Formulations: An Approach to Problem Solving using Mixture Response Surface Methodology@ *International Journal of Pharmaceutics* 23, pp. 195-207.

15.4 Narcy and Renaud[2] Studied mixtures of three components to see their effects on the viscosity and cold water clear point of a light-duty liquid detergent. Below is a list of some of their experiments and results.

#	Water x_1	Alcohol x_2	Urea x_3	Y_1 Viscosity	Y_2 Clear Point
1	1	0	0	362.5	35.0
2	0	1	0	78.0	11.3
3	0	0	1	1630.0	23.5
4	2	2	0	165	12.8
5	2	0	2	537.5	6.0
6	0	2	2	202.5	5.0
22	a	a	a	265.0	5.0

a) Fit the Scheffé linear model to the viscosity data

b) Fit the Scheffé linear model to the clear point data

c) Fit the Scheffé quadratic model to the viscosity data

d) Fit the Scheffé quadratic model to the clear point data

e) Using the additional data value for viscosity given in e) calculate the coefficients in the special cubic model for viscosity

15.5 a) Using the formulas in Table B.6-2, fit a quadratic model to the pseudo components of the data in table 15.6.

b) Using the transformation $X_i = (1 - \sum_{i=1}^{k} L_i) X_i' + L_i$ convert the equation you found in

a) relating the response to pseudo components (X_i') into an equation relating Y to the actual components (X_i).

---

[2]Narcy, J.P. and Renaud, J. (1972) AUse of Simplex Experimental Designs in Detergent Formulation@, *Journal of the American Oil Chemists= Society* 49, pp. 598-608.

15.6 Given the following experimental conditions and data:

#	X1	X2	X3	Y (replicates)
1	0.8	0.2	0.0	4, 6
2	0.3	0.7	0.0	8, 11
3	0.3	0.2	0.5	13, 9
4	0.55	0.45	0.0	15, 19
5	0.3	0.45	0.25	20, 23
6	0.55	0.2	0.25	18, 17
7	0.466	0.366	0.166	19, 22, 24, 17

a) Convert the components to pseudo (X_i') components and calculate the coefficients of the quadratic Scheffé model using the formulas shown in Table B.6-2

b) Using the transformation $X_i = (1 - \sum_{i=1}^{k} L_i) X_i' + L_i$ convert the equation you found in a) relating the response to pseudo components (X_i') into an equation relating Y to the actual components (X_i).

c) Set up an X-matrix or use a regression computer program, and calculate the coefficients for the model you found in b) directly from the data in the table.

d) Using a computer program, or by repeated calculation of predicted values, sketch the contours of your fitted surface and find the combination of X_1, X_2, and X_3 that maximizes Y.

15.7 Anik and Sukumar[3] modeled the soluability of an antifungal agent in terms of the proportions of X_1 = polysorbate 600, X_2 = polyethylene glycol 400, X_3 =glycerin, and X_4 = water. If the constraints on the proportions are given as follows:

$$0.0 \leq X_1 \leq 0.08$$
$$0.1 \leq X_2 \leq 0.40$$
$$0.1 \leq X_3 \leq 0.40$$
$$0.2 \leq X_4 \leq 0.80$$

a) Use the algorithm of Snee and Marquardt described in Section 15.3.3.2 to find the extreme verticies for this constrained region.

b) List the experiments that you would use to fit a linear model

c) Compute the face centroids.

d) List the experiments you would run to fit a quadratic model.

15.8 Given the 13 motor octane values determined in the mixture screening design in Table 15.16

Run	Motor Octane
1	78
2	86
3	98
4	94
5	86
6	92.7
7	74
8	90
9	77
10	79
11	92
12	93
13	88

---

[3]Anik, S. T. and Sukumar, L. (1981) "Extreme Vertexes Design in formulation Development: Solubility of Butoconazole Nitrate in a Multi-component System", *Journal of Pharmaceutical Sciences* 70, pp. 897-900.

a) Use the regression procedure to determine the coefficients in a linear Sheffé Model.

b) Make an effect plot of the six components and determine if any can be dropped from the study.

15.9 Make an effect plots using the quadratic equations fit to the data from problem 4 and explain in words the effects of the three components.

15.10 Given that six components of a mixture have the following constraints:

$$0.1 \leq X_1 \leq 0.2$$
$$0.1 \leq X_2 \leq 0.4$$
$$0.1 \leq X_3 \leq 0.5$$
$$0.0 \leq X_4 \leq 0.5$$
$$0.0 \leq X_5 \leq 0.6$$
$$0.0 \leq X_6 \leq 0.8$$

a) Set up a 12 run screening design

b) Set up a 16 run screening design

15.11 Soo, Sander, and Kess[4] Investigated the texture of shrimp patties made by blending proportions of isolated soy protein (X_1), sodium chloride (X_2), sodium tripolyphosphate (X_3), and Alaskan shrimp (X_4). The proportions of the ingredients were bounded by the constraints:

$$0.05 \leq X_1 \leq 0.10$$
$$0.01 \leq X_2 \leq 0.03$$
$$0.001 \leq X_3 \leq 0.005$$
$$0.85 \leq X_4 \leq 1.0$$

The experimental design and resulting data are shown below:

---

[4]Soo, H.M., Sander, E.H., and Kess, D.W. (1978) "Definition of a Prediction Model for Determination of the Effect of Processing and Compositional Parameters on the Textural Characteristics of Fabricated Shrimp", *Journal of Food Science*, 43 pp. 1165-1171.

Run	ISP X_1	NaCl X_2	STP X_3	Shrimp X_4	Texture Y
1	0.10	0.03	0.005	0.865	9.88
2	0.10	0.03	0.001	0.869	9.35
3	0.10	0.01	0.005	0.885	9.85
4	0.10	0.01	0.001	0.889	9.65
5	0.05	0.03	0.005	0.915	9.35
6	0.05	0.03	0.001	0.919	7.90
7	0.05	0.01	0.005	0.935	7.65
8	0.05	0.01	0.001	0.939	7.85
9	0.10	0.02	0.003	0.877	9.75
10	0.075	0.03	0.003	0.892	8.03
11	0.075	0.02	0.005	0.900	8.03
12	0.075	0.02	0.001	0.904	8.60
13	0.075	0.01	0.003	0.912	8.05
14	0.05	0.02	0.003	0.927	7.65
15	0.075	0.02	0.003	0.902	8.18
16	0.10	0.03	0.005	0.865	9.60
17	0.10	0.01	0.005	0.885	9.55
18	0.05	0.03	0.001	0.919	7.72
19	0.05	0.01	0.001	0.939	7.63
20	0.075	0.02	0.003	0.902	8.48

a) Fit the Scheffé quadratic model to the data using regression

b) Using the coding and scaling $X_1' = (X_1 - 0.075)/0.025$, $X_2' = (X_2 - 0.02)/0.01$, and $X_3' = (X_3 - 0.003)/0.002$, fit the slack variable model to the coded and scaled proportions.

c) Make effect plots for the 4 components and explain in words the effects of the components upon texture.

15.12 Bohl[5] studied the strength characteristics of a plastic compound as a function of X_1 = virgin resin plus two additives X_2 = glass fiber, and X_3 = glass micro-spheres. The data from his experiment is shown in the table below. Two responses are shown (tensile strength and cost in cents per inch).

	virgin resin	glass fiber	micro-spheres	tensile strength	material cost
run	X_1	X_2	X_3	Y_1	Y_2
1	1.0	0.00	0.0	100	100
2	0.9	0.00	0.10	177	100
3	0.9	0.10	0.0	86	94.1
4	0.86333	0.06666	0.06666	139	96.0
5	0.86333	0.06666	0.06666	137	96.0
6	0.8	0.0	0.2	217	100
7	0.8	0.10	0.10	148	94.1
8	0.8	0.20	0.0	79	88.1

a) Fit the quadratic Scheffé model for each response.

b) Perform a lack of fit test for the quadratic model of tensile strength.

c) Make an effect plot for each of the two responses.

d) Overlay contour plots of the two responses, or use a nonlinear optimization program like the Solver in Excel to predict the mixture proportions that would result in a tensile strength of at least 160 with minimum cost.

15.13   a) Include terms for X_1X_2, $X_1X_2Z_2$ and $X_1X_2Z_1Z_2$ to the X matrix in Section 15.6.3 and fit the expanded model by regression.

   b) Calculate the t-statistic for the terms

   c) Calculate the F-statistic for the expanded model.

---

[5]Bohl, A. H., "A Formulation Tool," *Chemtech*, May 1988, pp. 284 - 289.

Part V
Variability and Quality

In this part we cover the basic quantitative tools used in quality improvement efforts. Chapter 16 presents the basic methods for studying the measurement system. If something cannot be measured accurately, there is no basis for making improvements. Chapter 17 is an outline of the statistical methods used in online quality control to reduce variation and manufacturing imperfections. Chapter 18 presents Taguchi's offline quality control philosophy.

CHAPTER 16
Characterizing Variability in Data

16.1 Introduction

All data collected in engineering, manufacturing, or scientific investigations is subject to variability. Variability in data makes decision making more difficult. However, precise results are rare in process research, and variability is a fact of life. We must learn to make informed decisions in the face of uncertain or variable information. Already in this book we have discussed much about variability, and decision making in the presence of variability. In Chapter 4 we discussed some ways of summarizing data (histograms, sample variances, etc.) to reveal the amount of variability present. In Chapter 3 we discussed some of the theory for variability. In Chapters 6 and 8 - 15 we learned to judge the significance of the effects of experimental factors by comparing their magnitude to experimental noise variability. In Chapter 2 we discussed the connection between product quality and variability, and we introduced the use of control charts for detecting and removing assignable causes for variability. In Chapter 17 we will discuss, in more depth, the various types of control charts that can be used to reduce variability in manufacturing processes. But first, in this chapter, we will explain how variability can be classified into its component sources. This will help us to focus efforts to reduce variability on the major sources.

We can study sources of variability in two ways that we will call the *bottom up method* and the *top down method*. In the bottom up method we can synthesize the variability or variance in product characteristics or process performance measures if we know the variances of the component sources. For example we may synthesize the variance in the dimension of an assembled product, if we know the variances of the dimensions of the component parts. Likewise we may synthesize the variance of a process performance if we have an equation relating the process performance to process factors, and we know the variances of the individual factors. In the top down method we deduce the variances of component sources, that normally are not measured directly, from an overall variance. For example, the total variance in assays of a chemical product may be apportioned into variance due to product variability and laboratory error.

In this section we will discuss the additivity law of variances, and then some aspects of the bottom up method of synthesizing variances in Section 16.1.2. In Sections 16.2 through 16.5 we will discuss the top down method, and various types of experiments and data analysis techniques that will allow us to apportion variances into their component sources.

16.1.1 Additivity of Variances

As we saw in Section 3.7 uncertainty or variability in data can usually be expressed as a sum of uncertainties or errors from various sources. If Y represents a variable quantity, and we can write:

$$Y = A + B + C$$

where A, B, and C are each independent random variables. This is just a special case of Equation 3.14 in Section 3.7 and therefore the variance in Y can be expressed as a sum of variances of the three random variables that we will call *sources of variability*. In mathematical terms:

$$\sigma_Y^2 = \sigma_A^2 + \sigma_B^2 + \sigma_C^2$$

where σ_Y^2 represents the total variability, while the variances on the right side of the equation represent what we call the *components of variance*. σ_A^2 represents variability from the first source, σ_B^2 represents the variability from the second source, etc. Note that it is the variances which are additive; the standard deviations are not additive, although that is what most people intuitively assume. In general we can write

$$\sigma_S^2 = \sum_{i=1}^{n} \sigma_i^2 \qquad (16.1)$$

when $X_s = \sum_{i=1}^{n} X_i$ and the X_i's are independent or uncorrelated random variables and σ_s^2 is the variance of X_s, and the variances of the X_i's are σ_i^2. We call this the *law of additivity of variances*. Consider the following three examples.

> **Example 1:** If Y represents the thickness of a layered assembly made up of three component parts, then Y = A + B + C, where A, B, and C represent the thicknesses of the component parts. If the three component parts are manufactured and subject to manufacturing variability in thickness, then the variability in the thickness of the layered assembly, σ^2, is equal to the sum of the variances of the component parts, i.e.,
>
> $$\sigma^2 = \sigma_A^2 + \sigma_B^2 + \sigma_C^2$$
>
> This is the classical tolerance stacking problem. Often in this situation σ, the standard deviation, is mistakenly expressed as a sum of the three standard deviations. But, this is the so-called worst case formulation, and it usually motivates engineers to

586

request more stringent tolerance limits on the individual components than needed, in their attempt to meet a given tolerance on the assembly. The additivity of variance law shows us that we can actually get away with wider tolerances or standard deviations for component parts and still meet the assembly tolerance.

Example 2: Laboratory analyses are made on each batch of a powdered chemical product so that information about the purity can be passed to customers. It is hoped that the variability in laboratory analysis will reflect actual differences in average batch purity. However it is known that the material within a batch is not completely homogeneous, and that there is variability in the laboratory procedure. If σ^2 represents the variability among laboratory analysis from different batches, then

$$\sigma^2 = \sigma_B^2 + \sigma_S^2 + \sigma_L^2$$

where σ_B^2 represents the batch-to-batch variability, σ_S^2 represents the sample-to-sample variability within a batch, and σ_L^2 represents the laboratory error.

Example 3: The length of parts are measured by a micrometer. It is hoped that the majority of variability in these measurements is due to actual differences in the lengths of parts. If that is the case, decisions made after plotting the results on control charts will be effective in determining whether assignable causes have affected the part lengths.

However, if repeat measurements are made on the same part, we know they will vary due to assorted reasons. Changes in the actual length of the part could be caused by thermal expansion, or by deformation due to force exerted on the specimen by the measuring instrument. Also, several sources of variation may exist within the measuring instrument such as gear backlash, and sensitivity to temperature. All these sources of measurement error may be lumped into one category called *gage error*.

Finally the operator may induce error by making repeat measurements at slightly different positions, or by making errors in reading the graduated scale. If σ^2 represents the variance in measured lengths of a set of parts, we may write:

$$\sigma^2 = \sigma_P^2 + \sigma_G^2 + \sigma_O^2$$

where: σ_P^2 is the variance in actual part lengths from part to part, σ_G^2 is the gage error or variability of repeat measurements on the same part, and σ_O^2 is the variance induced by operator error. These latter two error sources are normally called repeatability and reproducibility. We will study them in more detail in Section 16.5.

If the components of the variance are known or can be estimated, then we can calculate the variance in the measured lengths by summing. For example, suppose the components of variance for the lengths of parts in Example 3 were ($\sigma_P^2 = 12.1$ $\sigma_G^2 = 4.2$ $\sigma_O^2 = 5.6$) then $\sigma^2 = 21.9$, and the variability in actual part lengths $\sigma_P^2 / (\sigma_P^2 + \sigma_G^2 + \sigma_O^2) = \sigma_P^2 / \sigma^2 = .552$ is only slightly more than 50% of the total. The remaining variability is due to measurement error. If this were the situation, control charts of measured length would not be very useful because the large measurement error would be confused with changes in actual part dimensions. On the other hand, if $\sigma_P^2 / \sigma^2 = .96$ control charts would accurately reflect changes in actual part dimensions. As a rule of thumb in a production environment, we would like the measurement error, in terms of a standard deviation, to be no more than 5% of the specification range for the product characteristic being measured. If measurement error is greater than that, we will need to find a way to reduce it.

16.1.2 Propagation of Errors

Often we need to generalize the additive law of variances, Equation 16.1, to functions of random variables other than sums. For example, suppose we know the variance of X_1 and X_2, and we would like to determine or synthesize the variance of a function of X_1 and X_2 like the product $X_1 \times X_2$, or the ratio X_1 / X_2? The simplest function we might consider is the weighted sum $Z = aX_1 + bX_2$ where X_1 and X_2 are independent or uncorrelated random variables. In this case we learned in Section 3.7 that:

$$\sigma_Z^2 = a^2 \sigma_1^2 + b^2 \sigma_2^2$$

and, in general, if $Z = c + \sum_{i=1}^{n} w_i X_i$ where c is a constant and the X_i's are uncorrelated

random variables , then

$$\sigma_Z^2 = \sum_{i=1}^{n} w_i^2 \sigma_i^2 \qquad (16.2)$$

Consider the following example: If an assembly requires a peg with outside diameter X_1, be placed in a hole with inside diameter X_2, the clearance is $Z = X_2 - X_1$, and the variance of the clearance will be $\sigma^2_Z = (-1)^2 \sigma^2_1 + (1)^2 \sigma^2_2 = \sigma^2_1 + \sigma^2_2$.

If Z is a function of X_1 and X_2 other than a weighted sum, we can always approximate Z within the vicinity of a point $(X_1 = \mu_1, X_2 = \mu_2)$ as a weighted sum of X_1 and X_2, using the first two terms of a Taylor's series expansion. In mathematical terms if $Z = f(X_1, X_2)$ then

$$Z \approx f(\mu_1, \mu_2) + \frac{\partial f(X_1, X_2)}{\partial X_1}(X_1 - \mu_1) + \frac{\partial f(X_1, X_2)}{\partial X_2}(X_2 - \mu_2) \text{ . This is a weighted sum in } X_1 \text{ and } X_2,$$

therefore, applying Equation 16.2 the variance of Z can also be approximated by

$$\left(\frac{\partial f(X_1, X_2)}{\partial X_1} \right)^2 \sigma^2_1 + \left(\frac{\partial f(X_1, X_2)}{\partial X_2} \right)^2 \sigma^2_2 \Big|_{X_1 = \mu_1, \ X_2 = \mu_2}$$

In general, if $Z = f(X_1, X_2, \ldots, X_n)$, where X_i's are uncorrelated random variables, then we can approximate the variance of Z with the following formula:

$$\sigma^2_Z = \sum_{i=1}^{n} \left(\frac{\partial f(X_1, X_2, \ldots, X_n)}{\partial X_i} \right)^2 \sigma^2_i \tag{16.3}$$

This is called the Variance Propagation formula. Consider two examples of its use:

Example 1: If $Z = X_1/X_2$ then $\dfrac{\partial(X_1/X_2)}{\partial X_1} = \dfrac{1}{X_2}$ and $\dfrac{\partial(X_1/X_2)}{\partial X_2} = -\dfrac{X_1}{X_2^2}$ and

$$\sigma^2_Z \approx \left(\frac{1}{X_2} \right)^2 \sigma^2_1 + \left(-\frac{X_1}{X_2^2} \right)^2 \sigma^2_2 \Big|_{X_1 = \mu_1, \ X_2 = \mu_2}$$

Example 2: In the design of a thin-film redistribution layer (discussed by Lawson and Madrigal[1]), the circuit impedance (Z) is a function of three design factors. These factors are the insulator

---

[1]Lawson, J. S., and Madrigal, J .L., "Robust Design Through Optimization Techniques," *Quality Engineering*, 6(4), (1994), pp. 593–608.

589

thickness (A), the line-width (B) and the line-height (C) as shown in Figure 1. From engineering first principles, it can be shown that the impedance is given by the following equation:

$$Z = f(A,B,C) = \frac{87.0}{\sqrt{\varepsilon + 1.41}} \log_e[(5.98A)/(0.8B+C)]$$

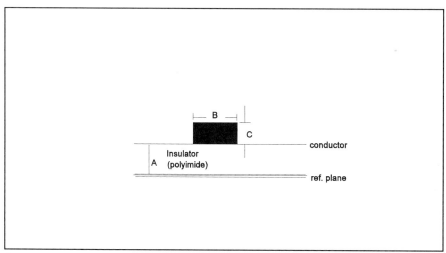

Figure 16.1 Thin Film Redistribution Layer

The ϵ in the equation is the di-electric constant of the insulator and is assumed to be constant at 3.10. The nominal, or mean, values of A, B, and C can be specified by the design engineer, but in actual circuits these characteristics will vary from the nominal due to manufacturing imperfections and wear during use. From process control and inspection data it was known that the low cost standard deviations in A, B, and C, were σ_A=0.333, σ_B=0.222, and σ_C=0.111. To approximate how much variability there will be in the impedence of circuits produced at the nominal settings for A, B, and C, the variance propagation formula results in:

$$\sigma_Z^2 \simeq \left[\left(\frac{\partial Z}{\partial A} \right)^2 \sigma_A^2 + \left(\frac{\partial Z}{\partial B} \right)^2 \sigma_B^2 + \left(\frac{\partial Z}{\partial C} \right)^2 \sigma_C^2 \right]\Bigg|_{A=\mu_A,B=\mu_B,C=\mu_C}$$

$$\simeq \frac{186.47}{\mu_a^2} + \frac{73.76}{(0.8\mu_B+\mu_C)^2}$$

For functions that cannot be differentiated analytically or are time consuming to compute, such as those defined by finite element analysis or Monte Carlo Simulation, the variance propagation formula cannot be used directly. However the partial derivatives can be approximated by secants and the same principle applied (see Nelson[2] for examples and a computer program in BASIC).

16.2 Simple Experiments for Estimating Two Sources of Variance

In this section we will begin to discuss the top down method of apportioning variance to component sources. We do this by collecting and analyzing data. We will start with the simplest case of two sources of variability. Suppose we are studying a batch chemical process where two reactants A and B are combined in a pot and stirred until the reaction is complete in forming a product C. Lab assays show there is variability over time in yields of the product C. Suppose we want to determine what part of the variability in measured yields is due to uncontrollable operating condition changes (such as ambient temperature, and purity of reactants that occur between batches), and what part is due to measurement error (such as sampling and analytical lab error). We can think of the measured response (yield) for the jth measurement of the ith operating condition or batch as the sum of two sources of variability. We can write a mathematical model for yield as:

$$Y_{ij} = \mu + B_i + M_{ij} \tag{16.4}$$

where μ represents the overall mean yield, B_i represents the deviation from the mean caused by the ith operating condition or batch , and M_{ij} is the deviation from $\mu + B_i$ caused by the jth measurement.

Model 16.4 in form looks exactly like Model 11.5, but the interpretation is different. In the experiments we discussed in Chapter 10, deliberate changes were made and the effects, A_i , represented the difference between the average response in the ith level of Factor A and the overall average of the experiment. In statistical terminology the A_i's are called *fixed effects*. In Chapter 10, we were interested in the values of the fixed effects, A_i , as a measures of the effects of the deliberate changes. On the other hand, in Model 16.4 B_i 's represent the differences from the overall average caused by random changes that we have no control over. In statistical terminology the B_i's are called *random effects*. We are not specifically interested in the values of the B_i's themselves, but rather the variance of the B_i's which is σ^2_B . In the same way, we were interested in the variance of the random effects ε_{ij} in Model 10.5, that represented our inability to repeat experimental results exactly. The random effects M_{ij} in Model 16.4 represent the effects of measurement error such as sampling and laboratory analytical error as discussed in Example 2. The M_{ij} in model 16.4 plays the same role as the ε_{ij} effects in Model 11.5. We are specifically interested in the variance of the M_{ij}'s which is σ^2_M .

Y_{ij} is a sum of a constant μ, and two random variables B_i and M_{ij}. Therefore the variance of Y_{ij} , σ^2 , is equal to the sum of the variances of the two random variables. In other words, $\sigma^2 = \sigma^2_B + \sigma^2_M$. We call these two variances the *variance components*. We can estimate these two variance components separately if we conduct an experiment similar to the one-factor experiment described in Section 11.5. We will have to make a number (n_B) of batches (using the same reactor, raw

---

[2]Nelson, L. S., "Propagation of Error," *Journal of Quality Technology*, 24(4), (1992), pp. 232–235.

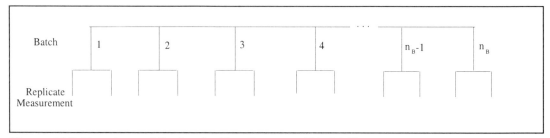

Figure 16.2 Experimental Design for Two Sources of Variability

material(s) and operator(s)) and measure the yield of each batch a number (n_M) of times. The experimental design or sampling design is shown schematically in Figure 16.2 and reproduced in the top of Table B.7.1-1. The only decisions to be made are the selection of the number of batches to make, n_B (denoted as n_A in the appendix table), and the number of measurements per batch, n_M. It is recommended that n_B be chosen to be as large as is feasibly possible, normally 25 to 30, in order to get a reliable estimate of σ^2_B. The recommended value for n_M is 2, which is the value shown in Figure 16.2. If more than two measurements were made for each batch, the design will be very "unbalanced" in the sense that there would be much more information available to estimate σ^2_M, than there would be to estimate σ^2_B.

16.2.1 Estimation of Variance Components Using ANOVA

One simple way to estimate the two variances from the experimental data is by the Analysis of Variance (ANOVA) method. The calculations needed to complete the ANOVA table are exactly the same as those shown in Table 11.5, the ANOVA for a one-factor experiment. However, additional information is needed to make the ANOVA table useful for estimating the components of variance. This additional information is the expected values of the mean squares. The mean squares, like sample variances are random variables. Their expected values are linear combinations of the variance components. For the present two sources of variability case, the complete ANOVA table with expected mean squares is shown symbolically in Table 16.1. Derivation of the expected mean square values is beyond the scope of this book, but can be found in Graybill[3]. To estimate the variance components, set up two linear equations by equating the mean squares (which are numerical constants after the ANOVA calculations are complete) to their expected values, and solving simultaneously. In the next section we will show a numerical example of these calculations.

---

[3]Graybill, F. A., *Theory and Application of The Linear Model* (North Scituate, MA: Duxbury Press, 1976), Chapter 15.

Table 16.1 ANOVA for Two Sources of Variability

Source	Degrees of Freedom	Sum of Squares	Mean Square	Expected Mean Square
Batch	n_B-1	$SSB = \sum_{i=1}^{n_B} (\bar{Y}_{i\cdot} - \bar{Y}_{\cdot\cdot})^2$	$SSB/(n_B-1)$	$\sigma^2_M + 2\,\sigma^2_B$
Measurement Error	n_B	$SSE = \sum_{i=1}^{n_B} \sum_{j=1}^{2} (Y_{ij} - \bar{Y}_{i\cdot})^2$	SSE/n_B	σ^2_M
Total	$2 \times n_B - 1$	SST=SSE+SSB		

16.2.2 Numerical Example of ANOVA for Two Sources

We will illustrate estimating variance components for the chemical batch yield example using the data in Table 16.2. In this table $n_B = 10$ for illustrative purposes, although in practice a larger number of batches (>25) should be sampled to get a reasonable estimate of σ^2_B. $n_M = 2$ (the recommended value). In the table sample data are given, along with the batch averages, and within batch sum of squares. These summary statistics can be easily computed with a spreadsheet program like EXCEL and are used to calculate the sums of squares in the ANOVA. The complete ANOVA table with expected mean squares is shown in Table 16.3. To estimate the variance components we solve the two simultaneous linear equations:

$$0.000180 = \sigma^2_M$$

$$0.071822 = \sigma^2_M + 2\,\sigma^2_B.$$

where the left side of each equation is the mean square and the right side is the expected mean square given for each line of the ANOVA in Table 16.3. This is an easy set of equations to solve since the first equation is already solved for σ^2_M. If we substitute the solution to the first equation into the second we get:

$$0.071822 = 0.000180 + 2\,\sigma^2_B$$

The solution to both equations is then:

$$\hat{\sigma}_M^2 = 0.000180$$

$$\hat{\sigma}_B^2 = 0.035821$$

Table 16.2 Yields of a Batch Chemical Process

i Batch	j Meas.	Y_{ij}	$\bar{Y}_{i\cdot}$	$\sum_{j=1}^{2}(Y_{ij}-\bar{Y}_{i\cdot})^2 = (Y_{i1}-Y_{i2})^2/2$
1	1	95.24		
1	2	95.20	95.22	0.0008
2	1	95.00		
2	2	95.03	95.066	0.00045
3	1	95.05		
3	2	95.05	95.05	0.00
4	1	95.13		
4	2	95.14	95.135	0.00005
5	1	94.70		
5	2	94.68	94.69	0.0002
6	1	94.97		
6	2	94.97	94.97	0.000
7	1	95.14		
7	2	95.13	95.135	0.00005
8	1	94.66		
8	2	94.68	94.67	0.0002
9	1	95.16		
9	2	95.17	95.165	0.00005
10	1	94.95		
10	2	94.95	94.95	0.000
Totals				SSE=0.0018

where the hats over the symbols indicate they are the estimated values. The estimated variance of a single measurement on one batch of material is:

$$\hat{\sigma}^2 = \hat{\sigma}_B^2 + \hat{\sigma}_M^2 = 0.035821 + 0.00018 = 0.036001$$

and the proportion of variation that is due to changes in operating conditions is:

$$\frac{\hat{\sigma}_B^2}{\hat{\sigma}^2} = \frac{0.035821}{0.036001} = 0.995$$

The measurement error due to sampling and laboratory error is a very small proportion (i.e., 1.0-0.995 =0.005) of the total variability and we should feel very comfortable using control charts to detect shifts in process operating conditions, or using designed experiments to determine the effects of various process operating factors upon yield.

Table 16.3 ANOVA Table of Data in Table 16.2

Source	Degrees of Freedom	Sum of Squares	Mean Square	Expected Mean Square
Batch	9	0.6464	0.71822	$\sigma^2_M + 2\,\sigma^2_B$
Measurement Error	10	0.0018	0.000180	σ^2_M
Total	19	0.6482		

One problem with the ANOVA method of estimating the variance components is that occasionally the mean square for measurement error is larger than the mean square for batch in the ANOVA table. This will result in a negative estimate for σ_B^2 when the two linear equations involving the expected mean squares are solved for the variance components. If this occurs, the estimate, $\hat{\sigma}_B^2$, is set equal to zero, since it is impossible, by definition, to have a negative variance.

16.2.3 Estimating Variance Components Using Ranges

For data from simple experiments like that shown in Figure 16.2 and Table 16.2, the variance components σ^2_B, and σ^2_M, can be estimated using simple ranges rather than using the ANOVA calculations. The range, which is the difference between the highest and lowest values in a group of observations, as described in Chapter 4, can be converted to an estimate of the standard deviation by dividing by an appropriate factor. In general, the range estimates are not considerd to be as accurate as the ANOVA estiamtes but they are simpler to compute by hand. Range estimates of variance components are normally used in quality control measurement error studies such as the Gage Capability experiments, that will be described in Section 16.5. In this section, we will illustrate the use of the range in estimating the two variance components, σ^2_B and σ^2_M, from the data in Table 16.2.

Since the observations Y_{i1}, and Y_{i2} in Table 16.2 represent repeat measurements of yield from the same batch of material, the ranges $R_i = |Y_{i1} - Y_{i2}|$ represent measurement variability as does $\bar{R} = \sum_{i=1}^{10} R_i$. These ranges are shown in the fourth column of Table 16.4 that is labeled Range of Reps. The fifth column of Table 16.4, that is labeled Range of 4, includes variability from batch to batch as well as variability within a batch, since there are two batches of material represented in each group of four observations and repeat measurements from both batches. Average ranges, $\bar{R}$, for groups of two and four are shown at the bottom of the Table. These quantities can be easily computed in a spreadsheet program.

An average range, $\bar{R}$, can be converted to an estimate of the standard deviation by dividing by the factor d_2 found in Table A.9. In this table, values of d_2 are found for converting ranges of two values to ranges of 10 values. They are indexed by the left column in the table by, n, which represents the number of data values in the range, and by the row heading, k, that represents the number of ranges averaged. For values of n greater than 10, the range is not an accurate estimator of variability. To convert the average Range of Reps in Table 16.4 to a standard deviation, we divide by the range factor $d_2=1.160$ for n=2, and k=10. This results in an estimate of $\hat{\sigma}_M$, i.e., $\hat{\sigma}_M = 0.014/1.160 = 0.012068$. To convert the average Range of 4 to a standard deviation we divide by the range factor $d_2=2.101$ for n=4, k=5. This results in an estimate of $\hat{\sigma}$, i.e., $\hat{\sigma} = 0.378/2.101 = 0.1799$. To get an estimate of the batch to batch variance $\hat{\sigma}_B$, we use the relationship:

$$\hat{\sigma}^2 = \hat{\sigma}^2_B + \hat{\sigma}^2_M$$

or
$$0.1799^2 = \hat{\sigma}^2_B + 0.012068^2$$

which results in the estimates:

$$\hat{\sigma}^2 = 0.03236$$

$$\hat{\sigma}_B^2 = 0.03222$$

$$\hat{\sigma}_M^2 = 0.0001456$$

These estimates are reasonably close to those obtained using the ANOVA method in the last section.

In general, this method of estimation can be extended to data from designs like those shown in Figure 16.2 as long as n_B is chosen to be an even number. Another more direct method of estimating σ_B^2, is to convert the range of the batch averages, $\bar{Y}_{i\cdot}'s$ in Table 16.2, to a standard deviation. The range of the batch averages is 95.22 - 94.67 = 0.55. Dividing this by the d_2 factor from Table A.9 with n=10 and k=1, we get the estimate $\hat{\sigma}_B$ = 0.55/3.185 = .17268. Squaring this we get the estimate $\hat{\sigma}_B^2$ =.0298 which again is only slightly different than the two previous estimates given. This alternate method of estimating σ_B^2 has the added advantage that it will not produce a negative estimate of σ_B^2, and will work as long as $n \le 10$ so that we can find the appropriate d_2 factor in Table A.9.

Table 16.4 Yields of a Batch Chemical Process

i Batch	j Meas.	Y_{ij}	Range of Reps.	Range of 4
1	1	95.24		
1	2	95.20	0.04	
2	1	95.00		
2	2	95.03	0.03	0.24
3	1	95.05		
3	2	95.05	0.00	
4	1	95.13		
4	2	95.14	0.01	0.09
5	1	94.70		
5	2	94.68	0.02	
6	1	94.97		
6	2	94.97	0.00	.29
7	1	95.14		
7	2	95.13	0.01	
8	1	94.66		
8	2	94.68	0.02	0.48
9	1	95.16		
9	2	95.17	0.01	
10	1	94.95		
10	2	94.95	0.00	0.79
Average Ranges			0.014	0.378

16.2.4 Graphical Representation of Data from Simple Experiments for Estimating Two Sources of Variance

One odd or unrepresentative value can have a major influence on variance estimates. When the variability in measurement error calculated from a simple experiment for estimating two sources of variability is a large >25% of the total, it is useful to know whether this measurement error is approximately equal in all batches or parts measured, or whether a majority of the measurement error occurred on one batch or part. If the majority of measurement error occurred on one batch or part, it may be possible to repeat the lab analysis to determine if something out of the ordinary has occurred. Graphical displays (see Snee[1]) allow us to quickly view data from a simple experiment to determine if there are any strange values or outliers. A

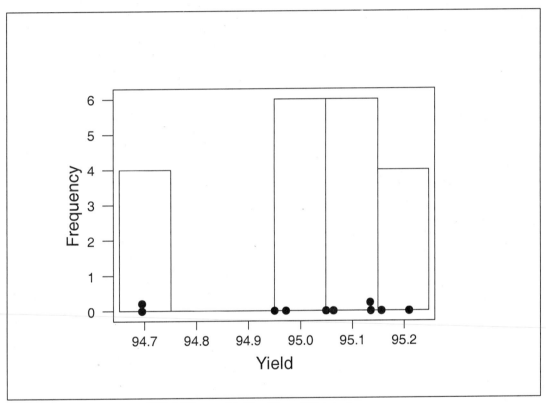

Figure 16.3 Histogram of Yield Data with Dot Plot of Batch Averages

histogram of all the data, possibly overlaid with the batch averages, will allow us to detect clusters of atypical results and peculiar distribution shapes. For example, Figure 16.3 shows a

[1]Snee, R. D., "Graphical Analysis of Process Variation Studies," *Journal of Quality Technology*, 15(2), (1983) , pp. 76 - 88.

histogram of the data in Table 16.2. From this graph we can see that the distribution shape is not normal, and that the majority of the variability among batches is due to the two batches with low yields around 94.7.

Dot frequency diagrams in which all data points are shown in a single plot with notations indicating the sources of variation are useful for visualizing all the sources of variation in one diagram. For example, the data of Table 16.2 are graphed in this manner in Figure 16.4. In this figure, boxes are drawn around replicate measurements from the same batch. The vertical length of the box indicates the measurement variability, and the variability of placement of the boxes around the horizontal line representing the grand average shows the variability from batch to batch. The near equality of the box lengths shows the variability within batch is fairly consistent, while the batch-tobatch variability is dominated by batches 5 and 8.

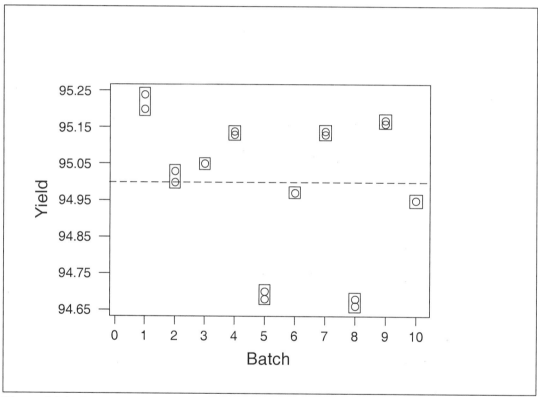

Figure 16.4-- Dot Frequency Diagram of Yield Data from Table 16.2

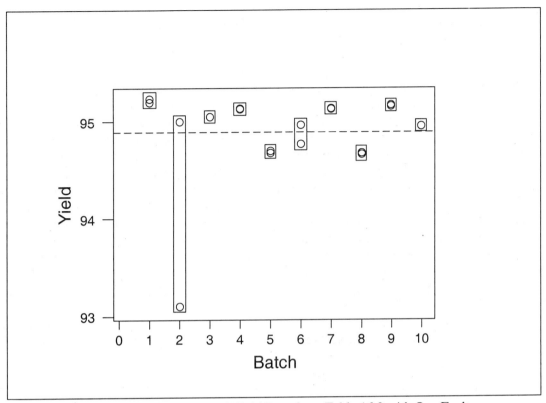

Figure 16.5 Dot Frequency Diagram of Yield Data from Table 16.2 with One Faulty Measurement

To visualize the effect of one faulty measurement on variance component estimates, consider changing the value of the second measurement on the second batch in Table 16.2, from 95.03 to 93.10. The estimate of measurement error would change from $\hat{\sigma}_M^2 = .00018$ (calculated using the ANOVA method) to $\hat{\sigma}_M^2 = 0.18264$; three orders of magnitude! However, Figure 16.5 would easily draw your attention to the fact that the majority of the measurement error is occurring within the second batch, and leads you to investigate the possibility of a mistake. In general, graphical analysis should be done in conjunction with any variance component estimation in order to lead to correct interpretation of the data.

16.3 Components of Variance Multiple Sources

16.3.1 Introduction

In most industrial measurements used in research or quality control, there are multiple sources of variability that should be understood. For example, in the chemical and process industries typical components of measurement error are sampling error and analytical error as described in Example 2 of Section 16.1. In the piece parts manufacturing industries, repeatability and reproducibility of gauges is normally of interest. Repeatability represents the error in measurements taken under the same conditions, while reproducibility represents the error variability caused by different gauges or different operators. The simple experiments described in the last section, are only adequate for estimating one source of error variability by separating it from process or piece-to-piece variation. In this section, we will describe nested sampling experiments that are often used in the chemical and process industries for estimating multiple sources of variability. In Section 16.4 we will describe staggered nested designs which, like fractional factorial designs, allow us to estimate the necessary quantities with less experimentation. In Section 16.5 we will describe the repeatability reproducibility studies used in piece parts manufacturing.

16.3.2 Nested Designs for Estimating Multiple Components of Variance

There are two common purposes for estimating multiple sources of variability. One is to characterize or classify process variability so that quality control efforts can focus on reducing the largest sources of variability. The second is to characterize sampling and laboratory error, so that efforts can be directed toward getting the most precise information for a given cost.

Tables B.7.1-2 through B.7.1-3 schematically represent nested experimental designs that are used for estimating three, four or five different sources of variability. At the bottom of the tables are the formulas for the ANOVA calculations used to estimate the variance components.

To illustrate their use, let's consider an example experiment and analysis for three sources of variability. In the semiconductor industry, several integrated circuits, called dies, are manufactured on wafers of silicon. Routine tests are conducted to measure various electrical properties of the individual circuits. For illustrative purposes the data in Table 16.5 represent a typical set of test values collected from the production line during one week. In practice, 25 or 30 days of data would be preferred. Two wafers were chosen randomly from each day's production. Two circuits (or dies) were measured on each wafer. The data in Table 16.5 follow the pattern shown schematically in Table B.7.1-2. Mathematically we can represent the data by the equation:

$$Y_{ijk} = \mu + D_i + W_{ij} + C_{ijk}$$

where μ represents the overall average, D_i represents the difference from the overall average in the ith day, W_{ij} represents the difference from the ith day average in the jth wafer, and C_{ijk}

represents the difference of the kth circuit or die from the average of the jth wafer in the ith day. This is called the *nested model* for three sources. We would like to estimate the variance of the D_i, W_{ij}, and the C_{ijk} in the previous equation, namely σ^2_D, σ^2_W, and σ^2_C in order to characterize the process variability.

      The summary statistics in the right five columns of Table 16.5 can be easily computed with a spreadsheet, and are used for calculating the sums of squares for the ANOVA table shown in the formulas at the bottom of Table B.7.1-2. Using these formulas and the summary statistics in Table 16.5, we produced the ANOVA Table 16.6. For example, the sum of squares for the top source of variability (days) is computed using the formula for SS_A, as:

$$SS_A = (\sum_{i=1}^{4} \frac{T_i^2}{4}) - \frac{T^2}{4n_A} = \frac{479.61}{4} + \frac{492.84}{4} + \frac{462.25}{4} + \frac{561.69}{4} - \frac{(89.3)^2}{4 \cdot 4} = .6919$$

where T=89.3 is the grand total of all the observed values.

Table 16.5 Experimental Data and Summary Statistics for Semiconductor Measurements

Day	Wafer	Circuit Die	Y_{ijk}	(Y_{ijk}^2)	T_{ij}	(T_{ij}^2)	T_i	(T_i^2)
i	j	k						
1	1	1	5.1	26.01				
1	1	2	5.4	29.16	10.5	110.25		
1	2	1	5.5	30.25				
1	2	2	5.9	34.81	11.4	129.96	21.9	479.61
2	1	1	5.1	26.01				
2	1	2	5.7	32.49	10.8	116.64		
2	2	1	5.5	30.25				
2	2	2	5.9	34.81	11.4	129.96	22.2	492.84
3	1	1	4.9	24.01				
3	1	2	5.5	30.25	10.4	108.16		
3	2	1	5.4	29.16				
3	2	2	5.7	32.49	11.1	123.21	21.5	462.25
4	1	1	6.0	36.0				
4	1	2	5.6	31.36	11.6	134.56		
4	2	1	5.8	33.64				
4	2	2	6.3	39.69	12.1	146.41	23.7	561.69

Table 16.6 ANOVA for Semiconductor Data in Table 16.5

Source	Sum of Squares	df	Mean Square	Expected Mean Square
Days	.6919	3	0.2306	$\sigma^2_C + 2\sigma^2_W + 4\sigma^2_D$
Wafer	.4775	4	0.1194	$\sigma^2_C + 2\sigma^2_W$
Circuit (Die)	.8150	8	0.1019	σ^2_C

To estimate the variance components, we solve the three equations obtained by equating the calculated mean squares in Table 16.6 to the expected mean squares.

$$0.2306 = \sigma^2_C + 2\sigma^2_W + 4\sigma^2_D$$

$$0.1194 = \sigma^2_C + 2\sigma^2_W$$

$$0.1019 = \sigma^2_C$$

The solution is:

$$\hat{\sigma}^2_D = 0.02781$$
$$\hat{\sigma}^2_W = 0.00875$$
$$\hat{\sigma}^2_C = 0.10187$$

where the hats over the letters indicate these are estimates calculated from the data. The total variability of measurements is estimated to be:

$$\hat{\sigma}^2 = \hat{\sigma}^2_D + \hat{\sigma}^2_W + \hat{\sigma}^2_C = 0.13843$$

Now it can be seen that most of the variability is due to variation from circuit to circuit within a single wafer, i.e.,

$$100\% \times (.10187/.13843) = 73.58\%$$

The day-to-day and wafer-to-wafer variability is a minor part. Therefore quality control efforts to reduce variability should concentrate on determining causes for variability of individual circuits or dies within a wafer, rather than looking for assignable causes for differences that affect entire wafers or an entire day's production.

Again simple graphs like those shown in Section 16.2.4 can help to make sure no odd

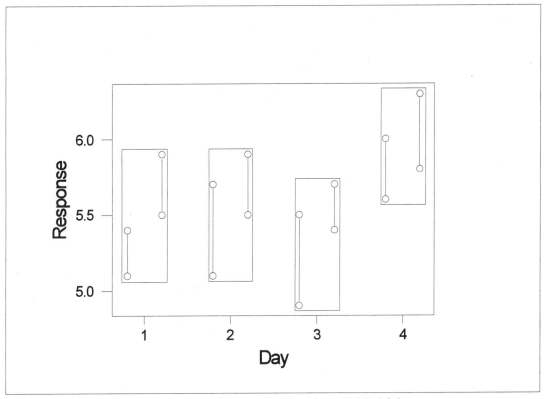

Figure 16.6 Dot Frequency Diagram of Response Data from Table 16.5

measurements are influencing the estimates. Figure 16.6 shows a dot frequency diagram of the data in Table 16.4. Here the boxes contain the four measurements for each day and vertical lines connect the two measurements from the two circuits made on each wafer. In the figure it can be seen that the variability or difference between the two circuits is fairly uniform from wafer to wafer. There are no problem measurements like those shown in Figure 16.5. The variability from wafer to wafer within a day also appears to be fairly consistent. The only odd characteristic that can be seen is the fact that the measurements were always higher in the second wafer measured each day than they were in the first. Normally in an actual experiment like this, we should have 25 or more days of data in order to have a good estimate of σ^2_D. If we were to see a pattern of the second wafer always being higher with 25 or more days of data, there would be cause for concern. However with only four days of data we wouldn't be too concerned. After all it is possible to toss four heads in a row while coin tossing.

Next let's consider an example for four sources of variability. The data in Table 16.7 are from Bennett[2] who describes a sampling study to separate the process variability from three sources of measurement error in lots of metallic oxide. In this example two samples were taken from each of 14 different lots of material. Two chemists (not necessarily the same for each lot) each made

---

[2]Bennett, C. A., "Effect of Measurement Error on Chemical Process Control," *Industrial Quality Control*, Vol. 11, No. 6, (1954), pp.17–20.

duplicate analysis from each sample. The figures reported in the table are the metal content minus 80. Mathematically we can represent the measured metal content as a sum of five different quantities, i.e.,

$$Y_{ijkl} = \mu + L_i + S_{ij} + C_{ijk} + A_{ijkl} \tag{16.5}$$

where μ represents the grand average

L_i represents the deviation of the average of Lot i from the grand average

S_{ij} represents the deviation of the average in the jth sample in Lot i from the average in Lot i

C_{ijk} represents the deviation of the average for the kth chemist for the jth sample in Lot i from the average in the jth sample in Lot i

A_{ijkl} represents the deviation of the lth analysis by the kth chemist of the jth sample in Lot i from the average for the kth chemist on the jth sample in Lot i

This is called a nested model for four sources. The variance of Y_{ijkl}, σ^2, likewise can be written as the sum of four different variances, i.e.,

$$\sigma^2 = \sigma^2_L + \sigma^2_S + \sigma^2_C + \sigma^2_A$$

These variance components can be estimated by the ANOVA method. We can complete the ANOVA calculations by hand using the formulas at the bottom of Table B.7.1-3 as in the last example, but with the availability of modern computer software this is rarely done. Most statistical software programs, such as MINITAB have procedures for making the ANOVA calculations and estimating the variance components.

Table 16.7 Metallic Oxide Data

Lot	Sample	Chemist 1		Chemist 2	
1	1	3.4	3.4	3.6	3.5
	2	3.7	3.5	3.1	3.4
2	1	4.2	4.1	4.3	4.2
	2	4.2	4.2	4.3	4.2
3	1	3.5	3.5	4.2	4.5
	2	3.4	3.7	3.9	4.0
4	1	3.4	3.3	3.5	3.1
	2	4.2	4.2	3.3	3.1
5	1	3.2	2.8	3.1	2.7
	2	3.0	3.0	3.2	2.7
6	1	0.2	0.7	0.8	0.7
	2	0.3	0.4	0.2	-0.1
7	1	0.9	0.6	0.3	0.6
	2	1.0	1.1	0.7	1.0
8	1	3.3	3.5	3.5	3.4
	2	3.9	3.7	3.7	3.7
9	1	2.9	2.6	2.8	2.9
	2	3.1	3.1	2.9	2.7
10	1	3.8	3.8	3.9	3.8
	2	3.4	3.6	4.0	3.8
11	1	3.8	3.4	3.6	3.8
	2	3.8	3.6	3.9	4.0
12	1	3.2	2.5	3.0	3.5
	2	4.3	3.5	3.8	3.8
13	1	3.4	3.4	3.3	3.3
	2	3.5	3.5	3.2	3.3

Table 16.8 shows the first few lines of the data format required by the balanced ANOVA procedure in MINITAB. The nested model for the data in Equation 16.2 is specified to the program as:

$$\%\text{Metal} = \text{Lot Sample(Lot) Chemist(Sample)}$$

The last source of variability (analysis) is not specified to the computer program since it automatically computes a term it labels "error" which is always the last source of variability. The output appears as Table 16.9.

Table 16.8 Data Format for MINITAB Balanced ANOVA

Row	Lot	Sample	Chemist	%Metal
1	1	1	1	3.4
2	1	1	1	3.4
3	1	1	2	3.6
4	1	1	2	3.5
5	1	2	1	3.7
6	1	2	1	3.5
7	1	2	2	3.1
8	1	2	2	3.4
9	2	1	1	4.2
10	2	1	1	4.1
11	2	1	2	4.3
12	2	1	2	4.2
13	2	2	1	4.2
14	2	2	1	4.2
15	2	2	2	4.3
16	2	2	2	4.2

.
.
.

Table 16.9 ANOVA for Metallic Oxide Data

Analysis of Variance (Balanced Designs)

```
Factor                      Type Levels Values
Lot                       random     13    1    2    3    4    5    6    7
                                           8    9   10   11   12   13
Sample(Lot)               random      2    1    2
Chemist(Lot Sample) random           2    1    2


Analysis of Variance for %Metal

Source                      DF        SS        MS      F      P
Lot                         12   128.8235   10.7353  54.84  0.000
Sample(Lot)                 13     2.5450    0.1958   1.72  0.115
Chemist(Lot Sample)         26     2.9550    0.1137   3.16  0.000
Error                       52     1.8700    0.0360
Total                      103   136.1935


Source                    Variance Error Expected Mean Square
                          component  term (using unrestricted model)
  1 Lot                    1.31744    2   (4) + 2(3) + 4(2) + 8(1)
  2 Sample(Lot)            0.02053    3   (4) + 2(3) + 4(2)
  3 Chemist(Lot Sample)    0.03885    4   (4) + 2(3)
  4 Error                  0.03596        (4)
```

The computer output shows the ANOVA followed by the estimates of the variance components and expected mean squares. In the expected mean squares table at the bottom, σ^2_A is labeled (4), σ^2_C is labeled (3), σ^2_S is labeled (2), and σ^2_L is labeled (1). In this example the measurement error is the sum of three components $\sigma^2_S + \sigma^2_C + \sigma^2_A$, and appears to be a small proportion of the total variability, i.e.,

$$(\sigma^2_S + \sigma^2_C + \sigma^2_A)/(\sigma^2_L + \sigma^2_S + \sigma^2_C + \sigma^2_A) = (0.02053+0.03885+0.03596)/(1.31744+0.0.02053+0.03885+0.03596)$$

$$= 0.0675$$

However, a simple histogram of the data shown in Figure 16.7 shows that the majority of the lot to lot variability can be explained by the fact that lots 6 and 7 have much lower metal content than the other 11 lots. If an assignable cause can be found for the low values found in these two lots, they should not be included in a study to identify and quantify the relative importance of measurement error. If lots 6 and 7 are removed from the data and the ANOVA calculations redone, the lot-to-lot variance, σ^2_L, is reduced substantially causing the proportion of total variance due to measurment to increase to over 50% of the total. This is much different than the results from all the data. If a larger study was conducted with 25 or more lots, the conclusions would be less dependent on the results of one or two lots. As in the last section, we always recommend the use of simple graphical displays of the data in conjunction with the variance component estimation.

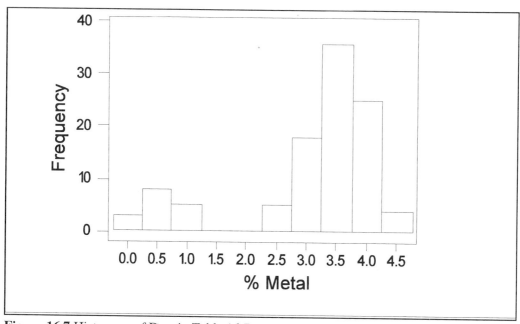

Figure 16.7 Histogram of Data in Table 16.7

16.4 Staggered Nested Designs

It can be seen in the ANOVA Table 16.9 from the last examples, that the degrees of freedom for the mean squares are unequal or unbalanced when we have more than two components of variance. In the last example, with four components, fully half of the information we collected went into estimating the variance due to repeat analysis by the same chemist (the least important component). This seems like a waste, and it is! If we would have collected 25-30 lots, as recommended, we would have needed to make as many as 30×8=240 lab analysis. In cases like this, there are designs that we can employ to reduce the number of lab analysis by 50% or more and make things more balanced. These designs are called *staggered nested designs*[3].

16.4.1 Design and Analysis with Staggered Nested Designs

As with the nested designs for estimating multiple sources discussed in Section 16.3, there are tables in Appendix B.7 that give the appropriate staggered nested designs for three, four, and five sources of variation (Tables B.7.2-1, B.7.2-2, and B.7.2-3 respectively). As with the other tables, the formulas for the ANOVA table are included below each design.

---

[3]Smith, J. R. and Beverly, J. M., "The Use and Analysis of Staggered Nested Factorial Designs," *Journal of Quality Technology*, Vol 13, No. 3, (1981), pp.166–173.

16.4.2 Staggered Design Example with Three Sources

Let us revisit the first example discussed in Section 16.3 to illustrate the use of the staggered nested designs. The appropriate design corresponding to three sources of variation is given in Table B.7.2-1. It is the same as the nested design already used, except that for one of the wafers from each day only one circuit is measured, while for the other wafer two circuits are measured, as before. The reduced set of data is given in Table 16.10. This table looks much like Table 16.5 (for the full nested design) with some of the rows removed. The missing rows correspond to the circuits that were not measured.

Table 16.10 also has several additional columns of summary statistics that were used to compute the sums of squares in the ANOVA Table 16.11. These columns consist of totals over one or more subscripts, the number of analyses totaled, and the total squared divided by the number of analyses in the total. It should be noted that these are the same summary statistics used for the full nested design, except that the number of analyses in any column was constant for the full nested design. Therefore the step of dividing by the number of analyses in the total was done later. The number of analyses totaled was not even listed in Table 16.5 but was included in the calculations for the ANOVA table given in the appendix. For the staggered nested design, the number of analyses in the total is not constant (except for T_i where n_i is always 3), and so n is listed and the step of dividing by n is done when squaring the total. These totals and counts can be easily computed with a spreadsheet.

The calculations needed to estimate the components of variance were carried out using the formulas in Appendix B, Table B.7.2-1. They are summarized in the ANOVA Table 16.11. The sum of squares for days was calculated as $SS_A = 376.036 - (67.1)^2/(3 \times 4) = .8355$; the sum of squares for wafers was calculated as $376.8 - 376.036 = .764$, and the sum of squares for circuit was calculated as $377.13 - 376.8 = .3300$. From the mean squares, the components of variance were estimated as shown in the bottom of Table 16.11. It can be seen by comparing Table 16.11 to Table 16.6 that the variance estimate for wafers is a bit different due to the small number of days in this example. For larger studies the variance component estimates obtained from a staggered nested design are normally quite comparable to those that would be obtained from a full nested design with a considerable savings in sampling and measurements. For the three-component study the reduction in number of observations was only 1/4 (12 in Table 16.10, compared to 16 in Table 16.5), however the savings are much greater for larger studies (i.e., ½ for four components and 9/16 for five components).

Table 16.10 Semiconductor Measurements—Staggered Nested Design

(Note that the data were coded: $Y_{ijk} = Y_{ijk} - 95$)

D	W	C	Y_{ijk}	$(Y_{ijk})^2$	T_{ij}	n_{ij}	$(T_{ij})^2/n_{ij}$	T_i	n_i	$(T_i)^2/n_i$
1	1	1	5.1	26.01	5.1	1	26.01			
1	2	1	5.5	30.25						
1	2	2	5.9	34.81	11.4	2	64.98	16.5	3	90.75
2	1	1	5.1	26.01						
2	2	1	5.5	30.25						
2	2	2	5.9	34.81	11.4	2	64.98	16.5	3	90.75
3	1	1	4.9	24.01	4.9	1	24.01			
3	2	1	5.4	29.16						
3	2	2	5.7	32.49	11.1	2	61.605	16.0	3	85.333
4	1	1	6.0	36.0	6.0	1	36.0			
4	2	1	5.8	33.64						
4	2	2	6.3	39.69	12.1	2	73.205	18.1	3	109.203
Totals			67.1	377.13	67.1			376.8	67.1	376.036

Table 16.11 ANOVA Table for Example 6.4.2

Source	Sum of Squares	DF	Mean Squares	Expected MS
Day	.8355	3	0.27851	$\sigma_C^2 + (5/3)\sigma_W^2 + 3\sigma_D^2$
Wafer	.7637	4	0.19092	$\sigma_C^2 + (4/3)\sigma_D^2$
Circuit (Die)	.3300	4	0.0825	σ_C^2

$\sigma_D^2 = [0.27851 - 0.0825 - (5/3)0.08125]/3 = .02199$ (10.98%) $\sigma_D^2 = 0.020199$

$\sigma_W^2 = [0.19092 - 0.0825]/(4/3)$ $= 0.08125$ (44.17%) $\sigma_W^2 = 0.08125$

$\sigma_C^2 = 0.0825$ $= 0.0825$ (44.85%) $\sigma_C^2 = 0.0825$

$\sigma^2 = \sigma_R^2 + \sigma_O^2 + \sigma_M^2$ $= 0.1839$ (100%)

16.4.3 Staggered Design Example with Four Sources

As mentioned in Section 16.3, there are two main motives for characterizing components of variance: (1) quality improvement (i.e., variance reduction) and (2) sampling efficiency. The example just discussed in Section 16.4.2 was aimed at quality improvement. The example to be discussed in this section will deal with sampling efficiency - getting the most precise number you can for a given amount of effort (money).

Suppose you are a manufacturer of dye and you are interested in measuring the strength of a batch of dye so that you can formulate it properly for a consistent strength product. The measurement process involves taking a grab sample from the batch, making up a solution of the dye, dying a piece of white cloth, and finally measuring the strength of the color of the cloth with a spectrophotometer. Each step in the procedure introduces variability and incurs a cost. In order to get the best assessment of dye strength for a given expenditure we must know both.

For illustrative purposes, Table 16.12 shows the data from a staggered nested design with 10 batches (again 25 or more batches should be used in a practical application). The design is patterned after Table B.7.2-2. In this table, as with the previous examples, the data were coded (by subtracting 100) to minimize the number of digits needed to be written down. There is great efficiency in this design since only four measurements are made per batch, rather than eight that would be required by a full nested design.

The components of variance were estimated using the ANOVA method. The formulas in Table B.7.2-2 along with the summary statistics in Table 16.12 were used to produce the ANOVA Table 16.13. For example, the sum of squares for grab samples was calculated to be $[1884.34 - (246.8)^2/(4)(10)]=[1884.34-1522.756]=361.584$.

Normally the computations necessary to complete Table 16.13 would be carried out on a computer. However, because the staggered nested designs are incomplete, a standard balanced Analysis of Variance program like the MINITAB procedure illustrated in Table 16.9 will not work. PROC Varcomp in SAS will produce the correct sums of squares and variance component estimates, and Nelson[4] provides a BASIC program that computes both the sums of squares and variance component estimates for staggered nested designs with three to six sources. This program can be incorporated as a VBA MACRO if computations are made with Microsoft's EXCEL spreadsheet. If neither of the programs are available, any general linear models program that will calculate the "sequential sum" of squares (called the type I sum of squares by SAS) can be used and will give the same sums of squares and degrees of freedom as those shown in Table 16.13. These sums of squares and degrees of freedom can be used with the expected mean squares from the appendix table to set up the linear equations necessary to solve for the variance components as shown in Table 16.13. Table 16.14 shows the output from the general linear model procedure in MINITAB. The sequential sums of squares can be seen to match hand-calculated sums of squares in Table 16.13.

---

[4]Nelson, L. S., "Variance Estimation Using Staggard Nested Design," *Journal of Quality Technology*, Vol 15, No. 4, (1983), pp.195–198.

Table 16.12 Strengths of Batches of Dye—Staggered Nested Design
(Note that the data were coded: $Y_{ijk} = y_{ijk} - 100$)

G	S	C	M	y_{ijkl}	Y_{ijkl}	Y^2_{ijkl}	T_{ijk}	n_{ijk}	$\dfrac{T^2_{ijk}}{n_{ijk}}$	T_{ij}	n_{ij}	$\dfrac{T^2_{ij}}{n_{ij}}$	T_i	n_i	$\dfrac{T^2_i}{n_i}$
1	1	1	1	101.3	1.3	1.69	1.3	1	1.69	1.3	1	1.69	13.6	4	46.24
1	2	1	1	103.7	3.7	13.69	3.7	1	13.69	12.3	3	50.43			
1	2	2	1	104.0	4.0	16.00	8.6	2	36.98						
1	2	2	2	104.6	4.6	21.16									
2	1	1	1	107.8	7.8	60.84	7.8	1	60.84	7.8	1	60.84	26.4	4	174.24
2	2	1	1	105.4	5.4	29.16	5.4	1	29.16	18.6	3	115.32			
2	2	2	1	107.0	7.0	49.00	13.2	2	87.12						
2	2	2	2	106.2	6.2	38.44									
3	1	1	1	106.1	6.1	37.21	6.1	1	37.21	6.1	1	37.21	19.6	4	96.04
3	2	1	1	106.1	6.1	37.21	6.1	1	37.21	13.5	3	60.75			
3	2	2	1	103.7	3.7	13.69	7.4	2	27.38						
3	2	2	2	103.7	3.7	13.69									
4	1	1	1	105.8	5.8	33.64	5.8	1	33.64	5.8	1	33.64	10.0	4	25.00
4	2	1	1	102.4	2.4	5.76	2.4	1	5.76	4.2	3	5.88			
4	2	2	1	101.2	1.2	1.44	1.8	2	1.62						
4	2	2	2	100.6	0.6	0.36									
5	1	1	1	108.3	8.3	68.89	8.3	1	68.89	8.3	1	68.89	35.6	4	316.94
5	2	1	1	107.1	7.1	50.41	7.1	1	50.41	27.3	3	248.43			
5	2	2	1	110.5	10.5	110.25	20.2	2	204.02						
5	2	2	2	109.7	9.7	94.09									
6	1	1	1	101.2	1.2	1.44	1.2	1	1.44	1.2	1	1.44	8.4	4	17.64
6	2	1	1	101.4	1.4	1.96	1.4	1	1.96	7.2	3	17.28			
6	2	2	1	103.0	3.0	9.00	5.8	2	16.82						
6	2	2	2	102.8	2.8	7.84									
7	1	1	1	103.3	3.3	10.89	3.3	1	10.89	3.3	1	10.89	21.0	4	110.25
7	2	1	1	106.3	6.3	39.69	6.3	1	39.69	17.7	3	104.43			
7	2	2	1	106.0	6.0	36.00	11.4	2	64.98						
7	2	2	2	105.4	5.4	29.16									
8	1	1	1	107.0	7.0	49.00	7.0	1	49.00	7.0	1	49.00	27.4	4	187.69
8	2	1	1	106.4	6.4	40.96	6.4	1	40.96	20.4	3	138.72			
8	2	2	1	107.5	7.5	56.25	14.0	2	98.00						
8	2	2	2	106.5	6.5	42.25									
9	1	1	1	112.7	12.7	161.29	12.7	1	161.29	12.7	1	161.29	47.2	4	556.96
9	2	1	1	109.9	9.9	98.01	9.9	1	98.01	34.5	3	396.75			
9	2	2	1	112.7	12.7	161.29	24.6	2	302.58						
9	2	2	2	111.9	11.9	141.61									
10	1	1	1	109.4	9.4	88.36	9.4	1	88.36	9.4	1	88.36	37.6	4	353.44
10	2	1	1	108.4	8.4	70.56	8.4	1	70.56	28.2	3	265.08			
10	2	2	1	110.2	10.2	104.04	19.8	2	196.02						
10	2	2	2	109.6	9.6	92.16									
		Totals		246.8		1938.38			1936.18			1916.32			1884.34

Table 16.13 ANOVA Table for Example 16.12

Source	Sum of Squares	DF	Mean Squares	Expected Mean Squares
Grab Samples (G)	361.584	9	40.176	$\sigma_M^2 + (3/2)\sigma_C^2 + (5/2)\sigma_S^2 + 4\sigma_G^2$
Solutions (S)	31.980	10	3.198	$\sigma_M^2 + (7/6)\sigma_C^2 + (3/2)\sigma_S^2$
Cloths (C)	19.860	10	1.988	$\sigma_M^2 + (4/3)\sigma_C^2$
Measurements	2.200	10	0.220	σ_M^2
Total	415.624	39		

$\sigma_G^2 = [40.176 - 0.220 - (3/2)(1.324) - (5/2)(0.977)] / 4$ $\quad = 8.895 \quad (78.1\%) \quad \sigma_G = 2.983$

$\sigma_S^2 = [3.198 - 0.220 - (7/6)(1.324)] / (3/2)$ $\quad = 0.955 \quad (8.4\%) \quad \sigma_S = 0.977$

$\sigma_C^2 = [1.986 - 0.220] / (4/3)$ $\quad = 1.324 \quad (11.6\%) \quad \sigma_C = 1.151$

$\sigma_M^2 =$ $\quad = 0.220 \quad (1.9\%) \quad \sigma_M = 0.469$

$\sigma^2 = \sigma_G^2 + \sigma_S^2 + \sigma_C^2 + \sigma_M^2$ $\quad = 11.394 \quad (100.0\%) \quad \sigma = 3.375$

Table 16.14 MINITAB General Linear Model Analysis

```
Factor   Levels Values
G           10    1     2     3     4     5     6     7     8     9     10
S(G)         2    1     2
C(G S)       2    1     2

Analysis of Variance for Strength

Source   Model DF  Reduced DF   Seq SS
G            9          9       361.584
S(G)        10         10        31.980
C(G S)      20         10+       19.860
Error        0         10         2.200
Total       39         39       415.624

+ Rank deficiency due to empty cells, unbalanced nesting or collinearity.
  No storage of results or further analysis will be done.
```

615

Before we can formulate a good strategy for measuring dye strength, we need to consider cost information as well as variability. Let us say that we found the cost of collecting a grab sample to be $5, the cost of making a solution of dye to be $4, the cost of adding a piece of cloth to the solution to be $1, and the cost of making the spectrophotometer measurement to be $2. Then, the total cost of a single measurement would be $12 (= $5 + $4 + $1 + $2). The total variability of a single measurement, as quantified by the variance, is simply the sum of all the components as indicated in Table 16.13. In this case the total variance, σ^2 is 11.39 (or the standard deviation is 3.38).

So for $12 we get a measurement of dye strength that has a standard deviation of 3.38. If that is precise enough, then we are happy and no further measurements are needed. If is not precise enough, then we need to improve it the only way we can (short of some technical change) via replication. The concept, which we have encountered many times before, is that if we take several measurements and average them, the average will have a variance which is smaller than that of a single measurement by the factor $1/n$, where n is the number of measurements averaged. Thus if we are willing to duplicate the whole procedure, the cost will be $24 and the variance of the average of two measurements will be half of what it was with only one measurement (or the standard deviation will be 2.39). If we do everything in triplicate the cost will be $36 and the variance will be one-third (or the standard deviation will be 1.95). And if we do everything in quadruplicate the cost will be $48 and the variance will be one-fourth (or the standard deviation will be 1.69). These numbers are all summarized in Table 16.15.

Table 16.15 Dye Batch Strength Measurement—Base Case

Source	Variance Component	Cost	Schemes with Various Amounts of Replication (with Resultant Cost and Variability)						
Sample (G)	8.895	$2	1	2	1	2	3	3	4
Solution (S)	0.955	$4	1	2	1	2	3	3	4
Cloth (C)	1.324	$1	1	2	3	4	3	3	4
Measure (M)	0.220	$5	1	2	3	4	3	6	4
Total Cost			$12	$24	$24	$36	$36	$51	$48
Total Variance			11.39	5.70	10.36	5.31	3.80	3.76	2.85
Total Standard Deviation			3.38	2.39	3.22	2.30	1.95	1.94	1.69

The main point of this example, however, is that we do not have to replicate everything the same number of times. So, for example, we could take one grab sample, make up only one dye solution from it, but dye three pieces of cloth and make three measurements (one on each piece of cloth) for a total cost of $24 (see Table 16.15). This is the same cost as duplicating everything, but the variance will be different. The way to calculate the variance when the various pieces are not replicated equally is straightforward—you divide each variance component by the number of times it is replicated and then add them together to get the total, as given in Equation 16.6.

$$\sigma^2 = \frac{\sigma_G^2}{n_G} + \frac{\sigma_S^2}{n_S} + \frac{\sigma_C^2}{n_C} + \frac{\sigma_M^2}{n_M} \tag{16.6}$$

The total variance of our example with one grab sample and solution but three cloths and

$$\sigma^2 = \frac{8.895}{1} + \frac{0.955}{1} + \frac{1.324}{3} + \frac{0.220}{3} = 10.36$$

measurements is 10.36 (or the standard deviation is 3.22) as shown in detail in previous equation and summarized in Table 16.15. This scheme gives a larger variance than duplicating everything (for the same cost) and so it would not be wise to use it. From the table it can be seen that the most efficient use of resources is to replicate the whole procedure as many times as is necessary to get the precision needed. This is not generally true, but it is quite reasonable in this case. If one inspects the components of variance, the largest component by far is the variability from one grab sample to another. So, to put a significant dent in the total variance, we must replicate the grab samples. When we take cost into account, we can see that the cost of taking a sample is small, and so it is a good strategy to spend our money replicating grab samples as much as possible.

As an aside, it should be noted that all of our discussions in this section assume that when a particular source of variation is replicated, it is necessary to replicate all the sources of variation downstream of it (below it in the variance pyramid). In other words, it is assumed that we cannot measure the strength of another grab sample without making a solution, dying a piece of cloth and measuring the color of the cloth. This is the typical case, but it is not always true. For example, it may be feasible to take a number of grab samples (from different parts of the batch of dye) and combine them to form one composite sample which is analyzed further. This would make it possible to replicate grab samples without replicating solutions, etc. Since this is a relatively unusual circumstance, is not discussed in detail here. Suffice it to say that the basic formula for
total variance with replication, Equation 16.6, is still valid.

Before we leave this example, let us look at a case where replicating the whole analysis procedure is not the most efficient scheme to get better precision. Let us pretend that the variance components and costs were as given in Table 16.16. These are the same values as previously determined, but several of them are associated with different components. Specifically the values of the variances for grab samples and cloths are interchanged, and the costs for grab samples and measurements are interchanged. The total variance and cost for a single measurement are the same as before. And if we wanted to replicate the whole procedure a second, third or fourth time, the costs and variances would also not change. What is different is that, for a given expenditure, we can find a better scheme than total replication. Several competing schemes are listed in Table 16.16, which give lower total variance for a fixed cost. For example, if we were willing to spend $36, the best scheme is NOT to replicate the whole procedure three times, but rather to take two grab samples, make one solution from each (two solutions total), dye three cloths in each of the two solutions (six cloths total) and measure the color of each cloth (six measurements total). This would give us a total variance of 2.66 rather than the 3.80 value for three total replicates.

If we look at the components of variance we can see why this would be true; the largest source of variability is the cloth, so we should try to dye as many pieces of cloth as we can for a given expenditure. Of course, the other components cannot be ignored. It turns out that we reach a state of diminishing returns if we try to replicate cloths too heavily relative to the other components. For example, we could spend the same $36 by taking only one grab sample and making one solution from it, dying nine pieces of cloth in the solution, and measuring the color of each. But we see that the total variance is bigger, even though we have nine pieces of cloth instead of six. The fact that we have only have one grab sample and solution instead of two became important.

Table 16.16 Dye Batch Strength Measurement—Case II

Source	Variance Component	Cost	Schemes with Various Amounts of Replication (with Resultant Cost and Variability)							
Sample (G)	1.324	$5	1	2	1	3	2	1	4	2
Solution (S)	0.955	$4	1	2	1	3	2	1	4	2
Cloth (C)	8.895	$1	1	2	5	3	6	9	4	10
Measure (M)	0.220	$2	1	2	5	3	6	9	4	10
	Total Cost		$12	$24	$24	$36	$36	$36	$48	$48
	Total Variance		11.39	5.70	4.10	3.80	2.66	3.29	2.85	2.05
	Total Standard Deviation		3.38	2.39	2.0	1.95	1.63	1.81	1.69	1.43

16.5 Classifying Sources of Measurement Error

Measurements are necessary in all aspects of engineering work. Most all data that are collected in engineering work are the results of measurements. It would be difficult for us to improve product designs, process designs, or maintenance and service schedules, if we can't measure product and process performance in some way. However, in the real world measurements are not exact. They are always subject to variability. Repeat measurements of the same part or component characteristic often yield differing results. The quality of a measurement is analogous to the quality of the output of a production process. A measurement process includes elements such as operational definitions, physical instruments, environmental conditions, and observers. It is beneficial to characterize and quantify the various sources of measurement variability.

Although this is not the first chapter in the book, identifying sources of variability in measurements should be the first step taken in any study involving data collection. Before using control charts and before conducting designed experiments, we should understand how much of the variability in data we collect is due to assignable causes, and how much represent errors in measurement. In this section we describe the traditional type of study that is used to classify measurement error.

16.5.1 Classical Gage Capability Studies

The most common type of study used in industry to classify measurement error, or variability, is called a Gage Capability Study. These studies are used to quantify two sources of measurement variability that are called repeatability error and reproducibility error. Repeatability refers to the variation obtained when one person, using the same measuring instrument, measures the same dimension two or more times. Reproducibility refers to the variation among different operators measuring the same part. The Gage Capability Study consists of having a single characteristic on each of ten different parts measured three times with the same gage by each of three operators. The number of parts, repeat measurements, and operators may be varied, although the numbers cited: 10, 3, and 3, are considered to be optimum[1] . The total measurement error or variability, is the sum (in terms of variances) of the repeatability error and reproducibility error. In general, if the ratio of the standard deviations of the total measurement error is greater than 5% of the tolerance range for the product characteristic in question, then the measurement error could have a large affect on decisions made from data. In this case, appropriate steps should be taken to reduce measurement error. For example, if most of the measurement error is due to reproducibility, better training of operators may reduce the error. If repeatability is the major source of error, higher precision gages may be required or averaging repeat measurements must be done routinely.

In this section we will describe a small gage capability study with five parts and three operators to illustrate the study design and methods of data analysis. Table 16.17 shows the steps

---

[1]*Measurement Systems Analysis Reference Manual*, Copyrighted by Automotive Industry Action Group, Developed by A.I.A.G. in conjunction with American Society for Quality Control.

routinely performed in the conduct of a Gage Capability Study.

Table 16.17 Procedure for Gage Capability Studies

1. Randomly select five or more parts and prepare them for gaging (i.e., wash, de-burr and number so that the numbers are not visible to operators making the measurements).

2. Choose two or more operators or appraisers (those that use the equipment routinely are preferred).

3. Calibrate the gage and make sure that it is defect free.

4. Each operator or appraiser measures each of the parts in random order without seeing each other's readings and records the values. Repeat the cycle two more times.

Table 16.18 shows the data resulting from a typical study along with summary statistics (ranges and averages) that are used to calculate the gage repeatability and reproducibility error. In the table rows 1–3 , 6–8 and 11–13 we see the measurement data collected by operators A, B, and C. In rows 4 and 5 we see the average and range of the three measurements made by operator A on each part, along with the overall average, and average range for operator A. Similar summary statistics are shown in rows 9 and 10 for operator B and 14 and 15 for operator C. The overall average (of all measurements from all operators) for each part are shown in row 16, along with the range, $\bar{R}_p$, in these part averages. In row 17 the three average ranges for each operator are averaged to get the overall range $\bar{R}$. In row 18 the range in the three operator averages (labeled $\bar{X}$ Diff) is calculated.

Table 16.18 Data from Gage Capability Study

Operator/ Trial #	Part					Average
	1	2	3	4	5	
1. A 1	6.387	6.239	6.334	6.493	6.277	
2. 2	6.370	6.243	6.342	6.459	6.248	
3. 3	6.398	6.244	6.332	6.456	6.265	
4. AVG.	6.385	6.242	6.336	6.469	6.263	$\bar{X}_a$ =6.339
5. RNG.	0.028	0.005	0.010	0.037	0.029	$\bar{R}_a$ =.0218
6. B 1	6.378	6.223	6.329	6.448	6.228	
7. 2	6.425	6.237	6.350	6.482	6.245	
8. 3	6.389	6.232	6.349	6.457	6.240	
9. AVG.	6.397	6.231	6.343	6.462	6.238	$\bar{X}_b$ =6.334
10. RNG.	0.047	0.014	0.007	0.034	0.017	$\bar{R}_b$ =.0238
11. C 1	6.391	6.243	6.387	6.345	6.259	
12. 2	6.447	6.224	6.420	6.478	6.208	
13. 3	6.412	6.241	6.335	6.482	6.213	
14. AVG.	6.417	6.236	6.381	6.435	6.227	$\bar{X}_c$ =6.339
15. RNG.	0.056	0.019	0.085	0.137	0.051	$\bar{R}_c$ =.0696
16. PART AVG. $\bar{X}_p$	6.400	6.236	6.353	6.455	6.243	$\bar{R}_p$ =0.219
17. [R_a = .0218 +R_b =.0238 + R_c =.0696]/[#operators=3]=						$\bar{R}$ = .0385
18. [Max ($\bar{X}$) =6.339 - Min ($\bar{X}$) = 6.334] =						$\bar{X}$ Diff = .005

16.5.2 Range Estimates of Repeatability and Reproducibility Errors

The traditional way of estimating the reproducibility and repeatability variation from the study data is through simple range calculations. The repeatability error is called the equipment variation or EV. The standard deviation of repeatability variation (SD_{EV}) is calculated by taking the grand average range $\bar{R}$, shown in Table 16.18, and dividing by the appropriate $d_2(k,n)$ factor, from Table A.2, to convert the range into a standard deviation. The reproducibility error is called appraiser variation or AV. The standard deviation of reproducibility variation (SD_{AV}) is calculated by converting the range between operators, $\bar{X}$ Diff, into a standard deviation by dividing by the appropriate $d_2(k,n)$ factor. The SD_{AV} calculated in this way contains some repeatability error. A better estimate can be obtained by subtracting a fraction of the repeatbility

error in terms of variances as: $\sqrt{SD_{AV}^2 - \dfrac{SD_{EV}^2}{p \cdot r}}$,where p is the number of parts and r is the

number of repeat measurements made of each part by each operator. If the radical is negative set it to zero as we did with the ANOVA estimates of variance components. The standard deviation of the repeatability and reproducibility variation is called SD_{RR} and is calculated as:

$SD_{RR} = \sqrt{SD_{EV}^2 + SD_{AV}^2}$. The standard deviation in the process (SD_{PV}), or part-to-part error,

can also be estimated from the Gage Capability Study by dividing the range of part averages,

$\bar{R}_p$, by the appropriate conversion factor, $d_2(k,n)$. Combining the repeatability and

reproducibility variation and the process variation we obtain the total variation. In terms of a

standard deviation this is: $SD_{TV} = \sqrt{SD_{RR}^2 + SD_{PV}^2}$. However, if SD^2_{PV} were obtained from

the Gage Capability Study with ten or less parts, it would not be very reliable. Normally it would be better to have an estimate of SD_{TV} from a more extensive process capability study like those to be described in Chapter 17.

Once the SD_{EV}, SD_{AV}, and SD_{RR} are obtained they are usually expressed as a percentage

of the SD_{TV} by the formulas: $EV\% = \dfrac{SD_{EV}}{SD_{TV}}$, $AV\% = \dfrac{SD_{AV}}{SD_{TV}}$, $RR\% = \dfrac{SD_{RR}}{SD_{TV}}$. These

percentages will not sum to 100. Normally if RR% is less than 10, the measurement system is considered acceptable. If RR% is between 10 and 30 the measurement system maybe acceptable depending on the application, but if RR% > 30 the measurement system should be deemed unacceptable.

Example calculations are shown below using the data in Table 16.18.

Repeatability: (k=#ranges averaged =15, n =#values in each range=3)

$$SD_{EV} = \bar{R} \; / \, d_2(15,3) = .0384/1.689 =- .02273$$

Reproducibility: (k=#ranges averaged =1, n =#values in each range=3)

$$SD_{AV} = \sqrt{\left(\frac{\bar{X} \; Diff}{d_2(k,n)}\right)^2 - \frac{SD_{EV}^2}{p \cdot r}} = \sqrt{\left(\frac{.005}{1.908}\right)^2 - \frac{.02273^2}{5 \cdot 3}} = \sqrt{-.00002758} \cong 0.0$$

Repeatability and Reproducibility:

$$SD_{RR} = \sqrt{.02273^2 + 0.0^2} = 0.02273$$

The total process variation in terms of a standard deviation was known, from a more extensive process capability studies, to be $SD_{TV}= .493$. Therefore the total measurement error, or repeatability and reproducibility, is acceptable since $RR\% = 100 \times (SD_{RR}/SD_{TV})\% = 4.6\% < 5\%$. If the estimate of SD_{TV} determined from the Gage Capability Study (calculations shown below) was used instead of the more accurate estimate, $RR\% = 24.94\%$ and the acceptability of the measurement system would be questionable. However, SD_{PV} is frequently underestimated in Gage Capability Studies because of the small number of parts studied. The effort to obtain a more accurate estimate of SD_{PV} should taken whenever possible.

Part and Total Variation:

$$SD_{PV} = \bar{R}_p \; / \, d_2(1,5) = 0.219/2.481 = 0.08827$$

$$SD_{TV} = \sqrt{SD_{PV}^2 + SD_{RR}^2} = \sqrt{.08827^2 + .02273^2} = .09115$$

If the RR% is greater than 30% and unacceptable, the quality engineer should investigate ways to reduce it. The first step would be to look at the separate components due to equipment variation (EV%), and appraiser variation (AV%). If AV% is much larger than EV%, better training of operators that will allow them to perform more alike may produce a satisfactory reduction in measurement error. If the major source of variability is equipment variation (EV%), either more precise gages must be obtained or as a temporary method of improvement repeated measurements could be made and averaged as shown in the sampling efficiency example in Section 16.4.3. In this case, the standard deviation in repeatability and reproducibility error can be reduced as the number of repeat measurements r is increased, i.e.,

$$SD_{RR} = \sqrt{\frac{SD_{EV}^2}{r} + SD_{AV}^2} < \sqrt{SD_{EV}^2 + SD_{AV}^2} \qquad \text{if } r > 1$$

16.5.3 ANOVA Estimates of Repeatability and Reproducibility Errors

Although the range estimates shown in the last section have been traditionally used to estimate the standard deviation of gage repeatability and reproducibility, theoretically more accurate estimates can be obtained using the Analysis of Variance (ANOVA) method. If a computer program like the MINITAB Balanced ANOVA procedure is available, the estimation of the repeatability and reproducibility standard deviation can be automated and an additional piece of error can be estimated if there is any interaction between operator and part.

To use the ANOVA method we write a model for the data in exactly the same form as Equation 11.6, i.e.,

$$Y_{ijk} = \mu + O_i + P_j + OP_{ij} + \varepsilon_{ijk} \tag{16.7}$$

The effects of operator O_i, part P_j, and interaction OP_{ij}, are defined similar to the effects in the two factor model in Chapter 10. However, in this case they are *random effects*, as defined in Section 16.2, and we are not interested in estimating their values, but rather their variances. To estimate their variances we need in addition to the ANOVA table, the expected values of the mean squares. Table 16.19 shows the general form for the expected mean squares. In this table o represents the number of operators in the study, p is the number of parts, and r is the number of repeat measurements of each part. The formulas for the sums of squares and means squares for the two-factor Model 16.7 are the same as those given in Equations 11.7 to 11.10. To estimate the four variances σ^2_O, σ^2_P, σ^2_{OP}, and σ^2_E, the mean squares in the ANOVA table are equated to their respective expected mean squares, and the four linear equations are solved simultaneously, as shown in Sections 16.2 and 16.3 with the nested models.

Table 16.19 Expected Mean Squares for Gage Capability Studies

Source	Degree of Freedom	Mean Square	Expected Mean Square
Operator	o-1	MS_O	$\sigma^2_E + r\,\sigma^2_{OP} + op\,\sigma^2_O$
Part	p-1	MS_P	$\sigma^2_E + r\,\sigma^2_{OP} + op\,\sigma^2_P$
Operator×Part	$(o$-1$)\times(p$-1$)$	MS_{OP}	$\sigma^2_E + r\,\sigma^2_{OP}$
Repeatability Error	$o\times p\times(r$-1$)$	MS_E	σ^2_E

We will illustrate the ANOVA method using the data from the last section. The measurement data from Table 16.18 is copied into a single column in Table 16.19 with indicators for operator and part added as columns to the left. This is the format required by most ANOVA programs. Table 16.20 shows the output of the MINITAB Balanced Analysis of Variance procedure. The model was specified to the program in the form:

measure = operator + part + operator*part

and the two factors operator and part were designated as random factors. The upper part of

Table 16.20 Measurement Data from Table 16.18 in Format for ANOVA Program

	Data Display		
Row	Operator	Part	Measure
1	1	1	6.387
2	1	1	6.370
3	1	1	6.398
4	1	2	6.239
5	1	2	6.243
6	1	2	6.244
7	1	3	6.334
8	1	3	6.342
9	1	3	6.332
10	1	4	6.493
11	1	4	6.459
12	1	4	6.456
13	1	5	6.277
14	1	5	6.248
15	1	5	6.265
16	2	1	6.378
17	2	1	6.425
18	2	1	6.389
19	2	2	6.223
20	2	2	6.237
21	2	2	6.232
22	2	3	6.329
23	2	3	6.350
24	2	3	6.349
25	2	4	6.448
26	2	4	6.482
27	2	4	6.457
28	2	5	6.228
29	2	·5	6.245
30	2	5	6.240
31	3	1	6.391
32	3	1	6.447
33	3	1	6.412
34	3	2	6.243
35	3	2	6.224
36	3	2	6.241
37	3	3	6.387
38	3	3	6.420
39	3	3	6.335
40	3	4	6.345
41	3	4	6.478
42	3	4	6.482
43	3	5	6.259
44	3	5	6.208
45	3	5	6.213

Table 16.21 shows the sums of squares, degrees of freedom, and mean squares. The lower part of the table shows the expected mean squares and variance component estimates. In the table, σ^2_O is called (1), σ^2_P is called (2), σ^2_{OP} is called (3), and σ^2_E is called (4). Since the estimate of σ^2_O is negative (i.e., -.00007) we would assume that it is zero. The standard deviation for repeatability or equipment variation would then be:

$$SD_{EV} = \sqrt{\sigma^2_E} = \sqrt{.00078} = .02793$$

The standard deviation for reproducibility or appraiser variability is the square root of the sum of the operator and operator×part variance components, i.e.,

$$SD_{AV} = \sqrt{\sigma^2_O + \sigma^2_{OP}} = \sqrt{-.00007 + .00012} = .00707$$

Finally the standard deviation of the total measurement or repeatability and reproducibility error is

$$SD_{RR} = \sqrt{SD^2_{EV} + SD^2_{AV}} = \sqrt{.02793^2 + .00707^2} = 0.02881$$

Table 16.21 MINITAB ANOVA Table for Measurement Data in Table 16.19

Analysis of Variance (Balanced Designs)

Factor	Type	Levels	Values				
Operator	random	3	1	2	3		
Part	random	5	1	2	3	4	5

Analysis of Variance for Measure

Source	DF	SS	MS	F	P
Operator	2	0.000244	0.000122	0.11	0.899
Part	4	0.335854	0.083963	74.13	0.000
Operator*Part	8	0.009061	0.001133	1.46	0.215
Error	30	0.023331	0.000778		
Total	44	0.368489			

Source	Variance component	Error term	Expected Mean Square (using unrestricted model)
1 Operator	-0.00007	3	(4) + 3(3) + 15(1)
2 Part	0.00920	3	(4) + 3(3) + 9(2)
3 Operator*Part	0.00012	4	(4) + 3(3)
4 Error	0.00078		(4)

16.5.4 Graphical Analysis of Gage Capability Studies

In Section 16.2.4 we emphasized the need to examine results of variance component studies graphically to make sure that one or two odd data values did not have a large influence on the variance estimates. For the same reason, similar graphical analysis should be made of data from Gage Capability Studies. A simple graph that can be utilized for this purpose is the Multi-vari plot shown in Figure 16.8. In this plot the nine measurements for each part are shown as points on the graph. The individual measurements are labeled by the operator A, B or C, and lines connect the measurements made by the same operator. From this plot we can see that most of the repeatability and reproducibility error is due to variation within repeat measurements made by operator C, and an apparent difference between the average measurement by operator C and the other two operators on parts 3 and 4. It is possible that error could be reduced if operator C were better trained. Other peculiar differences that might occur in Gage Studies such as higher or lower measurements always being made on the first cycle of measurements are easily identified on a Multi-vari plot that might occur in Gage Studies. Another type of graph that is normally used in Gage Capability Studies, to verify the consistency of the repeatability error over different operators and parts, is the range control chart. These were described briefly in Chapter 2 and will be described in more detail in Chapter 17.

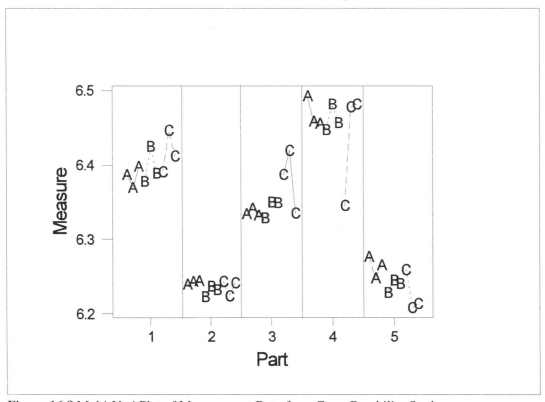

Figure 16.8 Multi-Vari Plot of Measurement Data from Gage Capability Study

16.6 Summary

In this Chapter we have studied ways to characterize variability. We have examined both the bottom up method, for synthesizing the variance of the dimension of an assembly from the known variances of the dimensions of each component, and the top down method for estimating variance components from experimental data. We have described the variance propogation formula for synthesizing the variance of a function. We have shown that the variance components can be calculated using simple ranges or the ANOVA method. We have presented specific experimental designs for collecting data to estimate variance components. These include nested designs, staggered nested designs, and classical Gage Capability Studies. The difference in fixed and random effects was explained, and some simple graphical displays that help in interpreting data used to estimate variance components.

The list below are the important terms and concepts covered in this chapter:

16.7 Exercises for Chapter 16

16.1 An assembly consists of three stacked parts. Due to manufacturing imperfections the parts are not identical, and their thicknesses follow a Normal distribution with a range of natural variation of ±.006 inches (which represents six standard deviations). What tolerance range (i.e., ±3σ limits) could you hope to hold on the thickness of the stacked assembly?

16.2 A finish on a metal consist of three coats. A primer coat, a finish coat, and a clear coat. The mean and standard deviation of the thicknesses of the three coats are $\mu_P = 0.0025$, $\sigma_P = 0.0013$, $\mu_F = 0.0021$, $\sigma_F = 0.0011$, $\mu_C = 0.0017$, $\sigma_C = 0.0008$. What is the mean and standard deviation of the finish?

16.3 Finish part dimensions are measured on a micrometer, and the variability in measurements can be assumed to be a sum of the part to part variance, the variance due to gage, and finally the variance of repeat measurements. Symbolically this is expressed as:

$$\sigma_T^2 = \sigma_P^2 + \sigma_G^2 + \sigma_R^2$$

If the three variance components are respectively: 2.73, 0.034, and 0.132.

    a) Calculate the total variance in measured part dimensions.

    b) Calculate the percentage or proportion of the total variation in measures that is actually due to part-to-part differences in dimensions.

    c) Would micrometer measurements taken in this way be accurate enough to use in control charts? Why or why not?

16.4 Injection molded caps are produced for cylindrical metal parts that are purchased from a supplier. The standard deviation of the clearance between the inside diameter of the cap, and the outside diameter of the metal cylinders must be no more than .02 inches if the cap is to fit snugly. If the standard deviation of the outside diameter of the cylinders is .005 inches, what must the standard deviation of the cap inside diameters be in order to ensure that the caps fit snugly?

16.5 A voltage is connected across a resistor. Due to manufacturing imperfections there is variability in both the voltage and the resistance. The resulting current is a random variable whose value is given by Ohm's law as a function of the random voltage and resistance as:

$$I = V\!/\!R$$

If the variance of the voltage is $\sigma_V^2 = 23$, and the variance of the resistor is $\sigma_R^2 = 95$, the mean or nominal voltage is 120V, and the mean or nominal resistance is 900Ω. Use the variance propagation formula to find the variance in the current.

16.6 The thermal expansion, δ, is given by the equation $\delta = \alpha \times L \times \Delta T$, where α is a constant,

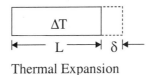

Thermal Expansion

L is the length, and ΔT is the change in temperature. Assuming $\alpha = 0.031$, and the change in temperature, and length are random variables with means and standard deviations given by: $\mu_L = 54.15$, $\sigma_L = 0.13$, $\mu_{\Delta T} = 150.0$, $\sigma_{\Delta T} = 2.0$. Use the variance propagation formula to calculate the variance of the thermal expansion.

16.7 Wernimont[1] studied the variability in the measured melting point of hydroquinone. There were four different thermometers used in the study, and two repeat experiments were performed with each thermometer. The data is shown below:

Thermometer	Measurement	Melting Point
A	1	174.0
A	2	173.0
B	1	173.0
B	2	172.0
C	1	171.5
C	2	171.0
D	1	173.5
D	2	171.0

[1]Werimont, G. "Quality Control in the Chemical Industry II. Statistical Quality Control in the Chemical Laboratory," *Industrial Quality Control*, (1947), p. 8.

a) Complete the ANOVA table similar to Table 16.3, and compute the variance components for thermometer and repeat measurement.

b) Repeat the calculation of variance components for thermometer and measurement using the range method, dividing the measurements into groups of four as shown in Table 16.4.

c) Repeat the calculation of variance components for thermometer and measurement using the range method, and the range of thermometer averages similar to the example in Section 16.2.3.

d) Make a dot frequency diagram of the data and spot any outliers.

16.8 In a production process, a web of material is extruded and wound on a reel. Periodically, a sample of the web is punched out as the material is wound and the thickness is measured and recorded. QA personnel are not sure how much of the variability in recorded thicknesses are actually due to product variability and how much is due to measurement error. The measurement error could be due to differences in gages used to measure the thickness, or due to the inability to get exactly the same value when repeating the measurement with the same gage. Describe how you would collect the data needed to calculate estimates of these three sources of variability.

16.9 Suppose that in the preceding problem the cost of taking a sample from the web is $5, the cost of selecting a gage for measurement is $1, and the cost of measuring the thickness with a gage is $2. If no more than $16 can be spent on getting a measure of thickness, and the variance due to sampling $\sigma^2_S = 0.0326$, the variance due to gages $\sigma^2_G = 0.0017$, and the variance due to repeat measurement with the same gage $\sigma^2_M = 0.0724$, how can the most accurate measure be made (i.e., smallest variance):

a) Take two samples from the web, measure each sample twice with two different gages and average the result as the measurement.

b) Take one sample, measure it five times with one gage and average the results as the measure.

Would your conclusion change if the variances were $\sigma^2_S = 0.0537$, $\sigma^2_G = 0.0021$, and $\sigma^2_M = 0.0224$?

16.10 Bennett and Franklin[2] presented data showing the variability in the concentration of iron in a standard solution as determined by different analysts. A portion of the data is shown below:

Analyst	Repeat Determination	Concentration of iron
1	1	2.963
1	2	2.996
2	1	2.958
2	2	2.964
3	1	2.956
3	2	2.945
4	1	2.948
4	2	2.960
5	1	2.953
5	2	2.961
6	1	2.941
6	2	2.940
7	1	2.963
7	2	2.928
8	1	2.987
8	2	2.989
9	1	2.946
9	2	2.950
10	1	2.956
10	2	2.947

[2]Bennett, C. A. and Franklin, N. L., *Statistical Analysis in Chemistry and the Chemical Industry*, (New York: John Wiley and Sons,1954), p. 331.

a) Complete the ANOVA table similar to Table 16.3, and compute the variance components for analyst and repeat determination.

b) Repeat the calculation of variance components for analyst and determination using the range method, dividing the determinations into groups of four as shown in Table 16.4.

c) Repeat the calculation of variance components for analyst and determination using the range method, and the range of determination averages similar to the example in Section 16.2.3.

d) Make a dot frequency diagram of the data and spot any outliers.

16.11 Duncan[3] presents data from a rubber-tread experiment. Two mixes were made for each of six different tread formulas. Two samples were taken from the slab cured from each mix and duplicate tests run. The samples were nested in mixes and the mixes within tread formulas. The data is shown below:

Tread Formula	Mix 1		Mix 2	
	Sample 1	Sample 2	Sample 1	Sample 2
1	98	52	86	122
2	75	96	64	50
3	34	7	34	26
4	32	19	7	-22
5	138	113	53	108
6	102	74	204	223

a) Write the model for this data. Define each term in your model.

b) With the help of Table B.7.1-2 complete the ANOVA table and estimate the three components of variance.

c) Make a dot frequency diagram of the data, note any unusual data points and describe how they might affect the estimates.

---

[3]Duncan, A .J., *Quality Control and Industrial Statistics*, (Homewood, IL: Richard D. Irwin Inc, 1952), p. 614.

16.12 Wernimont[4] presented data from an experiment designed to test the homogeneity of the copper content of a series of bronze castings from the same pour. Two samples were taken from each of 11 castings and each sample was analyzed in duplicate.

Casting	Sample	Analysis 1	Analysis 2
1	1	85.54	85.56
1	2	85.51	85.54
2	1	85.54	85.60
2	2	85.25	85.25
3	1	85.72	85.77
3	2	84.94	84.95
4	1	85.48	85.50
4	2	84.98	85.02
5	1	85.54	85.57
5	2	85.84	85.84
6	1	85.72	85.86
6	2	85.81	85.91
7	1	85.72	85.76
7	2	85.81	85.84
8	1	86.13	86.12
8	2	86.12	86.20
9	1	85.47	85.49
9	2	85.75	85.77
10	1	84.98	85.10
10	2	85.90	85.90
11	1	85.12	85.17
11	2	85.18	85.24

a) Write the model for this data. Define each term in your model.

b) With the help of Table B.7.1-2 complete the ANOVA table and estimate the three components of variance.

---

[4]Wernimont, G. Ibid.

c) Make a dot frequency diagram of the data, note any unusual data points and describe how they might affect the estimates.

16.13 Eliminating the second sample for mix 2 (i.e. the last column), in the data for Problem 11 results in a staggered nested design in three components. Use the formulas for the sums of squares in Table B.7.2-1, or a statistics program with a GLM ANOVA routine, and the coefficients for the expected mean squares given in Table B.7.2-1 to estimate the three variance components from the reduced data set. How do they compare to the estimates you made in Problem 11?

16.14 The following data are the results of a Gage Capability Study with five parts, two repeat measurements and three operators.

Operator trial		Part				
		1	2	3	4	5
A	1	2.0	2.0	1.5	3.0	2.0
A	2	1.0	3.0	1.0	3.0	1.5
B	1	1.5	2.5	2.0	2.0	1.5
B	2	1.5	2.5	1.5	2.5	.5
C	1	1.0	1.5	2.0	2.5	1.5
C	2	1.0	2.5	1.0	3.0	.5

a) Set up a worksheet similar to Table 16.18 and compute the average and range of the repeat measurements made by each operator, the overall part averages, the range in part averages, and the average and range of all measurements made by each operator.

b) Use the summary statistics you calculated in a) and the formulas in Section 16.5.2 to calculate the SD_{EV}, SD_{AV}, SD_{RR}, SD_{PV} and SD_{TV}. Convert your estimates SD_{EV}, SD_{AV}, and SD_{RR} to percentages. Should the measurement system be deemed acceptable in this situation?

c) Calculate the sums of squares and mean squares in a two-way ANOVA using the formulas in Chapter 11 or the ANOVA routine in a statistical program or spreadsheet. Use the expected mean square formulas in Table 16.19 along with your mean squares to solve for estimates of σ^2_E, σ^2_{OP}, σ^2_O, σ^2_P.

d) Make a Multi-vari plot of the data and determine if there are any unusual patterns.

CHAPTER 17
Shewhart Control Charts

17.1 Introduction

Shewhart and Deming have taught us that reducing variability in manufacturing processes is tantamount to improving quality. Walter Shewhart was the first to propose the use of control charts to reduce variability and improve quality in manufacturing processes. In Chapter 2, we described his concepts of chance or common causes for variation in a process, and special or assignable causes. We also showed the value of control charts with two brief examples. In this chapter we will expand our description of Shewhart style control charts. First, we will explain the philosophy of process control and the theoretical basis for Shewhart style control charts. Next, we will describe the different types of variables and attribute control charts that can be used, and finally provide guidance on how control charts should be used in practice.

Before getting into details, one point should be clarified. The use of control charts to improve performance is not restricted to manufacturing. They can be applied to any area where work is performed and outputs are variable. Examples could include bookkeeping errors, transit times for materials or paperwork in process, or performance characteristics of computer networks.

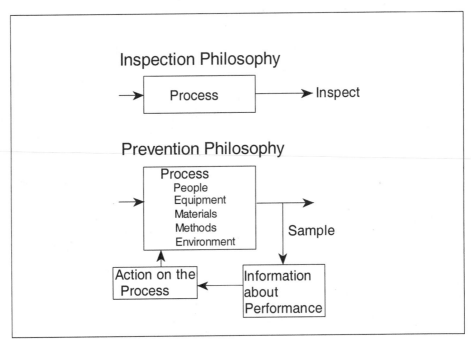

Figure 17.1 Differing Quality Philosophies

Figure 17.1 (adapted from Ford[1]) shows the contrast between the traditional and control chart approaches to quality control. The top of the figure shows the traditional inspection philosophy, where the output of a process is inspected at the end of the line to separate the usable from unusable. This applies to both manufacturing processes where product is inspected as well as administrative situations where work is often checked and rechecked to prevent errors. Shewhart said that this approach is inherently inefficient, and instead proposed the prevention philosophy depicted in the lower half of the figure. Using this philosophy, a feedback loop is established to improve the process performance. Any work process can be defined as the combination of people, equipment, input materials, methods or procedures, and environment that in combination work together to produce outputs. Through the prevention or control chart philosophy, in-process inspections are made to obtain information about the process. This information can be characteristics of the process output or intermediate outputs such as operating temperature, cycle times, etc. This information shows us when adjustments are needed to keep the process on track, and therefore helps us to avoid making unusable output altogether. Since there will always be variability or noise in the results of in-process inspections, the control chart is an invaluable tool for separating true signals for adjustments (special causes) from random process noise (common causes).

Outputs of processes are always subject to variability. Differences among outputs may be large such as those due to improper setup of a machine or use of an incorrect procedure in an administrative process. Other differences among output may be small, such as those due to gear backlash or part centering in metal machining processes, or digit transposition in the single dollar column of a bookkeeping process. Some causes of variation could be short term such as those due to a temporary power surge, or long term such as those due to a worn tool or catalyst degradation in a chemical process.

When aggregated, most (85%) of the minor causes for variation will result in output that appears like a constant system of chance results similar to what would be seen in a casino gambling game. Thus, Shewhart called these aggregate causes chance causes or common causes. As a group these causes will form a stable distribution that is consistent over time as shown in the top half of Figure 17.2 (again adapted from Ford[2]). When a process is influenced only by common causes, we say that it is in a state of statistical control and is predictable within limits.

Special causes for variability occur less frequently (15%) and cannot be characterized by a single distribution or density function, as would be the case if the process were in a state of statistical control. Unless these special causes can be identified and removed, the results of the process will be unpredictable as shown in the lower half of Figure 17.2.

Statistical control is usually not the natural state for any process, and it can only be achieved by determined effort to identify and eliminate, one by one, the special causes for

---

[1]*Continuing Process Control and Process Capability Improvement*, Published by Statistical Methods Office Operations Support Staff, Ford Motor Company, Dearborn MI, 1983.

[2]*Continuing Process Control and Process Capability Improvement*, Ibid.

variation[3]. The discovery of special causes and their correction should be the responsibility of someone who is directly involved with the day-to-day operation of the process (a local operator). The control chart is a useful tool that allows local operators to be aware that special or assignable causes are present. Given the knowledge that special causes are present, the means of identifying and removing these causes may be obvious in some situations. In other cases this may not be so, and could require extensive trial and error or use of experimental designs to discover the source and remedy.

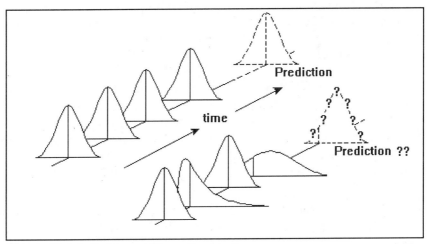

Figure 17.2 A Stable and Unpredictable Process

Once a process is in a state of statistical control, local operators are powerless to further reduce variation. Only management, who has the authority to change things such as established procedures, equipment, training of operators, raw material suppliers, etc. has the ability to further reduce variation.

17.2 Statistical Theory for Control Charts

A control chart is basically a graph upon which we plot a sequence of process measurements. Control limits and a center line are calculated and placed upon the chart, like the examples shown in Chapter 2. When the points on a control chart fall within the calculated control limits in a random pattern, it is an indication that all variation seen on the chart is due to common or chance causes. On the other hand, when points fall outside the control limits, or non-random patterns appear in the data, it is a signal that special or assignable causes for variation are present. Control charts are used in two basic situations. First is to study past process data to determine whether the process has been in a state of statistical control, and to develop control

---

[3]Deming, W. E., "On Some Statistical Aids Toward Economic Production," *Interfaces*, Vol. 5., No. 4. 1975.

limits. Second is to control future process output by extending previously developed control limits for real time decisions.

Control limits should be calculated from a sequence of data that includes only common or chance cause variation. In Shewhart's definition, chance cause variation appears like that experienced when rolling dice or flipping spinners. Therefore if a process measurement is a continuous value like a finished part length, or an in-process temperature, then variability due to common causes can be described by a probability density function like those described in Section 3.4. In a practical situation we may not know which density function (i.e., Normal, Exponential, etc.) would best describe the data. However if we plot sample averages, $\bar{X}$, of a set of consecutive measurements (like those shown in Table 2.4) on a control chart, we know the density function of the averages, $\bar{X}$, will be approximately normal, due to the Central Limit Theorem we described in Section 3.6. Therefore, the control limits are based upon the Normal Distribution.

From our description of the Normal Distribution in Section 3.4.4, we know that 99.73% of all observations should fall within ± 3 standard deviations (σ) of the mean (μ) (e.g., see Figure 3.11). This is the rationale for calculating control chart center line and limits. The center line for a variables control chart is calculated as the best estimate of the process mean μ, namely $\bar{\bar{X}}$

the average of the subgroup averages. The control limits are calculated as $\bar{\bar{X}} \pm 3\dfrac{\hat{\sigma}}{\sqrt{n}}$, where

$\hat{\sigma}$ is an estimate of the standard deviation. In a sense the control chart is nothing more than turning the Normal Distribution on its side as shown in Figure 17.3. If no assignable causes are present in the process the chance of an individual mean, $\bar{X}$, falling outside the limits is very small (i.e., 1-.9973 = .0027). However, if assignable causes shift the process mean, as shown in the dotted density curves in Figure 17.3, the probability of a sample mean, $\bar{X}$, falling outside the limits are greatly increased. Therefore, an assignable cause would be signaled on the control chart when a point falls outside the limits or when a series of points fall close to the limits.

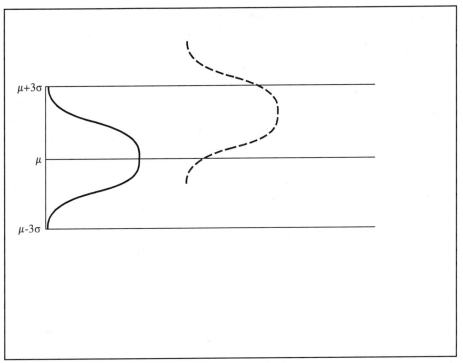

Figure 17.3 Normal Distribution Turned on Its Side.

Control charts help in distinguishing common cause variation from special or assignable cause variation. When points on a control chart fall outside the control limits or exhibit a nonrandom pattern, we say the process is *out of control*. By out of control, we do not mean the operator cannot control the process, but rather we mean the process is providing signals to the operators that indicate the presence of a special causes. Being aware of these signals, and having the know how to remove special causes when they occur, allows process operators to maintain consistent results.

If the points on a control chart fall within the control limits and exhibit nothing more than random patterns, the process is said to be in *a state of statistical control*. When a process is in a state of statistical control, there is nothing the operators can do to reduce variability. However, management can review the control charts from processes that are in statistical control, to quantify the level of common cause variability present. If common cause variation is excessive, management should institute changes to improve the process. Viewing data on a control chart, before and after changes to the process like the casting example in Chapter 2, allows management to visualize the effects of their process changes. In a sense this is like a graphical analysis of a simple comparative experiment conducted in the production environment.

The accuracy of the control limits depends on the assumption that all variability in data used to calculate the limits, was due to common causes. How do we guarantee this critical assumption in practice? If we knew the appropriate control limits, we could plot our data on the control chart and remove any data where the chart signaled assignable causes were present. But, since we don't know the limits, how can we guarantee no assignable causes were present during data collection and calculation of control limits? There is no simple way to answer this question. We will start by defining a *rational subgroup*.

Definition: A *rational subgroup* of measurements is defined to be a sample that covers a period of time that is long enough to exhibit common cause variation, but short enough to make the possibility of assignable cause variation occurring within a subgroup very remote.

In piece part manufacturing, rational subgroups are normally taken to be a sample of consecutive parts like the shank dimensions shown in Table 2.4. In that example, special or assignable causes for variation would be things such as a worn or broken tool, while common causes for variation would be things such as the variable hardness of the raw materials (bar stock) and slight changes in position or placement of the bar stock in the lathe fixture. It is unlikely that the cutting tool would suddenly break or wear significantly within a set of five consecutive parts. Therefore it is unlikely that assignable causes for variation will occur within a subgroup. This justifies the use of a set of consecutively produced parts as a rational subgroup in many situations.

17.3 Procedure for Determining Control Limits

With this definition of a rational subgroup in mind, we can proceed to describe the recommended procedure for computing control limits for variables control charts. First, a set of 25 or more rational subgroups of data are collected during normal operation of the process. Next, an estimate of the mean (μ) and the standard deviation (σ) is calculated from each subset of data separately. Next the separate estimates of μ and σ are combined by averaging, and preliminary control chart limits are calculated with the resulting combined estimates. Next, two control charts are made. One for charting changes in the process mean over time, and one for charting changes in a measure of the process variability over time. If there are no out of control signals on either of these two charts, it may be assumed that only common cause variation is present and the control limits may be finalized. If, on the other hand, out of control signals appear on either chart, the assignable cause for these signals must be determined. Once the assignable causes are determined, affected subgroups of data should be removed, or replaced by additional data from the process and the control limits should be recalculated as before. The two control charts should again be constructed and examined. If no out of control signals occur after this second iteration, the control limits may be finalized. Otherwise, the procedure should be repeated. The entire process is shown in the form of a flow diagram in Figure 17.4.

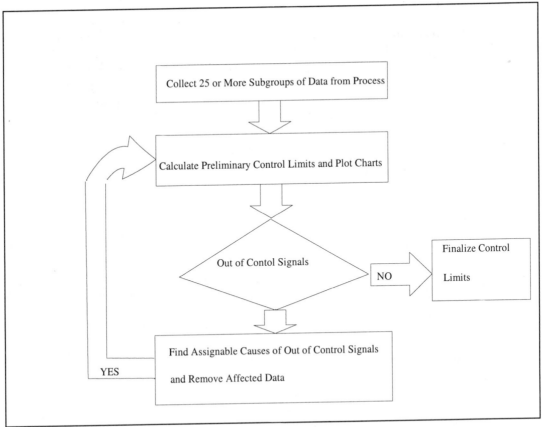

Figure 17.4 Procedure for Establishing Control Limits

The bottom step in the loop, namely finding the causes for out of control signals, sounds simple. But in practice this is the most difficult step. Failure to complete it properly is the reason that many control chart applications are ineffective. In some processes, such as the metal removal process discussed in the Chapter 2 example, causes and corrections for out of control signals are obvious. In other more complicated processes the causes may not be so obvious. Many company bulletin boards are covered with control charts containing unexplained out of control signals. Unless the causes for these signals are identified and prevented (or quickly remedied when they arise), control charts will not help to reduce variation and improve process performance.

17.4 Variables and Attribute Data

Measurements of length, weight, temperature, pressure, etc. normally result in values that may take on any real number in a continuous interval. When only common or chance causes for variability are present, the results of measurements will be continuous random variables as discussed in Chapter 3. In quality control language the results of this type of measurements are called *variables data*. Traditionally, $\bar{X}$ and R charts are the control charts used for variables data. With the advent of calculators and computers, $\bar{X}$ and s, or $\bar{X}$ and s^2 charts are also used for variables data.

In some quality control applications it is difficult or impossible to obtain variables data, because nothing is being measured; rather products are classified as to whether they have certain attributes or not. In this case we can obtain what is called *attribute data* in quality control language. Attribute data consists of counts. They may be counts of defective parts, counts of defects per unit area, or counts of some other attribute of a process output. Since attribute data are counts they can only take on integer values. When only common causes for variation are present in a process, defect counts or attribute data will appear like the discrete random variables discussed in Chapter 3.

There are two types of attribute data that we will encounter in practice. The first type is usually called the *number of defectives* and is restricted to be less than or equal to some known value. For example, if we are inspecting samples of 50 injection molded parts and counting the defective ones, the count can never exceed 50. The second type is called the *number of defects* and there is no restriction on these counts. For example, if we are counting the number of specks in an automobile paint job or the number of contaminant particles on a silicon wafer, the count could be 0, 1, 2, . . ., with no apparent upper limit. In terms of random variables, the number of defectives is equivalent to counts from a Binomial Distribution as described in Chapter 3, while the number of defects is equivalent to Poisson Distribution counts. P and NP control charts are used for the number of defectives, and C and U charts are used for the number of defects.

In some situations where control charts are to be used, we may have a choice whether to use variables data or attribute data. For example, if we are trying to control the length of discrete part lengths, we could measure the lengths with a caliper and record the variables data. Or, we could use a go/no-go gage and simply report attribute data in the form of counts of the number of parts within the spec limits. In these situations, the variables data would be preferred for three reasons. First, smaller rational subgroups (normally 3 - 6) are required for variables data.

Attribute data, on the other hand, requires rational subgroups of 100 to 200 or more in order to have counts greater than zero to chart. Second, we may be able to avoid defects entirely if we use variables data, while attribute type control charts imply defects are being made. The shank diameter study in Chapter 2 was an example of reducing the number of parts outside the specification limits to zero by bringing the process into statistical control. Finally, it is usually easier to identify assignable causes when working with variables. If, for example, we could see that part lengths are trending up and indicating an out of control signal, it would probably be easier to identify the cause than if we only knew the percentage of defectives was increasing (oblivious to whether the defects were due to long or short parts). If at all possible we would like to work with variables data, but attribute control charts also have important uses.

In some cases, there are no measurements to be made and we are stuck with attribute data. Attribute control charts are also a convenient way of introducing the use of control charts, and they are easily understood, since they directly graph counts of defects or defectives that are familiar to most managers and workers. In many situations attribute data is already being collected from end of the line inspection. This data is usually collected to make some decision about the disposition of product already produced (i.e., scrap it, rework it, or send it on to the customer), but it is not used to control or improve the production process. However, the same attribute data collected from end of the line inspection can be used to construct attribute control charts. The attribute control charts can be used to separate common causes and special causes, and get management thinking about the proper course of action to reduce defects in general. Many attributes or defects can be grouped on one chart, reducing the need for maintaining multiple charts. Finally, attribute charts are applicable in any process where outputs can be classified as conforming or nonconforming. For example, in administrative areas, it is possible to use attribute control charts to study and reduce errors in paper work, etc.

From a process control viewpoint, rather than an inspection viewpoint, a stricter inspection order is usually held. Defects and defectives are often counted that would not normally result in rejection of the product or output, such as abnormalities in a paint job on the unseen underside of a product. This would not be a defect from the customer point of view, yet the same process phenomena that caused this abnormality could cause similar abnormalities on the upper surface which would be a defect to the customer. Therefore, for purposes of process control it should be counted as a defect.

17.5 Variables Control Charts

In this section we will discuss the most common variables control charts that are used and show examples of the use of each type.

17.5.1 $\bar{X}$ - R Charts

$\bar{X}$ - R charts were the earliest variables control charts used. These two charts are used simultaneously to control the mean level and variability in a process. Figure 17.5 illustrates how the charts are intended to work. Increases in process variability are usually not visible on the $\bar{X}$ chart, but become quite apparent on the R chart as shown in the left side of Figure 17.5. Shifts in the process mean are not visible on the R chart, but are obvious on the $\bar{X}$ Chart as shown on the right side of Figure 17.5.

Ranges are used as the measure of variability because they were simple to calculate by hand in 1924 when Walter Shewhart first proposed the charts. The steps in constructing and interpreting $\bar{X}$ - R charts are:

1. Arrange sample or historical data in natural subgroups of 3 - 6 (there should be at least 25 subgroups).

2. Calculate the average, $\bar{X}$, and range R for each subgroup.

3. Calculate the estimates of population characteristics, (μ, and σ) as the grand average, $\bar{\bar{X}}$, and the average range , $\bar{R}$.

4. Calculate the control chart limits.

5. Plot the points on the charts and identify any out of control signals.

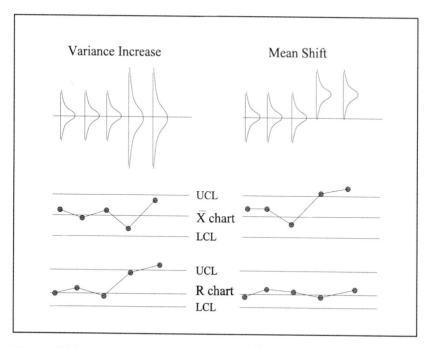

Figure 17.5 Process Changes Detected by $\bar{X}$ - R Charts

To calculate control limits for the $\bar{X}$ chart, using the formula $\bar{\bar{X}} \pm 3(\hat{\sigma}/\sqrt{n})$, we use the range as an estimate of the common cause variability by tradition. When using the range, $\bar{R}$, we must convert it to an estimate of the standard deviation σ. To do this we divide the average range, $\bar{R}$, by the factor d_2 as discussed in Chapter 16, i.e., $\hat{\sigma} = \bar{R}/d_2$. The factor $A_2 = 3/d_2\sqrt{n}$ in Table A.2 can be used to directly calculate the control limits as $\bar{\bar{X}} \pm A_2\bar{R}$. The values for A_2 are tabulated in the table by n, the number of values in a rational subgroup. The table covers subgroup sizes from 2 to 10. For subgroups greater than 10 the range becomes a poor estimator of variability. If subgroup sizes larger than 10 are used, $\bar{X}$ and s charts should be used rather than $\bar{X}$ and R charts.

The control limits for the R chart are based on the distribution of ranges from a Normal Distribution. The factors D_3 and D_4 are used for converting the average range of a set of rational subgroups into the lower and upper control limits for the R chart. These are factors also tabulated in Table A.2. The lower control limit for the R chart is $D_3\bar{R}$, and the upper limit is $D_4\bar{R}$. In Table A.2, notice that the lower limit is always zero unless the subgroup size is greater than 6. The center line for the R chart is drawn at $\bar{R}$, but this is always closer to the lower control limit than the upper because the distribution of ranges is skewed to the right. An example of computing the chart limits is shown in the next section. Table 17.1 summarizes the formulas used to calculate the control limits for $\bar{X}$ and R charts.

Table 17.1 Control Limit Formulas for $\bar{X}$ and R Charts

	$\bar{X}$ Chart	R chart
Central Line	$\bar{\bar{X}}$	$\bar{R}$
Lower Control Limit (LCL)	$\bar{\bar{X}} - A_2\bar{R}$	$D_3\bar{R}$
Upper Control Limit (UCL)	$\bar{\bar{X}} + A_2\bar{R}$	$D_4\bar{R}$

17.5.2 An Example of the Use of $\bar{X}$ - R Charts

Consider for illustrative purposes, the data presented in Table 17.2 which represents the chemical assays of the percent active product in 50 gallon drums of nominally 25% active material. This product is produced in a continuous chemical processing plant and four drums are selected per shift to be assayed by the laboratory. The four results per day represent rational subgroups of the data, and are displayed as rows in Table 17.2. In an actual application we would require 25 subgroups or more for accurate estimation of the control limits.

The specification limits for the product are 20 to 30 and the plant would like to know whether it can meet these limits, based on a sample of all product from this production unit. The first thing that can be noticed about the sample data in Table 17.2 is the variability. Some of this variability may have been caused by common causes, and some may have been caused by differences that we can recognize in the way the process was run (i.e., assignable causes).

Table 17.2 Chemical Assay Data

Day	Assays, X				$\overline{X}$	R
1	28	31	27	25	27.75	6
2	22	26	29	28	26.25	7
3	29	28	27	22	26.50	7
4	26	23	21	29	24.75	8
5	28	26	29	28	27.75	3
6	25	21	22	27	23.75	6
7	21	22	21	19	20.75	3
8	26	26	27	27	26.50	1
9	27	23	28	25	25.75	5
10	29	32	35	28	31.00	7
11	22	23	26	22	23.25	4
12	18	22	19	27	21.50	9
13	27	25	22	23	24.25	5
14	24	18	27	21	22.50	9
15	22	26	23	18	22.25	8
16	31	26	24	31	28.00	7
17	27	26	23	28	26.00	5
18	33	36	27	27	30.75	9
19	30	28	23	38	29.75	15
20	37	19	26	27	27.25	18
			Totals:		516.25	142

$$\overline{\overline{X}} = \frac{516.25}{20} = 25.81, \quad \overline{R} = \frac{142}{20} = 7.10$$

$\overline{X}$ Chart Limits : UCL $25.81+(.729)7.10 = 30.98$, LCL $= 25.81 - 5.17 = 20.64$

R - Chart Limits : UCL $= (2.282)(7.10) = 16.20$, LCL $= (0)(7.10) = 0$.

Figure 17.6 shows the $\overline{X}$ chart and R chart of these data . Normally the R chart is studied first. In this example there is one subgroup (20) which falls well above the upper control limit of the R chart, along with a run of seven points below the $\overline{R}$ line. Investigation of these points resulted in the discovery that on days 19 and 20 a new operator had started work and was probably the cause of the wide variability on those days. Also, from the lab reports it was noticed that the quality of the test solution was in doubt on day 10. No reasons for the run of consistent assays (i.e., low ranges) from days 5 to 11 could be found.

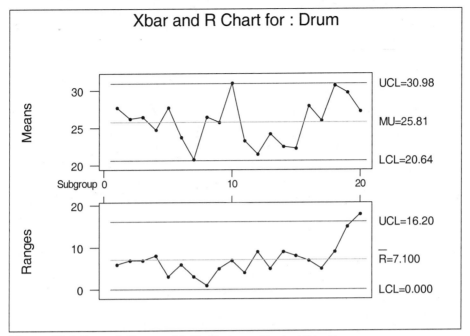

Figure 17.6 Control Charts with Preliminary Control Limits

Nothing could be done, at this point, about the product produced on days 10, 19, and 20. However, since assignable causes were found for these days they could be eliminated from the data that we use to calculate control limits. Control limits should only represent common cause variation. After elimination of subgroups 10, 19, and 20, new control limits were calculated as shown in Figure 17.7. In practice, additional subgroups of data should be collected when elimination of some groups results in less than 25 subgroups for computation of control limits.

After the elimination of subgroups with assignable causes, the R chart now appears to be *in control* with only random or common causes affecting the variability within subgroups, but the $\bar{X}$ chart still appears to be out of control with subgroups 7 and 18 beyond the control chart limits. Further investigation revealed that on day 18 the process was not run as normal because the new operator was being trained and several unusual procedures were demonstrated on that day, but no cause for the low assay on day 7 could be found. After elimination of the day 18 data, the $\bar{X}$ chart was reconstructed as shown in Figure 17.8, and this time all points are within the limits and no unusual runs or patterns are seen.

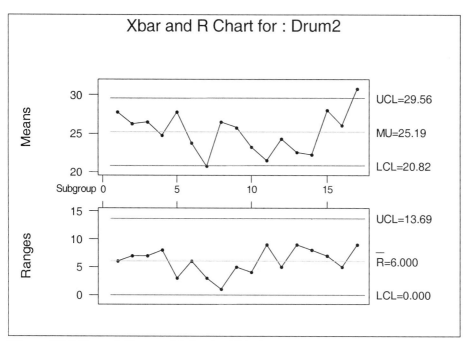

Figure 17.7 Control Charts with Revised Limits

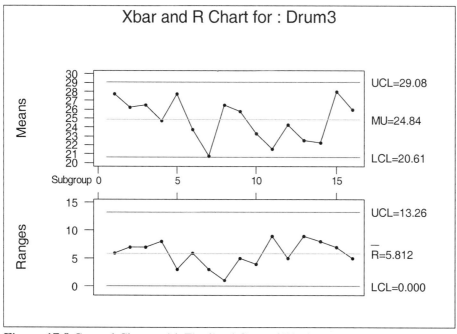

Figure 17.8 Control Charts with Finalized Control Limits

With both the $\bar{X}$ chart and R chart showing no assignable causes, the control limits shown in Figure 17.8 can be considered accurate, and will be useful for monitoring the process in the future to detect when assignable causes enter. Additionally, $\hat{\mu}=\bar{\bar{X}}$, and $\hat{\sigma}=\bar{R}/d_2$ can be considered good estimates of the mean and standard deviation of chemical assays that can be achieved from the process (assuming that the operator is fully trained, the test solution at the lab is not faulty, and no unusual changes are being made to the process such as those made for demonstration purposes on day 18). These estimates are useful to management for comparing to specification limits in what we call a *process capability study*. These studies will be described in Section 17.7.

Typically the process of establishing the control chart limits will involve one or more iterations between calculating limits and plotting data, and investigating potential causes for aberrant points, as demonstrated in this example. Further description of this iteration will be described in Section 17.7.

17.5.3 Comparison of Control Limits and Specification Limits

Control limits and specification limits have no relationship to one another. Specification limits are defined by a design engineer as he attempts to define what is needed for functionality of the product. Control limits are based on natural or common cause variation in the production process. Control limits are used by process operators to prevent assignable causes for variation, and they are used by management to judge the effectiveness of improvement programs they institute. Specification limits cannot help to improve manufacturing processes in any way.

A common misunderstanding results when specification limits are drawn on a control chart. The range of variability on $\bar{X}$ charts is narrower than the range of variation for individual measurements, because the standard deviation of an average is the standard deviation of an individual divided by the square root of n, were n is the subgroup size. Therefore, when specification limits are drawn on the control chart, the picture is always too good to be true. For example, Figure 17.9 shows a plot of individual measurements of part lengths from 20 subgroups of data. The dotted lines in the graph represent the specification limits. It can be seen that individual parts are being produced that are outside both the lower and upper specification limits. However, when the specification limits are drawn on the $\bar{X}$ chart, the results look like Figure 17.10 which is very misleading.

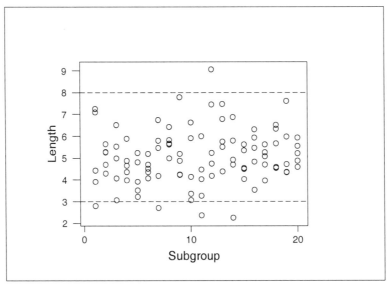

Figure 17.9 Individual Points and Specification Limits

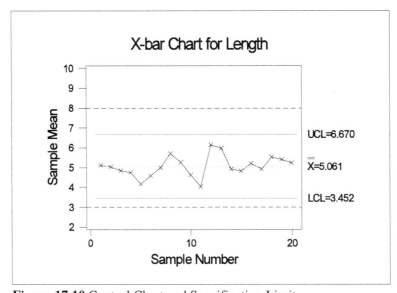

Figure 17.10 Control Chart and Specification Limits

652

Therefore, to determine the proportion within or out of spec, we must use the individual data and not the subgroup means. Let's reconsider the chemical assay data presented in Table 17.2. We could get a crude estimate of the percent out of spec by simply counting the proportion of assays exceeding the upper or lower spec limit (after eliminating subgroups identified with special causes). However, we can gain a more accurate estimate if we can assume that they follow a Normal Distribution, as was shown in Chapter 3. The upper specification limit of 30 is 1.83 standard deviations above the mean. We determine this by calculating the z-ratio

$$z = \frac{30-24.84}{2.82} = 1.83$$

where 24.84= $\overline{\overline{X}}$, our estimate of the process mean, and 2.82 = $\overline{R}/d_2$ is our estimate of the process standard deviation. Next we locate the z-value we have calculated in the margin of Table A.3, in Appendix A, and read the upper tail area in the body of the table. We see the proportion to the right of 1.83 is .0336 which represents the proportion of individual drum assays that will exceed the upper specification limit as shown in Figure 17.11.

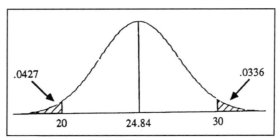

Figure 17.11 Proportion of Drums Exceeding Specification Limits

To determine the proportion of individual drum assays that will fall below the lower spec of 20 we again calculate the z-ratio

$$z = \frac{20-24.84}{2.82} = -1.72$$

This z-value is negative and cannot be found in Table A.3; however, since the Normal Distribution is symmetrical around zero, we know the left tail area below -1.72 is the same as the

653

right tail area above 1.72. Therefore, we locate 1.72 in the margins of the table and read the tail area as .0427. Therefore, management would predict that .0336 + .0427 = .0763 or 7.6% of individual drums would fall outside the specification limits.

   This prediction is very much based on the assumption that the individual drum assays follow a symmetric Normal Distribution. If the distribution were lopsided or skewed the predictions could be inaccurate. To check the normality we should construct histograms and or Normal Probability plots of the data. Figures 17.12 and 17.13 show a histogram and Normal plot of the data from Table 17.2 after eliminating subgroups 10, 18, 19, and 20. The Normal plot shows that the normality assumption is justified, and the addition of specification limits to the histogram is a visual way of verifying the proportion out of spec calculated by the Normal Distribution by comparison to the crude proportion counting method. This is a much better way of looking at the data for comparison to spec limits than putting spec limits directly on the control chart like Figure 17.10.

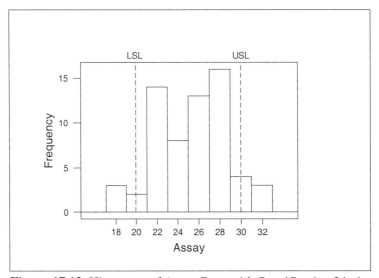

Figure 17.12 Histogram of Assay Data with Specification Limits

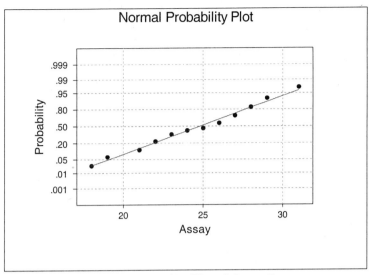

Figure 17.13 Normal Probability Plot of Assay Data

17.5.4 $\bar{X}$ -s Charts

When Walter Shewhart developed control charts in 1924, there were no computers and all computations and graphing were done by hand. The range became the standard measure of variability because of its computational simplicity and the fact that for small sample sizes (i.e., subgroup sizes less than 10) it is nearly as efficient as the sample standard deviation (s) as an estimate of the population standard deviation (σ). Since the advent of the computer, there are many software programs such as MINITAB that can be used to perform control chart calculations. As was illustrated in Section 17.5.2 it may take several iterations of calculating control limits, searching out assignable causes, deleting subgroups of data, and recomputing control limits before the final control limits are determined. Therefore, it is most convenient to use a computer for these repetitive calculations. If a computer is being used rather than hand computations, there is no reason to use the range as a measure of variability instead of the more efficient sample standard deviation, s.

Formulas for the control limits for $\bar{X}$ and s charts are very similar to those formulas used for the $\bar{X}$ and R charts, given previously, except R is replaced by s, and different constants A_3, B_3, and B_4 are used in place of A_2, D_3, and D_4. The constants are found in Table A.2 in Appendix A, and the control chart formulas are summarized in Table 17.3.

Table 17.3 Control Limit Formulas for $\bar{X}$ and s Charts

	$\bar{X}$ Chart	s chart
Central Line	$\bar{\bar{X}}$	$\bar{s}$
Lower Control Limit (LCL)	$\bar{\bar{X}} - A_3\bar{s}$	$B_3\bar{s}$
Upper Control Limit (UCL)	$\bar{\bar{X}} + A_3\bar{s}$	$B_4\bar{s}$

As an example of the use of $\bar{X}$ and s charts, consider the thrust washer thickness data shown in Table 17.4 taken from Chaudhry and Higbie[4] . These parts are used in the transmissions of Ford Taurus-Sable automobiles. Specifications for the thickness on these washers is .07550 to .07950. The production process for these parts consists of five basic steps: sanding, stripping, punching, baking, and shipping. Ford Motor Company, through their Q101 program, requires that their suppliers ensure that their products meet all requirements and specifications. Demonstrating that processes are in a state of statistical control is a vital step in doing this.

Table 17.4 Thrust Washer Thickness Data

Subgroup	Thickness Measurements					$\bar{X}$	$\underline{s}$
1	0.0775	0.0770	0.0770	0.0775	0.0770	0.0772	0.0002739
2	0.0770	0.0770	0.0770	0.0770	0.0770	0.0770	0.0000000
3	0.0775	0.0775	0.0780	0.0775	0.0775	0.0776	0.0002236
4	0.0770	0.0775	0.0780	0.0770	0.0770	0.0773	0.0004472
5	0.0775	0.0775	0.0770	0.0775	0.0770	0.0773	0.0002739
6	0.0770	0.0770	0.0775	0.0770	0.0775	0.0772	0.0002739
7	0.0770	0.0770	0.0780	0.0770	0.0775	0.0773	0.0004472
8	0.0775	0.0770	0.0770	0.0780	0.0770	0.0773	0.0004472
9	0.0775	0.0775	0.0770	0.0775	0.0775	0.0774	0.0002236
10	0.0780	0.0780	0.0780	0.0770	0.0780	0.0778	0.0004472

$$\bar{\bar{X}} = .07734 \qquad \bar{s} = 3.17E{-}04$$

---

[4]Chaudhry, S. S. and Higbie, J. R. 'Quality Improvement Through Statistical Process Control," *Quality Engineering*, Vol. 2, No. 4, (1990), pp. 411 - 419.

The $\bar{X}$ chart limits are calculated as:

$$UCL = \bar{\bar{X}} + A_3 \bar{s} = .07734 + (1.43)(3.17E\text{-}04) = .07779$$

$$LCL = \bar{\bar{X}} - A_3 \bar{s} = .07734 - (1.43)(3.17E\text{-}04) = .07689$$

and for the s Chart as:

$$LCL = B_4 \bar{s} = (0)(3.17E\text{-}04) = 0$$

$$UCL = B_3 \bar{s} = (2.09)(3.17E\text{-}04) = 6.63E\text{-}04$$

where the constants A_3, B_3, and B_4 are taken from Table A.2 with subgroup size n = 5. The $\bar{X}$ and s charts are plotted in Figure 17.14. Here it can be seen that for the last subgroup the mean thickness was above the upper control limit. The cause of the out of control group was determined to be the raw material thickness which was at the upper end of the specifications.

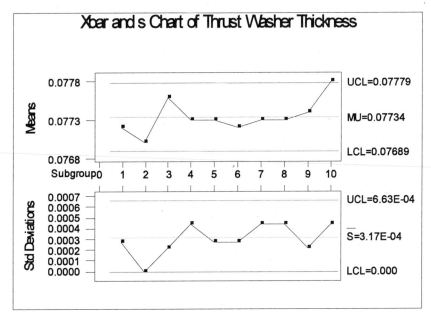

Figure 17.14 $\bar{X}$ and s Charts of Thrust Washer Thickness Data

17.5.5 Individuals Charts

Control charts allow us to identify special causes of variability and determine the magnitude of common cause variability. This is a critical step in improving any process, because the countermeasures for common and special causes of variation are different. Reacting to common causes by making adjustments, similar to what would be done to counteract special causes, is what Deming has classified as *tampering* with the process. Tampering may actually increase the process variability, and demoralize the workforce. Deming has demonstrated this with his colored bead and funnel experiments[5]. The concepts of special and common causes and appropriate countermeasures are applicable to much more than measured characteristics of manufactured parts. For example, we may be interested in monitoring in-process manufacturing variables such as the viscosity of a paint solution, or characteristics of administrative or professional processes such as the time or labor costs for completing a particular task. Unfortunately variables of this kind cannot be conveniently grouped into rational subgroups, and the $\bar{X}$, R, and s charts will be of little value. Individuals control charts are often used in this situation.

Individuals charts consist of plotting individual observations along with appropriate control limits. Because individuals charts can be applied to a wide variety of problems, they are potentially one of the most useful types of variables control charts. To illustrate the construction and interpretation of an individuals chart, refer to the data in column 2 of Table 17.5, which represents the pH of a chemical solution used in an industrial process.

Computing control limits for individuals charts using the standard $\mu \pm 3\sigma$ formula requires an estimate of the process mean and standard deviation of common causes, along with evidence of the normality of the data (since there is no averaging to invoke the Central Limit Theorem). As we have seen from previous examples, it is the rule rather than the exception to have special causes present in the data when calculating the preliminary control limits. When special causes are present in the data, the usual estimate of the process standard

deviation, $s = \sqrt{\sum_{i=1}^{n} (x_i - \bar{x})^2}$, will be inflated due to the special cause points. Therefore an

alternate method of estimating the standard deviation of common causes is required.

---

[5]Latzko, W. J. and Saunders, D .M., *Four Days with Dr. Deming: A Strategy for Modern Methods of Management*, (ASQC Quality Press, 1995).

Table 17.5 pH of Chemical Solution

Time	pH =x_i	Moving Range=R_i	
08:00	8.0	*	
10:00	8.5	0.5	
12:00	7.4	1.1	
14:00	10.5	3.1	
16:00	9.3	1.2	
18:00	11.1	1.8	
20:00	10.4	0.7	
22:00	10.4	0.0	
24:00	9.0	1.4	
02:00	10.0	1.0	$\bar{R}$ = 1.135
04:00	11.7	1.7	
06:00	10.3	1.4	
08:00	14.2	5.9	$\bar{R}/d_2$ =1.135/1.128 = 1.006
10:00	11.6	4.6	
12:00	11.5	0.1	$\bar{X}$ = 10.48
14:00	11.0	0.5	
16:00	12.0	1.0	
18:00	11.0	1.0	
20:00	10.2	0.8	
22:00	10.1	0.1	
24:00	10.5	0.4	
02:00	10.3	0.2	
04:00	11.5	1.2	
06:00	11.1	0.4	

The traditional way of doing this is by using the mean of moving ranges of 2. This assumes that the closer together individual values are in time the less likely that they will differ due to special causes. For the data in Table 17.5, the moving ranges of two are shown in column 3. They are computed as follows: the first moving range, R_1=|8.0-8.5| =0.5, is the absolute value of the difference in the first and second individual values; the second moving range, R_2=|8.5-7.4|=1.1, is the absolute value of the difference in the second and third individual values, and so forth. All of the moving ranges are shown in the third column of Table 17.5. There is one less moving range than there are data values. The estimate of the standard deviation of the common causes is calculated in the normal way by taking the average of these moving ranges divided by the factor d_2 taken from Table A.2. This calculation is shown in Table 17.5 where the estimate of the common cause standard deviation (σ) is seen to be 1.006. Subgroup size n=2 is used in determining the value of d_2 from Table A.2, since the ranges represent moving groups of two individual observations. The estimate of the process mean is the usual $\bar{X}$ = 10.48.

The upper control limit is calculated as:

$$UCL = 10.48 + 3(1.006) = 13.50$$

and the lower control limit is:

$$LCL = 10.48 - 3(1.006) = 7.46$$

The individuals control chart is seen in Figure 17.15 with point 13 clearly outside the upper control limit, and point 3 below the lower control limit. Because of these special cause points, the normal estimate of σ, s, is inflated and the control limit formed using this estimate does not show any special causes. Demonstration of this potentially costly mistake will be left as an exercise.

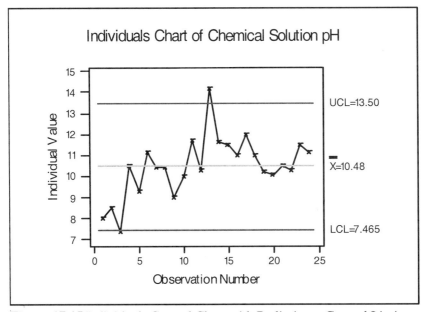

Figure 17.15 Individuals Control Chart with Preliminary Control Limits

Assuming special causes were found for points 1–3, and 13, we will construct revised control limits eliminating these points. The revised mean $\bar{X} = 10.68$, and the revised average moving range $\bar{R} = 0.8526$ results in a smaller variance estimate $\bar{R}/d_2 = .8526/1.128 = 0.756$. The reconstructed control chart is shown in Figure 17.16, where all points appear to be due to common causes.

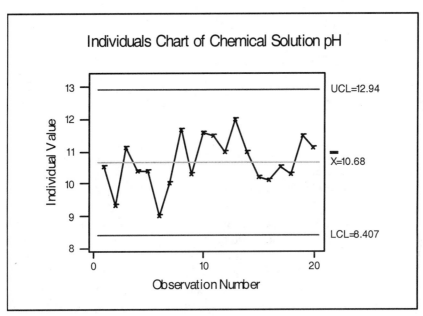

Figure 17.16 Individuals Control Chart Final Limits

To justify the use of the three sigma limits we should examine the normality of the data after eliminating all special cause points. Figure 17.17 shows a Normal Probability plot of the data in Table 17.5 after eliminating points 1–3 and 13. Here it can be seen that the normality assumption is reasonable, and it was justifiable to identify points 1–3 and 13 as potentially due to assignable causes.

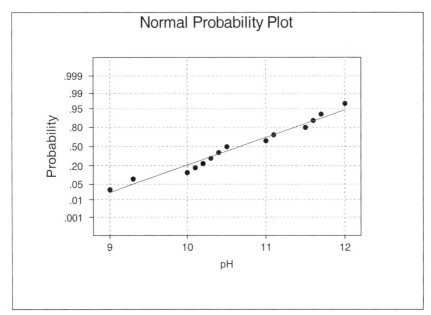

Figure 17.17 Normal Plot of pH Data

The last example shows that special causes which affect the mean of a process can inflate the estimate of the common cause process standard deviation (σ) on an individuals control chart. Each isolated special cause will result in two inflated values of the moving range and subsequently an inflated value of $\bar{R}$. Two iterations of calculating control limits, plotting the graph to identify special causes, eliminating special cause data, and recalculating limits were required to come up with a reasonable estimate of the process mean and standard deviation.

Bryce, Gaudard and Joiner[6] show that a better estimator of the standard deviation may be obtained in one step by computing the median of the moving ranges, $\tilde{R}$, rather than the mean, $\bar{R}$, of the moving ranges. The median, or 50th percentile of the moving ranges, is not inflated by a few large values due to special causes. Using the median range, an estimate of σ is formed as $\tilde{R}/d_4$, where the constant d_4 (from Ferrell[7]) is used in place of the d_2 constant normally used in variables control charts. Values of d_4 are given for subgroup sizes from 2 - 10 in Table A.2 along with the other control chart constants. The control limits for the individuals chart based on the median moving range are then given by :

[6]Bryce, G. R., Gaudard, M. A., and Joiner, B. L., "Computing Control Limits for the Individuals Control Chart" *American Statistical Association Proceedings of the Section on Quality and Productivity*, Alexandria, VA, 1995.

[7]Ferrell E. B., "Control Charts Using Midranges and Medians" *Industrial Quality Control*, (March 1953), pp. 30–34.

$$\text{UCL} = \bar{X} + 3(\tilde{R}/d_4) \quad \text{and} \quad \text{LCL} = \text{UCL} = \bar{X} - 3(\tilde{R}/d_4)$$

Bryce, Gaudard and Joiner show an example of the use of these formulas by making a control chart of weekly production costs from a manufacturing firm, which are reproduced in Table 17.6. The control chart is displayed in Figure 17.18. With this chart, they identified costs for week 18 to be out of the control limits, and identified the special cause that occurred. Interestingly, the traditional individuals chart based on the average moving range does not show any out of control signals, and is therefore misleading. For this reason, Bryce, Gaudard, and Joiner suggest using the median range based individual charts rather than the individual chart based on the mean range.

Table 17.6 Weekly Production Costs in ($1000) for a Manufacturing Firm

Week	Production Cost (in $1000)	Moving Range	Moving Range (Sorted)	
1	31.4	*	*	
2	22.1	9.3000	0.8000	
3	39.5	17.400	0.9000	
4	36.5	3.0000	1.0000	
5	33.6	2.9000	1.2000	
6	26.9	6.7000	1.5000	
7	31.0	4.1000	2.4000	
8	29.8	1.2000	2.4000	
9	30.6	0.8000	2.4000	
10	31.5	0.9000	2.5000	
11	29.1	2.4000	2.9000	$\bar{X} = 31.10$
12	31.5	2.4000	2.9000	
13	36.2	4.7000	3.0000	$\tilde{R} = 3.0$
14	27.6	8.6000	3.6000	
15	28.6	1.0000	4.1000	$\tilde{R}/d_4 = 3.0/.954 = 3.14$
16	31.5	2.9000	4.6000	
17	26.9	4.6000	4.7000	UCL= 31.10+3(3.14) = 40.53
18	44.0	17.100	5.5000	
19	32.7	11.300	6.7000	LCL= 31.10-3(3.14) = 21.67
20	29.1	3.6000	8.6000	
21	27.6	1.5000	9.3000	
22	25.1	2.5000	11.300	
23	30.6	5.5000	17.100	
24	33.0	2.4000	17.400	

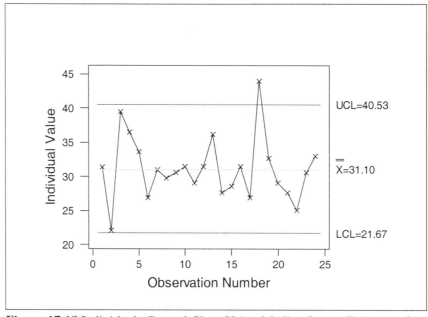

Figure 17.18 Individuals Control Chart Using Median Range Estimate of σ

A specific but common situation where individuals control charts are useful is in job shop type production, where only a few parts of any type are produced at one time. In this setting there are certainly not enough consecutive parts produced to chart subgroup means and ranges, but there may be enough to chart individual values. In some cases production runs may be so short that there is not even enough data to construct an individuals chart. In that case, measurements from production runs with different parts may be combined on a single chart called a *nominals chart*, which is a variation of the individuals control chart.

The purpose for using control charts is to give information about the process, so that the process can be improved by making in-process adjustments. In job shop production environments, many different types of parts may be produced on the same production equipment, with short runs of each type. For example, many different size bearings could be machined on the same NC lathe, using the same type bar stock. Since we would be interested in identifying special causes that affect performance of the lathe (such as a worn or broken cutting tool, improper set up, etc.) and not specific parts, we could combine the data taken from short production runs of different size bearings on the same control chart. In order to do this we would have to adjust the data by subtracting the nominal values to bring all measurements to a common mean of zero, and we would have to assume the variability was constant for different part types. If the different groups of parts are produced on the same manufacturing equipment and from the same type of raw material this may not be a bad assumption. The nominals chart is made by constructing an individuals chart of $(X_{ij} - N_i)$, where X_{ij} is the jth consecutive measurement made during the ith production run, and N_i represents the nominal value (i.e., target or mid-point of spec range) for the parts produced during the ith production run.

17.5.6 Moving Average Moving Range Charts

The advantage of individuals charts, discussed in the last section, is the fact that they can be used on a wide variety of variables measurements that can not be conveniently grouped into rational subgroups. One disadvantage is the fact that they are less sensitive to detecting small shifts in the process mean than are the traditional $\bar{X}$ charts. Recall that the control limits on an $\bar{X}$ chart are $\bar{\bar{X}} \pm 3\sigma/\sqrt{n}$ which are tighter than the individuals chart control limits $\bar{X} \pm 3\sigma$. Another disadvantage of the individuals charts is that we can't rely on the Central Limit Theorem to justify the normality of the plotted points and the validity of the 3σ limits. In attempt to preserve the advantage of the Individuals Chart and reduce its disadvantages, moving average, moving range charts have been proposed.

Moving average moving range charts are constructed from data that is not subgrouped, by simply creating moving subgroups in the data. For example, using the data in Table 17.5 and a moving subgroup of 4, the first subgroup would be composed of (8.0, 8.5, 7.4, and 10.5). The second subgroup would be composed of (8.5, 7.4, 10.5 and 9.3), etc. Table 17.7 shows all the data of Table 17.5 rearranged into moving subgroups of 4. The control chart limits are calculated using the normal formulas for $\bar{X}$-R charts shown in Table 17.1.

The moving average chart will be quicker to detect small shifts in the process mean than the individuals chart. However it is not as quick to detect large changes in the process mean, and due to averaging, it may not detect an abrupt but shor- lived shift in the process mean at all. Therefore, an individuals chart is often used in conjunction with a moving average moving range chart, to get the advantages of both.

Table 17.7 Data from Table 17.5 Rearranged into Moving Subgroups

Moving Subgroup Number	Data				Subgroup Mean	Range
1	8.0	8.5	7.4	10.5	8.600	3.1
2	8.5	7.4	10.5	9.3	8.925	3.1
3	7.4	10.5	9.3	11.1	9.575	3.7
4	10.5	9.3	11.1	10.4	10.325	1.8
5	9.3	11.1	10.4	10.4	10.300	1.8
6	11.1	10.4	10.4	9.0	10.225	2.1
7	10.4	10.4	9.0	10.0	9.950	1.4
8	10.4	9.0	10.0	11.7	10.275	2.7
9	9.0	10.0	11.7	10.3	10.250	2.7
10	10.0	11.7	10.3	14.2	11.550	4.2
11	11.7	10.3	14.2	11.6	11.950	3.9
12	10.3	14.2	11.6	11.5	11.900	3.9
13	14.2	11.6	11.5	11.0	12.075	3.2
14	11.6	11.5	11.0	12.0	11.525	1.0
15	11.5	11.0	12.0	11.0	11.375	1.0
16	11.0	12.0	11.0	10.2	11.050	1.8
17	12.0	11.0	10.2	10.1	10.825	1.9
18	11.0	10.2	10.1	10.5	10.450	0.9
19	10.2	10.1	10.5	10.3	10.275	0.4
20	10.1	10.5	10.3	11.5	10.600	1.4
21	10.5	10.3	11.5	11.1	10.850	1.2

17.6 Attribute Control Charts

Attribute control charts are essentially plots of counts or standardized counts augmented by appropriate control limits. There are four types of attribute charts that are normally used. Two of the four types are used for charting counts of defectives and the other two are used for charting counts of defects. The specific type of chart used will also depend on whether the subgroup or sample size is constant or variable. Table 17.8 summarizes the four types of charts.

Table 17.8 Attribute Control Charts

What Is Counted	Subgroup or Sample Size	Appropriate Chart	Distribution
Number of Defective items	Constant	NP Chart	Binomial
Proportion of Defective items	Variable	P Chart	Binomial
Defects on items	Constant	C Chart	Poisson
Defects on items	Variable	U Chart	Poisson

Counts are essentially sums of discrete events. Therefore, the justification for the control limits that are placed on attribute control charts is the same as that for variables control charts, namely the Central Limit Theorem. The Binomial Distribution and the Poisson Distribution describe the distribution of defective counts and defect counts, respectively, under a system of common causes. As explained in Chapter 3, both of these distributions can be approximated very closely by the Normal Distribution when their expected value is greater than or equal to 5. This can be seen by examining the Binomial probability mass function in the middle row of Figure 3.3 in Chapter 3, and the Poisson probability mass function in the last row of Figure 3.4. Both of these mass functions have expected values (or center of gravity) greater than or equal to 5 and have a bell-shaped Normal Distribution shape.

Recall that when constructing variables control charts with subgroups of data, two control charts are plotted side-by-side. The $\bar{X}$ chart is used to monitor shifts in the process mean, while the R or s chart is used to monitor shifts in the process variability. With attribute charts, there is no need to construct two charts because the mean and variance are inseparably connected. For example, the mean or expected value of a Binomial Distribution was shown in Chapter 3 to be np, while the variance is np(1-p). As p, the expected proportion defective in a sample of size n, increases both np and np(1-p) will increase. This is illustrated graphically in Figure 17.19 and Figure 17.20 which show graphs of the probability mass function for a Binomial Distribution with n=25, and p=.2, and for p=.5. We can see that as p increases both the mean or center of gravity and the variance or spread in the probability mass functions increases. A similar situation is true for Poisson counts. Because of this relationship between the mean and variance it is only necessary to construct one control chart for attribute data that will monitor both shifts in the mean and variability.

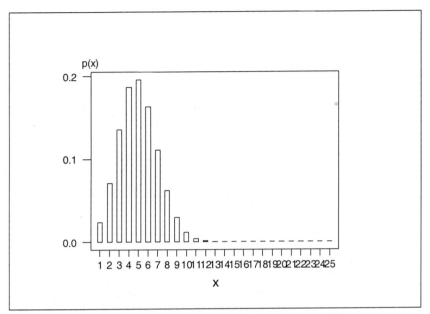

Figure 17.19 Probability Function for Binomial n=25, p=.20

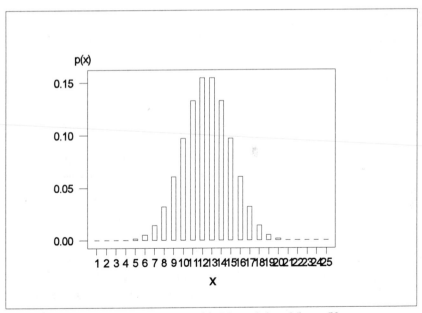

Figure 17.20 Probability Function for Binomial n=25, p=.50

17.6.1 NP Charts

NP charts are used to monitor counts of defectives when the subgroup size is constant. For example, consider the following situation. There are 250 integrated circuits on a silicon wafer, and each circuit, or die, is tested at an automated testing station and marked with a red ink spot if it is defective. Using a single wafer as a subgroup, the sample size is constant at n=250. If the process is influenced only by common causes, the distribution of the number of defective circuits on a wafer will follow the Binomial Distribution with mean np, and variance np(1-p). Since the Binomial Distribution can be approximated well by the Normal Distribution (if np>=5), three sigma limits can be used for control limits on the NP chart. The center line for the control chart and upper and lower control limits are given by:

$$\text{Center Line} = n\bar{p}$$

$$\text{LCL} = n\bar{p} - 3\sqrt{n\bar{p}(1-\bar{p})}$$

$$\text{UCL} = n\bar{p} + 3\sqrt{n\bar{p}(1-\bar{p})}$$

where $\bar{p}$ is the grand average proportion of defectives.

To illustrate the use of these formulas consider the data in Table 17.9. These data represent counts of red beads from one of Deming's famous red bead experiments[7]. In this experiment there is a bowl of 3750 beads of which 750 are red. Counts are determined by mixing the beads in the bowl and scooping up a sample of 50 beads with a paddle that contains 50

Table 17.9 Red Bead Counts

Sample No.	1	2	3	4	5	6	7	8	9	10	11	12	13
No. Red Beads	3	6	13	11	9	12	13	9	12	8	13	11	8

Sample No.	14	15	16	17	18	19	20	21	22	23	24	Total
No. Red Beads	8	7	10	8	7	9	10	10	15	11	15	238

beveled depressions. Thus the counts will approximate counts from a Binomial Distribution. In Table 17.9 there are a total of 238 defectives (i.e., red beads) and a total sample size of 1200. Thus $\bar{p} = 238/1200 = .198$, the center line and control limits for the control chart are calculated

---

[7]Deming, W. E., *Out of the Crisis*, MIT Center for Advanced Engineering Study, Cambridge Mass., (1986), p. 347.

below:

$$\text{Center Line} = 50(.198) = 9.92$$

$$\text{LCL} = 50(.198) - 3\sqrt{50(.198)(1-.198)} = 1.46$$

$$\text{UCL} = 50(.198) + 3\sqrt{50(.198)(1-.198)} = 18.38$$

The NP chart is shown in Figure 17.21. Here it can be seen that only common causes are

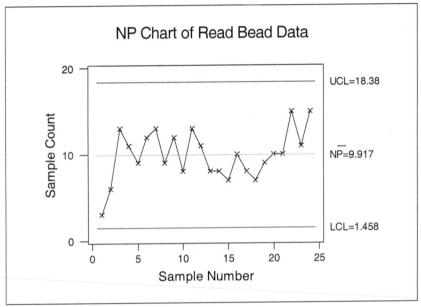

Figure 17.21 NP Chart

affecting the outcome. As Deming taught, the cause of the high number of defectives is the incoming raw material, and the willing workers are totally helpless to improve quality.

17.6.2 P Charts

In cases where the rational subgroup size or number of items inspected is not constant, NP charts will not work since neither the center line, $n\bar{p}$, nor the control limits will be constant. A P chart is a modification of the NP chart which makes the center line constant by standardizing or dividing each count to be plotted by its respective subgroup size. In the P chart we plot the fraction defectives in the ith subgroup, $\hat{p}_i = r_i/n_i$, rather than the count, r_i, of defectives. It can be shown that the mean and variance of the sample proportion, $\hat{p}_i$, are the population parameter p and $p(1-p)/n$, respectively, assuming no special causes are present. Therefore the control limits for ith sample proportion $\hat{p}_i$ in the P chart are given by:

$$\text{Center Line} = \bar{p}$$

$$\text{LCL}_{\hat{p}_i} = \bar{p} - 3\sqrt{\frac{\bar{p}(1-\bar{p})}{n_i}}$$

$$\text{UCL}_{\hat{p}_i} = \bar{p} + 3\sqrt{\frac{\bar{p}(1-\bar{p})}{n_i}}$$

where again $\bar{p}$ is the grand average fraction defective. For this chart, the centerline will be constant and horizontal, however, the control limits will vary depending on the number of items inspected, n_i, for each rational subgroup.

To illustrate the calculation of control limits for a P chart, consider Table 17.10 data on defective electric machine parts from Ishikawa[8]. In this table there are a total of 610 defectives and a total sample of 5925. Therefore, the grand average proportion defective is $\bar{p}=610/5925 = 0.103$, and the control limits for the ith proportion or fraction defectives are calculated as:

$$\text{Center Line} = 0.103$$

$$\text{LCL}_{\hat{p}_i} = 0.103 - 3\sqrt{\frac{0.103(1-0.103)}{n_i}}$$

$$\text{UCL}_{\hat{p}_i} = 0.103 + 3\sqrt{\frac{0.103(1-0.103)}{n_i}}$$

which will differ from subgroup to subgroup depending upon n_i. For example, in the first

[8]Ishikawa, K., *Guide to Quality Control*, Unipub-Krause International Publications, White Plains N.Y., (1982), p. 78.

Table 17.10 Defective Electric Motor Parts

Subgroup No. (i)	Subgroup Size n_i	Number Defective r_i	Proportion Defective $\hat{p}_i = r_i/n_i$	UCL	LCL
1	115	15	.130	.188	.018
2	220	18	.082	.165	.041
3	210	23	.109	.166	.040
4	220	22	.100	.165	.041
5	220	18	.082	.165	.041
6	255	15	.058	.160	.046
7	440	44	.100	.146	.060
8	365	47	.129	.151	.055
9	255	13	.051	.160	.046
10	300	33	.110	.156	.050
11	280	42	.146	.158	.048
12	330	46	.139	.153	.053
13	320	38	.119	.165	.041
14	225	29	.129	.164	.042
15	290	26	.089	.157	.049
16	170	17	.100	.173	.033
17	65	5	.077	.216	0
18	100	7	.070	.194	.012
19	135	14	.104	.182	.024
20	280	36	.128	.158	.048
21	250	25	.100	.161	.045
22	220	24	.109	.165	.041
23	220	20	.091	.165	.041
24	220	15	.068	.165	.041
25	220	18	.082	.165	.041
Total	5925	610	.103		

subgroup where n=115 the limits are:

$$\text{Center Line} = 0.103$$

$$\text{LCL}_{\hat{p}_1} = 0.103 - 3\sqrt{\frac{0.103(1-0.103)}{115}} = 0.018$$

$$\text{UCL}_{\hat{p}_1} = 0.103 + 3\sqrt{\frac{0.103(1-0.103)}{115}} = 0.188$$

and the remaining limits are shown in Table 17.10. Figure 17.22 shows the control chart, which does not exhibit any signs of special causes.

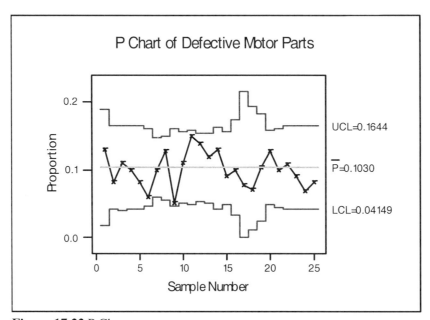

Figure 17.22 P Chart

17.6.3 C Charts

Counts of defects, such as the number of dust contamination spots on a silicon wafer, are not bounded above by the subgroup size, and are more appropriately modeled by the Poisson Distribution than the Binomial. In Table 3.4 of Chapter 3 it was shown that the variance of the Poisson Distribution is equal to the mean, and the standard deviation is the square root of the mean. If we represent the defect count from subgroup i with the letter c_i, then the estimate of the of the mean count per subgroup, λ, is

$$\hat{\lambda} = \bar{c} = \sum_{i=1}^{n} c_i/n$$

and the three sigma control limits are given by $LCL = \bar{c} - 3\sqrt{\bar{c}}, \quad UCL = \bar{c} + 3\sqrt{\bar{c}}.$

To illustrate a C chart consider the counts of complex circuit board assembly errors shown in Table 17.11 taken from DataMyte[9].

Table 17.11 Circuit Board Assembly Errors

Subgroup	1	2	3	4	5	6	7	8	9	10	11	12	13
Defect Count	1	7	3	6	4	2	3	10	3	5	4	1	7
Subgroup	14	15	16	17	18	19	20	21	22	23	24	25	Total
Defect Count	2	4	2	4	7	6	8	5	6	10	9	11	130

The subgroup size for this data is 10 circuit boards inspected each hour.

The average count $\bar{c} = 130/25 = 5.2$, and the control limits are calculated as:

$$UCL = 5.2 + 3\sqrt{5.2} = 12.0$$

$$LCL = 5.2 - 6.8 = -1.6 \text{ (Assume Zero)}$$

and the chart is shown in Figure 17.23. Here it can be seen that all the counts lie within the control limits, but there does appear to be a troubling upward trend after subgroup 16. Is there a

---

[9]*DataMyte Handbook Fourth Edition*, DataMyte Corporation, Minnetonka, Minn., (1989), p. 3 - 37.

need for concern, or is this trend within the possibility of common causes? In Section 17.7 we will discuss some rules for detecting out of control signals on charts with no points beyond the limits.

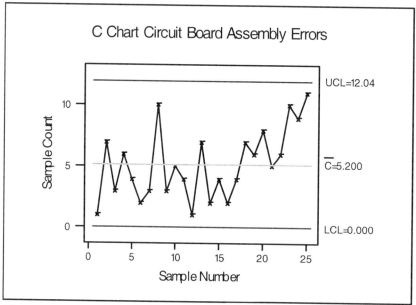

Figure 17.23 C Chart

17.6.4 U Charts

When rational subgroup sizes vary as defects are being counted the counts need to be standardized or divided by the subgroup size (or area inspected, for counts made over space or time) just as the counts of the number of defectives were standardized when constructing a P chart. A U chart is a standardized version of the C chart on which we plot standardized counts or *defects per unit*. If we represent the defect count for the ith subgroup as u_i, and the number of items or area of inspection as n_i, then the standardized counts are u_i/n_i. The center line for the control chart will be

$$\bar{u} = \frac{u_1 + u_2 + \ldots + u_k}{n_1 + n_2 + \ldots + n_k} = \frac{\sum_{i=1}^{k} u_i}{\sum_{i=1}^{k} n_i}$$

for k subgroups of counts.

The control limits for the U chart are given by

$$UCL_i = \bar{u} + 3\sqrt{\frac{\bar{u}}{n_i}}$$

$$LCL_i = \bar{u} - 3\sqrt{\frac{\bar{u}}{n_i}}$$

To illustrate the use of the U chart consider the data in Table 17.12 which represent leaks in daily samples of radiators tested at General Motors[10] . The center line for the U chart is

$$\bar{u} = \frac{116}{841} = 0.1379$$

and the control limits are given by the formulas

$$UCL_i = 0.1379 + 3\sqrt{\frac{0.1379}{n_i}}$$

$$LCL_i = 0.1379 - 3\sqrt{\frac{0.1379}{n_i}}$$

---

[10]Burns, W. L., "Quality Control Proves Itself in Assembly," *Quality Engineering* Vol. 2, No. 1, 1989 - 1990 --paper originally presented at the Second Midwest Quality Control Conference, June 6, 1947.

For example, the limits for the first subgroup where 39 radiators were inspected is:

$$UCL_i = 0.1379 + 3\sqrt{\frac{0.1379}{39}} = 0.3163$$

$$LCL_i = 0.1379 - 3\sqrt{\frac{0.1379}{39}} = -0.0404 \text{ (Assume Zero)}$$

Table 17.12 Radiator Leaks

Date	No. Tested n_i	No. Leaks in Sample u_i	Average per unit u_i / n_i	LCL_i	UCL_i
6/3	39	14	0.35	0	0.3163
6/4	45	4	0.08	0	0.3040
6/5	46	5	0.10	0	0.3022
6/6	48	13	0.27	0	0.2987
6/7	40	6	0.15	0	0.3140
6/10	58	2	0.03	0	0.2842
6/11	50	4	0.08	0	0.2955
6/12	50	11	0.22	0	0.2955
6/13	50	8	0.16	0	0.2955
6/14	50	10	0.20	0	0.2955
6/17	32	3	0.09	0	0.3349
6/18	50	11	0.22	0	0.2955
6/19	33	1	0.03	0	0.3319
6/20	50	3	0.06	0	0.2955
6/24	50	6	0.12	0	0.2955
6/25	50	8	0.16	0	0.2955
6/26	50	5	0.10	0	0.2955
6/27	50	2	0.04	0	0.2955
Total	841	116			

The remaining control limits are shown in Table 17.12. The U chart is displayed in Figure 17.24.

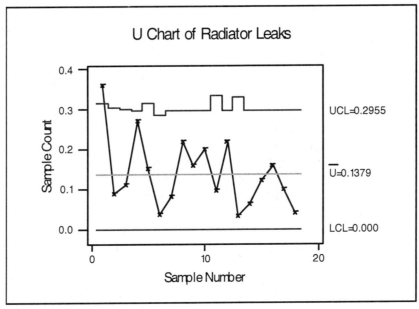

Figure 17.24 U Chart

Here it can be seen that the defect count in the first subgroup exhibits signs of special causes. Investigation revealed many leaks near the rear side of the lower tank to core joint. It was found that the high temperature required for assembly of this joint was disturbing the solder joints from a previous operation. A material change was instituted that required less heat at assembly, and the defect counts were immediately lowered.

17.7 Using Control Charts for Process Improvements

Process improvement using control charts is an iterative procedure. The fundamental phases that are used repeatedly are data collection, bringing the process into control, process capability assessment and process capability improvement. First, data is collected during process operation. The data could be variables data (such as the dimension of a critical part or the delivery times from subprocess A to B), or attribute counts (such as the number of errors). Next, preliminary control limits are calculated and used to access the presence of special causes in the process. The special causes should be identified by local operators with the help of engineers and management. These special causes should be removed, and the process run to collect additional data to show the process can be run in a state of statistical control. Once the process can be demonstrated to be in a state of statistical control, the level of common cause variation can be assessed. This is called a *process capability study*. If the variation from common causes is excessive and cannot meet the customer needs, the process itself must be investigated, and management must make the changes necessary to improve the system.

Use of control charts for process improvement has three major advantages. First, the charts can be used by local operators to keep a process in a state of statistical control by allowing them to identify when adjustments are needed. When a process is in a state of statistical control, it performs more efficiently and predictably with respect to quality of output and cost. Secondly, management can use the control charts to determine when process improvements are needed. Finally, the control charts can be used as a communication tool for discussing process performance and process improvement projects.

To use control charts efficiently any organization must first establish an environment suitable for action. By this we mean that local operators should be empowered to act in their area of responsibility, and management must recognize its duty for correcting common causes. Fear within the organization that prevents objective discussion of process problems and remedies must be eliminated. The process must be understood in terms of its relationship to other processes or operations upstream or downstream. Simple techniques such as flow diagrams and cause and effect diagrams presented in Chapter 1, are useful for making these relationships clear and pooling the knowledge of all individuals involved. Finally, if measurements are being charted, the measurement system needs to be understood. Variance characterization techniques from Chapter 16 are useful in this regard.

Not all characteristics need be subject to control charts, or all resources could be quickly exhausted. Pick those characteristics that are most critical to the customer (which could be the external customer, or the next processing step), and those that provide opportunities for the greatest potential improvements. Frequent communication with external and internal customers is essential. Consideration of problem areas such as waste, poor performance, scrap, rework, long cycle times, and missed targets, and potential problem areas such as upcoming changes to product or process can provide opportunities for improvement. Pareto Diagrams discussed in Chapter 1 are useful for ranking priorities.

When actually using control charts, special attention needs to be given to recognize special causes of variation by characteristic control chart patterns or signals. Typical special causes should be recognizable and procedures established for eliminating them. Finally, tools for accessing process capability should be defined and a feedback control process for management action to improve processes should be in place. We will address these items in more detail in the remainder of this section.

17.7.1 Examining Control Charts for Signs of Special Causes

When a process is subject only to common causes, we would expect the points on the control charts to appear as a sample from the Normal Distribution. From our discussion of random variables and the Normal Distribution in Chapter 3, we have a good idea of how the chart should look. The points should not follow any unusual trends or cycles, and from Figure 3.11, we know that about 2/3 of the points from a Normal Distribution should fall within one standard deviation. About 95% should fall within two standard deviations, and as stated before, it would be very rare to have a point outside the control limits.

When assignable or special causes are present in the process, and they shift the process mean or change the process variability, they will eventually be detected as points fall outside the control limits on the $\bar{X}$ or R charts. But, to be more sensitive in detecting special causes, the Western Electric Company developed some additional rules for examining control charts. These rules are applicable to all symmetric control charts such as $\bar{X}$, NP, P, C, and U charts. They do not apply to R or s charts with small subgroup sizes (i.e., $n \leq 10$) since these charts are not symmetric. Also, these rules do not apply to moving average moving range charts, since the points on these charts are necessarily correlated.

The first four rules are called *tests for instability*. To use these tests, we divide control charts with symmetric control limits into three equal zones on *each side* of the center line, as shown in Figure 17.25.

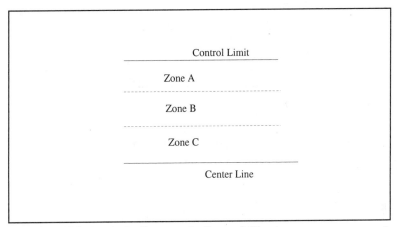

Figure 17.25 Zones in Symmetric Control Charts

The four rules for instability are:

1. One point outside Zone A (i.e., out of the control limits)

2. 2 out of 3 points in Zone A or beyond.

3. 4 out of 5 in Zone B or beyond.

4. 8 successive points in Zone C or beyond.

In applying these rules consider the upper and lower halves of the control chart separately. For example, Figure 17.26 shows an example of Rule 2 being violated at subgroup 7, Rule 3 violated at subgroup 9, and Rule 4 violated with subgroup 12 all on the lower half of the chart. It is very unlikely that any of these rules would be broken if only common cause variability was involved. The probabilities of breaking these rules, assuming only common causes and independence of subgroups, can be determined from areas under the Normal Distribution. When one of these rules is broken it indicates that the process mean has suddenly shifted to a new level, and we should look for the assignable cause. The cause for a sustained shift in the process level is usually something like tool breakage, improper set up at shift change, etc. Cause and effect diagrams followed up with verification tests can help in identifying these causes.

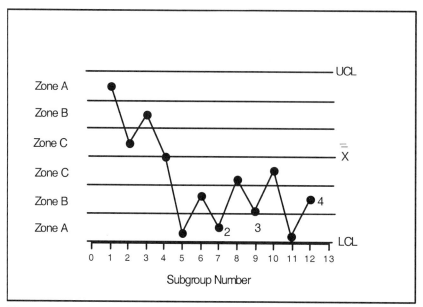

Figure 17.26 Violation of Stability Rules

Other unnatural patterns such as trends, systematic variation, over dispersion and under dispersion are sometimes evident on control charts. Rules 5 through 8, given below, are for detecting these patterns:

5. Trends—a consecutive run of 8 or more points up or down, or a long series of points without an apparent change in direction. A pattern like this on a control chart could be caused by something like the tool wear shown in example in Chapter 2.

6. Systematic Variation—a long series of points that are high, low, high, low without interruption, or an obvious cycle. A pattern on a control chart like this could be caused by a cycle in the process, like that caused by worn bearings in the pump for an extrusion process.

7. Over Dispersion—eight consecutive points avoiding Zone C on (either side of the chart). This might appear like Figure 17.27. This would indicate that two or more different distributions are being seen on the same chart. This can be caused when consecutive subgroups of data actually come from different processes feeding the same output stream (such as multiple

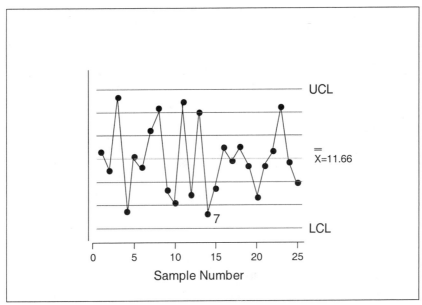

Figure 17.27 An Example of Over Dispersion Detected by Rule 7 on Subgroup 14

filling heads on a packaging line), or when over-control or tampering is taking place and local operators are adjusting the process too frequently. If the cause of over dispersion is merging outputs from several processes in the same stream, then the solution is to chart each process separately and calibrate them to the same level. If, on the other hand the over dispersion is due to tampering, or unnecessary adjustments, local operators should simply stop adjusting the process, unless the control chart gives a clear signal that such adjustment is required.

8. Under Dispersion—15 or more consecutive points within Zone C (either side of the chart). This usually indicates that σ has been overestimated. This usually appears like Figure 17.28, and is sometimes mistaken for good control of the process. This can happen if different distributions are represented within a subgroup, such as parts from different heads on an injection molding machine being grouped into the same rational subgroup, or thickness measurements within a subgroup taken from different positions across a nonuniform sheet, as illustrated in Figure 17.29 (adapted from[11]). When σ is overestimated due to grouping unequal items in the same subgroup, they should be separated into separate subgroups so that the differences will show up as assignable causes on the chart. This will identify dissimilarities that when corrected will make the process more consistent.

Most companies that use control charts use these rules for detecting assignable causes to some extent. Sometimes not all rules are applied, and sometimes the rules are modified

---

[11]DeVor, R .E., Chang, T., and Sutherland, J. W., *Statistical Quality Design and Control*, (New York: McMillan Publishing Co., 1992).

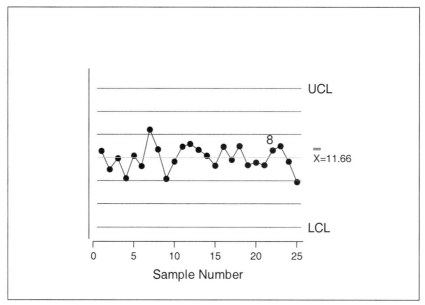

Figure 17.28 An Example of Under Dispersion Detected by Rule 8 on Subgroup 22

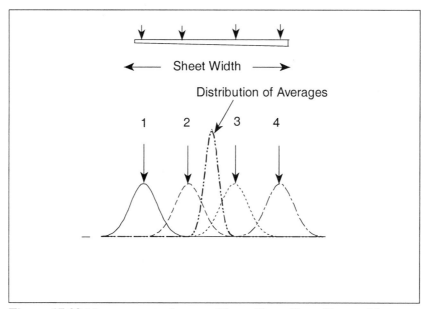

Figure 17.29 Measurements Across a Nonuniform Sheet Grouped into One Subgroup

to specific needs (e.g., runs of 7 rather than 8, etc.). Most software packages for plotting control charts also have these rules (or some modification of them) built in. The programs flag points that break the rules and provide a legend of which rules have been broken, similar to what has been shown in Figures 17.26 - 17.28.

17.7.2 Using Control Charts to Control Future Production

Once final control limits have been established, these limits can then be used to control future production similar to the way that control charts were used to reduce variation in cylindrical wheel balancer parts described in Example 1 of Chapter 2. To be successful in controlling future process outputs, not only must control charts be used, but there must be a firm understanding of special causes that may arise and process adjustments needed to eliminate these causes. In some cases these special causes are easy to identify and correct, but more often than not careful work must be done at the preliminary stages to develop a firm understanding.

Several iterations of data collection and preliminary control chart construction may be needed to confirm that assignable causes for variation have been removed from the process, allowing final chart limits to be calculated. During these preliminary steps, when data is being collected, notes and other qualitative information about the process should be recorded. The next two pages show a data recording form for and R chart used at Ford Motor Company. Similar forms are available through ASQ (American Society for Quality) and other organizations. As can be seen, there is space on the back side of the form for local operators to record comments which may be useful later in identifying special causes. With use of automatic data collectors and computer programs for charting, the same information collected on the paper form could be collected directly into an electronic database.

Engineers, with their theoretical knowledge, should be involved with operators in teams to perform the preliminary steps of establishing control limits and making lists of assignable causes and remedies. The scientific method should be the tool used to identify these causes. The preliminary control charts form the available data as described in Figure 1.1 of Chapter 1. Process flow diagrams, Pareto diagrams and cause and effect diagrams all can be used in team meetings to hypothesize the causes (and potential correcting adjustments) for out of control points and nonrandom patterns identified on the preliminary charts. Confirmation tests, or simple comparative experiments should be performed to confirm the hypothesized special causes and corrections.

During normal process operations, local process operators, without the help of engineers or management, should be responsible for maintaining and using the control charts to keep the process in a state of statistical control. When a firm understanding of special causes and their correction is established, and control limits are finalized at the preliminary stages, this information should be recorded for local operators to use in controlling the process. Control limits can be drawn directly on a paper, leaving room for potential out of limits points to be recorded during future data collection. The box in the upper right of the chart shows the special nonrandom signals Ford would like its operators to recognize while charting real time data. Finally, the second box on the right of the form leaves space to record instructions for operator adjustments necessary to correct special causes when they arise.

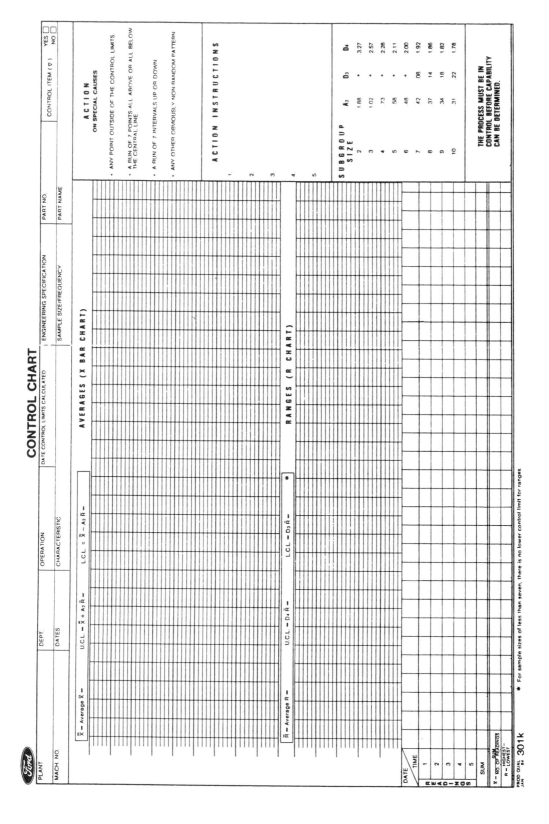

CONTROL CHART

684

CONTROL CHART

PROCESS LOG SHEET

ANY **CHANGE** IN PEOPLE, MATERIALS, ENVIRONMENT, METHODS OR MACHINES SHOULD BE NOTED. THESE NOTES WILL HELP YOU TO TAKE CORRECTIVE ACTION WHEN SIGNALED BY THE CONTROL CHART.

DATE	TIME	COMMENTS

DATE	TIME	COMMENTS

PROD QUAL
APR 83 301k (reverse)

When using control charts to control current production, the space on the back of the form can be used to record any necessary adjustments that were made to the process. The form then becomes a running record and valuable source of information concerning assignable causes identified, and corrective measures taken.

Separately, engineers may utilize the data collected by operators on control charts during normal operation to perform process capability studies, and to refine the action lists of instructions to operators placed on future charts. We will discuss the details of process capability studies in the next section.

In a modern factory, automatic data collection and computer based feedback control systems may replace the local operator's chore of plotting control charts and making manual adjustments. This does not negate the value of control charts. Automatic feedback control systems may over adjust, resulting in added variation or over dispersion if control chart algorithms are not built in to their set of instructions. Furthermore, manual work must be done in preliminary studies to identify assignable causes and their corrections in order to make the feedback controls work.

When processes are computer controlled, there are several newer forms of control charts that can be used that are more sensitive to minor process drifts. Some examples of these charts are Cusum charts, EWMA charts, and Multivariate charts. It has been reported that at DuPont over 10,000 chemical processes are run with automatic feedback control systems based on Cusum charts. Most of these charts are too complicated to be used in manual operations and they are beyond the scope of this text to present. Further information is available in texts such as those by Ryan[12], Montgomery[13], and Mitra[14].

17.7.3 Predicting Future Process Performance - A Process Capability Study

Capability is often thought of as the proportion of process output that meets the customer specifications. When a process is operating in a state of statistical control, its performance can be described by a predictable distribution, and the proportion of output within or outside of specifications can be determined from this distribution. Therefore, the process capability is determined from the variability due to common causes. If the process is not running in a state of statistical control, the proportion of output meeting the customer specifications will not be predictable, and process capability cannot be accessed.

Process capability studies are performed to determine to what degree a process is

---

[12]Ryan, T. P., *Statistical Methods for Quality Control*, (New York: John Wiley and Sons, 1989).

[13]Montgomery, D. C., *Introduction to Statistical Quality Control*, (NewYork: John Wiley and Sons, 1985).

[14]Mitra, A., *Fundamentals of Quality Control and Improvement*, (New York: Macmillan Publishing Company, 1993).

capable of producing results within the specifications. Capability studies can be used by management to determine if a process needs to be improved, and how much it needs to be improved. capability studies are also used to 1) determine if specific processing equipment will be able to meet design specifications for newly designed products, 2) evaluate new equipment purchases, 3) estimate costs of quality for contracts, and 4) set specification limits. Capability studies are based on predicting future process performance from sample process data. The results of capability studies are presented as graphs such as histograms, estimates of the proportion of products that will meet the specs, and other summary statistics that are called *capability indices*.

Performing a capability study requires several steps. First, the critical dimensions to be studied should be chosen. Next, data should be collected from the process and plotted on variables control charts. Assignable causes evident on the control charts should be identified, and it must be demonstrated that the process can be run in a state of statistical control. If the process is not in statistical control it will be impossible to predict future process performance accurately.

To perform a process capability study, we should have at least 25 rational subgroups, of four or more observations, that demonstrate the process to be in control when plotted on a control chart. The next step is to get estimates of the process mean and standard deviation. With 25 or more subgroups of data we can get good estimates of the *in-control* process mean and standard deviation, due to the Law of Large Numbers. The final step is to make predictions. The Central Limit Theorem helps us predict the *average* of future events. Just as we could predict that 68% of the *average* results on the toss of four dice would fall between 3.5 ± .8539, in Section 3.6, we can make predictions about future subgroup means. For example, using the chemical process discussed in Section 17.5.2, we can predict that 68% of the daily *average*

assays of the chemical product will fall between $\overline{\overline{X}} \pm \sigma_{\bar{X}}$ or $24.84 \pm \dfrac{2.82}{\sqrt{4}}$

The shape of the Normal Distribution shows us that the following will be true:

Range	% of Daily Average Assays	Limits
23.43 - 26.25	68.27%	1σ
22.02 - 27.66	95.45%	2σ
20.61 - 29.07	99.73%	3σ

Based on this information, the management of our example chemical plant would be reasonably (99.73%) confident that daily averages of four drums will meet the product specification limits of 20 - 30. However, individual drums vary more than daily averages. Before making predictions about the percentage of individual process outputs within specifications using the Normal Distribution, we should check the normality of the data as was shown in Section 17.5.3. There is no need to use the Normal Distribution to predict the percentage out of spec for attribute charts like NP and P charts because $\bar{p}$ is a direct estimate of the proportion defects.

When studying variables data rather than attributes, it is possible to predict very low percentages (e.g., parts per million or hundred thousand) of output outside the specifications if the process standard deviation is small with respect to the specification width. Thus when comparing capabilities of different processing equipment or suppliers of raw materials, it is desirable to develop a capability index rather than directly comparing small predicted percentages.

Capability indices are statistics calculated from the data and specification limits that can concisely summarize the results of a capability study in a single number. Capability indices are based on the Normal Distribution, just as predictions about individual outputs. Therefore, the normality of data should be checked before they are used. The most commonly used indices are C_p, and C_{pk}. The Cp index is the ratio of the tolerance width to 6σ, as shown in the equation below

$$C_p = \frac{USL - LSL}{6\sigma} = PCR$$

C_p can range between zero and infinity. The larger a value of C_p the better. Normally a process is considered capable if the C_p index is greater than or equal to one. If the process mean is centered between the specification limits, then a C_p index of 1.0 indicates that 99.73 percent of the output will be within specs. C_p values less than 1.0 indicate more than $.0027 = (1-.9973)$ will be out of spec and C_p values greater than one indicate less than .0027 will be out of spec. To ensure defect levels (i.e., percent out of specs) in the parts per 100,000 or less, some companies require C_p indices of 1.5 or greater.

The C_p index is misleading if the process mean is not centered between the specification limits. In that case, the C_{pk} index should be used instead. The C_{pk} index takes into account both the process spread and location. It is calculated using the formula below:

$$C_{pk} = \text{Minimum}\left[\left(\frac{USL - \overline{\overline{X}}}{3\sigma}\right), \left(\frac{\overline{\overline{X}} - LSL}{3\sigma}\right)\right] = PCR_k$$

When the process mean is centered between the spec limits, C_{pk} and C_p will be identical. When the process is not centered, however, the C_{pk} index is penalized by taking the shorter distance between the process mean and the upper or lower specification limit. When the process mean is actually outside the specification limits, the C_{pk} index will be negative.

The C_{pk} index for the chemical assay data, from Section 17.5.2, is:

$$(24.84 - 20)/3(2.82) = 0.572$$

This process would be considered less than capable since the value is less than 1.

Figure 17.30 shows representations of four different C_{pk} values in terms of the distribution of process results and specification limits. In the upper two distributions, the means are the same ($\mu=0$), but smaller standard deviation ($\sigma=0.5$) of the top distribution results in a larger C_{pk}. In the lower two distributions both distributions have the same unacceptable C_{pk}'s for different reasons.

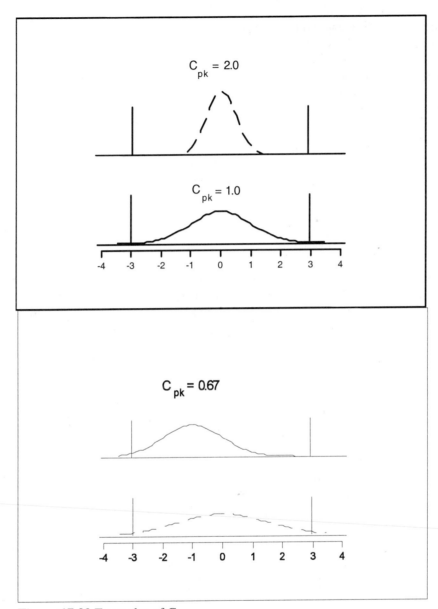

Figure 17.30 Examples of C_{pk}

In the third from the top, the mean is off-center ($\mu=-1$) which penalizes the C_{pk}, while in the bottom distribution the increased variance ($\sigma=1.5$) results in the same value of C_{pk}.

C_p and C_{pk} indexes are often used as a means of communication. Many companies in the automotive industry, and other industries, rate their suppliers by these indices or demand certain levels such as $C_{pk} \geq 1.33$ be obtained by all suppliers. By knowing that their suppliers process is in statistical control and has a high C_{pk} value, companies can eliminate inspection of incoming raw materials and associated costs.

When a process is less than capable, the values of the C_p and C_{pk} indices can help us to determine how the process can be improved. For example, if $C_{pk} \leq 1$, but $C_p \geq 1$, then the process can be made capable by centering the process mean. If both indices are less than one, however, both the process mean will have to be centered, and the process variance reduced in order to make the process capable.

17.7.4 PDCA for Process Improvement

When process capability studies indicate that a process is incapable of meeting customer specifications, what can management do to improve the situation? Possibly, the mean of the process output may need to be shifted or the process variance reduced. In some cases the solution may be simple, such as purchasing higher quality raw materials or more precise processing equipment. In other cases, the solution may not be so obvious or the obvious solution may be too expensive to pursue. For example, the obvious solution to reduce tile variation for the Ina Tile Company (mentioned in Chapter 2) was to install temperature sensors and controllers in their kiln. However, this option was too expensive to pursue. In these cases, management may be forced to use the same trial and error approach to find process improvements that local operators and engineers use to identify and correct special causes.

Figure 17.31 shows the PDCA cycle for process improvements. The idea for this cycle was originated by Shewhart and popularized by Deming. It is often called the Deming cycle. Deming recommended that this cycle be used by teams to improve the input and output of any process. To be effective, the team should be composed of individuals from different staff areas such as engineering, operations, accounting, and customer service. The PDCA cycle is really the same as the scientific method described in Chapter 1, and its success is therefore dependent on proper use of statistical methods. Let's examine in detail the four steps of this cycle to see where the statistical methods we have presented in this book fit into the scheme of things.

(1) Plan. This is essentially the hypothesis forming stage. What improvement is desired? What information is available that indicates fruitful approaches to achieving the desired improvement? Here we can examine available data from our own process or competitive benchmark data from alternate approaches. Flow diagrams, Pareto diagrams, and cause and effect diagrams completed by the cross functional team will lead to ideas on how the process may be improved. When consensus is reached on possible improvements, plan a change or test preferably on a small scale at first.

(2) Do. This is the experimentation phase or collection of additional data. Carry out the test or change planned. Here we need to be concerned about validity of the test. Concepts from experimental designs such as randomization and replication to avoid bias and ensure valid results

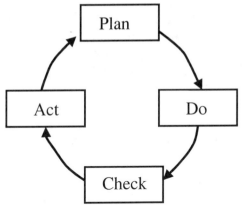

Figure 17.31 The Deming Cycle or
PDCA Cycle

are extremely pertinent at this stage. The effects of some changes may be studied on paper, or through computer simulation experiments avoiding actual experimentation. Probability calculations and cost analysis are sometimes all that is needed.

(3) Check. Sometimes this step is called study instead of check. It is the stage where we compare the results or data from our test to see if it confirms or contradicts our theory of improvement. What was learned? What can we predict? Here employment of objective methods of statistical data analysis are often appropriate. Humans tend to be optimistic in interpreting experimental data. Without objective analysis, false positive results abound. Deming has been quoted to say the only value of statistics and statisticians is to make predictions which form the basis for action.

(4) Act. In this stage we implement the change we predict will be beneficial. If the results of our test are favorable it might be wise to repeat it again under different environmental conditions to see if the result was spurious or valid (robust) under a variety of conditions. In experimental design terminology, this is equivalent to a blocked experiment. The change or difference in factor levels is tested in a number of blocks. This lends a broader base to our conclusion when a treatment effect is found.

If no beneficial change was found using the Deming cycle, the loop may be repeated several times trying different ideas, or combinations of ideas, as in multi-factor experiments to look for favorable interactions or synergism.

17.8 Summary

In this chapter we have presented the basic ideas for using control charts for process improvement. Engineers need to understand the theory and practice of control charts. Although they may not be directly involved in using control charts to control process outputs, they are usually involved with local operators in establishing the control limits and finding and eliminating special causes. They may often study the results of control charts in making process capability studies and process improvement studies using the Deming cycle. Some key concepts that should be understood from this chapter are:

Prevention vs Inspection Philosophy 637, 638
Special and Common Causes of Variability 638
Stable and Unpredictable Processes 639
Responsibility of Local Operators vs Management in Reducing Variability 639, 677
Variables and Attribute Data 644
Advantages and Disadvantages of Using Variables or Attribute Charts 644, 645
X and R Charts 645
X and s Charts 645
Charts for Individuals 658
Moving Average Moving Range Charts 665
NP Charts, P Charts, C Charts and U Charts 666
Iterative method for establishing Control Limits 642, 643
Interpreting Non-random Control Chart Patterns 678 to 682
Use of Control Charts to Control Future Process Output 683
Process Capability Studies, Capability Indices 686
PDCA Improvement Cycle 690

17.9 Exercises for Chapter 17

17.1 Walter Shewhart pioneered the idea of control charts and the control chart philosophy of quality control. What is the reason that he titled the book, that first introduced control charts, as: *The Economic Control of Quality of Manufactured Product?*

17.2 Given the table of fill weights for subgroups of size n=5

Subgroup	X_1	X_2	X_3	X_4	X_5	$\bar{X}$	R	s
1	48.85	47.22	45.95	43.04	55.17	48.046	12.13	4.514
2	43.35	54.73	53.82	37.35	43.51			
3	53.31	46.34	53.48	63.40	47.51			
4	39.83	60.09	54.35	50.53	51.16			
5	43.50	48.05	58.97	55.11	48.07			

a) Calculate the $\bar{X}$'s, the ranges, R, and sample standard deviations, s, for subgroups 2 through 5.

b) Calculate the grand mean $\bar{\bar{X}}$, $\bar{R}$, and $\bar{s}$.

c) Calculate the UCL and LCL for the $\bar{X}$ and R charts.

d) Calculate the UCL and LCL for the $\bar{X}$ and s charts.

17.3 The following data has been collected on the groove depth of a manufactured part:

Subgroup	Data				Operator Comments
1	12.2	12.6	12.2	12.5	
2	12.0	12.7	13.3	12.1	
3	12.9	13.4	13.1	12.4	
4	10.4	9.4	10.7	10.2	← New Cutting Tool before this group
5	10.0	10.9	11.4	11.3	
6	11.8	12.0	11.1	10.1	
7	11.6	11.0	11.5	12.0	
8	12.5	11.6	11.8	11.2	
9	12.9	12.0	10.5	12.2	
10	11.1	11.1	13.1	11.6	
11	11.7	13.6	12.5	11.8	
12	14.6	12.8	13.5	11.9	

13	11.7	10.0	10.5	12.2	← Adjusted Cutting Depth before this subgroup
14	11.6	10.6	11.0	10.8	
15	11.7	13.6	10.9	11.5	← New Operator on job
16	12.3	11.5	10.6	10.7	
17	11.6	12.0	12.5	11.9	
18	11.9	9.5	13.9	11.4	
19	12.3	14.6	9.6	11.7	Noticed high variability in raw material stock. Sent all parts back to previous step for 100% inspection.
20	12.6	12.8	13.2	11.5	

a) Construct $\bar{X}$ and R charts from the data using preliminary control limits calculated from all the data.

b) Check the charts for special causes and eliminate any data in this category.

c) Reconstruct the charts using the reduced data set, and estimate process mean and standard deviation, assuming the process will be kept in control.

d) Check the normality of the individual data in the reduced data set by constructing a Normal Probability plot.

e) Assuming the process can be kept in control, predict the percentage of finished parts that will fall within the specification limits of 10 - 13.

17.4 In a poultry processing plant in Merida, Mexico[1], 10 broilers were sampled every hour and a defect score was assigned based on the number and severity of defects based on various penalty factors. The data for July 1993 is shown below:

Subgroup		Scores				
1	14	13	14	13	15	14
2	15	14	13	16	19	13
3	13	3	4	14	15	16
4	7	16	14	14	16	15
5	3	3	13	6	16	6

[1]Elizundia, J. M. and Hau, G., "Use of SPC in two Mexican Poultry Processing Plants", *Proceedings of ASQC 49th Annual Quality Congress,"* (1995), pp. 579 - 587.

6	5	5	15	20	15	14
7	7	16	16	16	18	15
8	3	4	14	4	21	14
9	5	13	14	5	10	17
10	6	7	15	7	15	15
11	13	14	13	17	14	15
12	6	15	17	6	19	16
13	7	9	8	17	19	15
14	5	17	5	5	18	14
15	7	17	7	5	19	16
16	6	7	9	19	20	15
17	2	14	17	18	13	15
18	4	14	13	13	17	14
19	4	14	13	16	15	14
20	3	13	14	13	16	14
21	3	15	13	16	16	13
22	4	15	13	18	15	13
23	4	13	6	16	13	14
24	4	19	14	13	16	15

a) Calculate the limits for the $\bar{X}$, and s charts.

b) Plot the data and identify any special causes.

c) If the upper and lower specification limits for the defect scores are 0 - 24, calculate C_{pk}.

17.5 Match each of the time varying process behaviors (1)–(6) with one pair of $\bar{X}$, R charts (a)-(f) that appears most likely to result.

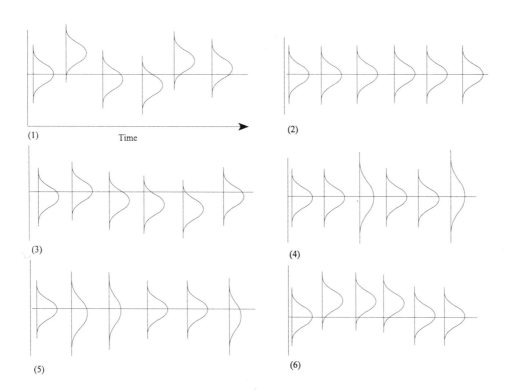

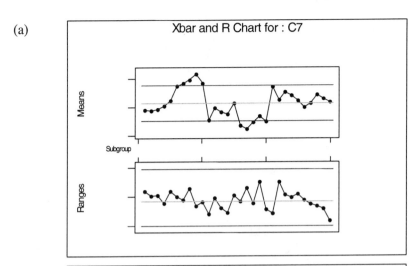

(a) Xbar and R Chart for : C7

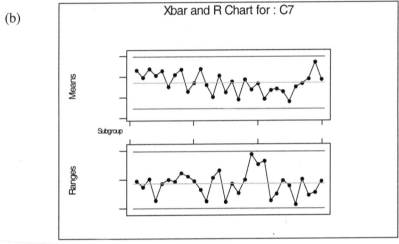

(b) Xbar and R Chart for : C7

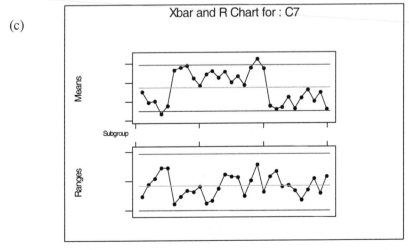

(c) Xbar and R Chart for : C7

697

(d)

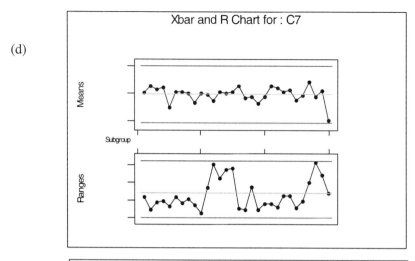

(e)

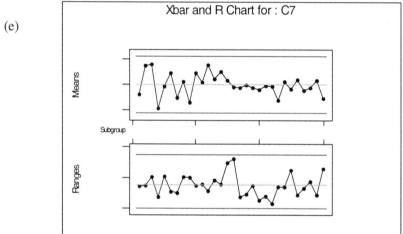

(f)

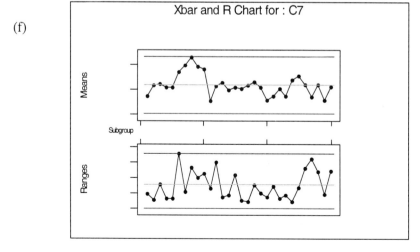

17.6 The following data represent the gas mileage obtained by a 1980 VW Scirocco for 60 consecutive fill-ups.

Fill Up #	MPG	Comments	Fill Up #	MPG	Comments
1	30.7		31	32.7	
2	31.5		32	28.8	
3	33.2		33	30.0	
4	33.1		34	28.4	
5	30.5		35	32.3	
6	37.9		36	29.9	
7	35.05		37	31.4	
8	36.6		38	27.4	Rear tires low re-inflated here
9	33.5		39	34.2	
10	36.4		40	34.09	
11	37.2		41	33.48	
12	32.1		42	32.04	
13	37.1		43	34.72	
14	36.3		44	32.5	
15	32.5		45	33.12	
16	37.3		46	33.9	
17	36.2		47	35.2	
18	34.5		48	32.2	
19	33.0		49	30.1	
20	34.4		50	39.9	Trip to Washington DC
21	33.2		51	38.6	
22	39.8	Highway Driving Upstate NY	52	36.2	
23	33.0		53	33.6	
24	37.3		54	36.3	
25	31.7		55	35.0	
26	32.4		56	35.1	
27	32.5		57	38.4	
28	29.7	Tune Up here	58	32.6	
29	30.2		59	35.7	
30	32.1		60	34.9	

a) Make an individuals chart using the mean of moving ranges of 2, and identify any special causes.

b) Make an individuals chart using the median of moving ranges of 2, and identify any special causes.

17.7 Defects, such as cracks, lips, warping, etc. in frozen hamburger patties produced at a meat processing plant are counted. These defects could cause incomplete or overcooking of patties at fast food restaurants. Below are defect counts for 30 subgroups of 100 patties. Construct an NP chart and identify any signals of assignable causes.

Subgroup	Sample Size	Number of Defects	Comments
1	100	5	
2	100	3	
3	100	6	
4	100	5	
5	100	5	
6	100	2	
7	100	12	
8	100	4	
9	100	3	
10	100	7	
11	100	3	
12	100	0	
13	100	6	
14	100	5	
15	100	3	
16	100	3	
17	100	9	
18	100	5	
19	100	3	
20	100	1	
21	100	23	
22	100	5	
23	100	12	
24	100	13	
25	100	31	Anti-lip bar changed
26	100	21	
27	100	4	
28	100	6	
29	100	4	
30	100	8	

17.8 At a meat processing plant, samples of incoming beef are analyzed for E. coli bacteria counts. Data for 20 samples are shown below. Construct a C chart and determine if there are any out of control signals.

Sample	E. Coli
1	5
2	5
3	10
4	20
5	5
6	5
7	10
8	5
9	10
10	30
11	30
12	5
13	30
14	5
15	50
16	10
17	5
18	10
19	20
20	5

17.9 In an automotive assembly plant, vehicles are tested for water leaks[2]. Assignable causes for leaks would have to be traced to the glass setting and welding processes. The data below is for 26 consecutive inspections in May and June:

Sample	Units Checked	Units Leaking
1	315	56
2	399	79
3	346	64
4	375	56
5	315	73
6	355	73
7	354	81
8	363	68

[2]Siegel J. C. "Managing with Statistical Methods," *Statistical Process Control SP-547*, Society of Automotive Engineers Inc., Warrendale, PA, June 1983.

9	362	76
10	275	51
11	298	61
12	312	58
13	359	78
14	321	80
15	329	80
16	282	57
17	344	60
18	332	82
19	395	76
20	358	89
21	451	90
22	358	65
23	417	85
24	413	57
25	396	90
26	404	75

Construct a P chart, and determine if the glass setters, welders, etc. are doing the best job they can with the process provided to them.

17.10 On the six control charts below, identify violations of the four rules for stability listed in Section 17.7.1 and the other rules for unnatural patterns listed in the same section.

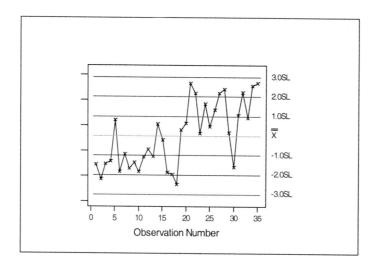

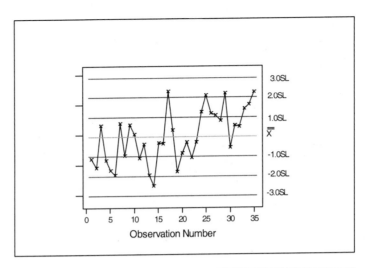

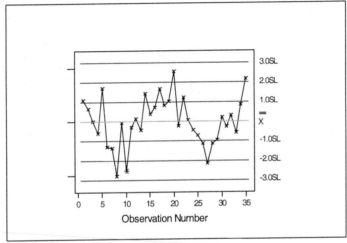

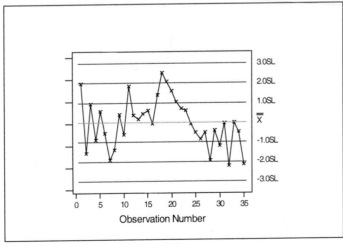

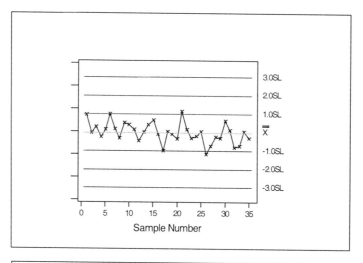

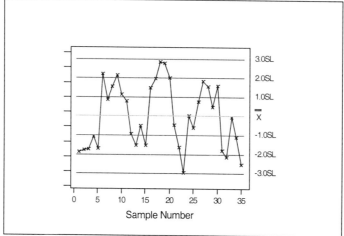

17.11 The control charts show a process to be in a state of statistical control with $\overline{\overline{X}}=.758$ and $\overline{R}= .149$ with subgroup size n=4. If the specification limits are .7 to .9,

a) calculate C_p

b) calculate C_{pk}

c) calculate C_{pk} assuming the process mean could be shifted to the center of the specification range

d) assuming the process mean could be adjusted to .8, to what value would the process standard deviation, σ, have to be reduced to in order to make $C_{pk} = 1.33$

CHAPTER 18
Off-line Quality Control and Robust Design

18.1 Introduction

As pointed out in Chapter 2, high quality means more than simply producing parts within the engineering specifications. In the example given in that chapter, we explained that by reducing variability of cylindrical component parts the total cost of producing and warranting a wheel balancer could be decreased, and customer satisfaction simultaneously increased. Deming said[1] that if he had to reduce his whole message about quality to a few words, he would say it all revolved around reducing variation. Chapter 17 presented various types of control charts that can be used to reduce variability in manufacturing. In this chapter we will discuss methods of reducing variation before manufacturing, at the engineering design stage. To understand why variation reduction is necessary during product design, let's further explore the relationship between quality and variation.

Some people would say that a Mercedes Benz is a higher quality automobile than a Geo Metro, or that a Bosch dishwasher is higher quality than a Hotpoint. But, in making a statement like that they are confusing the quality of the product with the features and intended utility of a product. The utility is the external appearance and function. Utility together with features defines what a product can do. An electric pencil sharpener, for example, has more utility than a manual pencil sharpener. That does not necessarily mean that it is higher quality. Consumers will buy a product, or service, when the price is less than or equal to the expected utility. Variety in price and utility is necessary in the market place because consumers differ in their budgets and tastes. Determining the particular market niche and price for a product are problems of marketing, not engineering.

Then what do we mean by the quality of a product? The price and intended function of a product are clear to customers at purchase, but the quality is not. Once the product is purchased and used, the quality becomes clear. How well did it perform its intended function? If a customer receives poor quality, he may complain and certainly won't buy the product again. Taguchi[2] has defined quality as the *loss a product causes society after it is shipped*. By loss, Taguchi specifically means loss caused by 1) variability in function or 2) loss due to harmful side effects. For example, losses to a customer could occur due to early failures, subpar performance, less power, or deterioration of a product, while losses to third parties (or society in general) could be caused by things such as auto exhaust, air conditioner noise, or toxic gas due to incinerating plastic.

Quality control does not concern itself with losses a product may cause society due to its intrinsic functions. For example, Taguchi[3] explained that children may become engrossed in TV viewing to the point where they sustain major losses by neglecting their studies and failing to develop necessary skills. That does not mean that TV programing should be made unappealing. Appeal is the intended function of the programming. The question of what functions and features a product should have is a marketing question and not an engineering problem. Once the price and intended function of a product are defined, the engineering problem is to design the product to prevent losses from functional variability and harmful side effects, and do this at the lowest possible cost. Therefore we are again led to the idea of reducing variability as the means for increasing quality, this time we are talking about variability from intended product function.

As we will explain in this chapter, the step where engineers can have the greatest effect in reducing functional variability is in the engineering design stage, not during production. The tools to accomplish this variation reduction will be the experimental design and analysis techniques discussed in Part III of this book. The activities used to reduce functional variability through engineering design are often called off-line quality control or quality engineering. The idea is to make product designs robust to the noise sources that cause functional variability.

18.2 Noise—Sources of Functional Variation

Variation in product function is caused by what is commonly called *noise*. Variation from nominal component dimensions and characteristics that occur during production is one class of noise. Taguchi defines three distinct classes of noise that cause functional variation in products. These classes are:

(1) inner or deterioration noise

(2) outer or environmental noise

(3) variational noise or piece to piece variation caused in manufacturing

Inner or deterioration noise refers to changes in product or components that occur as the product ages or wears out and prevent it from achieving its target function. Outer or environmental noise refers to external factors such as temperature, humidity, or human treatment that affect the performance of the product. Finally, variational noise refers to the manufacturing imperfections resulting from the special and common causes we have discussed previously. Phadke[4] described the important sources of noise in two common products:

Refrigerator temperature control:

Inner noise—the leakage of refrigerant and mechanical wear of compressor parts

Outer noise— number of times the door is opened and closed, the amount of food kept, initial temperature of food kept, variation in ambient temperature, supply, and voltage variation

Variational noise—tightness of door closure and amount of refrigerant used

Automobile brakes:

Inner noise—wear of the drums and break pads and leakage of brake fluid

Outer noise— road conditions, speed of the car, number of passengers

Variational noise—variation in the friction coefficient of the pads and drums

Variational noise or piece to piece variation in manufacturing is itself a combination of many sources of variability. Some of these sources are external factors like external temperature and humidity, and variation in raw materials. Other sources are internal to the process, such as within batch variation and process drift.

It is possible to reduce variational noise during production and thus reduce the functional loss caused by this noise source. For example, control charts were used to reduce variability of cylindrical part dimensions thus reducing functional variation and loss caused by wheel balancers. However, once a product is designed, component materials selected, and the manufacturing equipment designated, it is too late to do anything during production to prevent loss of function due to inner and outer noise, such as wear of internal parts during use or environmental factors. It is possible to mitigate these sources of noise during the product design.

The goal of off-line quality control is not necessarily to reduce variation of noise factors, but rather to find counter measures that will reduce their influence on product function. It is possible to find counter measures for inner and outer sources of noise during the product design stage. It is sometimes less expensive to find counter measures for variational noise during the design stage, than it is to control it during production. For example, variations in tile sizes described in Chapter 2, were caused by variational noise (a temperature gradient in the kiln). But, it was much cheaper to counter-act this temperature variation by changing the clay formulation (during the product design stage) than it was to control the kiln temperature gradient.

Table 18.1, adapted from Taguchi, summarizes the activities where counter measures to the three noise classes are possible. The critical importance of work done at the product design stage can be seen in this table. It is the only step where counter measures can be found for all noise sources. We will discuss specific ways to counteract inner and outer noise at the design stage in Sections 18.4 and 18.5. In the next section we will discuss loss functions which will quantify loss due to functional variation.

Table 18.1 Countermeasures

Activity \ Noise Source	Inner Noise	Outer Noise	Variational Noise
Manufacturing	X	X	O
Process Design	X	X	O
Product Design	O	O	O

X--No Countermeasure Possible   O--Countermeasures Possible

18.3 Loss Functions

Since the goal of quality improvement is to reduce variability of component dimensions and characteristics and product function, it is clear that simply counting the proportion of parts within engineering specifications will be a poor way to measure quality. For example, Figure 18.1 shows the distribution of the diameters of shanks (component part of the high speed wheel balancer), before and after the use of control charts. Here it can be seen that both distributions show essentially all

707

parts within the engineering specifications, but we know the more concentrated distribution represents higher quality. Capability indices like C_p and C_{pk}, described in Chapter 17, are a better way of measuring or quantifying quality, than counting the proportion within spec. Larger values of these indices relate directly to reduced variability around the nominal value. Taguchi defines another way of measuring quality which closely matches his definition of quality.

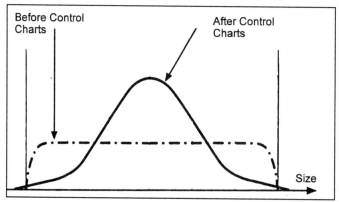

Figure 18.1 Comparison of Distribution of Wheel Balancer Components

Taguchi defines quality to be the loss to society after a product is shipped. Therefore, he measures quality with a loss function. If a component dimension or characteristic, y, is exactly at its nominal value, then it will cause the least possible loss. Therefore, Taguchi defines the loss function L(y) and its first derivative to be zero when the component is at its nominal value, i.e., L(m)=0 and L'(m)=0. Expanding the loss function through a Taylor's series around the nominal value m, the following equation results:

$$L(y) = L(m) + \frac{L'(m)}{1!}(y-m) + \frac{L''(m)}{2!}(y-m)^2 + \ldots$$
$$= \frac{L''(m)}{2!}(y-m)^2$$

Truncating this after two terms, we get the approximate quadratic loss function as:

$$L(y) \cong k(y-m)^2$$

and the expected loss is simply $E[L(y)] = k(E(y)-m)^2 + k\sigma^2$, which can be approximated by ks^2 when the component mean is adjusted to the nominal value.

When a component deviates from its nominal value the loss will increase. If the specific dollar loss could be determined for some value of y, then the unknown constant k could be determined. But, this may be difficult to do in practice, and we will discuss alternatives to loss functions later.

For some component characteristics there is no nominal value and the loss simply decreases as the characteristic increases or decreases. These are the so-called The Larger is Better and Smaller is Better situations. For example, if a component characteristic causes a harmful side effect, we would want to minimize it and this would be a Smaller is Better situation. Within our cost constraint, we would also like to maximize a product characteristic such as strength to support a load, and this would be a Larger is Better Situation. The loss functions Taguchi defines for Smaller is Better and Larger is Better are different from the quadratic loss function, shown above, which is appropriate for characteristics where the Nominal is Best. Table 18.2 shows the functional form of these loss functions and Figure 18.2 illustrates the difference between these loss functions and the traditional idea that no loss occurs when the engineering specification is met.

Table 18.2 Loss Functions

Characteristic	Average Loss from Sample Data
Nominal is Best	$\dfrac{k}{n}\sum_{i=1}^{n}(y_i - m)^2$
Smaller the Better	$\dfrac{k}{n}\sum_{i=1}^{n} y_i^2$
Larger the Better	$\dfrac{k}{n}\sum_{i=1}^{n}\left(\dfrac{1}{y_i^2}\right)$

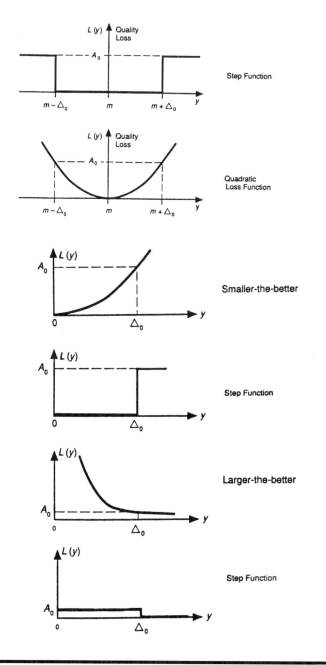

Figure 18.2 Comparison of Quality Loss Functions

18.4 Principles of Quality Engineering

The purpose of off-line quality control or quality engineering is to find ways, during the product and process design process, to reduce the effects of the three noise sources (inner, outer, and variational) upon intended product functionality. Taguchi has defined three steps where this can be accomplished. These steps are listed below:

1. System design

2. Parameter design

3. Tolerance design

System design is the step where a survey of pertinent technology is made to determine an architecture that is suitable for achieving the desired product function. This is the step where the skill and experience of the designer play an important role. Selection of an appropriate circuit design or sequence of manufacturing steps are examples of system design. Techniques such as Quality Function Deployment (QFD), see Hauser and Clausing[5], can help in organizing comparisons and making rational choices among competing technologies.

Once a system configuration is defined, parameter design is the step where individual values of the system parameters are chosen. System parameters are the controllable factors which affect the functionality of the product. Examples of system parameters are: nominal values of electrical components such a resistors and transistors in an electrical circuit, nominal dimensions and raw material types for component parts of mechanical devices, and temperatures and feed rates for steps in a manufacturing process.

Tolerance design is the final step where specifications or a range of tolerable deviations from the nominal values of each system parameter are defined. Since intended product function is affected by system parameters, defining tolerable ranges in addition to the nominal values of system parameters is necessary to ensure consistent function. However, at this step a tradeoff is usually made between cost and quality. Insisting on tighter specifications results in more consistent function and higher quality, but usually costs more. Tolerances should only be tightened after all possibilities for quality improvement at the parameter design stage have been exhausted.

Parameter design is the step where the designer can have the greatest effect in counteracting the noise sources that affect product performance without increasing cost. Figure 18.3, adapted from Phadke[3], illustrates the fact that product performance or function is affected by both system parameters that can be set by designers and noise factors (various noise sources). The values or levels chosen for some system parameters, called *control factors*, affect product performance directly and also indirectly by nullifying or enhancing the effect of noise factors on product performance. In the example in Chapter 2, the proportion of lime, in the clay formula for floor tiles, nullified the effect of the kiln temperature gradient on tile size, therefore it was a control parameter. Values of other system parameters, sometimes called *scale factors*, affect product performance independently of noise factors. Through appropriate choice of control and scale factor levels, or system parameter values, the designer may be able to produce a consistent product function in spite of various noise sources. The process of doing this is called a *robust system design*. System parameters are classified as either control or scale factors after the data is collected and interpreted.

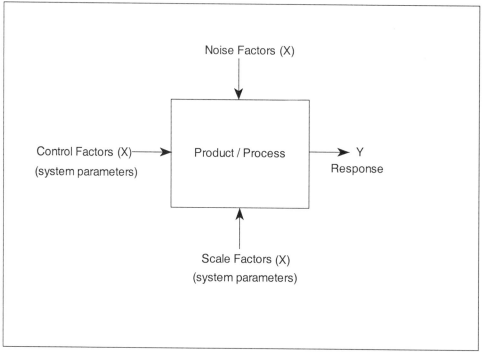

Figure 18.3 Block Diagram of Product or Process Showing Influence of Control, Scale and Noise Factors on Product Performance.

Despite its importance, parameter design is the step most often neglected in product development. Many engineers simply make adjustments during parameter design rather than making an organized study to increase robustness. They take the pertinent technology from the literature, then build and test a prototype. If product function in the prototype does not meet the desired target, they simply adjust the value for one (usually the most influential) system parameter until it can be achieved. Taguchi has made the analogy to a dyer, who makes a test batch of dye, then changes the mixing ratios or dying conditions until the target color is matched. This adjustment or calibration process should not be used in the design process because it will not lead to a robust system design.

Consider the example illustrated in Figure 18.4, again taken from Taguchi[2]. In this example a power circuit is designed. The desired product performance is an output voltage of 120V.

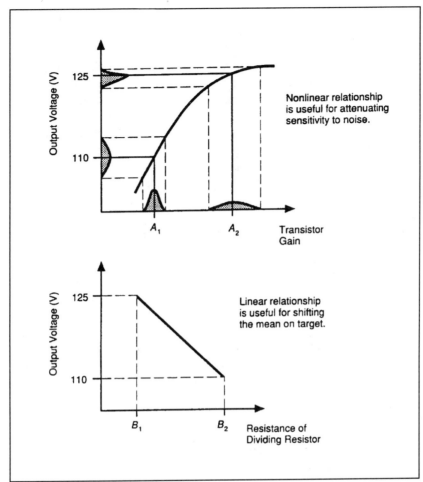

Figure 18.4 Parameter Design Values and Output Voltage

As the main effect plots in the figure show, the output voltage is affected by the nominal levels of two system parameters—the transistor gain and resistance of the dividing resistor. If an initial prototype circuit resulted in less than the desired 120V, it would be common to simply adjust the nominal value of the transistor gain or resistor to achieve the target. However, this ignores the possible influence of noise factors, and the opportunity to counteract them.

Variation in manufacture can result in 10–20% variation around nominal specifications for resistors and transistors. Deterioration and field use temperatures may cause additional variation from the nominal values. The distributions around the nominal transistor gains of A_1 and A_2 shown at the bottom of the top graph in Figure 18.4 illustrate the possible effect of these noise factors. The transistor gain, in this example, would be considered a control factor since its nonlinear relationship with output voltage causes differing sensitivity to variation in transistor gain (see Section 16.1.2). Choosing a nominal gain of A_2 would considerably reduce the effect of variability in transistors.

The resistor nominal value, on the other hand, would be called a scale factor since it directly affects output voltage, but its linear relationship with output voltage means the effect of resistor variability will be constant.

To produce the most robust design, the transistor nominal value should be set at A_2 to reduce variability in output voltage. But, this will increase the mean voltage output to something greater than 120V. To move the output voltage mean back to the target 120V, the dividing resistor nominal value should be set somewhere above its low level of B_1. Taguchi recommends, as this example illustrates, that values of the control factors should be selected in order to reduce variation in product performance as much as possible during the parameter design. Scale factor levels should then be adjusted to bring the product performance back to the target value.

To accomplish this recommendation we must either find and exploit nonlinear relationships between control factor levels and the product performance measure, as shown in Figure 18.4, or interactions between control factors and noise factors like the example shown in Figure 18.6. This figure, taken from Ross[6], shows the interaction between a noise factor (deflection) and a control factor (wall thickness) on the performance of the oil fill tube and associated plastic cap that is illustrated in Figure 18.5.

The product function in this case is to seal the tube against the engine crankcase pressure, yet be easy to install and remove. The force resisting installation or removal results from the rubber cap deflecting over the crimped ridge in the metal tube. The force can be increased by increasing the difference between the cap outside diameter (OD) and the inside diameter (ID) of the tube crimp, thus requiring a greater deflection of the cap. The force can also be increased by increasing the stiffness or wall thickness of the cap. As Figure 18.6 shows, the effect of deflection (or the difference between the cap OD and tube crimp ID) upon the installation force is greater with a thick-walled (high stiffness) cap than with a thin-walled cap (low stiffness).

The target force to install or remove the cap, as shown in Figure 18.6, can be achieved with stiff (thick) cap with less deflection (i.e., looser fit), or a more flexible (thinner) cap with more deflection (i.e., tighter fit). The stiffness or wall thickness is easy to control by choice of the injection molding die that is used to produce the cap, but the tightness of the fit (deflection) is a function of precise tolerances that must be held on the tube crimp (ID) and the cap (OD) as well as environmental factors such as temperature, which changes the thermal expansion coefficients. Thus when a target value for deflection is specified, variability about that target, indicated by the distributions in the figure, result. The variability in deflection is transferred to the force resisting installation or removal. As can be seen in Figure 18.6, the variability will affect the force to a lesser degree when using the flatter curve for a thin-walled cap. Thus, the designer should choose a thin-walled, tighter fitting cap in order to have a more consistent pull off force.

In order to make the choices of system parameter values in the last two examples, the information in the main effect plots in Figure 18.4 and the interaction plot in Figure 18.6 were necessary. Sometimes this type information is available from theory such as circuit approximation functions, but we also know that we can develop these plots in any situation by performing experiments like those described in Chapters 7 - 9, and 11 - 15. In the next section we will describe how experimental designs can be used to develop the information required to assign parameters in a robust system design.

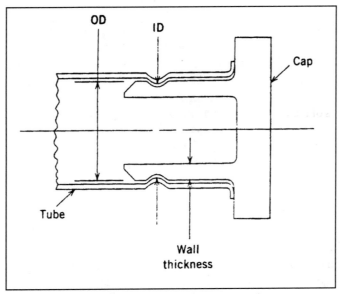

Figure 18.5 Oil Tube and Plastic Cap

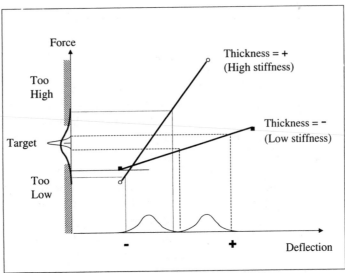

Figure 18.6 Interaction of Deflection and Stiffness

18.5 Parameter Design Experiments

The purpose of a parameter design experiment is to determine the relationship between the response, or measured quality characteristic, Y, and the control and scale parameters, X (as described in Figure 18.3). Once the relationship is determined, the settings for the control and scale parameters predicted to result in a value of the quality characteristic that is at the target value with minimum variation can be found. To determine a relationship between the response, Y, and parameters, X, an experiment similar to those described in Chapters 7–14 should be performed.

There are two approaches to conducting parameter design experiments. One method is called a combined array experiment, and the other method is called the inner outer array experiment. Both of these methods have advantages and disadvantages. We will describe and give examples of both of these methods, then discuss the situations where one method would be preferred over the other.

18.5.1 Combined Array Experiments

Recall the 2^3 factorial experiment described for voltmeter readings in Section 8.7, and interpreted in Section 8.8.3. There, an interaction between the circuit warm-up time and ambient temperature was discovered. In that situation, we can consider the circuit warm-up time a system parameter (in the voltage measurement process) while the ambient temperature would be considered a noise factor, since it cannot normally be controlled in the factory. By discovering this interaction, we could reduce the variability in meter readings by always choosing the high level (5 minute warm-up) of the circuit warm-up when making voltmeter readings. This is one simple example of a parameter design experiment using what is called the combined array approach.

In a *combined array experiment* we must be able to control the levels of both the system parameters (control and scale factors) and the noise factors. All are assigned as factors in designs like those we have studied in Chapters 8, 11, 12, and 13. What is needed from the experimental design is the ability to estimate all interactions between system parameters and noise factors. Using full factorials like the 2^k plans described in Chapter 8, we can estimate all two-factor interactions. Other fractions of factorials, like those described in Chapter 12, may reduce the number of experiments while still allowing us to estimate a required set of interactions. For example in Section 12.5.1, we determined an 8 run experiment that would allow estimation of four main effects A-D and three two-factor interactions AD, BD, and CD. If Factors A - C represented system parameters while Factor D was a noise factor, this 8 run design would be sufficient for a parameter design experiment.

We analyze combined array parameter design experiments as we would any experiment from a similar design. The optimal choice of nominal levels of the system parameters for robust design will be determined after interpreting the experimental results. Any system parameters that have significant interactions with noise factors may qualify as control factors, as did the transistor gain, oil cap stiffness, and circuit warm-up time parameters described above. Recall the choice of the nominal level of a control factor affects not only the product function but also the variability of product function. To determine which system parameters are control factors and set their nominal levels, we must examine all significant interactions between control and noise factors. System parameters that have significant main effects but do not interact with noise factors are scale factors that can be used to adjust system performance to the target level.

A procedure for conducting a combined array parameter design experiment is outlined below:

1. Select system parameters that are easily controllable.

2. Determine the sources of noise, and choose the most important noise sources to be represented as factors in the design. This sometimes requires some ingenuity to define factors that will represent the inner, outer, and variational noise that you are trying to mitigate.

3. Select an experimental design that will allow estimation of all main effects (system parameters and noise factors) and all two-way interactions between system parameters and noise factors.

4. Perform the experiment.

5. Determine the significant effects and interactions and interpret interactions between system parameters and noise factors in order to identify control factors.

6. Select nominal values of system parameters to reduce variation and meet target performance.

A common experimental plan used in parameter design experiments is the L_{18} design described in Section 12.5, and again in Section 13.7. This design allows for estimation of some two-factor interactions and allows for three levels on some factors in order to determine nonlinearities as shown in Figure 18.4. The next section describes an example of a combined array parameter design experiment using the L_{18} design.

18.5.2 Example of a Combined Array Parameter Design

The following example is patterned (and simplified) after a parameter design experiment that was conducted, with help of G. Taguchi, at a motorscooter manufacturer in Taiwan[7]. The purpose of the experiment was to select the nominal values for certain parameters in the electric starting motor. The desired function of the starting motor is to deliver a torque of .25 kgs or more with high probability. Four system parameters for the starting motor are the inside diameter (ID), the spring constant (SC), the R.C.L., and the type of weld (spot or line). To simulate noise, or variation from ideal conditions that could be caused by manufacturing imperfections and wear during use, a special noise factor was constructed which had two levels. At the first level of the noise factor, starting motors were assembled with loose, rusty springs, and motor casings with a rough surface finish. At the second level of the noise factor, starting motors were assembled with tight springs without rust and smooth, finished casings. These two levels were thought to represent the worst and best cases that could occur in use.

To study the four system parameters and the one two-level noise factor, a L_{18} combined array design was chosen. Three levels were used for all the system parameters except the one with qualitative levels (weld type). The noise factor was assigned to the first two-level column in the L_{18} design since its interaction with column 2 is unconfounded with all remaining columns. The three-level design is shown (with integers representing the coded levels) in Table 18.3. The factors ID, SC and RLC were assigned to columns 2, 3, and 4 in the L_{18} design shown in Table B.1-15. These factors all had quantitative levels, and three-levels in the design allowed for estimating quadratic (nonlinear) effects for these factors. The qualitative factor, weld type, was assigned to column 5 in Table B.1-15, and the third level was collapsed to a 2.

Following the experimental plan in Table 18.3, eighteen prototype starting motors were assembled and tested under various conditions. The response of interest was the torque generated by the starting motors. The torque generated in a locked position for each starting motor is recorded in kgs. in the seventh column of Table 18.3.

Table 18.3 L_{18} Design for Parameter Design Experiment

Run	Noise	ID	SC	RLC	Weld	Torque
1	1	1	1	1	1	0.225353
2	1	1	2	2	2	0.257185
3	1	1	3	3	2	0.276032
4	1	2	1	1	2	0.224485
5	1	2	2	2	2	0.242203
6	1	2	3	3	1	0.258791
7	1	3	1	2	1	0.208673
8	1	3	2	3	2	0.231972
9	1	3	3	1	2	0.257377
10	2	1	1	3	2	0.241026
11	2	1	2	1	1	0.259373
12	2	1	3	2	2	0.284073
13	2	2	1	2	2	0.266845
14	2	2	2	3	1	0.278359
15	2	2	3	1	2	0.307101
16	2	3	1	3	2	0.248256
17	2	3	2	1	2	0.270575
18	2	3	3	2	1	0.293259

The data was analyzed using the GLM ANOVA method. The ANOVA model included terms for each main effect and all interactions involving the noise factors. Since the interactions between factors in the L_{18} are only partially confounded with main effects, it is possible to estimate them using GLM ANOVA or regression, as long as the total degrees of freedom in the model do not exceed 18 (the number of experiments). The results of the analysis are shown in Table 18.4. There it can be seen that the significant main effects (at the .05 level of significance) are Noise, ID, and SC. There is one significant interaction between ID and the noise factor.

718

Table 18.4 Analysis of Variance for Torque

Source	DF	Seq SS	Adj SS	Adj MS	F	P
Noise	1	0.00395443	0.00342408	0.00342408	467.82	0.002
ID	2	0.00038171	0.00038171	0.00019085	26.08	0.037
SC	2	0.00572407	0.00320141	0.00160071	218.70	0.005
RLC	2	0.00002651	0.00002682	0.00001341	1.83	0.353
Weld	1	0.00009838	0.00004227	0.00004227	5.78	0.138
Noise*ID	2	0.00100678	0.00100678	0.00050339	68.78	0.014
Noise*SC	2	0.00000313	0.00000313	0.00000156	0.21	0.824
Noise*RLC	2	0.00000088	0.00000518	0.00000259	0.35	0.739
Noise*Weld	1	0.00002577	0.00002577	0.00002577	3.52	0.201
Error	2	0.00001464	0.00001464	0.00000732		
Total	17	0.01123629				

To determine if we have any control factors in a parameter design, we must examine any significant interactions between system parameters and noise factors. In this experiment there is only one significant interaction, and it can be interpreted by examining the interaction graph shown in Figure 18.7. Here it can be seen that by using the low level (=1) for the inside diameter (IC) appears to have a mitigating influence on the effect of the noise factor. Therefore IC appears to be a control factor and its nominal level should be set to its low level in order to reduce variation in torque caused by variation in inner and outer sources of noise.

The next step in the interpretation of the parameter design experiment is to search for scale factors that can be used to bring the product performance to the desired level. This is done by identifying significant main effects for system parameters that do not have significant interactions with noise factors. In this experiment there are two system parameters that have significant main effects: IC and SC. These main effects are shown in Figure 18.8. SC is the scale factor since in this experiment it has a significant main effect but does not interact with the noise factor.

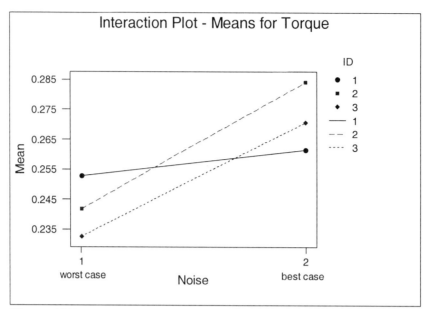

Figure 18.7 Interaction Between IC and Noise

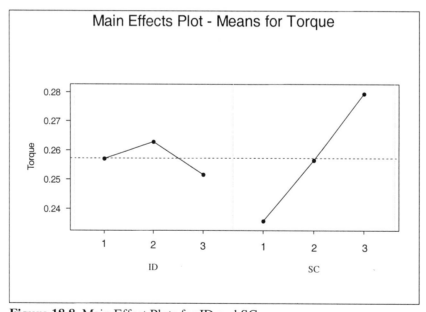

Figure 18.8 Main Effect Plots for ID and SC

The purpose of the experiment was to determine nominal levels of the system parameters that would make the torque consistently greater than .25 kg. Therefore, as can be seen in Figure 18.8, the spring constant (SC) should be set to its high level (=3). The main effect for ID indicates that a slightly higher average torque would at the mid (=2), but as can be seen in Figure 18.7 this would also cause increased variability. Therefore the levels of the parameters determined from the experiment would be ID = 1 and SC = 3. All other system parameters would be set at their least expensive levels since results of the experiment did not indicate that they have any effect.

Using a mathematical model developed from the data, predictions of the expected torque and variance of torque at the can be made at the recommended conditions. Table 18.5 shows the mean torque, at each level of the two significant factors, three-level factors, and the associated effects which were calculated as described in Chapter 11 by subtracting the overall mean torque of 0.2565.

Table 18.5 Means and Effects

Factor	Level	Mean	Effect
ID	1	.2564	-.0005
	2	.2622	.0057
	3	.2509	-.0056
SC	1	.2368	-.0197
	2	.2542	-.0023
	3	.2784	.0219

Because of the interaction, the effect of the noise factor depended upon the level of ID. Table 18.6 shows the means for both levels of noise at all three levels of ID. Since noise is a two-level factor the differences of the means at each level of ID was computed and divided by 2 to calculate the slope, as explained in Chapter 9.

Table 18.6 Means and Effects for Noise

	Noise Level			
Level of ID	Worst (-)	Best (+)	Difference (effect)	Slope (β)
1	.2515	.2613	.0098	.0049
2	.2405	.2839	.0434	.0217
3	.2313	.2705	.0392	.0196

The mathematical model is written below as a combination of the slope intercept form of the model for two-level factors shown in Chapter 9, and the subscripted effects form of the model shown in Chapter 11, for multilevel factors.

$$Y_{ijk} = \mu + A_i + B_j + \beta_i X + \varepsilon_{ijk} \qquad (18.1)$$

In this model, Y_{ijk} is the torque at the ith level of ID and the jth level of SC; μ is the grand average; A_i is the ith effect for ID; B_j is the jth effect for SC; β_i is the slope for the noise factor effect at the ith level of ID; X represents the coded level (- or + as in Chapter 9) of the noise factor; and ε_{ijk} represents the unexplained experimental error. The predicted torque at the recommended levels (i.e., ID=1 and SC=3) would be:

$$Y_{13} = \mu + A_1 + B_3 + \beta_1 X$$

$$= .2565 - .0005 + .0219$$

$$= 0.2779$$

which was found by substituting the grand average and estimated effects into model 18.1. The noise factor, X, was set to zero in this prediction since it should be somewhere between the worst (-) and best (+) case on the average.

Since Model 18.1 is a linear equation, the predicted variance of torque can be found using Formula 16.2 from Chapter 16. It is:

$$Var(Y_{ijk}) = \beta_i^2 \sigma_N^2 + \sigma^2 \qquad (18.2)$$

where σ_N^2 is the variance of the noise factor, and σ^2 is the variance of the experimental error. Using this equation, the variance of torque at the recommended conditions is

$$Var(Y_{13}) = \beta_1^2 \sigma_N^2 + \sigma^2$$

$$= .0049^2 \sigma_N^2 + \sigma^2$$

Evaluating this with the estimate $\hat{\sigma}^2 = .00000732$ obtained from the Error MS line in Table 18.4, and assuming that $\sigma_N^2 = 1$ (since the coded noise factor X varied between -1 and +1 in the experiment) the result is .00003133. Thus the predicted mean and standard deviation of torque at the recommended conditions of ID=1, SC=3 are: $\hat{\mu} = .2779$, $\hat{\sigma} = .0059$. A similar prediction made at ID=2, SC=1 results in a slightly higher predicted mean torque $\hat{\mu} = .2841$, but a greatly increased predicted standard deviation $\hat{\sigma} = .0218$. Thus, to ensure the maximum probability of the torque being greater than .25, we can see the wisdom of the recommended system parameter nominals.

Normally follow-up tests are done after the parameter design experiment to confirm the predicted results. This was done by building 100 more starting motors with the recommended levels of the system parameter levels determined above. The average torque of the starting motors in the confirmation test was $\bar{y}=0.278$ with a standard deviation of $s=0.007$; close to what was predicted from the parameter design experiment.

18.5.3 Inner Outer Array Experiments

An alternative to the combined array parameter design experiment is the *inner outer array experiment*. Using the inner outer array experiments, two experimental designs are employed. One design called the inner array is an experimental design in the system parameters. The other design called the *outer array* is an experimental design in the noise factors. The outer array, or noise array, experiment is repeated for each run in the inner array creating a set of quasi replicates for each run as shown in Figure 18.9. In this figure the inner array is a 2^{4-1} design in system factors A–D, and the outer array is a 2^2 factorial in the two noise factors, r and s. Thus, there are essentially four replicate experiments at each combination of levels of the system parameters for a total of 32 experiments. The replicate runs differ by changes in the levels of noise factors.

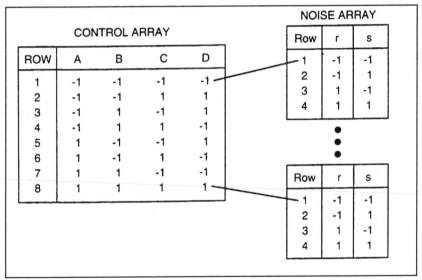

Figure 18.9 Diagram of Inner Outer Array Experiment

The data from an inner outer array experiment could be analyzed exactly as was shown in the last section for combined array experiments. Simply repeat each line of the inner array, one line corresponding to each run in the outer array, to create a combined array. Since each system parameter level will be repeated for each level of each noise factor, all interactions between system parameters and noise factors can be estimated and the analysis can proceed exactly like the examples in Chapter 9, if all factors have only two levels as shown in Figure 18.9, or like the examples in Chapters 11 if multilevel factors are used.

However, Taguchi has demonstrated a simpler method of analysis for inner outer array experiments. The procedure involves two steps. First, one or more summary, or performance, statistics, are calculated from the set of quasi replicates at each level of the inner array. Second, the summary statistics are treated as responses for the inner array and an appropriate analysis is made. One choice of summary statistics from each set of quasi replicates that is often used is the sample mean $\bar{y}$, and the sample variance, s^2. For example, consider the two copper plating experiments presented in Chapter 12. Both of these experiments were examples of inner outer arrays. The inner array for the printed circuit board example was a Plackett-Burman design, and the outer array was three levels of a single noise factor as illustrated in Figure 12.4. The levels of this noise factor simulated the variational noise in the manufacturing process. The summary statistics analyzed in this experiment were the mean, and log of the sample variance. In the example of copperplating ceramic substrate circuit boards at Motorola, the inner array was a 2^{3-1} design and the outer array consisted of replicate circuit boards coated under the same conditions. This again simulated the variational noise. In this experiment the summary statistic analyzed was the sample variance, s^2.

Other summary statistics that could be used in inner outer array experiments are the loss functions defined in Table 18.2. This would allow prediction of the system parameter nominal values that would minimize the appropriate loss function. The problem with using loss functions directly is that the constant k is unknown in most practical situations. To remedy the situation Taguchi, has devised other performance statistics called S/N ratios. These statistics are shown in Table 18.7.

Table 18.7 Taguchi's S/N Ratios

Characteristic	S/N Ratio
Nominal is Best	$10 \ \log_{10}(\bar{y}^2/s^2)$
Smaller the Better	$-10 \ \log_{10}(\sum y_i^2/n)$
Larger the Better	$-10 \ \log_{10}(\sum (1/y_i^2)/n)$

These statistics are expressed in standard units called decibels, and should be maximized as a function of the system parameter levels in order to satisfy the desired characteristic. For the Nominal is Best case, the sample mean is also analyzed separately. The two goals in that situation are to find the combination of system parameter levels that will maximize the S/N and bring the mean to the desired target value. In some cases these goals are contradictory and a compromise must be reached. To illustrate the simplicity of the analysis of inner outer array experiments we present an example in the next section.

18.5.4 An Example of an Inner Outer Array Experiment

This example of an inner outer array experiment is patterned after a report by Kackar and Shoemaker[8] of an experiment in integrated circuit fabrication. One of the first steps in an integrated circuit production process is the growth of added layers of silicon on top of silicon wafers. It is most important to grow layers of uniform thickness at this step, because electrical devices will be formed within these layers during later processing steps. The silicon layers are grown in a reactor where the wafers are mounted on the sides or facets of a rotating fixture called the *susceptor* that is shown in Figure 18.10.

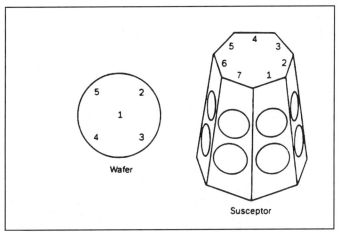

Figure 18.10 Susceptor for Holding Wafers in Reactor

The silicon layers are produced in the reactor when the temperature is elevated, and gasses are introduced. The gasses deposit silicon on the wafers and the layers grow until they reach their target thickness, then the process is stopped by lowering the temperature. There were four easily controllable system parameters in the production process. They were susceptor rotation method, deposition temperature, deposition time, and nozzle position. Deposition rates were not uniform within the reactor, therefore sampling different possible positions for wafers on the susceptor simulated within batch variational noise in the deposition process. The system parameters and noise factors and levels for the experiment are shown in Table 18.8.

Table 18.8 Factors and Levels for Silicon Deposition Experiment

Factor	Level	
System Parameters:	-	+
A. Rotation Method	Continuous	Oscillating
B. Deposition Time	Low	High
C. Deposition Temperature	1210	1220
D. Nozzle Position	2	6
Noise Factors:		
L. Susceptor Location	Top	Bottom
F. Facet	1	4

The inner array design for the system parameters was a 2^{4-1} fractional factorial with defining relation I=ABCD. The outer array is a 2^2 factorial in the two noise factors. The combined arrays are shown along with the response (silicon layer thickness) in Table 18.9.

Table 18.9 Inner Outer Array Design and Response Data

					Outer Array			
	Inner Array				L: -	+	-	+
Run	A	B	C	D	F: -	-	+	+
1	-	-	-	-	15.2957	14.3671	15.3472	14.1583
2	+	-	-	+	14.9497	14.7196	15.1063	14.9355
3	-	+	-	+	15.2179	14.9259	14.9439	14.6561
4	+	+	-	-	15.1984	13.9726	15.3102	14.2283
5	-	-	+	+	14.1731	14.1406	14.1711	13.9868
6	+	-	+	-	14.4655	13.2908	14.4904	13.4760
7	-	+	+	-	14.5811	13.4928	14.4518	13.1660
8	+	+	+	+	13.9321	13.9257	14.2042	14.0482

The four quasi-replicates for each run in the inner array were summarized by computing the mean and Nominal is Best S/N ratio, as shown in Table 18.10.

Table 18.10 Sample Means and S/N Ratios

Run	Mean	$S/N = 10\log_{10}(\bar{y}^2/s^2)$
1	14.7921	27.5872
2	14.9278	39.4571
3	14.9360	36.2703
4	14.6774	26.7359
5	14.1179	44.0411
6	13.9307	26.8034
7	13.9229	25.9703
8	14.0275	40.6270

The calculated effects for the inner array for each of the two performance statistics are shown in Table 18.11, and normal plots of the absolute effects are shown in Figure 18.11.

Table 18.11 Effects for Performance Statistics

Factor	Effect for Mean	Effect for S/N
A	-.0514	-.0614
B	-.0512	-2.0714
C	-.8336	1.8474
D	.1715	13.3246
AB+CD	-.0256	2.6225
AC+BD	.0101	-1.2291
AD+BC	.0021	-.0522

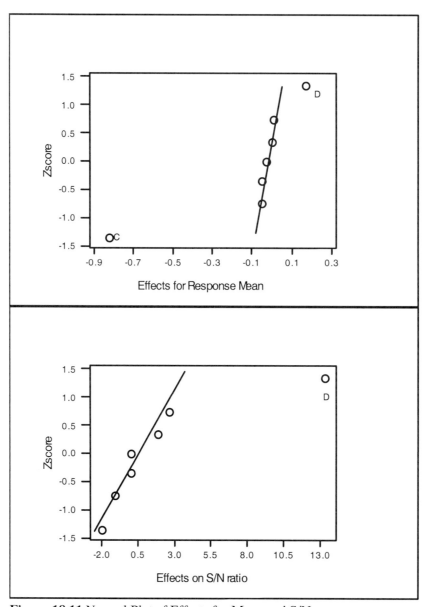

Figure 18.11 Normal Plot of Effects for Mean and S/N

It can be seen that main effect D (nozzle position) has a large positive effect on the S/N ratio. Therefore, nozzle position should be set to its high level (=6) to maximize the S/N ratio. It can also be seen that Factors D and C (deposition time) have significant effects on the average thickness. A mathematical model for the mean thickness using only significant effects is:

$$Y = 14.4165 - .4168\ X_C + .0858\ X_D$$

where 14.4165 is the grand mean, -.4168 is half the deposition temperature effect, .0858 is half the nozzle position effect, and X_C and X_D are the coded factor levels. If the high level of Factor D is chosen to maximize S/N ratio, this model simplifies to:

$$Y = 14.5023 - .4168((D.Temp-1215)/5)$$

and solving for D.Temp in this equation will determine the deposition temperature required to grow layers of a target thickness, Y.

18.5.5 Parameter Design Experiments on Paper

In some engineering designs no physical experimentation is necessary because there is already enough theoretical knowledge available. For example, in electrical engineering, circuit approximation equations exist. In other fields, computer simulation models or finite element analysis models can be used to obtain the information needed to optimize a product design. Even though no physical experiments are required in these situations, the concepts of parameter design can still be fruitfully employed. To understand the idea, consider the following simple example that is given by Taguchi[9] in an experimental design class.

If alternating current with a voltage of 100V and frequency of 50 Hz is applied to a circuit with a resistance of R ohms and self-inductance of L henrys, the current flow, y, in amps is given by:

$$y = \frac{100}{\sqrt{R^2+(2\pi\times50L)^2}} \qquad (18.3)$$

The customer requirements for the current flow are 10.0 ±4.0 amps. In this situation it is easy, using a spreadsheet or calculator solver program, to find a combination of R and L which will result in y=10 amps. In fact there are an infinite number of solutions to the equation y = 10, thus it is easy to meet the customer objective for nominal current. Nevertheless, this does not completely solve the problem. The design engineer specifies values for R and L, but when resistors and capacitors are manufactured and assembled in circuits, they will vary from the nominal values due to variational noise. When the circuits are in customer use, the values of R and L will drift further from the nominal values due to inner and outer noise. The differences from the nominal values of the components will alter the resulting current. The question is how to select the nominal R and L to ensure the current will remain within the specifications 10.0 ± 4.0 amps despite the effects of normal variational noise, and inner and outer noise. This question can be answered using the parameter design approach. For example, an inner outer array could be set up varying the nominal values of R and L over three levels in the inner array (such as 7, 9, and 11 for R and .003, .004, and .005 for L) . An outer array could be constructed by varying the deviations from the nominal specifications (i.e., ± 20%). Next, the resulting current can simply be calculated using Equation 18.3, rather than performing physical experiments (i.e., building and measuring prototype circuits). Table 18.12 shows the inner outer table. The body of the table shows the R and L values obtained from the marginal inner outer arrays along with the current computed from Equation 18.3.

Table 18.12 Inner Outer Array for Current Parameter Design

R	L	Nominal -20% / Nominal -20%	Nominal +20% / Nominal -20%	Nominal -20% / Nominal +20%	Nominal +20% / Nominal +20%	Mean	S/N
7	.003	R=5.6 L=.0024 y=17.7	R=8.4 L=.0024 y=11.9	R=5.6 L=.0036 y=17.5	R=8.4, L=.0036 y=11.8	14.725	-12.9357
9	.003	R=7.2 L=.0024 y=13.8	R=10.8 L=.0024 y=9.2	R=7.2 L=.0036 y=13.7	R=10.8 L=.0036 y=9.2	11.475	-12.8050
11	.003	R=8.8 L=.0024 y=11.3	R=13.2 L=.0024 y=7.6	R=8.8 L=.0036 y=11.3	R=13.2 L=.0036 y=7.5	9.425	-12.7747
7	.004	R=5.6 L=.0032 y=17.6	R=8.4 L=.0032 y=11.8	R=5.6 L=.0048 y=17.2	R=8.4 L=.0048 y=11.7	14.575	-12.9909
9	.004	R=7.2 L=.0032 y=13.8	R=10.8 L=.0032 y=9.2	R=7.2 L=.0048 y=13.6	R=10.8 L=.0048 y=9.2	11.45	-12.8788
11	.004	R=8.8 L=.0032 y=11.3	R=13.2 L=.0032 y=7.6	R=8.8 L=.0048 y=11.2	R=13.2 L=.0048 y=7.5	9.4	-12.8666
7	.005	R=5.6 L=.004 y=17.4	R=8.4 L=.004 y=11.8	R=5.6 L=.006 y=16.9	R=8.4 L=.006 y=11.6	14.425	-13.2044
9	.005	R=7.2 L=.004 y=13.7	R=10.8 L=.004 y=9.2	R=7.2 L=.006 y=13.4	R=10.8 L=.006 y=9.1	11.35	-12.9909
11	.005	R=8.8 L=.004 y=11.2	R=13.2 L=.004 y=7.5	R=8.8 L=.006 y=11.1	R=13.2 L=.006 y=7.5	9.325	-12.9167

Using the method of least squares the following two equations were fit to the performance statistics shown at the right of Table 18.14

$$\bar{y} = 11.439 - 2.596X_1 - .088X_2 + .554X_1^2 - .021X_2^2 + .050X_1X_2 \qquad (18.4)$$

$$S/N = -12.87 + .1X_1 - .1X_2 - .06X_1^2 - .03X_2^2 + .03X_1X_2 \qquad (18.5)$$

where $X_1 = (R-9)/2$ and $X_2 = (L-.004)/.001$ are the coded and scaled values for the systems parameters. Figure 18.12 shows contour plots of the fitted equations. In These figures it can be seen that the maximum S/N along a constant contour for the mean is near the bottom of the graph.

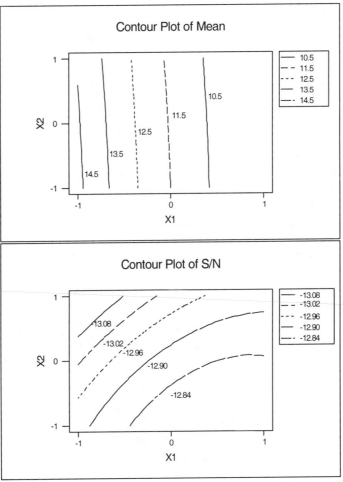

Figure 18.12 Contour Plots

Therefore, the most robust combination would be near $X_1 = .66$ or $R=10.32$, and $X_2 = -1$ or $L=.003$, that result in predicted values of $\bar{y}=10.06$, and $S/N=-12.678$ from Equations 18.4 and 18.5. The validity of this solution should be checked by plugging it back into Equation 18.3.

Lawson and Madrigal[10] show that with the help of a computer and appropriate software more accurate solutions to parameter design problems of this kind can be made using Nonlinear Optimization and Monte Carlo Simulation.

18.5.6 Planning Parameter Design Experiments

There are several things that should be considered when planning, analyzing and interpreting data from parameter design experiments. We will first discuss design considerations. The purpose of parameter designs is to find robust system designs. Products should function, and production processes should produce acceptable product despite normal changes in environmental factors and wear of internal components. Therefore the most crucial step in a parameter design experiment is to identify the appropriate system parameters and noise factors that can be used in the experiment. System parameters should be chosen that are easily varied and controlled in the product design specifications, or production process design. From these parameters the control factors and scale factors will be identified, and proper choice of their levels will robustify the design.

Identifying noise factors, and levels that can be varied and controlled during an experiment, may be more difficult than choosing system parameters. In some cases the noise factors are uncontrollable, yet their effects can be seen through a surrogate noise factor. For example in the integrated circuit example in Section 18.5.4, the cause of the within batch variation noise was unknown and uncontrollable, but by sampling from different locations and facets on the susceptor, a surrogate noise factor was created which would allow the effects of the true noise to be seen in the experiment.

The noise factors that will be used in a parameter design experiment, should guide the choice of a proper design. When changing the levels of the noise factors requires physically building additional prototypes, as was done in the motorscooter starting motor experiment, a combined array experiment may be the best design choice because, as Song and Lawson[11] explain, combined arrays usually require less experimentation. On the other hand, when noise factors consist of simply sampling over time or space as was done in the integrated circuit example, or in parameter design experiments conducted on paper or through computer simulation, the inner outer array design should be used because it will provide more information.

Estimation of interactions between system parameters is desirable when using either inner outer array designs or combined array designs. In the motoscooter example, the L_{18} design allowed estimation of interactions between system parameters and noise factors, but there were not enough runs to also estimate the interactions between system parameters. Therefore the particular L_{18} design used for that experiment was less desirable than the fractional factorial design used for the inner array in the silicon wafer experiment. The 2^{4-1} inner array for the wafer experiment did allow estimation of confounded pairs of two-factor interactions between system parameters. Fitting an additive model (that excludes interactions between control factors) when interactions do exist, will lead to suboptimal choices of nominal values of the system parameters as explained by Lawson and Helps[12]. Thus, inner arrays or combined array designs should be chosen that allow estimation of interactions between the system parameters. Good designs for doing this are the resolution V fractional factorials given in Tables B.1-9 or B.1-13, or any of the response surface designs given in Tables B.3-B.5. The L_{18} which allows estimation of quadratic and interaction effects is a popular choice for the inner array, in inner outer array experiments.

When analyzing data from parameter design experiments, the first consideration should be the characteristic you are trying to optimize. Is a target or nominal response best, or is consistently large or small results best? In inner outer array experiments the characteristic to be optimized will guide the choice of the appropriate performance statistics or S/N ratio. In combined array experiments, performance statistics are not calculated from the raw data because there is no outer array of quasi replicates. However, once a model is fit to the data the value of any performance statistic can be evaluated or predicted at any combination of system parameter settings. For example, in the combined array experiment presented in Section 18.5.2 , above, the mean torque and variance of torque was predicted (for IC=1 and 2 with SC=3) using the linear model fit to the raw data. Just as easily, a S/N performance statistic could have been predicted from a fitted model.

Once the appropriate performance statistics can be predicted as a function of the system parameters (control and scale factors) then the optimal combination can be obtained. Sometimes this optimization is a trivial task and can be accomplished by simple inspection or graphs of the prediction equation, as illustrated in the integrated circuit example above. In other cases where the models are more complicated, possibly involving quadratic and interaction terms, optimization can be accomplished through nonlinear programming methods. Most standard spreadsheet programs such as LOTUS 1-2-3, EXCEL, or QUATRO PRO have these algorithms built into a solver or optimizer tool.

One controversy that has been discussed frequently in the statistical literature is the validity of fitting models to performance statistics in inner outer array experiments. One of the assumptions that is made, when fitting a model to data by least squares, is that the variance of the response is constant at differing values of the independent variables. If a model can be fit to the $\log(s^2)$ or S/N ratio, it implies that the variance is not constant. Therefore, the least squares assumption is invalid when fitting a model to the mean response. A second assumption of normality of the data, justifies the use of t- and F-statistics in regression and ANOVA to determine significant effects. Although means would be considered Normally distributed due to the Central Limit Theorem, it is unclear whether S/N ratios can be considered Normally distributed. Residual plots should be checked when fitting these models and weighted least squares; or better yet Generalized Linear Models (see McCullagh and Nelder[13]) can be fit when the appropriate software is available to do so. When improved techniques like this are used for model fitting, often times additional significant control and scale factors may be identified leading to a better solution. For engineers working without the help of specialized statistical software, the simple approach of analyzing means and S/N ratios calculated directly from the quasi replicates is the simplest approach and usually leads to an adequate solution. An alternative simple approach is to analyze the data as a combined array fitting a model to the raw data. Then optimal system parameter values are determined through predicted performance statistics.

18.6 Tolerance Design

The choice of nominal values for the control and scale factors in parameter design reduces the variation in product function with little or no cost. If further reduction is required, a trade-off must be made between increased cost of the product and higher quality. This is the role of tolerance design. Tolerance design should only be conducted after parameter design is complete and S/N ratios have been maximized.

The first step in performing a tolerance design is to conduct an experiment. In some cases this could be an experiment on paper or through computer simulation, in other cases it will involve physical experiments. During the experiment, the control and scale factors are held constant (at their optimal settings determined during parameter design) and the design is created by varying the noise factors. From this experiment the contribution of the noise factors to variance in the product function response can be determined. This is done by expressing the variance in the response, σ^2, as a linear combination of variances of the noise factors as in Equation 18.6:

$$\sigma^2 = a\,\sigma^2_1 + b\,\sigma^2_2 + \qquad\qquad\qquad (18.6)$$

The next step in the tolerance design is to determine the proportion of σ^2 accounted for by each noise factor. For example, the proportion of σ^2 accounted for by noise factor 1 would be $(a\sigma^2_1/\sigma^2)$. To reduce σ^2, we must reduce the variance of the largest contributing noise factors, by holding tighter component tolerances or buying more expensive components. The added cost of doing this can sometimes be offset by substituting lower quality (i.e., with larger variances) for the noise factors that contribute little.

When noise factors cannot actually be controlled, and their effects are simulated by sampling, the nested designs presented in Section 16.3 can be used effectively for estimating the variance components in Equation 18.4. For example, once nominal values are chosen for the inside diameter (ID) and spring constant (SC) for the motorscooter starting motor, a tolerance design could be conducted by sampling I casings produced at the nominal, and IxJ springs. Next IxJ prototype starting motors could be assembled and tested according to the nested design plan like that shown in Figure 16.2. In this design, casing would replace batch on the top level, and spring would replace replicate measurement on the second level. Once the data is collected, an ANOVA should be performed, and the variance components estimated as shown in the example in Section 16.2.2.

When noise factors can be controlled during experimentation, simple two-level factorial or fractional factorial designs will allow fitting a standard slope intercept model to the data like Equation 18.7:

$$y = b_0 + b_1 X_1 + b_2 X_2 + \ldots + b_n X_n \qquad\qquad\qquad (18.7)$$

Coded noise factor levels, X_i, should be chosen one standard deviation above and below their nominal so that the variance in the response, y, can be expressed directly as a linear combination of the variances in the noise factors i.e.,

$$Var(y) = b_1^2 Var(X_1) + b_2^2\,Var(X_2) + \ldots + b_n^2\,Var(X_n)$$

$$= b_1^2\,\sigma^2_1 + b_2^2\,\sigma^2_2 + \ldots + b_n^2\,\sigma^2_n$$

as shown in Section 16.1.

18.7 Summary and Discussion

In this chapter we have introduced the Taguchi's definition of quality as loss caused by functional variation in products. We have presented the methods of system design, parameter design and tolerance design for reducing quality losses and we have emphasized the cost efficiency of counteracting noise sources through parameter design.

Taguchi's ideas for quality improvement through engineering design have been quite controversial since they were first introduced in the United States in the early 1980s. Generally, Taguchi is credited with bringing to the forefront the idea that it is more effective and efficient to accomplish quality improvement at the engineering design stage than it is in manufacturing. Also he is credited with drawing attention to the idea that product performance can be designed to be robust to environmental noise and manufacturing imperfections, and that experimental design methods such as those described in Chapters 7 - 14 can be used to find combinations of control factors that will minimize variation. However, some of the statistical experimental designs and methods of analysis first proposed by Taguchi have been criticized as inefficient, outdated, or incorrect. In our presentation of the ideas in this chapter some of these criticisms have already been mentioned, or improved methods have been presented instead. A few remaining criticisms are described below.

In terms of data analysis, use of Taguchi's Smaller the Better S/N ratio, and Larger the Better S/N ratio have been shown to be misleading. It is probably always better to analyze the mean and either the variance, or Nominal is Best S/N ratio, separately in all cases. In terms of design, inner outer array designs have been called inefficient in terms of the number of experiments that must be run, but as explained in Sections 18.5 and 18.6 when the outer array consists of replicate measurements or experiments on paper, the number of experiments is not a concern. Box has criticized Taguchi's experiment methods because the idea of a parameter design experiment seems to ignore the iterative nature of most experimental investigations, where we start at a screening stage and build up knowledge. It should always be remembered that things go wrong in experimentation, and we often need to repeat designs as an experimental program progresses. Therefore even in parameter design experiments no more than 25% of our resources should be expended on the initial experiment.

Below is a list of the important concepts in this chapter:

References

1. Deming, W. E. *Out of the Crisis*, (Cambridge MA: M.I.T. Press)

2. Taguchi, G., *Introduction to Quality Engineering*, Asian Productivity Organization, Tokyo 1986.

3. Taguchi, G., Ibid.

4. Phadke, M. S., *Quality Engineering Using Robust Design*, (Englewood Cliffs, N.J.: Prentice Hall, 1989).

5. Hauser, J. R. and Clausing, D., "The House of Quality", *Harvard Business Review*, Vol. 66, No. 3, 1988, pp. 63 - 73.

6. Ross, P., *Taguchi Techniques for Quality Engineering*, (New York: McGraw- Hill, 1988).

7. Wang, Y., from personal communication.

8. Kackar, R. and Shoemaker, A. "Robust Design: A Cost Effective Method for Improving Manufacturing Processes," *AT&T Technical Journal*, 1986.

9. Taguchi, G., Ibid.

10. Lawson, J. S. and Madrigal, J. L., "Robust Design Through Optimization Techniques," *Quality Engineering*, Vol 6, 1994, pp 593-608.

11. Song, J. and Lawson, J. "Use of 2^{k-p} Designs in Parameter design", *Quality and Reliability Engineering International*, Vol. 4 1988.

12. Lawson, J. S. and Helps, R,. "Detecting Undesirable Interactions in Robust Design Experiments," *Quality Engineering*, Vol. 8, No. 3., 1996, pp. 465 - 473.

13. McCullagh, P., and Nelder, J. A., *Generalized Linear Models* second edition, (London: Chapman & Hall, 1984).

18.8 Exercises for Chapter 18

18.1 Describe some of the important sources of noise (i.e., inner noise, outer noise, and variational noise) upon functionality of the following products:

   a) A VCR

   b) A pair of basketball shoes

   c) Mechanical pencil

18.2 Name the appropriate loss function that should be used to measure functional variation for the following systems:

   a) The response time of a computer network

   b) The bond strength of an adhesive product

   c) The output voltage of a power supply circuit used for a stereo. (Note: Half the circuits fail when the output voltage is outside the range 110 ± 20.)

18.3 In the late 1940s Taguchi[1] assisted a company that produced caramel candy. The plasticity (chewability) of the company's candy varied greatly with the ambient temperature. This temperature dependence was reduced by experimenting with the ten ingredients. Describe what step of the quality engineering process this was, and identify specifics such as noise factors, etc.

18.4   a) What is the purpose for studying three levels for each factor in parameter design experiments?

   b) What is the purpose for estimating interactions between noise factors and system parameters in parameter design experiments?

18.5 There are two approaches to parameter design experiments: combined array and inner outer array. Which would be appropriate in each of the following situations?

   a) Experiments on paper where levels of noise factors and control factors can be changed at will and the experimental result is obtained by evaluating a formula or running a computer algorithm.

---

[1]Taguchi, G. (1986) *Introduction to Quality Engineering*, Asian Productivity Organization, Tokyo.

b) Assembly and test experiments where components at both ends of the tolerance spectrum, and of different designs can be combined together, and the experimental result will be obtained by doing physical tests on the assembled prototype.

c) Manufacturing experiments where noise factors can be varied by sampling at different points of time or space.

18.6 List two strategies for selecting the control factor levels to reduce noise factor effects after completion of a parameter design experiment.

18.7 A manufacturer decides to buy springs from one of three suppliers: C_1, C_2 or C_3. The springs can be mounted using one of three different fixtures: F_1, F_2, and F_3. The objective is to control the extension of the springs in the manufacturer's product to a nominal or target value with minimum variation. Noise factors that can be controlled during a test are 1) Load (10 gram force to 20 gram force) 2) Temperature (-30°C , +70°C), and 3) Deterioration test on the springs (yes or no).

a) Using a 3×3 factorial for the inner array, set up the list of conditions for a inner outer array experiment to solve this problem (i.e., choose the supplier and fixture)

b) What S/N ratio do you propose to use with your inner outer array experiment?

c) Use the L18 orthogonal array to design a combined array to solve the problem.

18.8 The following inner outer array experiment is similar to one conducted by Zhou and Cao[2] to reduce the dimensional variation of various features on a stamped automobile inner door panel. The dimensions of the features were measured on CMM. Three system parameters in the stamping process were varied by using the L_{18} design.

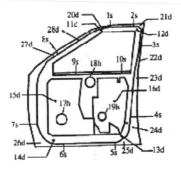

[2]Zhou, Z, and Cao, X., "Optimal Process Control in Stamping Operation," *Quality Engineering*, Vol. 6, No 4, (1994), pp. 621 - 631.

These parameters were the punch speed (PS), the inner tonnage that deforms the sheet metal to the form of the die (IT), and the outer tonnage (OT), which is responsible for holding the sheet metal in place while being deformed. The within run process variation (noise factor) was simulated by measuring one of the dimensions shown in Problem 18.8 on each of 9 stamped door panels through the course of a production run. This was the outer array. The results are shown below.

PS	IT	OT	Replicate Measurements	$\overline{Y}$	s
1	1	1	50.44 49.98 50.09 49.73 50.54 49.84 50.12 49.98 49.85	50.06	0.2728
1	1	2	50.00 49.99 49.99 50.00 49.99 49.99 49.99 50.00 49.99	49.99	0.0002
1	1	3	49.66 50.10 50.31 50.34 49.89 50.39 49.74 49.76 50.23	50.05	0.2874
1	2	1	50.45 49.91 51.41 49.26 49.26 49.60 49.91 51.11 48.59	49.94	0.9122
1	2	2	50.11 49.96 50.10 49.99 49.86 49.77 49.96 50.06 50.03	49.98	0.1122
1	2	3	49.45 51.00 48.93 51.47 48.31 50.40 49.93 48.57 50.16	49.80	1.0808
1	3	1	38.67 48.70 38.43 60.31 50.60 76.00 45.32 71.65 36.22	51.77	14.5732
1	3	2	48.52 50.96 45.58 52.54 41.75 54.18 54.62 51.86 51.82	50.20	4.2192
1	3	3	35.67 42.75 57.73 59.70 68.60 36.58 38.67 60.75 68.59	52.12	13.63
2	1	1	49.10 49.43 49.89 50.65 50.80 49.63 49.90 50.93 49.10	49.94	0.7061
2	1	2	49.99 50.08 49.96 49.92 49.97 49.94 49.86 49.95 49.99	49.96	0.0582
2	1	3	51.16 50.22 50.62 49.95 50.17 48.64 49.67 50.28 49.13	49.98	0.7590
2	2	1	45.30 50.31 52.71 52.26 52.27 49.22 53.38 53.04 47.50	50.67	2.8130
2	2	2	49.83 49.18 49.59 50.38 50.13 50.23 49.64 49.41 49.88	49.81	0.3937
2	2	3	49.57 50.07 47.11 51.98 50.89 50.96 50.33 48.39 50.96	50.03	1.4924
2	3	1	72.96 73.92 30.57 30.96 78.43 34.09 56.31 55.19 48.33	53.42	18.92
2	3	2	47.64 29.01 66.32 57.22 50.80 65.57 44.79 53.20 64.60	53.24	12.06
2	3	3	45.68 50.18 83.48 34.84 43.55 54.72 41.45 44.78 47.95	49.63	13.8581

If the goal is to achieve a target dimension with minimum variation,

a) Calculate the appropriate S/N ratio for each run in the inner array

b) Perform an ANOVA using a GLM program to fit the model:

$$y_{ijk} = \mu + P_i + I_j + O_k + PI_{ij} + PO_{ik} + IO_{jk}$$

to the data for $\overline{Y}$ and S/N. Determine which main effects and interactions are significant.

c) Using calculated means for each level of significant main effects, interpret the models you fit and select optimum levels for PS, IT, and OT to achieve target dimension with minimum variation.

18.9 Reconsider the impedance equation given in Chapter 16.

$$Z = \frac{87.0}{\sqrt{\varepsilon} + 1.41} \ln[(5.98A) / (0.8B + C)]$$

where Z is the impedance, A is the insulator thickness, B is the line width, and C is the line height as shown in Figure 16.1 This time consider the design implications. The customer requirement is to have the $Z = 85\ \Omega$ with minimum variation. There is an infinite number of combinations of A, B, and C that will result in $Z = 85\ \Omega$, but all will not result in the same variance of Z. The designer can choose values of A, B, and C from the table of feasible ranges given below. Once nominal values are chosen for A, B, and C the actual values for any circuit will vary due to variational noise in manufacture. The standard deviations shown in the table below represent this variational noise, and the low cost tolerance represents ± 3 standard deviations.

System Parameter	Feasible Range	Standard Deviation	Low Cost Tolerance
Thickness A	20 - 30 μm	0.333	$\pm 1.0\ \mu m$
Width B	12.5 - 17.5 μm	0.222	$\pm 0.67\ \mu m$
Height C	4 - 6 μm	0.111	$\pm 0.33\ \mu m$

a) Set up 2^3 design as in the feasible range of A, B, and C as an inner array.

b) Set up a 2^3 outer array in deviations (low cost tolerance) from the nominal settings in the inner array.

c) Perform a parameter design experiment by evaluating the impedance equation above, for each of the 64 combinations in the inner outer array you have created (similar to Table 18.14).

d) Calculate the mean impedance and appropriate S/N ratio for each run in your inner array.

e) Fit the full factorial model to your data for mean and S/N and use the methods of Chapter 8 to determine which main effects and interactions are significant.

f) Interpreting the significant terms for mean and S/N only, determine the nominal settings for A, B, and C to satisfy the customer requirements.

18.10 The diagram below[3] shows a simple circuit diagram for a temperature controller which uses the resistance thermometer R_T.

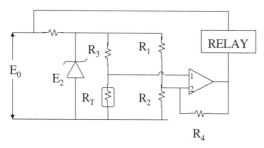

The resistance value of R_T at which the relay turns on is given by

$$R_{T-ON} = \frac{R_3 R_2 (E_2 R_4 + E_0 R_1)}{R_1 (E_2 R_2 + E_2 R_4 - E_0 R_2)}$$

The scale factor is R_3, this is known by the circuit relations. The control factors are $A = R_1$, $B = R_2/R_1$, $C = R_4/R_1$, $D = E_0$, and $F = E_2$. The objective is to minimize the variation in the R_{T-ON} for fixed values of the control factors, since the nominal value can be adjusted by the signal factor R_3.

a) Set up an inner array for the control factors using the L_{18} design and the levels of the control factors given below.

Factor	1	Levels 2	3
A	2.67	4.0	6.0
B	1.33	2.0	3.0
C	5.33	8.0	16.0
D	8.00	10.0	12.0
F	4.80	6.0	7.2

b) Set up an L_{18} design for an outer array, to be used for each row of the inner array. The noise factors will be deviations from the nominal tolerances given in the inner array similar to the example in Section 18.5.5. The three levels of the noise deviations are -2.04% of the nominal, nominal and +2.04% of nominal value for each control parameter in the inner array.

---

[3]Taguchi, T. and Phadke, M. S. "Quality Engineering Through Design Optimization," *The International Quality Forum*, Vol. 11 No. 11, (1985), pp. 27 - 48.

c) Evaluate the equation for R_{T-ON} for each of the 18×18 combinations in the inner outer array design, and summarize the 18 replicates in the outer array for each row in the inner array using the Nominal is Best S/N ratio.

d) Use a GLM program to compute the sums of squares an mean squares in and ANOVA of the control factors using the S/N ratios you computed in c).

e) Based on the significant factors and examination of means, determine the optimal settings of the control factors that will maximize S/N.

18.11 Song and Lawson[4] illustrate the combined array approach to a parameter design experiment. The objective was to modify the installation procedure for an elastometric connector to increase its separation pull-of-force at the same time as minimizing the variability in the pull-off-force. The experiment had four system parameters that could easily be controlled in the production process, and three noise factors that were normally hard to control in the production process, but could be controlled during an experiment. The system parameters were: A = interference (or allowance between the force fit rubber elastometric connector and the metal sleeve it is connected to), B = wall thickness of the connector, C = the insertion depth of the connector on to the metal sleeve, and D = percentage of adhesive in the solution the connector is dipped into before insertion on the metal sleeve.

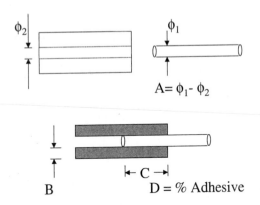

[4]Song, J. and Lawson, J., "Use of 2k-p Designs in Parameter Design," , *Quality and Reliability Engineering International*, Vol 4., (1988), pp.151 - 158.

The noise factors were E = the conditioning relative humidity, F = the conditioning temperature, and G = the conditioning time. The table below lists the levels for each factor:

Factor	-	+
A (interference)	Low	High
B (connecter wall thickness)	Thin	Thick
C (insertion depth)	Shallow	Deep
D (% Adhesive)	Low	High
E (conditioning Rel. Humid.)	25%	75%
F (conditioning temperature)	72°F	150°F
G (conditioning time)	24 hours	120 hours

The experiment and resulting Y = pull-off force, is shown in the table on the following page. This table is copied from Table B.1-11.

a) Calculate the effects and make a Normal plot as done in Chapter 8 to determine the significant effects. Use the table of confounding in Table B.1-11 to interpret the unassigned columns.

b) Determine if there are any control factors (i.e., system parameters that have significant interactions with noise factors), and if so, make an interaction plot to choose the level of the control factor that will minimize the effect of the noise factor.

c) Choose levels of the scale factors (i.e., system parameters that do not have significant interactions with noise factors) to be the level that will maximize pull-off-force.

Run	A	B	C	D	E	F	G	Pull Off Force
1	-	-	-	-	-	-	-	17.8
2	+	-	-	-	-	+	+	24.1
3	-	+	-	-	-	+	+	21.4
4	+	+	-	-	-	-	-	14.7
5	-	-	+	-	-	+	-	22.1
6	+	-	+	-	-	-	+	21.0
7	-	+	+	-	-	-	+	26.0
8	+	+	+	-	-	+	-	19.7
9	-	-	-	+	-	-	+	19.4
10	+	-	-	+	-	+	-	14.9
11	-	+	-	+	-	+	-	19.8
12	+	+	-	+	-	-	+	15.6
13	-	-	+	+	-	+	+	25.7
14	+	-	+	+	-	-	-	16.9
15	-	+	+	+	-	-	-	15.2
16	+	+	+	+	-	+	+	20.7
17	-	-	-	-	+	-	-	12.6
18	+	-	-	-	+	+	+	25.2
19	-	+	-	-	+	+	+	20.8
20	+	+	-	-	+	-	-	21.9
21	-	-	+	-	+	+	-	17.8
22	+	-	+	-	+	-	+	24.2
23	-	+	+	-	+	-	+	23.0
24	+	+	+	-	+	+	-	22.5
25	-	-	-	+	+	-	+	15.1
26	+	-	-	+	+	+	-	20.4
27	-	+	-	+	+	+	-	16.8
28	+	+	-	+	+	-	+	20.1
29	-	-	+	+	+	+	+	23.8
30	+	-	+	+	+	-	-	18.7
31	-	+	+	+	+	-	-	16.0
32	+	+	+	+	+	+	+	25.6

Appendix A

Table A.1
Random Digits

16776	65293	76698	06105	24268	72003
17808	63210	05595	89829	88641	62129
54380	82809	63356	87040	11948	35153
44412	52718	75468	94009	11502	29626
18816	50627	26698	46044	48161	74326
30852	52743	79058	64860	00971	49619
82751	02613	97581	02400	47933	14309
64242	53094	33950	51568	09629	92279
46943	95126	92782	23369	60545	11000
02374	07355	81220	03657	68532	90697
44228	99446	14850	42448	56114	62780
37972	05589	35879	55855	77491	56754
67575	33661	50787	20047	04736	12914
40014	28688	40823	32983	17624	23943
57700	49631	60187	21391	88843	71076
43381	37479	52430	43268	64290	32771
38873	15652	68444	10077	14383	07151
88548	74732	10924	31080	03551	99363
49820	60451	05670	47535	11457	95427
78863	83170	73623	28534	80261	32888
21902	69132	56940	88193	40029	12295
96028	46706	14210	20000	30933	92451
82677	54508	14488	35973	41345	15600
39397	89811	16139	22619	79488	99052
87982	26521	47483	52739	00675	58748
79746	32279	13310	10957	99021	43253
10649	36094	30006	34319	12010	17710
43103	55136	56132	94737	67713	54171
38999	25297	92861	37134	60340	34040

Table A.2

Factors for Control Charts

Subgroup Size	Divisors for Estimate of Standard Deviation		Factors for $\bar{X}$ Chart Control Limits		Factors for R Chart Control Limits		Factors for s Chart Control Limits	
n	d_2	d_4	A_2	A_3	D_3	D_4	B_3	B_4
2	1.128	0.954	1.880	2.659	--	3.267	--	3.267
3	1.693	1.588	1.023	1.954	--	2.574	--	2.568
4	2.059	1.978	0.729	1.628	--	2.282	--	2.266
5	2.236	2.257	0.577	1.427	--	2.114	--	2.089
6	2.534	2.472	0.483	1.287	--	2.004	0.030	1.970
7	2.704	2.645	0.419	1.182	0.076	1.924	0.118	1.882
8	2.847	2.791	0.373	1.099	0.136	1.864	0.185	1.815
9	2.970	2.916	0.377	1.032	0.139	1.816	0.239	1.761
10	3.078	3.024	0.308	0.975	0.223	1.777	0.284	1.716

* Parts of this table are reproduced with permission from Table B2 of the *A.S.T.M. Manual on Quality Control of Materials*, p. 115.

d_4 factor is from Ferrell, E. B., " Control Charts Using Midranges and Medians", *Industrial Quality Control*, March 1953, pp. 30-34.

Table A.3
Standard Normal Distribution
Tail Areas

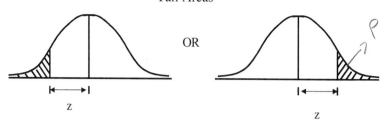

z	x.x0	x.x1	x.x2	x.x3	x.x4	x.x5	x.x6	x.x7	x.x8	x.x9
0.0	0.5000	0.4960	0.4920	0.4880	0.4840	0.4800	0.4760	0.4720	0.4681	0.4641
0.1	0.4601	0.4562	0.4522	0.4482	0.4443	0.4403	0.4364	0.4325	0.4285	0.4246
0.2	0.4207	0.4168	0.4129	0.4090	0.4051	0.4012	0.3974	0.3935	0.3897	0.3859
0.3	0.3820	0.3782	0.3744	0.3707	0.3669	0.3631	0.3594	0.3556	0.3519	0.3482
0.4	0.3445	0.3409	0.3372	0.3335	0.3299	0.3263	0.3227	0.3191	0.3156	0.3120
0.5	0.3085	0.3050	0.3015	0.2980	0.2945	0.2911	0.2877	0.2843	0.2809	0.2775
0.6	0.2742	0.2709	0.2676	0.2643	0.2610	0.2578	0.2546	0.2514	0.2482	0.2450
0.7	0.2419	0.2388	0.2357	0.2326	0.2296	0.2266	0.2236	0.2206	0.2176	0.2147
0.8	0.2118	0.2089	0.2061	0.2032	0.2004	0.1976	0.1948	0.1921	0.1894	0.1867
0.9	0.1840	0.1814	0.1787	0.1761	0.1736	0.1710	0.1685	0.1660	0.1635	0.1610
1.0	0.1586	0.1562	0.1538	0.1515	0.1491	0.1468	0.1445	0.1423	0.1400	0.1378
1.1	0.1356	0.1335	0.1313	0.1292	0.1271	0.1250	0.1230	0.1210	0.1190	0.1170
1.2	0.1150	0.1131	0.1112	0.1093	0.1074	0.1056	0.1038	0.1020	0.1002	0.0985
1.3	0.0968	0.0950	0.0934	0.0917	0.0901	0.0885	0.0869	0.0853	0.0837	0.0822
1.4	0.0807	0.0792	0.0778	0.0763	0.0749	0.0735	0.0721	0.0707	0.0694	0.0681
1.5	0.0668	0.0655	0.0642	0.0630	0.0617	0.0605	0.0593	0.0582	0.0570	0.0559
1.6	0.0547	0.0536	0.0526	0.0515	0.0505	0.0494	0.0484	0.0474	0.0464	0.0455
1.7	0.0445	0.0436	0.0427	0.0418	0.0409	0.0400	0.0392	0.0383	0.0375	0.0367
1.8	0.0359	0.0351	0.0343	0.0336	0.0328	0.0321	0.0314	0.0307	0.0300	0.0293
1.9	0.0287	0.0280	0.0274	0.0268	0.0261	0.0255	0.0249	0.0244	0.0238	0.0232
2.0	0.0227	0.0222	0.0216	0.0211	0.0206	0.0201	0.0196	0.0192	0.0187	0.0183
2.1	0.0178	0.0174	0.0170	0.0165	0.0161	0.0157	0.0153	0.0150	0.0146	0.0142
2.2	0.0139	0.0135	0.0132	0.0128	0.0125	0.0122	0.0119	0.0116	0.0113	0.0110
2.3	0.0107	0.0104	0.0101	0.0099	0.0096	0.0093	0.0091	0.0088	0.0086	0.0084
2.4	0.0081	0.0079	0.0077	0.0075	0.0073	0.0071	0.0069	0.0067	0.0065	0.0063
2.5	0.0062	0.0060	0.0058	0.0057	0.0055	0.0053	0.0052	0.0050	0.0049	0.0047
2.6	0.0046	0.0045	0.0043	0.0042	0.0041	0.0040	0.0039	0.0037	0.0036	0.0035
2.7	0.0034	0.0033	0.0032	0.0031	0.0030	0.0029	0.0028	0.0028	0.0027	0.0026
2.8	0.0025	0.0024	0.0024	0.0023	0.0022	0.0021	0.0021	0.0020	0.0019	0.0019
2.9	0.0018	0.0018	0.0017	0.0016	0.0016	0.0015	0.0015	0.0014	0.0014	0.0013
3.0	0.0013	0.0013	0.0012	0.0012	0.0011	0.0011	0.0011	0.0010	0.0010	0.0010
4.0	3.17E-5									
5.0	2.87E-7									
6.0	9.9E-10									
7.0	1.3E-12									

Table entries generated with MINITAB® Release 13

Table A.4
Right Tail Areas of the χ^2 Distribution with υ Degrees of Freedom

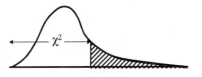

tail area probability

υ	0.99	0.975	0.95	0.9	0.5	0.1	0.05	0.025	0.01
1	0.000157	0.000982	0.003932	0.01579	0.45494	2.70554	3.84146	5.02389	6.63490
2	0.02010	0.05064	0.10259	0.21072	1.38629	4.60517	5.99146	7.37776	9.21034
3	0.1148	0.2158	0.3518	0.5844	2.3660	6.2514	7.8147	9.3484	11.3449
4	0.2971	0.4844	0.7107	1.0636	3.3567	7.7794	9.4877	11.1433	13.2767
5	0.5543	0.8312	1.1455	1.6103	4.3515	9.2364	11.0705	12.8325	15.0863
6	0.8721	1.2373	1.6354	2.2041	5.3481	10.6446	12.5916	14.4494	16.8119
7	1.2390	1.6899	2.1673	2.8331	6.3458	12.0170	14.0671	16.0128	18.4753
8	1.6465	2.1797	2.7326	3.4895	7.3441	13.3616	15.5073	17.5345	20.0902
9	2.0879	2.7004	3.3251	4.1682	8.3428	14.6837	16.9190	19.0228	21.6660
10	2.5582	3.2470	3.9403	4.8652	9.3418	15.9872	18.3070	20.4832	23.2093
11	3.0535	3.8157	4.5748	5.5778	10.3410	17.2750	19.6751	21.9200	24.7250
12	3.5706	4.4038	5.2260	6.3038	11.3403	18.5493	21.0261	23.3367	26.2170
13	4.1069	5.0088	5.8919	7.0415	12.3398	19.8119	22.3620	24.7356	27.6882
14	4.6604	5.6287	6.5706	7.7895	13.3393	21.0641	23.6848	26.1189	29.1412
15	5.2293	6.2621	7.2609	8.5468	14.3389	22.3071	24.9958	27.4884	30.5779
16	5.8122	6.9077	7.9616	9.3122	15.3385	23.5418	26.2962	28.8454	31.9999
17	6.4078	7.5642	8.6718	10.0852	16.3382	24.7690	27.5871	30.1910	33.4087
18	7.0149	8.2307	9.3905	10.8649	17.3379	25.9894	28.8693	31.5264	34.8053
19	7.6327	8.9065	10.1170	11.6509	18.3377	27.2036	30.1435	32.8523	36.1909
20	8.2604	9.5908	10.8508	12.4426	19.3374	28.4120	31.4104	34.1696	37.5662
21	8.8972	10.2829	11.5913	13.2396	20.3372	29.6151	32.6706	35.4789	38.9322
22	9.5425	10.9823	12.3380	14.0415	21.3370	30.8133	33.9244	36.7807	40.2894
23	10.1957	11.6886	13.0905	14.8480	22.3369	32.0069	35.1725	38.0756	41.6384
24	10.8564	12.4012	13.8484	15.6587	23.3367	33.1962	36.4150	39.3641	42.9798
25	11.5240	13.1197	14.6114	16.4734	24.3366	34.3816	37.6525	40.6465	44.3141
30	12.1981	13.8439	15.3792	17.2919	25.3365	35.5632	38.8851	41.9232	45.6417

Table entries generated with MINITAB® Release 13

Table A.5
Two-Sided Student's t-Statistic

t^*

Degrees of Freedom	Confidence Level		
	90%	95%	99%
1	6.314	12.706	63.656
2	2.920	4.303	9.925
3	2.353	3.182	5.841
4	2.132	2.776	4.604
5	2.015	2.571	4.032
6	1.943	2.447	3.707
7	1.895	2.365	3.499
8	1.860	2.306	3.355
9	1.833	2.262	3.250
10	1.812	2.228	3.169
11	1.796	2.201	3.106
12	1.782	2.179	3.055
13	1.771	2.160	3.012
14	1.761	2.145	2.977
15	1.753	2.131	2.947
16	1.746	2.120	2.921
17	1.740	2.110	2.898
18	1.734	2.101	2.878
19	1.729	2.093	2.861
20	1.725	2.086	2.845
21	1.721	2.080	2.831
22	1.717	2.074	2.819
23	1.714	2.069	2.807
24	1.711	2.064	2.797
25	1.708	2.060	2.787
26	1.706	2.056	2.779
27	1.703	2.052	2.771
28	1.701	2.048	2.763
29	1.699	2.045	2.756
30	1.697	2.042	2.750
40	1.684	2.021	2.704
60	1.671	2.000	2.660
120	1.658	1.980	2.617
∞	1.645	1.960	2.576

Table entries generated with TINV function in Excel

Table A.6
Upper 5% Points of F-Distribution with υ_1 and υ_2
Degrees of Freedom

υ_2	υ_1 1	2	3	4	5	6	7	8	9	10	12	15	20	24	30	40	60	120
1	161.5	199.5	215.7	224.6	230.2	234.0	236.8	238.9	240.5	241.9	243.9	246.0	248.0	249.0	250.1	251.1	252.2	253.2
2	18.51	19.00	19.16	19.25	19.30	19.33	19.35	19.37	19.38	19.40	19.41	19.43	19.45	19.45	19.46	19.47	19.48	19.49
3	10.13	9.55	9.28	9.12	9.01	8.94	8.89	8.85	8.81	8.79	8.74	8.70	8.66	8.64	8.62	8.59	8.57	8.55
4	7.71	6.94	6.59	6.39	6.26	6.16	6.09	6.04	6.00	5.96	5.91	5.86	5.80	5.77	5.75	5.72	5.69	5.66
5	6.61	5.79	5.41	5.19	5.05	4.95	4.88	4.82	4.77	4.74	4.68	4.62	4.56	4.53	4.50	4.46	4.43	4.40
6	5.99	5.14	4.76	4.53	4.39	4.28	4.21	4.15	4.10	4.06	4.00	3.94	3.87	3.84	3.81	3.77	3.74	3.70
7	5.59	4.74	4.35	4.12	3.97	3.87	3.79	3.73	3.68	3.64	3.57	3.51	3.44	3.41	3.38	3.34	3.30	3.27
8	5.32	4.46	4.07	3.84	3.69	3.58	3.50	3.44	3.39	3.35	3.28	3.22	3.15	3.12	3.08	3.04	3.01	2.97
9	5.12	4.26	3.86	3.63	3.48	3.37	3.29	3.23	3.18	3.14	3.07	3.01	2.94	2.90	2.86	2.83	2.79	2.75
10	4.96	4.10	3.71	3.48	3.33	3.22	3.14	3.07	3.02	2.98	2.91	2.85	2.77	2.74	2.70	2.66	2.62	2.58
11	4.84	3.98	3.59	3.36	3.20	3.09	3.01	2.95	2.90	2.85	2.79	2.72	2.65	2.61	2.57	2.53	2.49	2.45
12	4.75	3.89	3.49	3.26	3.11	3.00	2.91	2.85	2.80	2.75	2.69	2.62	2.54	2.51	2.47	2.43	2.38	2.34
13	4.67	3.81	3.41	3.18	3.03	2.92	2.83	2.77	2.71	2.67	2.60	2.53	2.46	2.42	2.38	2.34	2.30	2.25
14	4.60	3.74	3.34	3.11	2.96	2.85	2.76	2.70	2.65	2.60	2.53	2.46	2.39	2.35	2.31	2.27	2.22	2.18
15	4.54	3.68	3.29	3.06	2.90	2.79	2.71	2.64	2.59	2.54	2.48	2.40	2.33	2.29	2.25	2.20	2.16	2.11
16	4.49	3.63	3.24	3.01	2.85	2.74	2.66	2.59	2.54	2.49	2.42	2.35	2.28	2.24	2.19	2.15	2.11	2.06
17	4.45	3.59	3.20	2.96	2.81	2.70	2.61	2.55	2.49	2.45	2.38	2.31	2.23	2.19	2.15	2.10	2.06	2.01
18	4.41	3.55	3.16	2.93	2.77	2.66	2.58	2.51	2.46	2.41	2.34	2.27	2.19	2.15	2.11	2.06	2.02	1.97
19	4.38	3.52	3.13	2.90	2.74	2.63	2.54	2.48	2.42	2.38	2.31	2.23	2.16	2.11	2.07	2.03	1.98	1.93
20	4.35	3.49	3.10	2.87	2.71	2.60	2.51	2.45	2.39	2.35	2.28	2.20	2.12	2.08	2.04	1.99	1.95	1.90
21	4.32	3.47	3.07	2.84	2.68	2.57	2.49	2.42	2.37	2.32	2.25	2.18	2.10	2.05	2.01	1.96	1.92	1.87
22	4.30	3.44	3.05	2.82	2.66	2.55	2.46	2.40	2.34	2.30	2.23	2.15	2.07	2.03	1.98	1.94	1.89	1.84
23	4.28	3.42	3.03	2.80	2.64	2.53	2.44	2.37	2.32	2.27	2.20	2.13	2.05	2.01	1.96	1.91	1.86	1.81
24	4.26	3.40	3.01	2.78	2.62	2.51	2.42	2.36	2.30	2.25	2.18	2.11	2.03	1.98	1.94	1.89	1.84	1.79
25	4.24	3.39	2.99	2.76	2.60	2.49	2.40	2.34	2.28	2.24	2.16	2.09	2.01	1.96	1.92	1.87	1.82	1.77
26	4.23	3.37	2.98	2.74	2.59	2.47	2.39	2.32	2.27	2.22	2.15	2.07	1.99	1.95	1.90	1.85	1.80	1.75
27	4.21	3.35	2.96	2.73	2.57	2.46	2.37	2.31	2.25	2.20	2.13	2.06	1.97	1.93	1.88	1.84	1.79	1.73
28	4.20	3.34	2.95	2.71	2.56	2.45	2.36	2.29	2.24	2.19	2.12	2.04	1.96	1.91	1.87	1.82	1.77	1.71
28	4.18	3.33	2.93	2.70	2.55	2.43	2.35	2.28	2.22	2.18	2.10	2.03	1.94	1.90	1.85	1.81	1.75	1.70
30	4.17	3.32	2.92	2.69	2.53	2.42	2.33	2.27	2.21	2.16	2.09	2.01	1.93	1.89	1.84	1.79	1.74	1.68
40	4.08	3.23	2.84	2.61	2.45	2.34	2.25	2.18	2.12	2.08	2.00	1.92	1.84	1.79	1.74	1.69	1.64	1.58
60	4.00	3.15	2.76	2.53	2.37	2.25	2.17	2.10	2.04	1.99	1.92	1.84	1.75	1.70	1.65	1.59	1.53	1.47
120	3.92	3.07	2.68	2.45	2.29	2.18	2.09	2.02	1.96	1.91	1.83	1.75	1.66	1.61	1.55	1.50	1.43	1.35
∞	3.84	3.00	2.60	2.37	2.21	2.10	2.01	1.94	1.88	1.83	1.75	1.67	1.57	1.52	1.46	1.39	1.32	1.22

Table entries generated with FINV function in Excel

Table A.7
95% Confidence Values for the Studentized Range Statistic
q(.05, k,v)

Error Degress of Freedom v	Number of Means Compared k								
	2	3	4	5	6	7	8	9	10
1	18.0	27.0	32.8	37.2	40.5	43.1	45.4	47.3	49.1
2	6.09	8.33	9.80	10.89	11.73	12.43	13.03	13.54	13.99
3	4.50	5.91	6.83	7.51	8.04	8.47	8.85	9.18	9.46
4	3.93	5.04	5.76	6.29	6.71	7.06	7.35	7.60	7.83
5	3.64	4.60	5.22	5.67	6.03	6.33	6.58	6.80	6.99
6	3.46	4.34	4.90	5.31	5.63	5.89	6.12	6.32	6.49
7	3.34	4.16	4.68	5.06	5.35	5.59	5.80	5.99	6.15
8	3.26	4.04	4.53	4.89	5.17	5.40	5.60	5.77	5.92
9	3.20	3.95	4.42	4.76	5.02	5.24	5.43	5.60	5.74
10	3.15	3.88	4.33	4.66	4.91	5.12	5.30	5.46	5.60
11	3.11	3.82	4.26	4.58	4.82	5.03	5.20	5.35	5.49
12	3.08	3.77	4.20	4.51	4.75	4.95	5.12	5.27	5.40
13	3.06	3.73	4.15	4.46	4.69	4.88	5.05	5.19	5.32
14	3.03	3.70	4.11	4.41	4.64	4.83	4.99	5.13	5.25
15	3.01	3.67	4.08	4.37	4.59	4.78	4.94	5.08	5.20
16	3.00	3.65	4.05	4.34	4.56	4.74	4.90	5.03	5.15
17	2.98	3.62	4.02	4.31	4.52	4.70	4.86	4.99	5.11
18	2.97	3.61	4.00	4.28	4.49	4.67	4.83	4.96	5.07
19	2.96	3.59	3.98	4.26	4.47	4.64	4.79	4.92	5.04
20	2.95	3.58	3.96	4.24	4.45	4.62	4.77	4.90	5.01
24	2.92	3.53	3.90	4.17	4.37	4.54	4.68	4.81	4.92
30	2.89	3.48	3.84	4.11	4.30	4.46	4.60	4.72	4.83
40	2.86	3.44	3.79	4.04	4.23	4.39	4.52	4.63	4.74
60	2.83	3.40	3.74	3.98	4.16	4.31	4.44	4.55	4.65
120	2.80	3.36	3.69	3.92	4.10	4.24	4.36	4.47	4.56
∞	2.77	3.32	3.63	3.86	4.03	4.17	4.29	4.39	4.47

Adapted from *Biometrika Tables for Statisticians*, Vol. 1, 1966, edited by E.S. Pearson and H.O. Hartley, with permission of Biometrika Trustees.

Table A.8
Orthogonal Polynomials

Number of Factor Levels I=3

Factor Level	Linear c_{1i}	Quadratic c_{2i}
1	-1	1
2	0	-2
3	1	1

Number of Factor Levels I=4

Factor Level	Linear c_{1i}	Quadratic c_{2i}	Cubic c_{3i}
1	-3	1	-1
2	-1	-1	3
3	1	-1	-3
4	3	1	1

Number of Factor Levels I=5

Factor Level	Linear c_{1i}	Quadratic c_{2i}	Cubic c_{3i}	Quartic c_{4i}
1	-2	2	-1	1
2	-1	-1	2	-4
3	0	-2	0	6
4	1	-1	-2	-4
5	2	2	1	1

Adapted with permission from Table A 19 of *Statistical Methods* 7th Ed., by G. W. Snedecor and W.G. Cochran

Table A.9

Factors d_2^* for Converting the Average Range ($\overline{R}$) into a Standard Deviation (σ)

	k = 1	2	3	4	5	8	10	∞
n = 2	1.41	1.28	1.23	1.21	1.19	1.17	1.16	1.13
3	1.91	1.81	1.77	1.75	1.74	1.72	1.72	1.69
4	2.24	2.15	2.12	2.11	2.10	2.08	2.08	2.06
5	2.48	2.40	2.38	2.37	2.39	2.35	2.34	2.33
6	2.67	2.60	2.58	2.57	2.56	2.55	2.55	2.53
7	2.83	2.77	2.75	2.74	2.73	2.72	2.72	2.70
8	2.96	2.91	2.89	2.88	2.87	2.87	2.86	2.85
9	3.08	3.02	3.01	3.00	2.99	2.99	2.98	2.97
10	3.19	3.13	3.11	3.10	3.10	3.09	3.09	3.08

k = Number of Samples (Number of parts measured)
n = Sample Size (Number of times each part was measured)
*Based on d_2 factors, Table D3, *Quality Control and Industrial Statistics*, by A.J. Duncan, reproduced with permission.

Appendix B

TWELVE RUN PLACKETT-BURMAN DESIGN

Run	Mean	X_1	X_2	X_3	X_4	X_5	X_6	X_7	X_8	X_9	X_{10}	X_{11}
1	+	+	+	-	+	+	+	-	-	-	+	-
2	+	+	-	+	+	+	-	-	-	+	-	+
3	+	-	+	+	+	-	-	-	+	-	+	+
4	+	+	+	+	-	-	-	+	-	+	+	-
5	+	+	+	-	-	-	+	-	+	+	-	+
6	+	+	-	-	-	+	-	+	+	-	+	+
7	+	-	-	-	+	-	+	+	-	+	+	+
8	+	-	-	+	-	+	+	-	+	+	+	-
9	+	-	+	-	+	+	-	+	+	+	-	-
10	+	+	-	+	+	-	+	+	+	-	-	-
11	+	-	+	+	-	+	+	+	-	-	-	+
12	+	-	-	-	-	-	-	-	-	-	-	-

Table B.1-2
TWENTY RUN PLACKETT-BURMAN DESIGN

Run	Mean	X_1	X_2	X_3	X_4	X_5	X_6	X_7	X_8	X_9	X_{10}	X_{11}	X_{12}	X_{13}	X_{14}	X_{15}	X_{16}	X_{17}	X_{18}	X_{19}
1	+	+	+	−	−	+	+	+	+	−	+	−	+	−	−	−	−	+	+	−
2	+	+	−	−	+	+	+	+	−	+	−	+	−	−	−	−	+	+	−	+
3	+	−	−	+	+	+	+	−	+	−	+	−	−	−	−	+	+	−	+	+
4	+	−	+	+	+	+	−	+	−	+	−	−	−	−	+	+	−	+	+	−
5	+	+	+	+	+	−	+	−	+	−	−	−	−	+	+	−	+	+	−	−
6	+	+	+	+	−	+	−	+	−	−	−	−	+	+	−	+	+	−	−	+
7	+	+	+	−	+	−	+	−	−	−	−	+	+	−	+	+	−	−	+	+
8	+	+	−	+	−	+	−	−	−	−	+	+	−	+	+	−	−	+	+	+
9	+	−	+	−	+	−	−	−	−	+	+	−	+	+	−	−	+	+	+	+
10	+	+	−	+	−	−	−	−	+	+	−	+	+	−	−	+	+	+	+	−
11	+	−	+	−	−	−	−	+	+	−	+	+	−	−	+	+	+	+	−	+
12	+	+	−	−	−	−	+	+	−	+	+	−	−	+	+	+	+	−	+	−
13	+	−	−	−	−	+	+	−	+	+	−	−	+	+	+	+	−	+	−	+
14	+	−	−	−	+	+	−	+	+	−	−	+	+	+	+	−	+	−	+	−
15	+	−	−	+	+	−	+	+	−	−	+	+	+	+	−	+	−	+	−	−
16	+	−	+	+	−	+	+	−	−	+	+	+	+	−	+	−	+	−	−	−
17	+	+	+	−	+	+	−	−	+	+	+	+	−	+	−	+	−	−	−	−
18	+	+	−	+	+	−	−	+	+	+	+	−	+	−	+	−	−	−	−	+
19	+	−	+	+	−	−	+	+	+	+	−	+	−	+	−	−	−	−	+	+
20	+	−	−	−	−	−	−	−	−	−	−	−	−	−	−	−	−	−	−	−

759

Table B.1-3
TWENTY FOUR RUN PLACKETT-BURMAN DESIGN

Run	Mean	X₁	X₂	X₃	X₄	X₅	X₆	X₇	X₈	X₉	X₁₀	X₁₁	X₁₂	X₁₃	X₁₄	X₁₅	X₁₆	X₁₇	X₁₈	X₁₉	X₂₀	X₂₁	X₂₂	X₂₃
1	+	+	+	+	+	+	-	+	-	+	+	-	-	+	+	-	-	+	-	+	-	-	-	-
2	+	-	+	+	+	+	+	-	+	-	+	+	-	-	+	+	-	-	+	-	+	-	-	-
3	+	-	-	+	+	+	+	+	-	+	-	+	+	-	-	+	+	-	-	+	-	+	-	-
4	+	-	-	-	+	+	+	+	+	-	+	-	+	+	-	-	+	+	-	-	+	-	+	-
5	+	-	-	-	-	+	+	+	+	+	-	+	-	+	+	-	-	+	+	-	-	+	-	+
6	+	+	-	-	-	-	+	+	+	+	+	-	+	-	+	+	-	-	+	+	-	-	+	-
7	+	-	+	-	-	-	-	+	+	+	+	+	-	+	-	+	+	-	-	+	+	-	-	+
8	+	+	-	+	-	-	-	-	+	+	+	+	+	-	+	-	+	+	-	-	+	+	-	-
9	+	-	+	-	+	-	-	-	-	+	+	+	+	+	-	+	-	+	+	-	-	+	+	-
10	+	-	-	+	-	+	-	-	-	-	+	+	+	+	+	-	+	-	+	+	-	-	+	+
11	+	+	-	-	+	-	+	-	-	-	-	+	+	+	+	+	-	+	-	+	+	-	-	+
12	+	+	+	-	-	+	-	+	-	-	-	-	+	+	+	+	+	-	+	-	+	+	-	-
13	+	-	+	+	-	-	+	-	+	-	-	-	-	+	+	+	+	+	-	+	-	+	+	-
14	+	-	-	+	+	-	-	+	-	+	-	-	-	-	+	+	+	+	+	-	+	-	+	+
15	+	+	-	-	+	+	-	-	+	-	+	-	-	-	-	+	+	+	+	+	-	+	-	+
16	+	+	+	-	-	+	+	-	-	+	-	+	-	-	-	-	+	+	+	+	+	-	+	-
17	+	-	+	+	-	-	+	+	-	-	+	-	+	-	-	-	-	+	+	+	+	+	-	+
18	+	+	-	+	+	-	-	+	+	-	-	+	-	+	-	-	-	-	+	+	+	+	+	-
19	+	-	+	-	+	+	-	-	+	+	-	-	+	-	+	-	-	-	-	+	+	+	+	+
20	+	+	-	+	-	+	+	-	-	+	+	-	-	+	-	+	-	-	-	-	+	+	+	+
21	+	+	+	-	+	-	+	+	-	-	+	+	-	-	+	-	+	-	-	-	-	+	+	+
22	+	+	+	+	-	+	-	+	+	-	-	+	+	-	-	+	-	+	-	-	-	-	+	+
23	+	+	+	+	+	-	+	-	+	+	-	-	+	+	-	-	+	-	+	-	-	-	-	+
24	+	-	-	-	-	-	-	-	-	-	-	-	-	-	-	-	-	-	-	-	-	-	-	-

760

Table B.1-4
TWENTY-EIGHT RUN PLACKETT-BURMAN DESIGN

Run	Mean	X_1	X_2	X_3	X_4	X_5	X_6	X_7	X_8	X_9	X_{10}	X_{11}	X_{12}	X_{13}	X_{14}	X_{15}	X_{16}	X_{17}	X_{18}	X_{19}	X_{20}	X_{21}	X_{22}	X_{23}	X_{24}	X_{25}	X_{26}	X_{27}
1	+	+	−	+	+	+	+	−	−	−	−	+	+	−	−	+	+	−	+	+	+	−	+	−	+	+	−	+
2	+	+	+	−	+	+	+	−	−	−	−	−	+	+	+	−	+	−	−	−	+	+	+	+	−	+	+	−
3	+	−	+	+	+	−	+	−	+	−	+	−	+	+	+	+	−	+	−	+	+	+	+	+	+	−	+	+
4	+	−	−	+	+	−	+	+	+	+	+	−	−	+	+	−	−	+	−	+	+	−	+	+	−	+	−	+
5	+	−	−	−	+	+	−	+	+	+	+	+	−	+	−	+	−	+	−	+	+	+	−	+	+	+	+	−
6	+	+	+	+	+	−	−	+	+	+	−	+	+	−	+	−	+	+	+	+	−	+	+	+	−	−	−	+
7	+	+	+	+	−	+	−	+	+	−	+	+	+	+	+	−	−	−	+	−	−	+	+	+	+	+	+	−
8	+	+	+	+	−	+	+	+	−	+	−	+	+	+	−	+	+	−	−	+	+	+	+	−	+	−	+	+
9	+	−	+	−	+	+	+	−	−	+	+	+	−	−	+	+	+	+	−	−	+	−	+	+	+	+	+	+
10	+	+	−	+	−	+	+	+	+	−	−	−	+	+	+	−	+	+	+	+	−	−	+	−	+	−	+	+
11	+	+	−	+	−	−	−	+	+	+	+	+	+	+	−	+	−	−	+	+	−	−	−	+	−	+	−	−
12	+	+	+	+	+	−	−	+	+	+	−	+	+	+	+	+	−	+	+	+	+	+	+	+	−	−	−	+
13	+	+	+	−	+	+	+	+	−	+	+	+	−	+	+	+	+	+	−	−	+	+	+	+	+	+	−	−
14	+	+	+	−	−	+	−	+	+	+	−	−	+	−	+	+	+	−	−	+	+	+	+	−	+	+	−	−
15	+	−	+	+	−	−	+	−	−	+	−	−	−	−	+	−	+	+	+	−	+	+	+	+	+	+	+	−
16	+	+	−	+	+	+	+	−	+	−	−	+	+	−	+	+	+	+	+	+	+	−	+	−	+	−	+	+
17	+	+	−	+	+	−	−	+	+	+	+	+	+	−	+	−	+	+	+	−	−	+	+	+	+	−	−	−
18	+	−	+	−	−	+	+	−	−	+	+	+	+	+	+	−	+	+	+	−	+	−	+	+	+	+	+	−
19	+	−	+	−	−	−	−	+	−	+	+	−	+	+	+	+	−	−	+	+	+	+	+	−	+	−	−	−
20	+	+	−	+	−	−	+	+	+	−	+	−	−	−	−	+	+	+	−	−	−	−	−	+	−	−	−	−
21	+	−	−	−	−	+	−	−	−	−	+	+	+	+	+	+	+	+	+	−	+	+	+	+	+	+	+	+
22	+	+	−	+	+	−	+	−	−	+	+	+	+	+	+	+	+	+	+	+	+	+	+	+	+	+	+	+
23	+	−	+	+	−	+	−	+	+	−	−	+	+	+	+	+	+	+	+	−	+	−	+	−	−	+	+	+
24	+	+	−	−	+	+	−	+	+	+	+	+	+	−	+	+	−	+	+	+	−	+	+	+	+	+	−	+
25	+	−	+	+	−	+	−	−	+	−	−	−	+	−	−	+	+	−	+	+	+	+	+	−	−	+	+	−
26	+	+	+	+	+	+	+	+	+	+	+	+	+	+	+	−	−	−	+	−	+	+	+	−	−	−	+	+
27	+	−	−	−	−	−	−	−	−	−	−	−	−	−	−	−	−	−	−	−	−	−	−	−	−	−	−	−
28	+	−	−	−	−	−	−	−	−	−	−	−	−	−	−	−	−	−	−	−	−	−	−	−	−	−	−	−

Table B.1-5

EIGHT RUN FRACTIONAL FACTORIAL DESIGN FOR
FIVE TO SEVEN FACTORS

MAIN EFFECTS CONFOUNDED WITH TWO-FACTOR
INTERACTIONS

Run	Mean	X_1	X_2	X_3	X_4	X_5	X_6	X_7
1	+	-	-	-	+	+	+	-
2	+	+	-	-	-	-	+	+
3	+	-	+	-	-	+	-	+
4	+	+	+	-	+	-	-	-
5	+	-	-	+	+	-	-	+
6	+	+	-	+	-	+	-	-
7	+	-	+	+	-	-	+	-
8	+	+	+	+	+	+	+	+

CONFOUNDINGS

1 + 24 + 35 + 67
2 + 14 + 36 + 57
3 + 15 + 26 + 47
4 + 12 + 37 + 56
5 + 13 + 27 + 46
6 + 17 + 23 + 45
7 + 16 + 25 + 34

I = 124 = 135 = 236 = 1237

Table B.1-6
EIGHT RUN FRACTIONAL FACTORIAL DESIGN
FOR FOUR FACTORS

MAIN EFFECTS CLEAR OF TWO FACTOR INTERACTIONS

Run	Mean	X_1	X_2	X_3	X_4	E_1	E_2	E_3
1	+	-	-	-	-	+	+	+
2	+	+	-	-	+	-	-	+
3	+	-	+	-	+	-	+	-
4	+	+	+	-	-	+	-	-
5	+	-	-	+	+	+	-	-
6	+	+	-	+	-	-	+	-
7	+	-	+	+	-	-	-	+
8	+	+	+	+	+	+	+	+

CONFOUNDINGS

$$E_1 = 12 + 34$$
$$E_2 = 13 + 24$$
$$E_3 = 14 + 23$$

Table B.1-7
SIXTEEN RUN FRACTIONAL FACTORIAL DESIGN FOR NINE TO FIFTEEN FACTORS
(MAIN EFFECTS CONFOUNDED WITH TWO-FACTOR INTERACTIONS)

Run	Mean	X_1	X_2	X_3	X_4	X_5	X_6	X_7	X_8	X_9	X_{10}	X_{11}	X_{12}	X_{13}	X_{14}	X_{15}
1	+	-	-	-	-	-	-	-	-	+	+	+	+	+	+	+
2	+	+	-	-	-	+	-	+	+	-	-	-	-	+	+	+
3	+	-	+	-	-	+	+	-	+	-	-	+	+	-	-	+
4	+	+	+	-	-	-	+	+	-	+	+	-	-	-	-	+
5	+	-	-	+	-	+	+	+	-	-	+	-	+	-	+	-
6	+	+	-	+	-	-	+	-	+	+	-	+	-	-	+	-
7	+	-	+	+	-	-	-	+	+	+	-	-	+	+	-	-
8	+	+	+	+	-	+	-	-	-	-	+	+	-	+	-	-
9	+	-	-	-	+	-	+	+	+	-	+	+	-	+	-	-
10	+	+	-	-	+	+	+	-	-	+	-	-	+	+	-	-
11	+	-	+	-	+	+	-	+	-	+	-	+	-	-	+	-
12	+	+	+	-	+	-	-	-	+	-	+	-	+	-	+	-
13	+	-	-	+	+	+	-	-	+	+	+	-	-	-	-	+
14	+	+	-	+	+	-	-	+	-	-	-	+	+	-	-	+
15	+	-	+	+	+	-	+	-	-	-	-	-	-	+	+	+
16	+	+	+	+	+	+	+	+	+	+	+	+	+	+	+	+

CONFOUNDINGS

X_1 = 1 + 2,10 + 3,11 + 4,12 + 5,13 + 8,14 + 6,9 + 7,15

X_2 = 2 + 1,10 + 3,13 + 4,14 + 5,11 + 8,12 + 6,15 + 7,9

X_3 = 3 + 1,11 + 2,13 + 4,15 + 5,10 + 7,12 + 6,14 + 8,9

X_4 = 4 + 1,12 + 2,14 + 3,15 + 8,10 + 7,11 + 6,13 + 5,9

X_5 = 5 + 1,13 + 2,11 + 3,10 + 4,9 + 6,12 + 7,14 + 8,15

X_6 = 6 + 1,9 + 2,15 + 3,14 + 4,13 + 7,10 + 8,11 + 5,12

X_7 = 7 + 1,15 + 2,9 + 3,12 + 4,11 + 6,10 + 8,13 + 5,14

X_8 = 8 + 1,14 + 2,12 + 3,9 + 4,10 + 6,11 + 7,13 + 5,15

X_9 = 9 + 1,6 + 2,7 + 3,8 + 4,5 + 10,15 + 11,14 + 12,13

X_{10} = 10 + 1,2 + 3,5 + 4,8 + 11,13 + 12,14 + 9,15 + 6,7

X_{11} = 11 + 1,3 + 2,5 + 4,7 + 10,13 + 12,15 + 9,14 + 6,8

X_{12} = 12 + 1,4 + 2,8 + 3,7 + 10,14 + 11,15 + 9,13 + 5,6

X_{13} = 13 + 1,5 + 2,3 + 4,6 + 10,11 + 9,12 + 14,15 + 7,8

X_{14} = 14 + 1,8 + 2,4 + 3,6 + 10,12 + 9,11 + 13,15 + 5,7

X_{15} = 15 + 1,7 + 2,6 + 3,4 + 9,10 + 11,12 + 13,14 + 5,8

I = 1235 = 2346 = 1347 = 1248 = 12349 = 12(10) = 13(11) = 14(12) = 23(13) = 24(14) = 34(15)

SIXTEEN RUN FRACTIONAL FACTORIAL DESIGN FOR SIX TO EIGHT FACTORS
(MAIN EFFECTS CLEAR OF TWO-FACTOR INTERACTIONS)

Run	Mean	X_1	X_2	X_3	X_4	X_5	X_6	X_7	X_8	E_1	E_2	E_3	E_4	E_5	E_6	E_7
1	+	−	−	−	−	−	−	−	−	+	+	+	+	+	+	+
2	+	+	−	−	−	+	−	+	+	−	−	−	+	−	+	+
3	+	−	+	−	−	+	+	−	+	−	+	+	−	−	+	−
4	+	+	+	−	−	−	+	+	−	+	−	−	−	+	+	−
5	+	−	−	+	−	+	+	+	−	+	−	+	−	−	−	+
6	+	+	−	+	−	−	+	−	+	−	+	−	−	+	−	+
7	+	−	+	+	−	−	−	+	+	−	−	+	+	+	−	−
8	+	+	+	+	−	+	−	−	−	+	+	−	+	−	−	−
9	+	−	−	−	+	−	+	+	+	+	+	−	+	−	−	−
10	+	+	−	−	+	+	+	−	−	−	−	+	+	+	−	−
11	+	−	+	−	+	+	−	+	−	−	+	−	−	+	−	+
12	+	+	+	−	+	−	−	−	+	+	−	+	−	−	−	+
13	+	−	−	+	+	+	−	−	+	+	−	−	−	+	+	−
14	+	+	−	+	+	−	−	+	−	−	+	+	−	−	+	−
15	+	−	+	+	+	−	+	−	−	−	−	−	+	−	+	+
16	+	+	+	+	+	+	+	+	+	+	+	+	+	+	+	+

CONFOUNDINGS

$$E_1 = 12 + 35 + 48 + 67$$
$$E_2 = 13 + 25 + 47 + 68$$
$$E_3 = 14 + 28 + 37 + 56$$
$$E_4 = 15 + 23 + 46 + 78$$
$$E_5 = 16 + 27 + 38 + 45$$
$$E_6 = 17 + 26 + 34 + 58$$
$$E_7 = 18 + 24 + 36 + 57$$

$$I = 1235 = 2346 = 1347 = 1248$$

Table B.1-9
SIXTEEN RUN FRACTIONAL FACTORIAL DESIGN FOR FIVE FACTORS
(MAIN EFFECTS AND TWO-FACTOR INTERACTIONS CLEAR OF EACH OTHER)

Run	Mean	X_1	X_2	X_3	X_4	X_5	X_1X_2	X_1X_3	X_1X_4	X_1X_5	X_2X_3	X_2X_4	X_2X_5	X_3X_4	X_3X_5	X_4X_5
1	+	-	-	-	-	+	+	+	+	-	+	+	-	+	-	-
2	+	+	-	-	-	-	-	-	-	-	+	+	+	+	+	+
3	+	-	+	-	-	-	-	+	+	+	-	-	-	+	+	+
4	+	+	+	-	-	+	+	-	-	+	-	-	+	+	-	-
5	+	-	-	+	-	-	+	-	+	+	-	+	+	-	-	+
6	+	+	-	+	-	+	-	+	-	+	-	+	-	-	+	-
7	+	-	+	+	-	+	-	-	+	-	+	-	+	-	+	-
8	+	+	+	+	-	-	+	+	-	-	+	-	-	-	-	+
9	+	-	-	-	+	-	+	+	-	+	+	-	+	-	+	-
10	+	+	-	-	+	+	-	-	+	+	+	-	-	-	-	+
11	+	-	+	-	+	+	-	+	-	-	-	+	+	-	-	+
12	+	+	+	-	+	-	+	-	+	-	-	+	-	-	+	-
13	+	-	-	+	+	+	+	-	-	-	-	-	-	+	+	+
14	+	+	-	+	+	-	-	+	+	-	-	-	+	+	-	-
15	+	-	+	+	+	-	-	-	-	+	+	+	-	+	-	-
16	+	+	+	+	+	+	+	+	+	+	+	+	+	+	+	+

CONFOUNDINGS

I = 12345

766

Table B.1-10

THIRTY-TWO RUN FRACTIONAL FACTORIAL DESIGN FOR 17 TO 31 FACTORS
(MAIN EFFECTS CONFOUNDED WITH TWO-FACTOR INTERACTIONS)

Run	Mean	X_1	X_2	X_3	X_4	X_5	X_6	X_7	X_8	X_9	X_{10}	X_{11}	X_{12}	X_{13}	X_{14}	X_{15}	X_{16}	X_{17}	X_{18}	X_{19}	X_{20}	X_{21}	X_{22}	X_{23}	X_{24}	X_{25}	X_{26}	X_{27}	X_{28}	X_{29}	X_{30}	X_{31}
1	+	-	-	-	-	-	+	+	+	+	+	+	+	+	+	+	-	-	-	-	-	-	-	-	-	-	+	+	+	+	+	-
2	+	+	-	-	-	-	-	-	-	-	+	+	+	+	+	+	+	+	+	+	+	+	-	-	-	-	-	-	-	-	+	+
3	+	-	+	-	-	-	-	+	+	+	-	-	-	+	+	+	+	+	+	-	-	-	+	+	+	-	-	-	-	+	-	+
4	+	+	+	-	-	-	+	-	-	-	-	-	-	+	+	+	-	-	-	+	+	+	+	+	+	-	+	+	+	-	-	-
5	+	-	-	+	-	-	+	-	+	+	-	+	+	-	-	+	+	-	-	+	+	-	+	+	-	+	-	-	+	-	-	+
6	+	+	-	+	-	-	-	+	-	-	-	+	+	-	-	+	-	+	+	-	-	+	+	+	-	+	+	+	-	+	-	-
7	+	-	+	+	-	-	-	-	+	+	+	-	-	-	-	+	-	+	+	+	+	-	-	-	+	+	+	+	-	-	+	-
8	+	+	+	+	-	-	+	+	-	-	+	-	-	-	-	+	+	-	-	-	-	+	-	-	+	+	-	-	+	+	+	+
9	+	-	-	-	+	-	+	+	-	+	+	-	+	-	+	-	-	+	-	+	-	+	+	-	+	+	-	+	-	-	-	+
10	+	+	-	-	+	-	-	-	+	-	+	-	+	-	+	-	+	-	+	-	+	-	+	-	+	+	+	-	+	+	-	-
11	+	-	+	-	+	-	-	+	-	+	-	+	-	-	+	-	+	-	+	+	-	+	-	+	-	+	+	-	+	-	+	-
12	+	+	+	-	+	-	+	-	+	-	-	+	-	-	+	-	-	+	-	-	+	-	-	+	-	+	-	+	-	+	+	+
13	+	-	-	+	+	-	+	-	-	+	-	-	+	+	-	-	+	+	-	-	+	+	-	+	+	-	+	-	-	+	+	-
14	+	+	-	+	+	-	-	+	+	-	-	-	+	+	-	-	-	-	+	+	-	-	-	+	+	-	-	+	+	-	+	+
15	+	-	+	+	+	-	-	-	-	+	+	+	-	+	-	-	-	-	+	-	+	+	+	-	-	-	-	+	+	+	-	+
16	+	+	+	+	+	-	+	+	+	-	+	+	-	+	-	-	+	+	-	+	-	-	+	-	-	-	+	-	-	-	-	-
17	+	-	-	-	-	+	+	+	+	-	+	+	-	+	-	-	-	-	+	-	+	+	-	+	+	+	+	-	-	-	-	+
18	+	+	-	-	-	+	-	-	-	+	+	+	-	+	-	-	+	+	-	+	-	-	-	+	+	+	-	+	+	+	-	-
19	+	-	+	-	-	+	-	+	+	-	-	-	+	+	-	-	+	+	-	-	+	+	+	-	-	+	-	+	+	-	+	-
20	+	+	+	-	-	+	+	-	-	+	-	-	+	+	-	-	-	-	+	+	-	-	+	-	-	+	+	-	-	+	+	+
21	+	-	-	+	-	+	+	-	+	-	-	+	-	-	+	-	+	-	+	+	-	+	+	-	+	-	-	+	-	+	+	-
22	+	+	-	+	-	+	-	+	-	+	-	+	-	-	+	-	-	+	-	-	+	-	+	-	+	-	+	-	+	-	+	+
23	+	-	+	+	-	+	-	-	+	-	+	-	+	-	+	-	-	+	-	+	-	+	-	+	-	-	+	-	+	+	-	+
24	+	+	+	+	-	+	+	+	-	+	+	-	+	-	+	-	+	-	+	-	+	-	-	+	-	-	-	+	-	-	-	-
25	+	-	-	-	+	+	+	+	-	-	+	-	-	-	-	+	-	+	+	+	+	-	+	+	-	-	-	-	+	+	+	-
26	+	+	-	-	+	+	-	-	+	+	+	-	-	-	-	+	+	-	-	-	-	+	+	+	-	-	+	+	-	-	+	+
27	+	-	+	-	+	+	-	+	-	-	-	+	+	-	-	+	+	-	-	+	+	-	-	-	+	-	+	+	-	+	-	+
28	+	+	+	-	+	+	+	-	+	+	-	+	+	-	-	+	-	+	+	-	-	+	-	-	+	-	-	-	+	-	-	-
29	+	-	-	+	+	+	+	-	-	-	-	-	-	+	+	+	+	+	+	-	-	-	-	-	-	+	+	+	+	-	-	+
30	+	+	-	+	+	+	-	+	+	+	-	-	-	+	+	+	-	-	-	+	+	+	-	-	-	+	-	-	-	+	-	-
31	+	-	+	+	+	+	-	-	-	-	+	+	+	+	+	+	-	-	-	-	-	-	+	+	+	+	-	-	-	-	+	-
32	+	+	+	+	+	+	+	+	+	+	+	+	+	+	+	+	+	+	+	+	+	+	+	+	+	+	+	+	+	+	+	+

I = 126 = 137 = 148 = 159 = 23(10) = 24(11) = 25(12) = 34(13) = 35(14) = 45(15) = 123(16) = 124(17) = 125(18) = 134(19)=135(20)
= 145(21) = 234(22) = 235(23) = 245(24) = 345(25) = 1234(26) = 1235(27) = 1245(28) = 1345(29) = 2345(30) = 12345(31)

Table B.1-11

THIRTY-TWO RUN FRACTIONAL FACTORIAL DESIGN FOR SEVEN TO SIXTEEN FACTORS
(MAIN EFFECTS CLEAR OF TWO-FACTOR INTERACTIONS)

Run	Mean	X_1	X_2	X_3	X_4	X_5	X_6	X_7	X_8	X_9	X_{10}	X_{11}	X_{12}	X_{13}	X_{14}	X_{15}	X_{16}	E_1	E_2	E_3	E_4	E_5	E_6	E_7	E_8	E_9	E_{10}	E_{11}	E_{12}	E_{13}	E_{14}	E_{15}
1	+																															

CONFOUNDINGS FOR THIRTY-TWO RUN FRACTIONAL FACTORIAL DESIGN
WITH SEVEN TO SIXTEEN FACTORS

E_1	$= 1,2$	$+3, 6$	$+4, 7$	$+5, 8$	$+9,12$	$+10,13$	$+11,14$	$+15,16$
E_2	$= 1,3$	$+2, 6$	$+4, 9$	$+5,10$	$+7,12$	$+8,13$	$+11,15$	$+14,16$
E_3	$= 1,4$	$+2, 7$	$+3, 9$	$+5,11$	$+6,12$	$+8,14$	$+10,15$	$+13,16$
E_4	$= 1,5$	$+2, 8$	$+3,10$	$+4,11$	$+6, 13$	$+7,14$	$+ 9,15$	$+12,16$
E_5	$= 1,6$	$+2, 3$	$+4,12$	$+5,13$	$+7, 9$	$+8,10$	$+11,16$	$+14,15$
E_6	$= 1,7$	$+2, 4$	$+3,12$	$+5,14$	$+6, 9$	$+8,11$	$+10,16$	$+13,15$
E_7	$= 1,8$	$+2, 5$	$+3,13$	$+4,14$	$+6,10$	$+7,11$	$+9,16$	$+12,15$
E_8	$= 1,9$	$+2,12$	$+3, 4$	$+5,15$	$+6, 7$	$+8,16$	$+10,11$	$+13,14$
E_9	$=1,10$	$+2,13$	$+3, 5$	$+4,15$	$+6, 8$	$+7,16$	$+9,11$	$+12,14$
E_{10}	$=1,11$	$+2,14$	$+3,15$	$+4, 5$	$+6,16$	$+7, 8$	$+9,10$	$+12,13$
E_{11}	$=1,12$	$+2,9$	$+3,7$	$+4,6$	$+5,16$	$+8,15$	$+10,14$	$+11,13$	$(+1,2,3,4)$
E_{12}	$=1,13$	$+2,10$	$+3,8$	$+4,16$	$+5, 6$	$+7, 15$	$+9, 14$	$+11,12$	$(+1,2,3,5)$
E_{13}	$=1,14$	$+2,11$	$+3,16$	$+4,8$	$+5, 7$	$+6,15$	$+9,13$	$+10,12$	$(+1,2,4,5)$
E_{14}	$=1,15$	$+2,16$	$+3,11$	$+4,10$	$+5, 9$	$+6,14$	$+7,13$	$+ 8,12$	$(+1,3,4,5)$
E_{15}	$=1,16$	$+2,15$	$+3,14$	$+4,13$	$+5,12$	$+6,11$	$+7,10$	$+8, 9$	$(+2,3,4,5)$

$I = 123(6) = 124(7) = 125(8) = 134(9) = 135(10) = 145(11) = 234(12) = 235(13) = 245(14)$
$= 345(15) = 12345(16)$

Table B.1-13

THIRTY-TWO RUN FRACTIONAL FACTORIAL DESIGN FOR SIX FACTORS

(MAIN EFFECTS AND TWO FACTOR INTERACTIONS CLEAR OF EACH OTHER)

Run	Mean	X_1	X_2	X_3	X_4	X_5	X_6	X_1X_2	X_1X_3	X_1X_4	X_1X_5	X_1X_6	X_2X_3	X_2X_4	X_2X_5	X_2X_6	X_3X_4	X_3X_5	X_3X_6	X_4X_5	X_4X_6	X_5X_6	E_1	E_2	E_3	E_4	E_5	E_6	E_7	E_8	E_9	E_{10}
1	+																															

CONFOUNDINGS FOR THIRTY-TWO RUN FRACTIONAL
FACTORIAL DESIGN WITH SIX FACTORS

$$E_1 = 123 + 456$$

$$E_2 = 124 + 356$$

$$E_3 = 125 + 346$$

$$E_4 = 126 + 345$$

$$E_5 = 134 + 256$$

$$E_6 = 135 + 246$$

$$E_7 = 136 + 245$$

$$E_8 = 145 + 236$$

$$E_9 = 146 + 235$$

$$E_{10} = 156 + 234$$

$$I = 123456$$

Table B.1-15
L_{18} Orthogonal Array Design 2×3^7

	Column							
Run	1	2	3	4	5	6	7	8
1	1	1	1	1	1	1	1	1
2	1	1	2	2	2	2	2	2
3	1	1	3	3	3	3	3	3
4	1	2	1	1	2	2	3	3
5	1	2	2	2	3	3	1	1
6	1	2	3	3	1	1	2	2
7	1	3	1	2	1	3	2	3
8	1	3	2	3	2	1	3	1
9	1	3	3	1	3	2	1	2
10	2	1	1	3	3	2	2	1
11	2	1	2	1	1	3	3	2
12	2	1	3	2	2	1	1	3
13	2	2	1	2	3	1	3	2
14	2	2	2	3	1	2	1	3
15	2	2	3	1	2	3	2	1
16	2	3	1	3	2	3	1	2
17	2	3	2	1	3	1	2	3
18	2	3	3	2	1	2	3	1

Table B.2-1

The Factorial Pattern of Experiments

	Run	X1	X2	X3	X4	X5
	1	-	-	-	-	-
	2	+	-	-	-	-
	3	-	+	-	-	-
k = 2	4	+	+	-	-	-
	5	-	-	+	-	-
	6	+	-	+	-	-
	7	-	+	+	-	-
k = 3	8	+	+	+	-	-
	9	-	-	-	+	-
	10	+	-	-	+	-
	11	-	+	-	+	-
	12	+	+	-	+	-
	13	-	-	+	+	-
	14	+	-	+	+	-
	15	-	+	+	+	-
k = 4	16	+	+	+	+	-
	17	-	-	-	-	+
	18	+	-	-	-	+
	19	-	+	-	-	+
	20	+	+	-	-	+
	21	-	-	+	-	+
	22	+	-	+	-	+
	23	-	+	+	-	+
	24	+	+	+	-	+
	25	-	-	-	+	+
	26	+	-	-	+	+
	27	-	+	-	+	+
	28	+	+	-	+	+
	29	-	-	+	+	+
	30	+	-	+	+	+
	31	-	+	+	+	+
k = 5	32	+	+	+	+	+

Table B.2.-2
COMPUTATION TABLE FOR 2^2 DESIGN

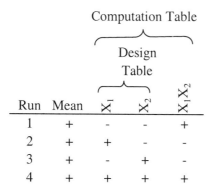

Computation Table — Design Table

Run	Mean	X_1	X_2	X_1X_2
1	+	-	-	+
2	+	+	-	-
3	+	-	+	-
4	+	+	+	+

Table B.2.-3
COMPUTATION TABLE FOR 2^3 DESIGN

Computation Table — Design Table

Run	Mean	X_1	X_2	X_3	X_1X_2	X_1X_3	X_2X_3	$X_1X_2X_3$
1	+	-	-	-	+	+	+	-
2	+	+	-	-	-	-	+	+
3	+	-	+	-	-	+	-	+
4	+	+	+	-	+	-	-	-
5	+	-	-	+	+	-	-	+
6	+	+	-	+	-	+	-	-
7	+	-	+	+	-	-	+	-
8	+	+	+	+	+	+	+	+

Table B.2-4
COMPUTATION TABLE FOR 2^4 DESIGN

Computation Table

Run	Mean	X_1	X_2	X_3	X_4	X_1X_2	X_1X_3	X_1X_4	X_2X_3	X_2X_4	X_3X_4	$X_1X_2X_3$	$X_1X_2X_4$	$X_1X_3X_4$	$X_2X_3X_4$	$X_1X_2X_3X_4$
		Design Table														
1	+	−	−	−	−	+	+	+	+	+	+	−	−	−	−	+
2	+	+	−	−	−	−	−	−	+	+	+	+	+	+	−	−
3	+	−	+	−	−	−	+	+	−	−	+	+	+	−	+	−
4	+	+	+	−	−	+	−	−	−	−	+	−	−	+	+	+
5	+	−	−	+	−	+	−	+	−	+	−	+	−	+	+	−
6	+	+	−	+	−	−	+	−	−	+	−	−	+	−	+	+
7	+	−	+	+	−	−	−	+	+	−	−	−	+	+	−	+
8	+	+	+	+	−	+	+	−	+	−	−	+	−	−	−	−
9	+	−	−	−	+	+	+	−	+	−	−	−	+	+	+	−
10	+	+	−	−	+	−	−	+	+	−	−	+	−	−	+	+
11	+	−	+	−	+	−	+	−	−	+	−	+	−	+	−	+
12	+	+	+	−	+	+	−	+	−	+	−	−	+	−	−	−
13	+	−	−	+	+	+	−	−	−	−	+	+	+	−	−	+
14	+	+	−	+	+	−	+	+	−	−	+	−	−	+	−	−
15	+	−	+	+	+	−	−	−	+	+	+	−	−	−	+	−
16	+	+	+	+	+	+	+	+	+	+	+	+	+	+	+	+

Table B.2-5
COMPUTATION TABLE FOR 2^5 DESIGN

Run	Mean	X_1	X_2	X_3	X_4	X_5	X_1X_2	X_1X_3	X_1X_4	X_1X_5	X_2X_3	X_2X_4	X_2X_5	X_3X_4	X_3X_5	X_4X_5	$X_1X_2X_3$	$X_1X_2X_4$	$X_1X_2X_5$	$X_1X_3X_4$	$X_1X_3X_5$	$X_1X_4X_5$	$X_2X_3X_4$	$X_2X_3X_5$	$X_2X_4X_5$	$X_3X_4X_5$	$X_1X_2X_3X_4$	$X_1X_2X_3X_5$	$X_1X_2X_4X_5$	$X_1X_3X_4X_5$	$X_2X_3X_4X_5$	$X_1X_2X_3X_4X_5$
1	+	−	−	−	−	−	+	+	+	+	+	+	+	+	+	+	−	−	−	−	−	−	−	−	−	−	+	+	+	+	+	−
2	+	+	−	−	−	−	−	−	−	−	+	+	+	+	+	+	+	+	+	+	+	+	−	−	−	−	−	−	−	−	+	+
3	+	−	+	−	−	−	−	+	+	+	−	−	−	+	+	+	+	+	+	−	−	−	+	+	+	−	−	−	−	+	−	+
4	+	+	+	−	−	−	+	−	−	−	−	−	−	+	+	+	−	−	−	+	+	+	+	+	+	−	+	+	+	−	−	−
5	+	−	−	+	−	−	+	−	+	+	−	+	+	−	−	+	+	−	−	+	+	−	+	+	−	+	−	−	+	−	−	+
6	+	+	−	+	−	−	−	+	−	−	−	+	+	−	−	+	−	+	+	−	−	+	+	+	−	+	+	+	−	+	−	−
7	+	−	+	+	−	−	−	−	+	+	+	−	−	−	−	+	−	+	+	+	+	−	−	−	+	+	+	+	−	−	+	−
8	+	+	+	+	−	−	+	+	−	−	+	−	−	−	−	+	+	−	−	−	−	+	−	−	+	+	−	−	+	+	+	+
9	+	−	−	−	+	−	+	+	−	+	+	−	+	−	+	−	−	+	−	+	−	+	+	−	+	+	−	+	−	−	−	+
10	+	+	−	−	+	−	−	−	+	−	+	−	+	−	+	−	+	−	+	−	+	−	+	−	+	+	+	−	+	+	−	−
11	+	−	+	−	+	−	−	+	−	+	−	+	−	−	+	−	+	−	+	+	−	+	−	+	−	+	+	−	+	−	+	−
12	+	+	+	−	+	−	+	−	+	−	−	+	−	−	+	−	−	+	−	−	+	−	−	+	−	+	−	+	−	+	+	+
13	+	−	−	+	+	−	+	−	−	+	−	−	+	+	−	−	+	+	−	−	+	+	−	+	+	−	+	−	−	+	+	−
14	+	+	−	+	+	−	−	+	+	−	−	−	+	+	−	−	−	−	+	+	−	−	−	+	+	−	−	+	+	−	+	+
15	+	−	+	+	+	−	−	−	−	+	+	+	−	+	−	−	−	−	+	−	+	+	+	−	−	−	−	+	+	+	−	+
16	+	+	+	+	+	−	+	+	+	−	+	+	−	+	−	−	+	+	−	+	−	−	+	−	−	−	+	−	−	−	−	−
17	+	−	−	−	−	+	+	+	+	−	+	+	−	+	−	−	−	−	+	−	+	+	−	+	+	+	+	−	−	−	−	+
18	+	+	−	−	−	+	−	−	−	+	+	+	−	+	−	−	+	+	−	+	−	−	−	+	+	+	−	+	+	+	−	−
19	+	−	+	−	−	+	−	+	+	−	−	−	+	+	−	−	+	+	−	−	+	+	+	−	−	+	−	+	+	−	+	−
20	+	+	+	−	−	+	+	−	−	+	−	−	+	+	−	−	−	−	+	+	−	−	+	−	−	+	+	−	−	+	+	+
21	+	−	−	+	−	+	+	−	+	−	−	+	−	−	+	−	+	−	+	+	−	+	+	−	+	−	−	+	−	+	+	−
22	+	+	−	+	−	+	−	+	−	+	−	+	−	−	+	−	−	+	−	−	+	−	+	−	+	−	+	−	+	−	+	+
23	+	−	+	+	−	+	−	−	+	−	+	−	+	−	+	−	−	+	−	+	−	+	−	+	−	−	+	−	+	+	−	+
24	+	+	+	+	−	+	+	+	−	+	+	−	+	−	+	−	+	−	+	−	+	−	−	+	−	−	−	+	−	−	−	−
25	+	−	−	−	+	+	+	+	−	−	+	−	−	−	−	+	−	+	+	+	+	−	+	+	−	−	−	−	+	+	+	−
26	+	+	−	−	+	+	−	−	+	+	+	−	−	−	−	+	+	−	−	−	−	+	+	+	−	−	+	+	−	−	+	+
27	+	−	+	−	+	+	−	+	−	−	−	+	+	−	−	+	+	−	−	+	+	−	−	−	+	−	+	+	−	+	−	+
28	+	+	+	−	+	+	+	−	+	+	−	+	+	−	−	+	−	+	+	−	−	+	−	−	+	−	−	−	+	−	−	−
29	+	−	−	+	+	+	+	−	−	−	−	−	−	+	+	+	+	+	+	−	−	−	−	−	−	+	+	+	+	−	−	+
30	+	+	−	+	+	+	−	+	+	+	−	−	−	+	+	+	−	−	−	+	+	+	−	−	−	+	−	−	−	+	−	−
31	+	−	+	+	+	+	−	−	−	−	+	+	+	+	+	+	−	−	−	−	−	−	+	+	+	+	−	−	−	−	+	−
32	+	+	+	+	+	+	+	+	+	+	+	+	+	+	+	+	+	+	+	+	+	+	+	+	+	+	+	+	+	+	+	+

776

Table B.3-1
CENTRAL COMPOSITE DESIGN FOR TWO FACTORS

Run	X_1	X_2
1	-1	-1
2	1	-1
3	-1	1
4	1	1
5	$-\alpha$	0
6	α	0
7	0	$-\alpha$
8	0	α
9	0	0
10	0	0
11	0	0
12	0	0
13	0	0

Note: $\alpha = \sqrt{2} = 1.41$

Table B.3-1
CENTRAL COMPOSITE FOR THREE FACTORS

Run	X_1	X_2	X_3	
1	-1	-1	-1	
2	1	-1	-1	
3	-1	1	-1	
4	1	1	-1	
5	-1	-1	1	
6	1	-1	1	Block 1
7	-1	1	1	
8	1	1	1	
9	0	0	0	
10	0	0	0	
11	0	0	0	
12	$-\alpha$	0	0	
13	α	0	0	
14	0	$-\alpha$	0	
15	0	α	0	
16	0	0	$-\alpha$	Block 2
17	0	0	α	
18	0	0	0	
19	0	0	0	
20	0	0	0	

Note: $\alpha = \sqrt{3} = 1.73$

Table B.3-3
CENTRAL COMPOSITE DESIGN FOR FOUR FACTORS

Run	X_1	X_2	X_3	X_4	
1	1	-1	-1	-1	
2	-1	1	-1	-1	
3	-1	-1	1	-1	
4	1	1	1	-1	
5	-1	-1	-1	1	Block 1
6	1	1	-1	1	
7	1	-1	1	1	
8	-1	1	1	1	
9	0	0	0	0	
10	0	0	0	0	
11	0	0	0	0	
12	-1	-1	-1	-1	
13	1	1	-1	-1	
14	1	-1	1	-1	
15	-1	1	1	-1	
16	1	-1	-1	1	Block 2
17	-1	1	-1	1	
18	-1	-1	1	1	
19	1	1	1	1	
20	0	0	0	0	
21	0	0	0	0	
22	0	0	0	0	
23	$-\alpha$	0	0	0	
24	α	0	0	0	
25	0	$-\alpha$	0	0	
26	0	α	0	0	
27	0	0	$-\alpha$	0	Block 3
28	0	0	α	0	
29	0	0	0	$-\alpha$	
30	0	0	0	α	
31	0	0	0	0	
32	0	0	0	0	

Note:

(1) $\alpha = \sqrt{4} = 2.00$

(2) If the experiments are not broken up into three blocks, only seven center points are needed not nine.

Table B.3-4

CENTRAL COMPOSITE DESIGN FOR FIVE FACTORS

Run	X_1	X_2	X_3	X_4	X_5	
1	-1	-1	-1	-1	1	
2	1	-1	-1	-1	-1	
3	-1	1	-1	-1	-1	
4	1	1	-1	-1	1	
5	-1	-1	1	-1	-1	
6	1	-1	1	-1	1	
7	-1	1	1	-1	1	
8	1	1	1	-1	-1	
9	-1	-1	-1	1	-1	
10	1	-1	-1	1	1	Block 1
11	-1	1	-1	1	1	
12	1	1	-1	1	-1	
13	-1	-1	1	1	1	
14	1	-1	1	1	-1	
15	-1	1	1	1	-1	
16	1	1	1	1	1	
17	0	0	0	0	0	
18	0	0	0	0	0	
19	0	0	0	0	0	
20	$-\alpha$	0	0	0	0	
21	α	0	0	0	0	
22	0	$-\alpha$	0	0	0	
23	0	α	0	0	0	
24	0	0	$-\alpha$	0	0	
25	0	0	α	0	0	Block 2
26	0	0	0	$-\alpha$	0	
27	0	0	0	α	0	
28	0	0	0	0	$-\alpha$	
29	0	0	0	0	α	
30	0	0	0	0	0	
31	0	0	0	0	0	
32	0	0	0	0	0	

Note: $\alpha = \sqrt{4} = 2.00$

Table B.3-5
CENTRAL COMPOSITE DESIGN FOR SIX FACTORS

BLOCK 1

Run	X_1	X_2	X_3	X_4	X_5	X_6
1	-1	-1	-1	-1	-1	-1
2	1	-1	-1	-1	-1	1
3	-1	1	-1	-1	-1	1
4	1	1	-1	-1	-1	-1
5	-1	-1	1	-1	-1	1
6	1	-1	1	-1	-1	-1
7	-1	1	1	-1	-1	-1
8	1	1	1	-1	-1	1
9	-1	-1	-1	1	-1	1
10	1	-1	-1	1	-1	-1
11	-1	1	-1	1	-1	-1
12	1	1	-1	1	-1	1
13	-1	-1	1	1	-1	-1
14	1	-1	1	1	-1	1
15	-1	1	1	1	-1	1
16	1	1	1	1	-1	-1
17	0	0	0	0	0	0
18	0	0	0	0	0	0
19	0	0	0	0	0	0

BLOCK 2

Run	X_1	X_2	X_3	X_4	X_5	X_6
20	-1	-1	-1	-1	1	1
21	1	-1	-1	-1	1	-1
22	-1	1	-1	-1	1	-1
23	1	1	-1	-1	1	1
24	-1	-1	1	-1	1	-1
25	1	-1	1	-1	1	1
26	-1	1	1	-1	1	1
27	1	1	1	-1	1	-1
28	-1	-1	-1	1	1	-1
29	1	-1	-1	1	1	1
30	-1	1	-1	1	1	1
31	1	1	-1	1	1	-1
32	-1	-1	1	1	1	1
33	1	-1	1	1	1	-1
34	-1	1	1	1	1	-1
35	1	1	1	1	1	1
36	0	0	0	0	0	0
37	0	0	0	0	0	0
38	0	0	0	0	0	0

BLOCK 3

Run	X_1	X_2	X_3	X_4	X_5	X_6
39	$-\alpha$	0	0	0	0	0
40	α	0	0	0	0	0
41	0	$-\alpha$	0	0	0	0
42	0	α	0	0	0	0
43	0	0	$-\alpha$	0	0	0
44	0	0	α	0	0	0
45	0	0	0	$-\alpha$	0	0
46	0	0	0	α	0	0
47	0	0	0	0	$-\alpha$	0
48	0	0	0	0	α	0
49	0	0	0	0	0	$-\alpha$
50	0	0	0	0	0	α
51	0	0	0	0	0	0
52	0	0	0	0	0	0
53	0	0	0	0	0	0

Note: $\alpha = \sqrt{5} = 2.24$

Table B.3-6
CENTRAL COMPOSITE DESIGN FOR SEVEN FACTORS

BLOCK 1									BLOCK 2							
Run	X1	X2	X3	X4	X5	X6	X7		Run	X1	X2	X3	X4	X5	X6	X7
1	-1	-1	-1	-1	-1	-1	1		19	1	-1	-1	-1	-1	-1	-1
2	-1	1	-1	-1	-1	-1	-1		20	1	1	-1	-1	-1	-1	1
3	1	-1	1	1	-1	-1	-1		21	-1	-1	1	1	-1	-1	1
4	1	1	1	1	-1	-1	1		22	-1	1	1	1	-1	-1	-1
5	1	-1	1	-1	1	-1	-1		23	-1	-1	1	-1	1	-1	1
6	1	1	1	-1	1	-1	1		24	-1	1	1	-1	1	-1	-1
7	-1	-1	-1	1	1	-1	1		25	1	-1	-1	1	1	-1	-1
8	-1	1	-1	1	1	-1	-1		26	1	1	-1	1	1	-1	1
9	1	-1	-1	-1	-1	1	1		27	-1	-1	-1	-1	-1	1	-1
10	1	1	-1	-1	-1	1	-1		28	-1	1	-1	-1	-1	1	1
11	-1	-1	1	1	-1	1	-1		29	1	-1	1	1	-1	1	1
12	-1	1	1	1	-1	1	1		30	1	1	1	1	-1	1	-1
13	-1	-1	1	-1	1	1	-1		31	1	-1	1	-1	1	1	1
14	-1	1	1	-1	1	1	1		32	1	1	1	-1	1	1	-1
15	1	-1	-1	1	1	1	1		33	-1	-1	-1	1	1	1	-1
16	1	1	-1	1	1	1	-1		34	-1	1	-1	1	1	1	1
17	0	0	0	0	0	0	0		35	0	0	0	0	0	0	0
18	0	0	0	0	0	0	0		36	0	0	0	0	0	0	0

BLOCK 3									BLOCK 4							
Run	X1	X2	X3	X4	X5	X6	X7		Run	X1	X2	X3	X4	X5	X6	X7
37	1	-1	1	-1	-1	-1	1		55	-1	-1	1	-1	-1	-1	-1
38	1	1	1	-1	-1	-1	-1		56	-1	1	1	-1	-1	-1	1
39	-1	-1	-1	1	-1	-1	-1		57	1	-1	-1	1	-1	-1	1
40	-1	1	-1	1	-1	-1	1		58	1	1	-1	1	-1	-1	-1
41	-1	-1	-1	-1	1	-1	-1		59	1	-1	-1	-1	1	-1	1
42	-1	1	-1	-1	1	-1	1		60	1	1	-1	-1	1	-1	-1
43	1	-1	1	1	1	-1	1		61	-1	-1	1	1	1	-1	-1
44	1	1	1	1	1	-1	-1		62	-1	1	1	1	1	-1	1
45	-1	-1	1	1	-1	1	-1		63	1	-1	1	1	-1	1	-1
46	-1	1	1	-1	-1	1	-1		64	1	1	1	-1	-1	1	1
47	1	-1	-1	1	-1	1	-1		65	-1	-1	-1	1	-1	1	1
48	1	1	-1	1	-1	1	1		66	-1	1	-1	1	-1	1	-1
49	1	-1	-1	-1	1	1	-1		67	-1	-1	-1	-1	1	1	1
50	1	1	-1	-1	1	1	1		68	-1	1	-1	-1	1	1	-1
51	-1	-1	1	1	1	1	1		69	1	-1	1	1	1	1	-1
52	-1	1	1	1	1	1	-1		70	1	1	1	1	1	1	1
53	0	0	0	0	0	0	0		71	0	0	0	0	0	0	0
54	0	0	0	0	0	0	0		72	0	0	0	0	0	0	0

BLOCK 3							
Run	X1	X2	X3	X4	X5	X6	X7
73	$-\alpha$	0	0	0	0	0	0
74	α	0	0	0	0	0	0
75	0	$-\alpha$	0	0	0	0	0
76	0	α	0	0	0	0	0
77	0	0	$-\alpha$	0	0	0	0
78	0	0	α	0	0	0	0
79	0	0	0	$-\alpha$	0	0	0
80	0	0	0	α	0	0	0
81	0	0	0	0	$-\alpha$	0	0
82	0	0	0	0	α	0	0
83	0	0	0	0	0	$-\alpha$	0
84	0	0	0	0	0	α	0
85	0	0	0	0	0	0	$-\alpha$
86	0	0	0	0	0	0	α
87	0	0	0	0	0	0	0

Note: $\alpha = \sqrt{7} = 2.65$

Table B.4-1

BOX-BEHNKEN DESIGN FOR THREE FACTORS

Run	X_1	X_2	X_3
1	-1	-1	0
2	1	-1	0
3	-1	1	0
4	1	1	0
5	-1	0	-1
6	1	0	-1
7	-1	0	1
8	1	0	1
9	0	-1	-1
10	0	1	-1
11	0	-1	1
12	0	1	1
13	0	0	0
14	0	0	0
15	0	0	0

Table B.4-2
BOX-BEHNKEN DESIGN FOR FOUR FACTORS

Run	X_1	X_2	X_3	X_4	
1	-1	-1	0	0	
2	1	-1	0	0	
3	-1	1	0	0	
4	1	1	0	0	
5	0	0	-1	-1	Block 1
6	0	0	1	-1	
7	0	0	-1	1	
8	0	0	1	1	
9	0	0	0	0	
10	-1	0	-1	0	
11	1	0	-1	0	
12	-1	0	1	0	
13	1	0	1	0	
14	0	-1	0	-1	Block 2
15	0	1	0	-1	
16	0	-1	0	1	
17	0	1	0	1	
18	0	0	0	0	
19	-1	0	0	-1	
20	1	0	0	-1	
21	-1	0	0	1	
22	1	0	0	1	Block 3
23	0	-1	-1	0	
24	0	1	-1	0	
25	0	-1	1	0	
26	0	1	1	0	
27	0	0	0	0	

Table B.4-3
BOX-BEHNKEN DESIGN FOR FIVE FACTORS

		BLOCK 1						BLOCK 2			
Run	X_1	X_2	X_3	X_4	X_5	Run	X_1	X_2	X_3	X_4	X_5
1	-1	-1	0	0	0	24	0	-1	-1	0	0
2	1	-1	0	0	0	25	0	1	-1	0	0
3	-1	1	0	0	0	26	0	-1	1	0	0
4	1	1	0	0	0	27	0	1	1	0	0
5	0	0	-1	-1	0	28	-1	0	0	-1	0
6	0	0	1	-1	0	29	1	0	0	-1	0
7	0	0	-1	1	0	30	-1	0	0	1	0
8	0	0	1	1	0	31	1	0	0	1	0
9	0	-1	0	0	-1	32	0	0	-1	0	-1
10	0	1	0	0	-1	33	0	0	1	0	-1
11	0	-1	0	0	1	34	0	0	-1	0	1
12	0	1	0	0	1	35	0	0	1	0	1
13	-1	0	-1	0	0	36	-1	0	0	0	-1
14	1	0	-1	0	0	37	1	0	0	0	-1
15	-1	0	1	0	0	38	-1	0	0	0	1
16	1	0	1	0	0	39	1	0	0	0	1
17	0	0	0	-1	-1	40	0	-1	0	-1	0
18	0	0	0	1	-1	41	0	1	0	-1	0
19	0	0	0	-1	1	42	0	-1	0	1	0
20	0	0	0	1	1	43	0	1	0	1	0
21	0	0	0	0	0	44	0	0	0	0	0
22	0	0	0	0	0	45	0	0	0	0	0
23	0	0	0	0	0	46	0	0	0	0	0

Table B.4-4
BOX-BEHNKEN DESIGN FOR SIX FACTORS

		BLOCK 1								BLOCK 2				
Run	X_1	X_2	X_3	X_4	X_5	X_6	Run	X_1	X_2	X_3	X_4	X_5	X_6
1	-1	-1	0	-1	0	0	28	1	-1	0	-1	0	0
2	1	1	0	-1	0	0	29	-1	1	0	-1	0	0
3	1	-1	0	1	0	0	30	-1	-1	0	1	0	0
4	-1	1	0	1	0	0	31	1	1	0	1	0	0
5	0	-1	-1	0	-1	0	32	0	1	-1	0	-1	0
6	0	1	1	0	-1	0	33	0	-1	1	0	-1	0
7	0	1	-1	0	1	0	34	0	-1	-1	0	1	0
8	0	-1	1	0	1	0	35	0	1	1	0	1	0
9	0	0	-1	-1	0	-1	36	0	0	1	-1	0	-1
10	0	0	1	1	0	-1	37	0	0	-1	1	0	-1
11	0	0	1	-1	0	1	38	0	0	-1	-1	0	1
12	0	0	-1	1	0	1	39	0	0	1	1	0	1
13	-1	0	0	-1	-1	0	40	1	0	0	-1	-1	0
14	1	0	0	1	-1	0	41	-1	0	0	1	-1	0
15	1	0	0	-1	1	0	42	-1	0	0	-1	1	0
16	-1	0	0	1	1	0	43	1	0	0	1	1	0
17	0	-1	0	0	-1	-1	44	0	1	0	0	-1	-1
18	0	1	0	0	1	-1	45	0	-1	0	0	1	-1
19	0	1	0	0	-1	1	46	0	-1	0	0	-1	1
20	0	-1	0	0	1	1	47	0	1	0	0	1	1
21	-1	0	-1	0	0	-1	48	1	0	-1	0	0	-1
22	1	0	1	0	0	-1	49	-1	0	1	0	0	-1
23	1	0	-1	0	0	1	50	-1	0	-1	0	0	1
24	-1	0	1	0	0	1	51	1	0	1	0	0	1
25	0	0	0	0	0	0	52	0	0	0	0	0	0
26	0	0	0	0	0	0	53	0	0	0	0	0	0
27	0	0	0	0	0	0	54	0	0	0	0	0	0

Table B.4-5
BOX-BEHNKEN DESIGN FOR SEVEN FACTORS

		BLOCK 1									BLOCK 2					
Run	X_1	X_2	X_3	X_4	X_5	X_6	X_7	Run	X_1	X_2	X_3	X_4	X_5	X_6	X_7
1	0	0	0	-1	-1	-1	0	32	0	0	0	1	-1	-1	0
2	0	0	0	1	1	-1	0	33	0	0	0	-1	1	-1	0
3	0	0	0	1	-1	1	0	34	0	0	0	-1	-1	1	0
4	0	0	0	-1	1	1	0	35	0	0	0	1	1	1	0
5	-1	0	0	0	0	-1	-1	36	1	0	0	0	0	-1	-1
6	1	0	0	0	0	1	-1	37	-1	0	0	0	0	1	-1
7	1	0	0	0	0	-1	1	38	-1	0	0	0	0	-1	1
8	-1	0	0	0	0	1	1	39	1	0	0	0	0	1	1
9	0	-1	0	0	-1	0	-1	40	0	1	0	0	-1	0	-1
10	0	1	0	0	1	0	-1	41	0	-1	0	0	1	0	-1
11	0	1	0	0	-1	0	1	42	0	-1	0	0	-1	0	1
12	0	-1	0	0	1	0	1	43	0	1	0	0	1	0	1
13	-1	-1	0	-1	0	0	0	44	1	-1	0	-1	0	0	0
14	1	1	0	-1	0	0	0	45	-1	1	0	-1	0	0	0
15	1	-1	0	1	0	0	0	46	-1	-1	0	1	0	0	0
16	-1	1	0	1	0	0	0	47	1	1	0	1	0	0	0
17	0	0	-1	-1	0	0	-1	48	0	0	1	-1	0	0	-1
18	0	0	1	1	0	0	-1	49	0	0	-1	1	0	0	-1
19	0	0	1	-1	0	0	1	50	0	0	-1	-1	0	0	1
20	0	0	-1	1	0	0	1	51	0	0	1	1	0	0	1
21	-1	0	-1	0	-1	0	0	52	1	0	-1	0	-1	0	0
22	1	0	1	0	-1	0	0	53	-1	0	1	0	-1	0	0
23	1	0	-1	0	1	0	0	54	-1	0	-1	0	1	0	0
24	-1	0	1	0	1	0	0	55	1	0	1	0	1	0	0
25	0	-1	-1	0	0	-1	0	56	0	1	-1	0	0	-1	0
26	0	1	1	0	0	-1	0	57	0	-1	1	0	0	-1	0
27	0	1	-1	0	0	1	0	58	0	-1	-1	0	0	1	0
28	0	-1	1	0	0	1	0	59	0	1	1	0	0	1	0
29	0	0	0	0	0	0	0	60	0	0	0	0	0	0	0
30	0	0	0	0	0	0	0	61	0	0	0	0	0	0	0
31	0	0	0	0	0	0	0	62	0	0	0	0	0	0	0

Table B.5-1
SMALL COMPOSITE DESIGN FOR TWO FACTORS

Run	X_1	X_2
1	1	-1
2	1	1
3	$-\alpha$	0
4	α	0
5	0	$-\alpha$
6	0	α

Note: $\alpha=1.19$, $n_0 = 2$ center points can be added to get an estimate of error

Table B.5-2
SMALL COMPOSITE DESIGN FOR THREE FACTORS

Run	X_1	X_2	X_3
1	-1	-1	1
2	1	-1	-1
3	-1	1	-1
4	1	1	1
5	$-\alpha$	0	0
6	α	0	0
7	0	$-\alpha$	0
8	0	α	0
9	0	0	$-\alpha$
10	0	0	α

Note: $\alpha=1.41$, $n_0 = 2$ center points can be added to get an estimate of error

Table B.5-3

SMALL COMPOSITE DESIGN FOR FOUR FACTORS

Run	X_1	X_2	X_3	X_4
1	1	-1	-1	-1
2	-1	1	-1	-1
3	-1	-1	1	-1
4	1	1	1	-1
5	1	-1	-1	1
6	-1	1	-1	1
7	-1	-1	1	1
8	1	1	1	1
9	$-\alpha$	0	0	0
10	α	0	0	0
11	0	$-\alpha$	0	0
12	0	α	0	0
13	0	0	$-\alpha$	0
14	0	0	α	0
15	0	0	0	$-\alpha$
16	0	0	0	α

Note: $\alpha=1.68$, $n_0 = 1$ center point can be added to get an estimate of error

Table B.5.4
SMALL COMPOSITE DESIGN FOR FIVE FACTORS

Run	X_1	X_2	X_3	X_4	X_5
1	1	1	1	-1	-1
2	-1	1	-1	-1	1
3	1	1	-1	1	-1
4	1	-1	-1	-1	1
5	1	-1	1	1	1
6	-1	-1	-1	1	-1
7	-1	1	1	-1	1
8	-1	-1	1	1	1
9	1	1	-1	1	1
10	1	-1	1	-1	-1
11	-1	-1	-1	-1	-1
12	$-\alpha$	0	0	0	0
13	α	0	0	0	0
14	0	$-\alpha$	0	0	0
15	0	α	0	0	0
16	0	0	$-\alpha$	0	0
17	0	0	α	0	0
18	0	0	0	$-\alpha$	0
19	0	0	0	α	0
20	0	0	0	0	$-\alpha$
21	0	0	0	0	α

Note: $\alpha=1.82$, $n_0 = 2$ center points can be added to get an estimate of error

Table B.5-5
SMALL COMPOSITE DESIGN FOR SIX FACTORS

Run	X_1	X_2	X_3	X_4	X_5	X_6
1	1	-1	-1	1	-1	-1
2	-1	1	-1	1	-1	-1
3	-1	-1	1	1	-1	-1
4	1	1	1	1	-1	-1
5	1	-1	-1	-1	1	-1
6	-1	1	-1	-1	1	-1
7	-1	-1	1	-1	1	-1
8	1	1	1	-1	1	-1
9	1	-1	-1	-1	-1	1
10	-1	1	-1	-1	-1	1
11	-1	-1	1	-1	-1	1
12	1	1	1	-1	-1	1
13	1	-1	-1	1	1	1
14	-1	1	-1	1	1	1
15	-1	-1	1	1	1	1
16	1	1	1	1	1	1
17	$-\alpha$	0	0	0	0	0
18	α	0	0	0	0	0
19	0	$-\alpha$	0	0	0	0
20	0	α	0	0	0	0
21	0	0	$-\alpha$	0	0	0
22	0	0	α	0	0	0
23	0	0	0	$-\alpha$	0	0
24	0	0	0	α	0	0
25	0	0	0	0	$-\alpha$	0
26	0	0	0	0	α	0
27	0	0	0	0	0	$-\alpha$
28	0	0	0	0	0	α

Note: $\alpha = 2.0$, $n_0 = 2$ center points can be added to get an estimate of error

Table B.5-6

SMALL COMPOSITE DESIGN FOR SEVEN FACTORS

Run	X_1	X_2	X_3	X_4	X_5	X_6	X_7
1	1	-1	-1	1	1	-1	1
2	1	-1	-1	-1	1	1	1
3	1	-1	1	1	-1	1	-1
4	1	1	1	1	1	-1	-1
5	1	1	-1	-1	1	1	1
6	-1	1	-1	1	-1	1	1
7	-1	1	1	1	1	-1	1
8	-1	-1	-1	-1	1	1	-1
9	-1	-1	1	1	-1	1	1
10	-1	1	-1	1	1	-1	1
11	1	-1	1	-1	1	1	1
12	1	-1	1	-1	-1	-1	1
13	-1	1	1	-1	-1	-1	1
14	-1	-1	1	1	1	1	-1
15	1	1	-1	1	1	1	-1
16	-1	1	1	-1	1	-1	-1
17	1	1	1	-1	-1	1	1
18	1	-1	-1	1	-1	-1	-1
19	1	1	-1	-1	-1	1	-1
20	1	1	-1	1	-1	-1	-1
21	-1	-1	-1	-1	1	-1	1
22	-1	1	1	-1	-1	1	-1
23	1	-1	1	1	-1	-1	1
24	1	1	1	-1	1	-1	-1
25	-1	-1	-1	-1	-1	-1	-1
26	$-\alpha$	0	0	0	0	0	0
27	α	0	0	0	0	0	0
28	0	$-\alpha$	0	0	0	0	0
29	0	α	0	0	0	0	0
30	0	0	$-\alpha$	0	0	0	0
31	0	0	α	0	0	0	0
32	0	0	0	$-\alpha$	0	0	0
33	0	0	0	α	0	0	0
34	0	0	0	0	$-\alpha$	0	0
35	0	0	0	0	α	0	0
36	0	0	0	0	0	$-\alpha$	0
37	0	0	0	0	0	α	0
38	0	0	0	0	0	0	$-\alpha$
39	0	0	0	0	0	0	α

Note: $\alpha = 2.24$

Table B.6-1

Simplex Design for Three Mixture Components *

Run No.	X_1	X_2	X_3	Symbolic Response	Coefficient Estimates for Special Cubic Scheffé Model
1	1	0	0	Y_1	$\beta_1 = Y_1$
2	0	1	0	Y_2	$\beta_2 = Y_2$
3	0	0	1	Y_3	$\beta_3 = Y_3$
4	1/2	1/2	0	Y_{12}	$\beta_{12} = 4Y_{12} - 2(Y_1 + Y_2)$
5	0	0	1/2	Y_{13}	$\beta_{13} = 4Y_{13} - 2(Y_1 + Y_3)$
6	1/2	1/2	1/2	Y_{23}	$\beta_{23} = 4Y_{23} - 2(Y_2 + Y_3)$
7	1/3	1/3	1/3	$Y_{123(1)}$	
8	1/3	1/3	1/3	$Y_{123(2)}$	$\beta_{123} = 27(\overline{Y}_{123}) - 12(Y_{12} + Y_{13} + Y_{23}) + 3(Y_1 + Y_2 + Y_3)$
9	1/3	1/3	1/3	$Y_{123(3)}$	

*Runs 1–3 are the linear design, adding 4–6 creates a quadratic design, and adding Runs 7–9 creates a special cubic design with a replicated center point.

Table B.6-2

Simplex Design for Four Mixture Components *

Run No.	X_1	X_2	X_3	X_4	Symbolic Response	Coefficient Estimates for Special Cubic Scheffé Model
1	1	0	0	0	Y_1	$\beta_1 = Y_1$
2	0	1	0	0	Y_2	$\beta_2 = Y_2$
3	0	0	1	0	Y_3	$\beta_3 = Y_3$
4	0	0	0	1	Y_4	$\beta_4 = Y_4$
5	½	½	0	0	Y_{12}	$\beta_{12} = 4Y_{12} - 2(Y_1 + Y_2)$
6	½	0	½	0	Y_{13}	$\beta_{13} = 4Y_{13} - 2(Y_1 + Y_3)$
7	½	0	0	½	Y_{14}	$\beta_{14} = 4Y_{14} - 2(Y_1 + Y_4)$
8	0	½	½	0	Y_{23}	$\beta_{23} = 4Y_{23} - 2(Y_2 + Y_3)$
9	0	½	0	½	Y_{24}	$\beta_{24} = 4Y_{24} - 2(Y_2 + Y_4)$
10	0	0	½	½	Y_{34}	$\beta_{25} = 4Y_{34} - 2(Y_3 + Y_4)$
11	1/3	1/3	1/3	0	Y_{123}	$\beta_{123} = 27Y_{123} - 12(Y_{12} + Y_{13} + Y_{23}) + 3(Y_1 + Y_2 + Y_3)$
12	1/3	1/3	0	1/3	Y_{124}	$\beta_{124} = 27Y_{124} - 12(Y_{12} + Y_{14} + Y_{24}) + 3(Y_1 + Y_2 + Y_4)$
13	1/3	0	1/3	1/3	Y_{134}	$\beta_{134} = 27Y_{134} - 12(Y_{13} + Y_{14} + Y_{34}) + 3(Y_1 + Y_3 + Y_4)$
14	0	1/3	1/3	1/3	Y_{234}	$\beta_{234} = 27Y_{234} - 12(Y_{23} + Y_{24} + Y_{34}) + 3(Y_2 + Y_3 + Y_4)$
15	1/4	1/4	1/4	1/4	$Y_{1234(1)}$	
16	1/4	1/4	1/4	1/4	$Y_{1234(2)}$	
17	1/4	1/4	1/4	1/4	$Y_{1234(3)}$	

*Runs 1–4 are the linear design, adding Runs 5–10 creates a quadratic design, adding Runs 11–14 creates a special cubic design, and adding Run 15–17 provides a replicated center point for checking goodness of fit.

Simplex Design for Five Mixture Components *

Run No.	X_1	X_2	X_3	X_4	X_5	Symbolic Response	Coefficient Estimates for Special Cubic Scheffé Model
1	1	0	0	0	0	Y_1	$\beta_1 = Y_1$
2	0	1	0	0	0	Y_2	$\beta_2 = Y_2$
3	0	0	1	0	0	Y_3	$\beta_3 = Y_3$
4	0	0	0	1	0	Y_4	$\beta_4 = Y_4$
5	0	0	0	0	1	Y_5	$\beta_5 = Y_5$
6	½	½	0	0	0	Y_{12}	$\beta_{12} = 4Y_{12} - 2(Y_1 + Y_2)$
7	½	0	½	0	0	Y_{13}	$\beta_{13} = 4Y_{13} - 2(Y_1 + Y_3)$
8	½	0	0	½	0	Y_{14}	$\beta_{14} = 4Y_{14} - 2(Y_1 + Y_4)$
9	½	0	0	0	½	Y_{15}	$\beta_{15} = 4Y_{15} - 2(Y_1 + Y_5)$
10	0	½	½	0	0	Y_{23}	$\beta_{23} = 4Y_{23} - 2(Y_2 + Y_3)$
11	0	½	0	½	0	Y_{24}	$\beta_{24} = 4Y_{24} - 2(Y_2 + Y_4)$
12	0	½	0	0	½	Y_{25}	$\beta_{25} = 4Y_{25} - 2(Y_2 + Y_5)$
13	0	0	½	½	0	Y_{34}	$\beta_{34} = 4Y_{34} - 2(Y_3 + Y_4)$
14	0	0	½	0	½	Y_{35}	$\beta_{35} = 4Y_{35} - 2(Y_3 + Y_5)$
15	0	0	0	½	½	Y_{45}	$\beta_{45} = 4Y_{45} - 2(Y_4 + Y_5)$
16	1/3	1/3	1/3	0	0	Y_{123}	$\beta_{123} = 27Y_{123} - 12(Y_{12} + Y_{13} + Y_{23}) + 3(Y_1 + Y_2 + Y_3)$
17	1/3	1/3	0	1/3	0	Y_{124}	$\beta_{124} = 27Y_{124} - 12(Y_{12} + Y_{14} + Y_{24}) + 3(Y_1 + Y_2 + Y_4)$
18	1/3	1/3	0	0	1/3	Y_{125}	$\beta_{125} = 27Y_{125} - 12(Y_{12} + Y_{15} + Y_{25}) + 3(Y_1 + Y_2 + Y_5)$
19	1/3	0	1/3	1/3	0	Y_{134}	$\beta_{134} = 27Y_{134} - 12(Y_{13} + Y_{14} + Y_{34}) + 3(Y_1 + Y_3 + Y_4)$
20	1/3	0	1/3	0	1/3	Y_{135}	$\beta_{135} = 27Y_{135} - 12(Y_{13} + Y_{15} + Y_{35}) + 3(Y_1 + Y_3 + Y_5)$
21	1/3	0	0	1/3	1/3	Y_{145}	$\beta_{145} = 27Y_{145} - 12(Y_{14} + Y_{15} + Y_{45}) + 3(Y_1 + Y_4 + Y_5)$
22	0	1/3	1/3	1/3	0	Y_{234}	$\beta_{234} = 27Y_{234} - 12(Y_{23} + Y_{24} + Y_{34}) + 3(Y_2 + Y_3 + Y_4)$
23	0	1/3	1/3	0	1/3	Y_{235}	$\beta_{235} = 27Y_{235} - 12(Y_{23} + Y_{25} + Y_{35}) + 3(Y_2 + Y_3 + Y_5)$
24	0	1/3	0	1/3	1/3	Y_{245}	$\beta_{245} = 27Y_{245} - 12(Y_{24} + Y_{25} + Y_{45}) + 3(Y_2 + Y_4 + Y_5)$
25	0	0	1/3	1/3	1/3	Y_{345}	$\beta_{345} = 27Y_{345} - 12(Y_{34} + Y_{35} + Y_{45}) + 3(Y_3 + Y_4 + Y_5)$
26	1/5	1/5	1/5	1/5	1/5	$Y_{12345(1)}$	
27	1/5	1/5	1/5	1/5	1/5	$Y_{12345(2)}$	
28	1/5	1/5	1/5	1/5	1/5	$Y_{12345(3)}$	

*Runs 1–5 are the linear design, adding Runs 6–15 creates a quadratic design, adding Runs 16–24 creates a special cubic design, and adding Runs 25–28 provides a replicated center point for checking goodness of fit.

Simplex Design for Six Mixture Components *

Run No.	X_1	X_2	X_3	X_4	X_5	X_6	Symbolic Response	Coefficient Estimates for Special Cubic Scheffé Models
1	1	0	0	0	0	0	Y_1	$\beta_1 = Y_1$
2	0	1	0	0	0	0	Y_2	$\beta_2 = Y_2$
3	0	0	1	0	0	0	Y_3	$\beta_3 = Y_3$
4	0	0	0	1	0	0	Y_4	$\beta_4 = Y_4$
5	0	0	0	0	1	0	Y_5	$\beta_5 = Y_5$
6	0	0	0	0	0	1	Y_6	$\beta_6 = Y_6$
7	1/2	1/2	0	0	0	0	Y_{12}	$\beta_{12} = 4Y_{12} - 2(Y_1 + Y_2)$
8	1/2	0	1/2	0	0	0	Y_{13}	$\beta_{13} = 4Y_{13} - 2(Y_1 + Y_3)$
9	1/2	0	0	1/2	0	0	Y_{14}	$\beta_{14} = 4Y_{14} - 2(Y_1 + Y_4)$
10	1/2	0	0	0	1/2	0	Y_{15}	$\beta_{15} = 4Y_{15} - 2(Y_1 + Y_5)$
11	1/2	0	0	0	0	1/2	Y_{16}	$\beta_{16} = 4Y_{16} - 2(Y_1 + Y_6)$
12	0	1/2	1/2	0	0	0	Y_{23}	$\beta_{23} = 4Y_{23} - 2(Y_2 + Y_3)$
13	0	1/2	0	1/2	0	0	Y_{24}	$\beta_{24} = 4Y_{24} - 2(Y_2 + Y_4)$
14	0	1/2	0	0	1/2	0	Y_{25}	$\beta_{25} = 4Y_{25} - 2(Y_2 + Y_5)$
15	0	1/2	0	0	0	1/2	Y_{26}	$\beta_{26} = 4Y_{26} - 2(Y_2 + Y_6)$
16	0	0	1/2	1/2	0	0	Y_{34}	$\beta_{34} = 4Y_{34} - 2(Y_3 + Y_4)$
17	0	0	1/2	0	1/2	0	Y_{35}	$\beta_{35} = 4Y_{35} - 2(Y_3 + Y_5)$
18	0	0	1/2	0	0	1/2	Y_{36}	$\beta_{36} = 4Y_{36} - 2(Y_3 + Y_6)$
19	0	0	0	1/2	1/2	0	Y_{45}	$\beta_{45} = 4Y_{45} - 2(Y_4 + Y_5)$
20	0	0	0	1/2	0	1/2	Y_{46}	$\beta_{46} = 4Y_{46} - 2(Y_4 + Y_6)$
21	0	0	0	0	1/2	1/2	Y_{56}	$\beta_{56} = 4Y_{56} - 2(Y_5 + Y_6)$
22	1/3	1/3	1/3	0	0	0	Y_{123}	$\beta_{123} + 27Y_{123} - 12(Y_{12} + Y_{13} + Y_{23}) + 3(Y_1 + Y_2 + Y_3)$
23	1/3	1/3	0	1/3	0	0	Y_{124}	$\beta_{124} + 27Y_{124} - 12(Y_{12} + Y_{14} + Y_{24}) + 3(Y_1 + Y_2 + Y_4)$
24	1/3	1/3	0	0	1/3	0	Y_{125}	$\beta_{125} + 27Y_{125} - 12(Y_{12} + Y_{15} + Y_{25}) + 3(Y_1 + Y_2 + Y_5)$
25	1/3	1/3	0	0	0	1/3	Y_{126}	$\beta_{126} + 27Y_{126} - 12(Y_{12} + Y_{16} + Y_{26}) + 3(Y_1 + Y_2 + Y_6)$
26	1/3	0	1/3	1/3	0	0	Y_{134}	$\beta_{134} + 27Y_{134} - 12(Y_{13} + Y_{14} + Y_{34}) + 3(Y_1 + Y_3 + Y_4)$
27	1/3	0	1/3	0	1/3	0	Y_{135}	$\beta_{135} + 27Y_{135} - 12(Y_{13} + Y_{15} + Y_{35}) + 3(Y_1 + Y_3 + Y_5)$
28	1/3	0	1/3	0	0	1/3	Y_{23}	$\beta_{136} + 27Y_{136} - 12(Y_{13} + Y_{16} + Y_{36}) + 3(Y_1 + Y_3 + Y_6)$
29	1/3	0	0	1/3	1/3	0	Y_{145}	$\beta_{145} + 27Y_{145} - 12(Y_{14} + Y_{15} + Y_{45}) + 3(Y_1 + Y_4 + Y_5)$
30	1/3	0	0	1/3	0	1/3	Y_{146}	$\beta_{146} + 27Y_{146} - 12(Y_{14} + Y_{16} + Y_{46}) + 3(Y_1 + Y_4 + Y_6)$
31	1/3	0	0	0	1/3	1/3	Y_{156}	$\beta_{156} + 27Y_{156} - 12(Y_{15} + Y_{16} + Y_{56}) + 3(Y_1 + Y_5 + Y_6)$
32	0	1/3	1/3	1/3	0	0	Y_{234}	$\beta_{234} + 27Y_{234} - 12(Y_{23} + Y_{24} + Y_{34}) + 3(Y_2 + Y_3 + Y_4)$
33	0	1/3	1/3	0	1/3	0	Y_{235}	$\beta_{235} + 27Y_{235} - 12(Y_{23} + Y_{25} + Y_{35}) + 3(Y_2 + Y_3 + Y_5)$
34	0	1/3	1/3	0	0	1/3	Y_{236}	$\beta_{236} + 27Y_{236} - 12(Y_{23} + Y_{26} + Y_{36}) + 3(Y_2 + Y_3 + Y_6)$
35	0	1/3	0	1/3	1/3	0	Y_{245}	$\beta_{245} + 27Y_{245} - 12(Y_{24} + Y_{25} + Y_{45}) + 3(Y_2 + Y_4 + Y_5)$
36	0	1/3	0	1/3	0	1/3	Y_{246}	$\beta_{246} + 27Y_{246} - 12(Y_{24} + Y_{26} + Y_{46}) + 3(Y_2 + Y_4 + Y_6)$
37	0	1/3	0	0	1/3	1/3	Y_{256}	$\beta_{256} + 27Y_{256} - 12(Y_{25} + Y_{26} + Y_{56}) + 3(Y_2 + Y_5 + Y_6)$
38	0	0	1/3	1/3	1/3	0	Y_{345}	$\beta_{345} + 27Y_{345} - 12(Y_{34} + Y_{35} + Y_{45}) + 3(Y_3 + Y_4 + Y_5)$
39	0	0	1/3	1/3	0	1/3	Y_{346}	$\beta_{346} + 27Y_{346} - 12(Y_{34} + Y_{36} + Y_{46}) + 3(Y_3 + Y_4 + Y_6)$
40	0	0	1/3	0	1/3	1/3	Y_{356}	$\beta_{356} + 27Y_{356} - 12(Y_{35} + Y_{36} + Y_{56}) + 3(Y_3 + Y_5 + Y_6)$
41	0	0	0	1/3	1/3	1/3	Y_{456}	$\beta_{456} + 27Y_{456} - 12(Y_{45} + Y_{46} + Y_{56}) + 3(Y_4 + Y_5 + Y_6)$
42	1/6	1/6	1/6	1/6	1/6	1/6	$Y_{123456(1)}$	
43	1/6	1/6	1/6	1/6	1/6	1/6	$Y_{123456(2)}$	
44	1/6	1/6	1/6	1/6	1/6	1/6	$Y_{123456(3)}$	

*Runs 1-6 are linear design, adding Runs 7-21 creates a quadratic design, adding Runs 22-41 creates a special cubic design, and adding Runs 42-44 provides a replicated center point for checking goodness of fit.

Table B.6-5
Quadratic Simplex Design for Seven Mixture Components *

Run No.	X_1	X_2	X_3	X_4	X_5	X_6	X_7	Symbolic Response	Coefficient Estimates for Special Cubic Scheffé Model
1	1	0	0	0	0	0	0	Y_1	$\beta_1 = Y_1$
2	0	1	0	0	0	0	0	Y_2	$\beta_2 = Y_2$
3	0	0	1	0	0	0	0	Y_3	$\beta_3 = Y_3$
4	0	0	0	1	0	0	0	Y_4	$\beta_4 = Y_4$
5	0	0	0	0	1	0	0	Y_5	$\beta_5 = Y_5$
6	0	0	0	0	0	1	0	Y_6	$\beta_6 = Y_6$
7	0	0	0	0	0	0	1	Y_7	$\beta_7 = Y_7$
8	½	½	0	0	0	0	0	Y_{12}	$\beta_{12} = 4Y_{12} - 2(Y_1 + Y_2)$
9	½	0	½	0	0	0	0	Y_{13}	$\beta_{13} = 4Y_{13} - 2(Y_1 + Y_3)$
10	½	0	0	½	0	0	0	Y_{14}	$\beta_{14} = 4Y_{14} - 2(Y_1 + Y_4)$
11	½	0	0	0	½	0	0	Y_{15}	$\beta_{15} = 4Y_{15} - 2(Y_1 + Y_5)$
12	½	0	0	0	0	½	0	Y_{16}	$\beta_{16} = 4Y_{16} - 2(Y_1 + Y_6)$
13	½	0	0	0	0	0	½	Y_{17}	$\beta_{17} = 4Y_{17} - 2(Y_1 + Y_7)$
14	0	½	½	0	0	0	0	Y_{23}	$\beta_{23} = 4Y_{23} - 2(Y_2 + Y_3)$
15	0	½	0	½	0	0	0	Y_{24}	$\beta_{24} = 4Y_{24} - 2(Y_2 + Y_4)$
16	0	½	0	0	½	0	0	Y_{25}	$\beta_{25} = 4Y_{25} - 2(Y_2 + Y_5)$
17	0	½	0	0	0	½	0	Y_{26}	$\beta_{26} = 4Y_{26} - 2(Y_2 + Y_6)$
18	0	½	0	0	0	0	½	Y_{27}	$\beta_{27} = 4Y_{27} - 2(Y_2 + Y_7)$
19	0	0	½	½	0	0	0	Y_{34}	$\beta_{34} = 4Y_{34} - 2(Y_3 + Y_4)$
20	0	0	½	0	½	0	0	Y_{35}	$\beta_{35} = 4Y_{35} - 2(Y_3 + Y_5)$
21	0	0	½	0	0	½	0	Y_{36}	$\beta_{36} = 4Y_{36} - 2(Y_3 + Y_6)$
22	0	0	½	0	0	0	½	Y_{37}	$\beta_{37} = 4Y_{37} - 2(Y_3 + Y_7)$
23	0	0	0	½	½	0	0	Y_{45}	$\beta_{45} = 4Y_{45} - 2(Y_4 + Y_5)$
24	0	0	0	½	0	½	0	Y_{46}	$\beta_{46} = 4Y_{46} - 2(Y_4 + Y_6)$
25	0	0	0	½	0	0	½	Y_{47}	$\beta_{47} = 4Y_{47} - 2(Y_4 + Y_7)$
26	0	0	0	0	½	½	0	Y_{56}	$\beta_{56} = 4Y_{56} - 2(Y_5 + Y_6)$
27	0	0	0	0	½	0	½	Y_{57}	$\beta_{57} = 4Y_{57} - 2(Y_5 + Y_7)$
28	0	0	0	0	0	½	½	Y_{67}	$\beta_{67} = 4Y_{67} - 2(Y_6 + Y_7)$
29	⅐	⅐	⅐	⅐	⅐	⅐	⅐	$Y_{1234567(1)}$	
30	⅐	⅐	⅐	⅐	⅐	⅐	⅐	$Y_{1234567(2)}$	
31	⅐	⅐	⅐	⅐	⅐	⅐	⅐	$Y_{1234567(3)}$	

*Runs 1–7 are the linear design, adding Runs 8–28 creates a quadratic design, and adding Runs 29–31 provides a replicated center point for checking goodness of fit.

Table B.7.1-1
NESTED DESIGN FOR TWO COMPONENTS OF VARIANCE

Source A

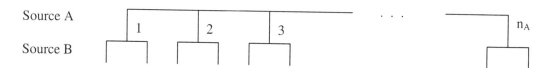

Source B

ANOVA Table for Two Components of Variance

Source	SS	ν	MS	Expected Mean Square (MS)
A	SS_A	$\nu_A = n_A - 1$	SS_A/ν_A	$\sigma_B^2 + 2\sigma_A^2$
B	SS_B	$\nu_B = n_A$	SS_B/ν_B	σ_B^2

where Y_{ij} = the observation for the *i*th value of A ($i = 1, 2, \ldots, n_A$) and the *j*th value of B ($j = 1, 2$)

$$T_i = \sum_j Y_{ij}$$

$$SS_B = \sum_i \left(\sum_i Y_{ij}^2 - T_i^2/2 \right)$$

$$T = \sum_i T_i$$

$$SS_B = \sum_i T_i^2/2 - T^2/2n_A$$

Table B.7.1-2

NESTED DESIGN FOR THREE COMPONENTS OF VARIANCE

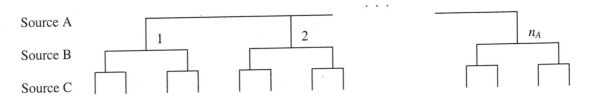

Source A

Source B

Source C

ANOVA Table for Three Components of Variance

Source	SS	v	MS	Expected Mean Square (MS)
A	SS_A	$v_A = n_{A-1}$	SS_A/v_A	$\sigma_C^2 + 2\sigma_B^2 + 4\sigma_A^2$
B	SS_B	$v_B = n_A$	SS_B/v_B	$\sigma_C^2 + 2\sigma_B^2$
C	SS_C	$v_C = 2n_A$	SS_C/v_C	σ_C^2

where Y_{ijk} = the observation for the ith value of A $(i = 1, 2, \ldots n_A)$ and the jth value of B $(j = 1,2)$ and the kth value of C $(k = 1,2)$

$$T_{ij} = \sum_k Y_{ijk}$$

$$SS_C = \sum_i \sum_j \left(\sum_k Y_{ijk}^2 - T_{ij}^2 / 2 \right)$$

$$T_i = \sum_j T_{ij}$$

$$SS_B = \sum_i \left(\sum_j T_{ij}^2 / 2 - T_i^2 / 4 \right)$$

$$T = \sum_i T_i$$

$$SS_A = \sum_i T_i^2 / 4 - T^2 / 4n_A$$

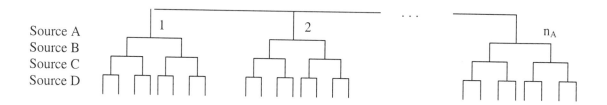

Source A					
Source B		1	2	$\cdots$	n_A
Source C					
Source D					

ANOVA Table for Four Components of Variance

Source	SS	v	MS	Expected Mean Square (MS)
A	SS_A	$v_A = n_A - 1$	SS_A/v_A	$\sigma_D^2 + 2\sigma_C^2 + 4\sigma_B^2 + 8\sigma_A^2$
B	SS_B	$v_B = n_A$	SS_B/v_B	$\sigma_D^2 + 2\sigma_C^2 + 4\sigma_B^2$
C	SS_B	$v_C = 2n_A$	SS_C/v_C	$\sigma_D^2 + 2\sigma_C^2$
D	SS_D	$v_D = 4n_A$	SS_D/v_D	σ_D^2

where Y_{ijkl} = the observation for the ith value of A ($i = 1, 2, \ldots n_A$) and the jth value of B ($j = 1,2$) the kth value of C ($k = 1,2$) and the lth value of D ($l = 1, 2$)

$$T_{ijk} = \sum_l Y_{ijkl}$$

$$SS_D = \sum_i \sum_j \sum_k \left(\sum_l Y_{ijkl}^2 - T_{ijk}^2/2 \right)$$

$$T_{ij} = \sum_k T_{ijk}$$

$$SS_C = \sum_i \sum_j \left(\sum_k T_{ijk}^2/2 - T_{ij}^2/4 \right)$$

$$T_i = \sum_j T_{ij}$$

$$SS_B = \sum_i \left(\sum_j T_{ij}^2/4 - T_i^2/8 \right)$$

$$T = \sum_i T_i$$

$$SS_A = \sum_i T_i^2/8 - T^2/8n_A$$

Table B.7.1-4
NESTED DESIGN FOR FIVE COMPONENTS OF VARIANCE

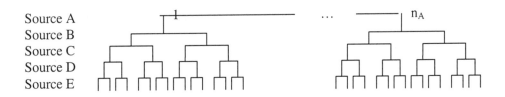

| Source A | | 1 | | ... | | n_A |

ANOVA Table for Five Components of Variance

Source	SS	ν	MS	Expected Mean Square (MS)
A	SS_A	$\nu_A = n_A - 1$	SS_A/ν_A	$\sigma_E^2 + 2\sigma_D^2 + 4\sigma_C^2 + 8\sigma_B^2 + 16\sigma_A^2$
B	SS_B	$\nu_B = n_A$	SS_B/ν_B	$\sigma_E^2 + 2\sigma_D^2 + 4\sigma_C^2 + 8\sigma_B^2$
C	SS_C	$\nu_C = 2n_A$	SS_C/ν_C	$\sigma_E^2 + 2\sigma_D^2 + 4\sigma_C^2$
D	SS_D	$\nu_D = 4n_A$	SS_D/ν_D	$\sigma_E^2 + 2\sigma_D^2$
E	SS_E	$\nu_E = 8n_A$	SS_E/ν_E	σ_E^2

where Y_{ijklm} = the observation for the ith value of A ($i = 1, 2, \ldots n_A$) and the jth value of B ($j = 1,2$) the kth value of C ($k = 1,2$) the lth value of D ($l = 1, 2$) and the mth value of E ($m=1,2$)

$$T_{ijkl} = \sum_m Y_{ijklm}$$

$$SS_E = \sum_i \sum_j \sum_k \sum_l \left(\sum_m Y_{ijklm}^2 - T_{ijkl}^2/2 \right)$$

$$T_{ijk} = \sum_l T_{ijkl}$$

$$SS_D = \sum_i \sum_j \sum_k \left(\sum_l T_{ijkl}^2/2 - T_{ijk}^2/4 \right)$$

$$T_{ij} = \sum_k T_{ijk}$$

$$SS_C = \sum_i \sum_j \left(\sum_k T_{ijk}^2/4 - T_{ij}^2/8 \right)$$

$$T_i = \sum_j T_{ij}$$

$$SS_B = \sum_i \left(\sum_j T_{ij}^2/8 - T_i^2/16 \right)$$

$$T = \sum_i T_{ij}$$

$$SS_A = \sum_i T_i^2/16 - T^2/16n_A$$

Source A

Source B

Source C

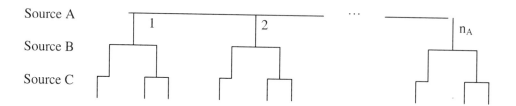

ANOVA Table for Three Components of Variance

Source	SS	v	MS	Expected Mean Square (MS)
A	SS_A	$v_A = n_{A-1}$	SS_A/v_A	$\sigma_C^2 + (5/3)\sigma_B^2 + 3\sigma_A^2$
B	SS_B	$v_B = n_A$	SS_B/v_B	$\sigma_C^2 + (4/3)\sigma_B^2$
C	SS_C	$v_C = n_A$	SS_C/v_C	σ_C^2

where Y_{ijk} = the observation for the ith value of A ($i = 1, 2, \ldots n_A$) and the jth value of B ($j = 1,2$) and the kth value of C ($k = 1,2$)

$$T_{ij} = \sum_k Y_{ijk}$$

$$SS_C = \sum_i \sum_j \left(\sum_k Y_{ijk}^2 - T_{ij}^2/n_{ij} \right)$$

$$T_i = \sum_j T_{ij}$$

$$SS_B = \sum_i \left(\sum_j T_{ij}^2/n_{ij} - T_i^2/3 \right)$$

$$T = \sum_i T_i$$

$$SS_A = \sum_i T_i^2/3 - T^2/3n_A$$

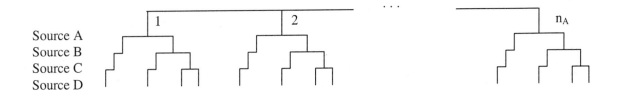

Source A
Source B
Source C
Source D

ANOVA Table for Four Components of Variance

Source	SS	ν	MS	Expected Mean Square (MS)
A	SS_A	$\nu_A = n_A - 1$	SS_A/ν_A	$\sigma_D^2 + (3/2)\sigma_C^2 + (5/2)\sigma_B^2 + 4\sigma_A^2$
B	SS_B	$\nu_B = n_A$	SS_B/ν_B	$\sigma_D^2 + (7/6)\sigma_C^2 + (3/2)\sigma_B^2$
C	SS_B	$\nu_C = n_A$	SS_C/ν_C	$\sigma_D^2 + (4/3)\sigma_C^2$
D	SS_D	$\nu_D = n_A$	SS_D/ν_D	σ_D^2

where Y_{ijkl} = the observation for the ith value of A ($i = 1, 2, \ldots n_A$) and the jth value of B ($j = 1,2$) the kth value of C ($k = 1,2$) and the lth value of D ($l = 1, 2$)

$$T_{ijk} = \sum_l Y_{ijkl}$$

$$SS_D = \sum_i \sum_j \sum_k \left(\sum_l Y_{ijkl}^2 - T_{ijk}^2/n_{ijk} \right)$$

$$T_{ij} = \sum_k T_{ijk}$$

$$SS_C = \sum_i \sum_j \left(\sum_k T_{ijk}^2/n_{ijk} - T_{ij}^2/n_{ij} \right)$$

$$T_i = \sum_j T_{ij}$$

$$SS_B = \sum_i \left(\sum_j T_{ij}^2/n_{ij} - T_i^2/4 \right)$$

$$T = \sum_i T_i$$

$$SS_A = \sum_i T_i^2/4 - T^2/4n_A$$

STAGGERED NESTED DESIGN FOR FIVE COMPONENTS OF VARIANCE

Source A
Source B
Source C
Source D
Source E

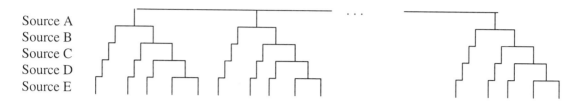

ANOVA Table for Five Components of Variance

Source	SS	ν	MS	Expected Mean Square (MS)
A	SS_A	$\nu_A = n_A - 1$	SS_A/ν_A	$\sigma_E^2 + (7/5)\sigma_D^2 + (11/5)\sigma_C^2 + (17/5)\sigma_B^2 + 5\sigma_A^2$
B	SS_B	$\nu_B = n_A$	SS_B/ν_B	$\sigma_E^2 + (11/10)\sigma_D^2 + (13/10)\sigma_C^2 + (8/5)\sigma_B^2$
C	SS_C	$\nu_C = n_A$	SS_C/ν_C	$\sigma_E^2 + (7/6)\sigma_D^2 + (3/2)\sigma_C^2$
D	SS_D	$\nu_D = n_A$	SS_D/ν_D	$\sigma_E^2 + (4/3)\sigma_D^2$
E	SS_E	$\nu_E = n_A$	SS_E/ν_E	σ_E^2

where Y_{ijklm} = the observation for the ith value of A ($i = 1, 2, \ldots n_A$) and the jth value of B ($j = 1,2$) the kth value of C ($k = 1,2$) the lth value of D ($l = 1, 2$) and the mth value of E ($m = 1,2$)

$$T_{ijkl} = \sum_m Y_{ijklm}$$

$$SS_E = \sum_i \sum_j \sum_k \sum_l \left(\sum_m Y_{ijklm}^2 - T_{ijkl}^2/n_{ijkl} \right)$$

$$T_{ijk} = \sum_l T_{ijkl}$$

$$SS_D = \sum_i \sum_j \sum_k \left(\sum_l T_{ijkl}^2/n_{ijkl} - T_{ijk}^2/n_{ijk} \right)$$

$$T_{ij} = \sum_k T_{ijk}$$

$$SS_C = \sum_i \sum_j \left(\sum_k T_{ijk}^2/n_{ijk} - T_{ij}^2/n_{ij} \right)$$

$$T_i = \sum_j T_{ij}$$

$$SS_B = \sum_i \left(\sum_j T_{ij}^2/n_{ij} - T_i^2/5 \right)$$

$$T = \sum_i T_{ij}$$

$$SS_A = \sum_i T_i^2/5 - T^2/5n_A$$

Index

Credits